Lecture Notes in Computer Sc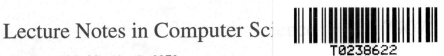

T0238622

Commenced Publication in 1973
Founding and Former Series Editors:
Gerhard Goos, Juris Hartmanis, and Jan van Leeuwen

Editorial Board

David Hutchison
Lancaster University, UK

Takeo Kanade
Carnegie Mellon University, Pittsburgh, PA, USA

Josef Kittler
University of Surrey, Guildford, UK

Jon M. Kleinberg
Cornell University, Ithaca, NY, USA

Alfred Kobsa
University of California, Irvine, CA, USA

Friedemann Mattern
ETH Zurich, Switzerland

John C. Mitchell
Stanford University, CA, USA

Moni Naor
Weizmann Institute of Science, Rehovot, Israel

Oscar Nierstrasz
University of Bern, Switzerland

C. Pandu Rangan
Indian Institute of Technology, Madras, India

Bernhard Steffen
TU Dortmund University, Germany

Madhu Sudan
Microsoft Research, Cambridge, MA, USA

Demetri Terzopoulos
University of California, Los Angeles, CA, USA

Doug Tygar
University of California, Berkeley, CA, USA

Gerhard Weikum
Max Planck Institute for Informatics, Saarbruecken, Germany

Marie-Pierre Béal   Olivier Carton (Eds.)

Developments in Language Theory

17th International Conference, DLT 2013
Marne-la-Vallée, France, June 18-21, 2013
Proceedings

 Springer

Volume Editors

Marie-Pierre Béal
Université Paris-Est Marne-la-Vallée
LIGM
5 Bd Descartes, Champs-sur-Marne, 77454 Marne-la-Vallée Cedex 2, France
E-mail: beal@univ-mlv.fr

Olivier Carton
Université Paris Diderot
LIAFA
UMR 7089, 75205 Paris cedex 13, France
E-mail: olivier.carton@liafa.univ-paris-diderot.fr

ISSN 0302-9743   e-ISSN 1611-3349
ISBN 978-3-642-38770-8   e-ISBN 978-3-642-38771-5
DOI 10.1007/978-3-642-38771-5
Springer Heidelberg Dordrecht London New York

Library of Congress Control Number: 2013939199

CR Subject Classification (1998): F.1, F.4, F.2, G.2, E.4

LNCS Sublibrary: SL 1 – Theoretical Computer Science and General Issues

© Springer-Verlag Berlin Heidelberg 2013
This work is subject to copyright. All rights are reserved, whether the whole or part of the material is
concerned, specifically the rights of translation, reprinting, re-use of illustrations, recitation, broadcasting,
reproduction on microfilms or in any other way, and storage in data banks. Duplication of this publication
or parts thereof is permitted only under the provisions of the German Copyright Law of September 9, 1965,
in its current version, and permission for use must always be obtained from Springer. Violations are liable
to prosecution under the German Copyright Law.
The use of general descriptive names, registered names, trademarks, etc. in this publication does not imply,
even in the absence of a specific statement, that such names are exempt from the relevant protective laws
and regulations and therefore free for general use.

Typesetting: Camera-ready by author, data conversion by Scientific Publishing Services, Chennai, India

Printed on acid-free paper

Springer is part of Springer Science+Business Media (www.springer.com)

Preface

The 17th International Conference on Developments in Language Theory (DLT 2013) was held at Université Paris-Est, Marne-la-Vallée, France, during June 18-21, 2013.

The DLT conference series is one of the major international conference series in language theory. It started in Turku, Finland, in 1993. It was held initially once every two years. Since 2001, it has been held every year, odd years in Europe and even years on other continents.

The scope of the conference includes, among others, the following topics and areas: combinatorial and algebraic properties of words and languages; grammars, acceptors and transducers for strings, trees, graphs, arrays; algebraic theories for automata and languages; codes; efficient text algorithms; symbolic dynamics; decision problems; relationships to complexity theory and logic; picture description and analysis; polyominoes and bidimensional patterns; cryptography; concurrency; cellular automata; bio-inspired computing; quantum computing.

The papers submitted to DLT 2013 were from 23 countries including Belgium, Canada, China, Czech Republic, Estonia, Finland, France, Germany, Hungary, India, Italy, Japan, Republic of Korea, Latvia, Poland, Portugal, Russian Federation, Slovakia, South Africa, Sweden, Turkey, UK, and USA.

There were 63 submissions. Each submission was reviewed by at least three referees and discussed by the Program Committee for presentation at the conference. The Committee decided to accept 34 papers. There were five invited talks given by Rūsiņš Freivalds (University of Latvia), Artur Jeż (University of Wrocław and Max Planck Institut für Informatik, Saarbrücken), Raphaël Jungers (UCL Louvain), Christof Löding (RWTH Aachen), and Jean-Éric Pin (CNRS LIAFA Paris). This volume of *Lecture Notes in Computer Science* contains the papers that were presented at DLT 2013 including the abstracts or full papers of the invited lectures.

We warmly thank all the invited speakers and all the authors of the submitted papers. We would also like to thank all the members of the Program Committee and the external referees (listed in the proceedings) for their hard work in evaluating the papers. We also thank all members of the Organizing Committee, and specially Arnaud Carayol who headed this Committee. We wish to express our sincere appreciation to the conference sponsors: the University Paris-Est Marne-la-Vallée, ESIEE Paris, the LabEx Bézout, the Computer Science Research Laboratory Gaspard-Monge, the CNRS and the European Association for Theoretical Computer Science. Finally, we wish to thank the editors of the *Lecture Notes in Computer Science* series and Springer.

March 2013

Marie-Pierre Béal
Olivier Carton

Organization

Program Committee

Marie-Pierre Béal (Co-chair)	Université Paris-Est, France
Arnaud Carayol	Université Paris-Est, CNRS, France
Olivier Carton (Co-chair)	Université Paris Diderot, France
Alessandro De Luca	Università di Napoli Federico II, Italy
Volker Diekert	Universität Stuttgart, Germany
Anna Frid	Sobolev Institute in Mathematics, Russia
Nataša Jonoska	University of South Florida, USA
Jarkko Kari	University of Turku, Finland
Michal Kunc	Masaryk University Brno, Czech Republic
Martin Kutrib	Universität Giessen, Germany
Slawomir Lasota	Warsaw University, Poland
Pierre McKenzie	Université de Montréal, Canada
Giovanni Pighizzini	Università degli Studi di Milano, Italy
Benjamin Steinberg	Carleton University, Canada
Klaus Sutner	Carnegie Mellon University, USA
Mikhail Volkov	Ural State University, Russia
Hsu-Chun Yen	National Taiwan University, Taiwan
Zoltán Ésik	University of Szeged, Hungary

Organizing Committee

Yasmina Abdeddaïm	Université Paris-Est, ESIEE, France
Arnaud Carayol (Head)	Université Paris-Est, CNRS, France
Olivier Carton	Université Paris Diderot, France
Didier Caucal	Université Paris-Est, CNRS, France
Claire David	Université Paris-Est, France
Philippe Gambette	Université Paris-Est, France
Matthew Hague	Université Paris-Est, France
Antoine Meyer	Université Paris-Est, France
Cyril Nicaud	Université Paris-Est, France
Corinne Palescandolo	Université Paris-Est, CNRS, France
Carine Pivoteau	Université Paris-Est, France
Chloé Rispal	Université Paris-Est, France

Steering Committee

Marie-Pierre Béal	Université Paris-Est Marne-la-Vallée, France
Véronique Bruyère	University of Mons, Belgium

Cristian S. Calude	University of Auckland, New Zealand
Volker Diekert	Universität Stuttgart, Germany
Juraj Hromkovic	ETH Zurich, Switzerland
Oscar H. Ibarra	University of California, Santa Barbara, USA
Masami Ito	Kyoto Sangyo University, Japan
Nataša Jonoska	University of South Florida, USA
Juhani Karhumäki (Chair)	University of Turku, Finland
Antonio Restivo	University of Palermo, Italy
Grzegorz Rozenberg	Leiden University, The Netherlands
Wojciech Rytter	Warsaw University, Poland
Arto Salomaa	University of Turku, Finland
Kai Salomaa	Queen's University, Canada
Mikhail Volkov	Ural State University, Russia
Takashi Yokomori	Waseda University, Japan

Additional Reviewers

Marcella Anselmo	Hermann Gruber
Sergey Avgustinovich	Christoph Haase
Nicolas Bedon	Matthew Hague
Francine Blanchet-Sadri	Vesa Halava
Guillaume Blin	Benjamin Hellouin de Menibus
Michael Blondin	Ulrich Hertrampf
Janusz Brzozowski	Mika Hirvensalo
Michelangelo Bucci	Piotr Hofman
Michaël Cadilhac	Štěpán Holub
Alan Cain	Markus Holzer
Julien Cassaigne	Juha Honkala
Giusi Castiglione	Hendrik Jan Hoogeboom
Didier Caucal	Dag Hovland
Alessandra Cherubini	Oscar Ibarra
Matthieu Constant	Szabolcs Iván
Erzsébet Csuhaj-Varjú	Sebastian Jakobi
Wojciech Czerwiński	Emmanuel Jeandel
Claire David	Galina Jirásková
Aldo de Luca	Mark Kambites
Frank Drewes	Juhani Karhumäki
Joost Engelfriet	Jonathan Kausch
Thomas Fernique	Sergey Kitaev
Gabriele Fici	Ines Klimann
Emmanuel Filiot	Bartek Klin
Achille Frigeri	Ondřej Klíma
Dora Giammarresi	Eryk Kopczyński
Amy Glen	Steffen Kopecki

Andreas Krebs
Manfred Kufleitner
Alexander Lauser
Ranko Lazić
Peter Leupold
Markus Lohrey
Christof Löding
Maria Madonia
Andreas Malcher
Roberto Mantaci
Sabrina Mantaci
Tomáš Masopust
Ian McQuillan
Katja Meckel
Carlo Mereghetti
Antoine Meyer
Nelma Moreira
Judit Nagy-György
Christos Nomikos
Dirk Nowotka
Zoltán Németh
Pascal Ochem
Alexander Okhotin
Vincent Penelle
Dominique Perrin
Joni Pirnot
Alberto Policriti
Libor Polák
Damien Pous
Elena Pribavkina
Julien Provillard
Svetlana Puzynina

Narad Rampersad
Bala Ravikumar
Rogério Reis
Antonio Restivo
Gwenaël Richomme
Aleksi Saarela
Ville Salo
Kai Salomaa
Aristidis Sapounakis
Shinnosuke Seki
Frédéric Servais
Arseny Shur
Pedro V. Silva
Michał Skrzypczak
Frank Stefan
Howard Straubing
K.G. Subramanian
Tony Tan
Szymon Toruńczyk
Ilkka Törmä
Antti Valmari
György Vaszil
Stéphane Vialette
Tobias Walter
Pascal Weil
Armin Weiss
Matthias Wendlandt
Ryo Yoshinaka
Luca Zamboni
Georg Zetzsche
Charalampos Zinoviadis

Sponsoring Institutions

University Paris-Est Marne-la-Vallée
ESIEE Paris
LabEx Bézout
Laboratoire d'informatique Gaspard-Monge UMR 8049
CNRS
European Association for Theoretical Computer Science

Table of Contents

Invited Talks

Regular Papers

Ultrametric Finite Automata
and Turing Machines*

Rūsiņš Freivalds

Institute of Mathematics and Computer Science, University of Latvia,
Raiņa bulvāris 29, Riga, LV-1459, Latvia
Rusins.Freivalds@lu.lv

Abstract. We introduce a notion of ultrametric automata and Turing machines using p-adic numbers to describe random branching of the process of computation. These automata have properties similar to the properties of probabilistic automata but complexity of probabilistic automata and complexity of ultrametric automata can differ very much.

1 Introduction

Pascal and Fermat believed that every event of indeterminism can be described by a real number between 0 and 1 called *probability*. Quantum physics introduced a description in terms of complex numbers called *amplitude of probabilities* and later in terms of probabilistic combinations of amplitudes most conveniently described by *density matrices*.

String theory [18], chemistry [15] and molecular biology [3, 12] have introduced p-adic numbers to describe measures of indeterminism.

Popularity of usage of p-adic numbers can be explained easily. There is a well-known difficulty to overcome the distinction between *continuous* and *discrete* processes. For instance, according to Rutherford's model of atoms, the electrons can be situated only on specific orbits. When energy of an electron increases, there is a quantum leap. Niels Bohr proposed, in 1913, what is now called the Bohr model of the atom. He suggested that electrons could only have certain classical motions:

1. Electrons in atoms orbit the nucleus.

2. The electrons can only orbit stably, without radiating, in certain orbits (called by Bohr the "stationary orbits"): at a certain discrete set of distances from the nucleus. These orbits are associated with definite energy levels. In these orbits, the electron's acceleration does not result in radiation and energy loss as required by classical electromagnetics.

---

* The research was supported by Project 271/2012 from the Latvian Council of Science.

M.-P. Béal and O. Carton (Eds.): DLT 2013, LNCS 7907, pp. 1–11, 2013.
© Springer-Verlag Berlin Heidelberg 2013

3. Electrons can only gain and lose energy by jumping from one allowed orbit to another, absorbing or emitting electromagnetic radiation with a frequency determined by the energy difference of the levels according to the Planck relation.

One of the methods to model such quantum leaps is to consider p-adic numbers and there norms. The p-adic numbers can have continuum distinct values but their norms can have only denumerable values. If a variable gradually changes taking p-adic values, its norm performs quantum leaps. Hence usage of p-adic numbers as measures of indeterminism provides a mechanism which is similar to probabilistic model but mathematically different from it.

There were no difficulties to implement probabilistic automata and algorithms practically. Quantum computation [11] has made a considerable theoretical progress but practical implementation has met considerable difficulties. However, prototypes of quantum computers exist, some quantum algorithms are implemented on these prototypes, quantum cryptography is already practically used. Some people are skeptical concerning practicality of the initial spectacular promises of quantum computation but nobody can deny the existence of quantum computation.

We consider a new type of indeterministic algorithms called *ultrametric* algorithms. They are very similar to probabilistic algorithms but while probabilistic algorithms use real numbers r with $0 \leq r \leq 1$ as parameters, ultrametric algorithms use *p-adic* numbers as the parameters. Slightly simplifying the description of the definitions one can say that ultrametric algorithms are the same probabilistic algorithms, only the interpretation of the probabilities is different.

Our choice of p-adic numbers instead of real numbers is not quite arbitrary. In 1916 Alexander Ostrowski proved that any non-trivial absolute value on the rational numbers Q is equivalent to either the usual real absolute value or a p-adic absolute value. This result shows that using p-adic numbers is not merely one of many possibilities to generalize the definition of deterministic algorithms but rather the only remaining possibility not yet explored.

Moreover, Helmut Hasse's local-global principle states that certain types of equations have a rational solution if and only if they have a solution in the real numbers and in the p-adic numbers for each prime p.

There are many distinct p-adic absolute values corresponding to the many prime numbers p. These absolute values are traditionally called *ultrametric*. Absolute values are needed to consider *distances* among objects. We have used to rational and irrational numbers as measures for distances, and there is a psychological difficulty to imagine that something else can be used instead of irrational numbers. However, there is an important feature that distinguishes p-adic numbers from real numbers. Real numbers (both rational and irrational) are linearly ordered. p-adic numbers cannot be linearly ordered. This is why *valuations* and *norms* of p-adic numbers are considered.

The situation is similar in Quantum Computation. Quantum amplitudes are complex numbers which also cannot be linearly ordered. The counterpart of valuation for quantum algorithms is *measurement* translating a complex number $a + bi$ into a real number $a^2 + b^2$. Norms of p-adic numbers are rational numbers.

Ultrametric finite automata and ultrametric Turing machines are reasonably similar to probabilistic finite automata and Turing machines.

Below we consider ultrametric versus deterministic Turing machines with one input tape which can be read only 1-way and a work tape which is empty at the beginning of the work.

The problem of more adequate mechanism to describe indeterminism in finite automata and Turing machines has been raised many times (see, e.g. [1, 5–7]). Ultrametric finite automata and Turing machines were introduced in [8]. This paper contains more explicit motivation of the research, more examples and more complete proofs.

2   p-adic Numbers

Let p be an arbitrary prime number. We will call *p-adic digit* a natural number between 0 and $p - 1$ (inclusive). A *p-adic integer* is by definition a sequence $(a_i)_{i \in N}$ of p-adic digits. We write this conventionally as

$$\cdots a_i \cdots a_2 a_1 a_0$$

(that is, the a_i are written from left to right).

If n is a natural number, and

$$n = \overline{a_{k-1} a_{k-2} \cdots a_1 a_0}$$

is its p-adic representation (in other words $n = \sum_{i=0}^{k-1} a_i p^i$ with each a_i a p-adic digit) then we identify n with the p-adic integer (a_i) with $a_i = 0$ if $i \geq k$. This means that natural numbers are exactly the same thing as p-adic integer only a finite number of whose digits are not 0. The number 0 is the p-adic integer all of whose digits are 0, and that 1 is the p-adic integer all of whose digits are 0 except the right-most one (digit 0) which is 1.

To have p-adic representations of all rational numbers, $\frac{1}{p}$ is represented as $\cdots 00.1$, the number $\frac{1}{p^2}$ as $\cdots 00.01$, and so on. For any p-adic number it is allowed to have infinitely many (!) digits to the left of the "decimal" point but only a finite number of digits to the right of it.

However, p-adic numbers is not merely one of generalizations of rational numbers. They are related to the notion of *absolute value* of numbers.

If X is a nonempty set, a distance, or metric, on X is a function d from pairs of elements (x, y) of X to the nonnegative real numbers such that

1. $d(x, y) = 0$ if and only if $x = y$,
2. $d(x, y) = d(y, x)$,
3. $d(x, y) \leq d(x, z) + d(z, y)$ for all $z \in X$.

A set X together with a metric d is called a *metric space*. The same set X can give rise to many different metric spaces.

The *norm* of an element $x \in X$ is the distance from 0:

1. $\| x \| = 0$ if and only if $x = y$,
2. $\| x.y \| = \| x \| . \| xy \|$,
3. $\| x + y \| \le \| x \| + \| y \|$.

We know one metric on Q induced by the ordinary absolute value. However, there are other norms as well.

A norm is called *ultrametric* if the third requirement can be replaced by the stronger statement: $\| x + y \| \le \max\{\| x \|, \| y \|\}$. Otherwise, the norm is called *Archimedean*.

Definition 1. *Let $p \in \{2, 3, 5, 7, 11, 13, \cdots\}$ be any prime number. For any nonzero integer a, let the p-adic ordinal (or valuation) of a, denoted $\mathrm{ord}_p a$, be the highest power of p which divides a, i.e., the greatest m such that $a \equiv 0(mod p^m)$. For any rational number $x = a/b$, denote $\mathrm{ord}_p x$ to be $\mathrm{ord}_p a - \mathrm{ord}_p b$. Additionally, $\mathrm{ord}_p x = \infty$ if and only if $x = 0$.*

Definition 2. *Let $p \in \{2, 3, 5, 7, 11, 13, \cdots\}$ be any prime number. For arbitrary rational number x, its p-norm is:*

$$\|x\|_p = \begin{cases} \frac{1}{p^{\mathrm{ord}_p x}}, & \text{if} \quad x \neq 0, \\ \neg p_i, & \text{if} \quad x = 0 \; ; \end{cases}$$

Rational numbers are p-adic integers for all prime numbers p. The nature of irrational numbers is more complicated. For instance, $\sqrt{2}$ just does not exist as a p-adic number for some prime numbers p. More precisely, $\sqrt{a}$ can be represented as a p-adic number if and only if a is a quadratic residue modulo p, i.e. if the congruence $x^2 = a(mod p)$ has a solution. On the other hand, there is a continuum of p-adic numbers not being real numbers. Moreover, there is a continuum of 3-adic numbers not being 5-adic numbers, and vice versa.

p-adic numbers are described in much more detail in [10, 13].

3   First Examples

The notion of p-adic numbers widely used in mathematics but not so much in Computer Science. The aim of our next sections is to show that the notion of ultrametric automata and ultrametric Turing machines is natural.

In mathematics, a stochastic matrix is a matrix used to describe the transitions of a Markov chain. A *right stochastic matrix* is a square matrix each of whose rows consists of nonnegative real numbers, with each row summing to 1. A *stochastic vector* is a vector whose elements consist of nonnegative real numbers which sum to 1. The *finite probabilistic automaton* is defined as an extension of a non-deterministic finite automaton $(Q, \Sigma, \delta, q_0, F)$, with the initial state q_0 replaced by a stochastic vector giving the probability of the automaton being in a given initial state, and with stochastic matrices corresponding to each symbol in the input alphabet describing the state transition probabilities. It is important to note that if A is the stochastic matrix corresponding to the input symbol

a and B is the stochastic matrix corresponding to the input symbol b, then the product AB describes the state transition probabilities when the automaton reads the input word ab. Additionally, the probabilistic automaton has a threshold λ being a real number between 0 and 1. If the probabilistic automaton has only one *accepting state* then the input word x is said to be accepted if after reading x the probability of the accepting state has a probability exceeding λ. If there are several accepting states, the word x is said to be accepted the total of probabilities of the accepting states exceeds λ.

Ultrametric automata are defined exactly in the same way as probabilistic automata, only the parameters called *probabilities of transition from one state to another one* are real numbers between 0 and 1 in probabilistic automata, and they are p-adic numbers called *amplitudes* in the ultrametric automata. Formulas to calculate the amplitudes after one, two, three, $\cdots$ steps of computation are exactly the same as the formulas to calculate the probabilities in the probabilistic automata. Following the example of finite quantum automata, we demand that the input word x is followed by a special end-marker. At the beginning of the work, the states of the automaton get *initial amplitudes* being p-adic numbers. When reading the current symbol of the input word, the automaton changes the amplitudes of all the states according to the transition matrix corresponding to this input symbol. When the automaton reads the end-marker, the *measurement* is performed, and the amplitudes of all the states are transformed into the p-norms of these amplitudes. The norms are rational numbers and it is possible to compare whether or not the norm exceeds the threshold λ. If total of the norms for all the accepting states of the automaton exceeds λ, we say that the automaton accepts the input word.

Paavo Turakainen considered various generalizations of finite probabilistic automata in 1969 and proved that there is no need to demand in cases of probabilistic branchings that total of probabilities for all possible continuations equal 1. He defined generalized probabilistic finite automata where the "probabilities" can be arbitrary real numbers, and that languages recognizable by these generalized probabilistic finite automata are the same as for ordinary probabilistic finite automata. Hence we also allow usage of all possible p-adic numbers in p-ultrametric machines. Remembering the theorem by P.Turakainen [17] we start with the most general possible definition hoping to restrict it if we below find examples of not so natural behavior of ultrametric automata. (Moreover, we do not specify all the details of the definitions in Theorems 1-4, and make the definition precise only afterwards. The reader may consider such a presentation strange but we need some natural examples of ultrametric automata before we concentrate on one standard definition.)

However, it is needed to note that if there is only one accepting state then the possible probabilities of acceptance are discrete values $0, p^1, p^{-1}, p^2, p^{-2}, p^3, \cdots$. Hence there is no natural counterpart of *isolated cut-point* or *bounded error* for ultrametric machines. On the other hand, a counterpart of Turakainen's theorem for probabilistic automata with isolated cut-point still does not exist. We also did not succeed to prove such a theorem for ultrametric automata. Most probably, there are certain objective difficulties.

Theorem 1. *There is a continuum of languages recognizable by finite ultrametric automata.*

Proof. Let $\beta = \cdots 2a_3 2a_2 2a_1 2a_0 2$ be an arbitrary p-adic number (not p-adic integer) where $p \geq 3$ and all $a_i \in \{0, 1\}$. Denote by B the set of all possible such β. Consider an automaton A_β with 3 states, the initial amplitudes of the states being $(\beta, -1, -1)$. The automaton is constructed to have the following property. If the input word is $2a_0 2a_1 2a_2 2a_3 2 \cdots 2a_n 2$ then the amplitude of the first state becomes $\cdots 2a_{n+4} 2a_{n+3} 2a_{n+2} 2a_{n+1} 2$. To achieve this, the automaton adds -2, multiplies to p, adds $-a_n$ and again multiplies to p.

Now let β_1 and β_2 be two different p-adic numbers. Assume that they have the same first symbols $a_m \cdots 2a_3 2a_2 2a_1 2a_0 2$ but different symbols a_{m+1} and b_{m+1}. Then the automaton accepts one of the words $a_{m+1} 2a_m \cdots 2a_3 2a_2 2a_1 2a_0 2$ and rejects the other one $b_{m+1} 2a_m \cdots 2a_3 2a_2 2a_1 2a_0 2$. Hence the languages are distinct.   □

Definition 3. *Finite p-ultrametric automaton is called* **integral** *if all the parameters of it are p-adic integers.*

Automata recognizing nonrecursive languages cannot be considered natural. Hence we are to restrict our definition.

Theorem 2. *There exists a finite integral ultrametric automaton recognizing the language $\{0^n 1^n\}$.*

Proof. When the automaton reads 0 it multiplies the amplitude to 2, and when it reads 1 it multiplies it to $\frac{1}{2}$. The norm of the amplitude equals p^0 iff the number of zeros is equal to the number of ones.   □

We consider the following language.

$$L = \{w | w \in \{0, 1\}^* \text{ and } w = w^{rev}\}$$

Theorem 3. *For every prime number $p \geq 5$, there is an integral p-ultrametric automaton recognizing L.*

Proof. The automaton has two special states. If the input word is

$$a(1)a(2) \cdots a(n)a(n+1)a(n+2) \cdots a(2n+1)$$

then one of these states has amplitude

$$a(1)p^n + \cdots + a(n)p^{+1} + a(n+1)p^0 + a(n+2)p^{-1} + \cdots + a(2n)p^{-n+1} + a(2n+1)p^{-n}$$

and the other one has amplitude

$$-a(1)p^{-n} - \cdots - a(n)p^{-1} - a(n+1)p^0 - a(n+2)p^{+1} - \cdots - a(2n)p^{+n-1} + a(2n+1)p^{+n}$$

If the sum of these two amplitudes equals 0 then the input word is a palindrome. Otherwise, the sum of amplitudes has a norm removed from p^0.   □

Definition 4. *A square matrix with elements being p-adic numbers is called* **balanced** *if for arbitrary row of the matrix the product of p-norms of the elements equals 1.*

Definition 5. *A finite ultrametric automaton is called* **balanced** *if all the matrices in its definition are balanced.*

Theorem 4. *If a language M can be recognized by a finite ultrametric automaton then M can be recognized also by a balanced finite ultrametric automaton.*

Proof. For every state of the automaton we add its duplicate. If the given state has an amplitude γ then its duplicate has the amplitude $\frac{1}{\gamma}$. Product of balanced matrices is balanced.   □

Definition 6. *A balanced finite ultrametric automaton is called* **regulated** *if there exist constants λ and c such that $0 < c < 1$ and for arbitrary input word x the norm $c\lambda <\parallel \gamma \parallel_p< \frac{\lambda}{c}$. We say that the word x is accepted if $\parallel \gamma \parallel_p> \lambda$ and it is rejected if $\parallel \gamma \parallel_p\leq \lambda$.*

Theorem 5. *(1) If a language M is recognized by a regulated finite ultrametric automaton then M is regular.*
(2) For arbitrary prime number p there is a constant c_p such that if a language M is recognized by a regulated finite p-ultrametric automaton with k states then there is a deterministic finite automaton with $(c_p)^{k.logk}$ states recognizing the language M.

4   Non-regulated Finite Automata

Since the numbers 1 and 0 are also p-adic numbers, every deterministic finite automaton can be described in terms of matrices for transformation of amplitudes. Hence every regular language is recognizable by a regulated p-ultrametric automaton. There is a natural problem : are there languages for which regulated p-ultrametric automata can have smaller complexity, i.e. smaller number of states.

The following 3 theorems seem to present such an example but there is a catch: these automata are not regulated because the norm of the amplitude to be measured can be arbitrary small (for lengthy input words).

Theorem 6. *For arbitrary prime number $p \geq 3$ the language*

$$L_{p-1} = \{1^n \mid n \equiv p - 1(\ mod\ p)\}$$

is recognizable by a p-ultrametric finite automaton with 2 states.

Proof. A primitive root modulo n is any number g with the property that any number coprime to n is congruent to a power of g modulo n. In other words, g is a generator of the multiplicative group of integers modulo n. Existence of primitive roots modulo prime numbers was proved by Gauss. The initial amplitude 1 of a special state in our automaton is multiplied to an arbitrary primitive root modulo p. When the end-marker is read the amplitude -1 of the other state is added to this amplitude. The result has p-norm p^0 iff $n \equiv p - 1$.   □

Theorem 7. *For arbitrary prime number $p \geq 3$ the language*

$$L_p = \{1^n \mid n \equiv p(\bmod\ p)\}$$

is recognizable by a p-ultrametric finite automaton with 2 states.

Proof. The value 1 of the amplitude of the second state is added to the amplitude of the accepting state at every step of reading the input word. The result has p-norm p^0 iff $n \equiv p$. □

Theorem 8. *For arbitrary natural number m there are infinitely many prime numbers p such that the language*

$$L_m = \{1^n \mid n \equiv 0(\bmod\ m)\}$$

is recognizable by a p-ultrametric finite automaton with 2 states.

Proof. Dirichlet prime number theorem, states that for any two positive coprime integers m and d, there are infinitely many primes of the form $m + nd$, where $n \geq 0$. In other words, there are infinitely many primes which are congruent to m modulo d. The numbers of the form $mn + d$ form an arithmetic progression

$$d,\ m + d,\ 2m + d,\ 3m + d,\ \ldots,$$

and Dirichlet's theorem states that this sequence contains infinitely many prime numbers.

Let p be such a prime and g be a primitive root modulo p. Then the sequence of remainders $g, g^2, g^3, \cdots$ modulo p has period m and $n \equiv 0(\bmod\ m)$ is equivalent to $g^n \equiv d(\bmod\ p)$. Hence the automaton multiplies the amplitude of the special state to g and adds $-d$ when reading the end-marker. □

5   Regulated Finite Automata

We wish to complement Theorem 5 by a proof showing that the gap between the complexity of regulated finite ultrametric automata and the complexity of deterministic finite automata is not overestimated. It turns out that this comparison is related to well-known open problems.

First, we consider a sequence of languages where the advantages of ultrametric automata over deterministic ones are super-exponential but the advantages are achieved only for specific values of the prime number p.

It is known that every p-permutation can be generated as a product of sequence of two individual p-permutations:

$$a = \begin{pmatrix} 1\ 2\ 3 \cdots p-1\ p \\ 2\ 3\ 4 \cdots\ \ \ p\ \ 1 \end{pmatrix}$$

$$b = \begin{pmatrix} 1\ 2\ 3 \cdots p-1\ p \\ 2\ 1\ 3 \cdots p-1\ p \end{pmatrix}$$

A string $x \in \{a, b\}^*$ is in the language M_p if the product of these p-permutations equals the trivial permutation.

Theorem 9. *(1) For arbitrary prime p, the language M_p is recognized by a p-ultrametric finite automaton with $p + 2$ states.*
(2) If a deterministic finite automaton has less than $p! = c^{p \cdot \log p}$ states then it does not recognize M_p.

Idea of the Proof. The ultrametric automaton gives initial amplitudes $0, 1, 2, \cdots, p - 1$ to p states of the automaton and after reading any input letter only permutes these amplitudes. After reading the endmarker from the input the automaton subtracts the values $0, 1, 2, \cdots, p - 1$ from these amplitudes. □

6   1-Way Pushdown Automata

Let $A = \{a, b, c, d, e, f, g, h, k, l, m, p, q, r, s, t, u, v\}$. Now we consider a language T in the alphabet $A \cup \{\#\}$ for which both the deterministic and probabilistic 1-way pushdown automata cannot recognize the language but there exists an ultrametric 1-way pushdown automaton recognizing it.

The language T is defined as the set of all the words x in the input alphabet such that either x is in all 9 languages T_i described below or in exactly 6 of them or in exactly 3 of them or in none of them where

$$T_1 = \{x\#y \mid x \in A^* \wedge y \in A^* \wedge proj_{ab}(x) = proj_{ab}(y)\},$$

$$T_2 = \{x\#y \mid x \in A^* \wedge y \in A^* \wedge proj_{cd}(x) = proj_{cd}(y)\},$$

$$T_3 = \{x\#y \mid x \in A^* \wedge y \in A^* \wedge proj_{ef}(x) = proj_{ef}(y)\},$$

$$T_4 = \{x\#y \mid x \in A^* \wedge y \in A^* \wedge proj_{gh}(x) = proj_{gh}(y)\},$$

$$T_5 = \{x\#y \mid x \in A^* \wedge y \in A^* \wedge proj_{kl}(x) = proj_{kl}(y)\},$$

$$T_6 = \{x\#y \mid x \in A^* \wedge y \in A^* \wedge proj_{mp}(x) = proj_{mp}(y)\},$$

$$T_7 = \{x\#y \mid x \in A^* \wedge y \in A^* \wedge proj_{qr}(x) = proj_{qr}(y)\},$$

$$T_8 = \{x\#y \mid x \in A^* \wedge y \in A^* \wedge proj_{st}(x) = proj_{st}(y)\},$$

$$T_9 = \{x\#y \mid x \in A^* \wedge y \in A^* \wedge proj_{uv}(x) = proj_{uv}(y)\}.$$

Theorem 10. *For the language T we have the following properties.*

(1) *There is a regulated 3-ultrametric 1-way pushdown automaton recognizing the language T.*
(2) *No deterministic 1-way pushdown automata can recognizing the language T.*
(3) *No probabilistic 1-way pushdown automata can recognizing the language T can have bounded error.*

7   Turing Machines

We denote by pUP the class of all languages recognizable by p-ultrametric Turing machines in a polynomial time. This is a large class of languages.

Theorem 11. *If a language M is recognizable by a probabilistic Turing machine in a polynomial time then for arbitrary $p \geq 3$ there is a p-ultrametric Turing machine recognizing M in a polynomial time.*

Proof. The class PP of all languages recognizable in a polynomial time has natural complete problems, for example, $MAJSAT$. $MAJSAT$ is a decision problem in which one is given a Boolean formula F. The answer must be YES if more than half of all assignments $x_1, x_2, \cdots, x_n$ make F true and NO otherwise. Hence M is reducible to $MAJSAT$ in deterministic polynomial time. On the other hand, $MAJSAT$ is recognizable by a p-ultrametric Turing machine in a polynomial time. This machine considers in parallel all possible assignments for $x_1, x_2, \cdots, x_n$ and adds a p-adic number 2^{-n} to the amplitude α of a special state. F is in $MAJSAT$ iff the resulting amplitude α has p-norm 0.   □

Definition 7. *A discrete Riemann surface on the rectangle $[a, b] \times [c, d]$ is a map from (x, y, z) (where $x \in [a, b]$, $y \in [c, d]$ and z is a string of symbols from a finite alphabet Σ whose length equals $y - c$) to a finite alphabet Δ. For each triple its neighbors are defined as the triples:*
(1) (x, y', z) where either $y' = y + 1$ or $y' = y + 1$,
(2) (x', y, z') where either $x' = x - 1$ and z' is z with the last symbol omitted, or $x' = x + 1$ and z' is z with the one symbol attached at its end.

Definition 8. *A discrete Dirichlet condition is a 5-tuple consisting of: (1) a map from (x, y) where $y = c$ to Δ, (2) a map from (x, y) where $y = d$ to Δ, (3) (x, y) where $x = a$ to Δ, (4) (x, y) where $x = b$ to Δ, and (5) neighboring conditions that may forbid some simultaneous maps of neighboring triples.*

Definition 9. *The discrete Dirichlet problem is whether or not it is possible a Riemann surface consistent with the given discrete Dirichlet condition.*

Theorem 12. *For arbitrary prime number $p \geq 3$, there is a pUP-complete language.*

Idea of the Proof. The language consists of all discrete Dirichlet conditions such that the discrete Dirichlet problem has a positive answer. The map in the Riemann surface can be used to describe the work of a ultrametric Turing machine. The symbols of Δ for all possible values of x for a fixed y and z describe the configuration of the tape at the moment y with the choices z made before the moment y and the amplitudes accumulated. The discrete Dirichlet problem asks whether the ultrametric machine accepts the input word. The difference $d - c$ represents the computation time allowed.   □

References

1. Ablayev, F.M., Freivalds, R.: Why Sometimes Probabilistic Algorithms Can Be More Effective. In: Wiedermann, J., Gruska, J., Rovan, B. (eds.) MFCS 1986. LNCS, vol. 233, pp. 1–14. Springer, Heidelberg (1986)
2. Artin, E.: Beweis des allgemeinen Reziprozitätsgesetzes. Mat. Sem. Univ. Hamburg, B.5, 353–363 (1927)
3. Dragovich, B., Dragovich, A.: A p-Adic Model of DNA Sequence and Genetic Code. p-Adic Numbers, Ultrametric Analysis, and Applications 1(1), 34–41 (2009)
4. Ershov, Y.L.: Theory of numberings. In: Griffor, E.R. (ed.) Handbook of Computability Theory, pp. 473–503. North-Holland, Amsterdam (1999)
5. Freivalds, R.: Complexity of Probabilistic Versus Deterministic Automata. In: Barzdins, J., Bjorner, D. (eds.) Baltic Computer Science. LNCS, vol. 502, pp. 565–613. Springer, Heidelberg (1991)
6. Freivalds, R.: How to Simulate Free Will in a Computational Device. ACM Computing Surveys 31(3), 15 (1999)
7. Freivalds, R.: Non-Constructive Methods for Finite Probabilistic Automata. International Journal of Foundations of Computer Science 19(3), 565–580 (2008)
8. Freivalds, R.: Ultrametric automata and Turing machines. In: Voronkov, A. (ed.) Turing-100. EPiC Series, vol. 10, pp. 98–112. EasyChair (2012)
9. Garret, P.: The Mathematics of Coding Theory. Pearson Prentice Hall, Upper Saddle River (2004)
10. Gouvea, F.Q.: p-adic Numbers: An Introduction (Universitext), Springer, 2nd edn. Springer (1983)
11. Hirvensalo, M.: Quantum Computing. Springer, Heidelberg (2001)
12. Khrennikov, A.Y.: Non Archimedean Analysis: Quantum Paradoxes, Dynamical Systems and Biological Models. Kluwer Academic Publishers (1997)
13. Koblitz, N.: p-adic Numbers, p-adic Analysis, and Zeta-Functions, 2nd edn. Graduate Texts in Mathematics, vol. 58. Springer (1984)
14. Kolmogorov, A.N.: Three approaches to the quantitative definition of information. Problems in Information Transmission 1, 1–7 (1965)
15. Kozyrev, S.V.: Ultrametric Analysis and Interbasin Kinetics. p-Adic Mathematical Physics. In: Proc. of the 2nd International Conference on p-Adic Mathematical Physics, American Institute Conference Proceedings, vol. 826, pp. 121–128 (2006)
16. Madore, D.A.: A first introduction to p-adic numbers, http://www.madore.org/~david/math/padics.pdf
17. Turakainen, P.: Generalized Automata and Stochastic Languages. Proceedings of the American Mathematical Society 21(2), 303–309 (1969)
18. Vladimirov, V.S., Volovich, I.V., Zelenov, E.I.: p-Adic Analysis and Mathematical Physics. World Scientific, Singapore (1995)
19. Weyl, H.: The concept of a Riemann surface. Dover Publications, New York (2009)

Recompression: Word Equations and Beyond

Artur Jeż[1,2,*]

[1] Max Planck Institute für Informatik,
Campus E1 4, DE-66123 Saarbrücken, Germany
[2] Institute of Computer Science, University of Wrocław
ul. Joliot-Curie 15, 50-383 Wrocław, Poland
aje@cs.uni.wroc.pl

Abstract. We present the technique of local recompression on the example of word equations. The technique is based on local modification of variables (replacing X by aX or Xa) and replacement of pairs of letters appearing in the equation by a 'fresh' letter, which can be seen as a bottom-up building of an SLP (Straight-Line Programme) for the solution of the word equation, i.e. a *compression*.

Using this technique we give a simple proof that satisfiability of word equations is in **PSPACE**. Furthermore we sketch the applications for some problems regarding the SLP compressed strings.

1   Introduction

Local Recompression. We demonstrate the technique of local recompression. It was developed for compressed membership problem for finite automata [6] and it was later found that it is applicable also to other problems for compressed data [4,5]. However, its most unexpected and surprising application is in the area of word equations, for which it gives simple proofs for many previously known results. In this paper we explain the technique using the word equations as the working example. In this way we obtain a nondeterministic algorithm that works in time $\mathcal{O}(\log N \mathsf{poly}(n))$ and in $\mathcal{O}(n^2)$ space, where n is the size of the input equation and N the size of the smallest solution. Furthermore, for $\mathcal{O}(1)$ variables a more careful analysis yields that the space consumption is $\mathcal{O}(n)$, thus showing that this case is context-sensitive. Lastly, the algorithm can be easily generalised to a generator of all solutions.

Word Equations. The problem of *word equations* is one of the oldest in computer science: given words U and V, consisting of letters (from Σ) and variables (from $\mathcal{X}$) we are to check the *satisfiability*, i.e. decide whether there is a substitution for variables that turns this formal equation into an equality of strings. It is useful to think of a solution S as a homomorphism $S : \Sigma \cup \mathcal{X} \mapsto \Sigma^*$, which is an identity on Σ. In the more general problem of *solving* the equation,

---

* Supported by NCN grant number 2011/01/D/ST6/07164, 2011–2014 and by Humboldt Foundation Research Fellowship.

M.-P. Béal and O. Carton (Eds.): DLT 2013, LNCS 7907, pp. 12–26, 2013.
© Springer-Verlag Berlin Heidelberg 2013

we are to give representation of all solutions of the equation. This problem was first solved by Makanin [13] and the currently best PSPACE algorithm is due to Plandowski [16].

Local Recompression and Word Equations. The idea of the technique is easily explained in terms of solutions of the equations rather than the equations themselves: consider a solution $S(U) = S(V)$ of the equation $U = V$. In one phase we first list all pairs of different letters ab that appear as substrings in $S(U)$ and $S(V)$. For a fixed pair ab of this kind we greedily replace all appearances of ab in $S(U)$ and $S(V)$ by a 'fresh' letter c. (A slightly more complicated action is performed for pairs aa, for now we ignore this case to streamline the presentation of the main idea). There are possible conflicts between such replacements for different types of pairs (consider string aba, in which we try to replace both pairs ab and ba), we resolve them by introducing some arbitrary order on types of pairs and performing the replacement for one type of pair at a time, according to the order. When all such pairs are replaced, we obtain another equal pair of strings $S'(U')$ and $S'(V')$ (note that the equation $U = V$ may have changed, say into $U' = V'$). Then we iterate the process. In each phase the strings are shortened by a constant factor, hence after $\mathcal{O}(\log N)$ phases we obtain constant-length strings. The original equation is solvable if and only if the obtained strings are the same.

The problematic part is that the operations are performed on the solutions, which can be large. If we simply guess S and then perform the compressions, the running time is polynomial in N. Instead we perform the compression directly on the equation (the *recompression*): the pairs ab appearing in the solution are identified using only the equation and the compression of the solution is two-fold: the pairs ab from U and V are replaced explicitly and the pairs fully within some $S(X)$ are replaced implicitly, by changing S (which is not stored). However, not all pairs of letters can be compressed in this way, as some of them appear on the 'crossing' between a variable and a constant: consider for instance $S(X) = ab$, a string of symbols Xc and a compression of a pair bc. This is resolved by *local decompression*: when trying to compress the pair bc in the example above we first replace X by Xb (implicitly changing $S(X)$ from ab to a), obtaining the string of symbols Xbc, in which the pair bc can be easily compressed.

Example 1. Consider an equation $aXca = abYa$ with a solution $S(X) = baba$ and $S(Y) = abac$. In the first phase, the algorithm wants to compress the pairs ab, ca, ac, ba, say in this order. To compress ab, it replaces X with bX, thus changing the substitution into $S(X) = aba$. After compression we obtain equation $a'Xca = a'Ya$. Notice, that this implicitly changed to solution into $S(X) = a'a$ and $S(Y) = a'ac$. To compress ca (into c'), we replace Y by Yc, thus implicitly changing the substitution into $S(Y) = a'a$. Then, we obtain the equation $a'Xc' = a'Yc'$ with a solution $S(X) = a'a$ and $S(Y) = a'a$. Remaining pairs no longer appear in the equation, so we proceed to the next phase.

Related Techniques. While the presented method of recompression is relatively new, some of its ideas and inspirations go quite back. It was developed in order to deal with fully compressed membership problem for NFA and the previous work on this topic by Mathissen and Lohrey [12] already implemented the idea of replacing strings with fresh letters as well as modifications of the instance so that this is possible. Furthermore they identified the importance of maximal blocks of a single letter and dealt with them appropriately. However, the replacement was not iterated, and the newly introduced letters could not be further compressed.

The idea of replacing short strings by a fresh letter and iterating this procedure was used by Mehlhorn et. al [14] in their work on data structure for equality testing for dynamic strings.

A similar technique, based on replacement of pairs and blocks of the same letter was proposed by Sakamoto [19] in the context of constructing a smallest grammar generating a given word. His algorithm was inspired by a practical grammar-based compression algorithm RePair [9]. It possessed the important features of the method: iterated replacement, and ignoring letters recently introduced. However, the analysis that stressed the modification of the variables was not introduced and it was done in a different way.

2   Toy Example: Equality of Strings

In this section we introduce the first 'half' of the recompression technique and apply it in the trivial case of equality testing of two explicit strings (say $u, v \in \Sigma^*$), i.e. their representation is not compressed. This serves as an easy introduction.

In case of equality testing, our approach is similar to the one of Mehlhorn et al. [14] from their work on equality testing for dynamic strings. The proposed method was based on iterative replacement of strings: they defined a schema, which replaced a string s with a string s' (where $|s'| \leq c|s|$ for some constant $c < 1$) and iterated the process until a length-1 string was obtained. Most importantly, the replacement is injective, i.e. if $s_1 \neq s_2$ then they are replaced with different strings.[1] In this way, for each string we calculate its unique *signature* and two strings are equal if and only if their signatures are.

The second important property of this schema is that the replacement is 'local': s is partitioned into blocks of $\mathcal{O}(1)$ letters and each of them is replaced independently.

The recompression, as presented in this section, is a variant of this approach, in which a different replacement schema is applied. To be more specific, our algorithm is based on two types of 'compressions' performed on strings:

pair compression of ab For two different letters ab appearing in v or u replace each of ab in v and u by a *fresh* letter c.

a**'s block compression** For each maximal block a^ℓ, with $\ell > 1$, that appears in v replace all a^ℓs in v and u by a fresh letter a_ℓ.

---

[1] This is not a information-theory problem, as we replace only strings that appear in the instance and moreover can reuse original letters.

By a *maximal block* we denote a substring of the form a^k that cannot be extended by a to the left nor to the right. By a fresh letter we denote any letter that does not appear in v or u. We adopt the following notational convention throughout rest of the paper: whenever we refer to a letter a_ℓ, it means that the last block compression was done for a and a_ℓ is the letter that replaced a^ℓ.

Clearly, both compressions operations preserve the equality of strings.

Lemma 1. *Let v', u' be obtained from v and u by a pair compression (or block compression). Then $v = u$ if and only if $v' = u'$.*

Using those two operations, we can define SimpleEqualityTesting, which tests the equality of two strings.

Its crucial property is that in each iteration of the main loop the lengths of v and u shorten by a constant factor.

Algorithm 1. SimpleEqualityTesting

1: **while** $|v| > 1$ and $|u| > 1$ **do**
2:     $L \leftarrow$ list of letters appearing in u and v
3:     **for** each $a \in L$ **do** compress blocks of a
4:     $P \leftarrow$ list pairs appearing in u and v
5:     **for** each $ab \in P$ **do** compress pair ab
6: Naively check the equality

Lemma 2. *When $|v|, |u| > 1$ then one iterations of the main loop of SimpleEqualityTesting shortens those lengths by a constant factor. In particular, there are $\mathcal{O}(\log(\min(|v|, |u|)))$ such iterations.*

Proof. We call one iteration of the main loop a *phase*. We first show that for two consecutive letters of v (the proof for u is the same) at least one of them is compressed in this phase.

Claim 1. *Consider any two consecutive letters a and b in v or u at the beginning of the phase. Then at least one of those letters is compressed till the end of the iteration of the .*

Proof. Suppose for the sake of contradiction that both of them are not compressed. If they are the same, then they are compressed during the blocks compression, contradiction. So suppose that they are different. Since none of them is compressed during the block compression $ab \in P$ and we try to compress this appearance during the pair compressions. This fails if and only if one of letters from this appearance was already compressed when we considered ab during the pair compression. □

So each uncompressed letter can be associated with a letter to the left and to the right, which were compressed (the first and last letter can be only associated with a letter to the right/left, respectively). Since when a substring is compressed, it is of length at least two, this means that no compressed letter is associated with two uncompressed letters. So, for a pattern v there are at most $\frac{|v|+2}{3}$ uncompressed letters and at least $\frac{2|v|-2}{3}$ compressed ones. Hence, the length of the pattern at the end of a phase is at most

$$\frac{|v|+2}{3} + \frac{1}{2} \cdot \frac{2|v|-2}{3} = \frac{2|v|+1}{3} \le \frac{5}{6}|v| \ ,$$

where the last inequality holds in the interesting case of $|v| > 1$. □

Now, by iterative application of Lemma 1 each compression performed by SimpleEqualityTesting preserves the equality of strings, so it returns a proper answer.

Theorem 1. SimpleEqualityTesting *properly tests the equality of* v *and* u.

3   Word Equations

We shall now adopt the local recompression to word equations. Observe that in this problems we also want to test equality of two strings, which are unfortunately given implicitly. We start with some necessary definitions.

By Σ we denote the set of letters appearing in the equation $U = V$ or are used for representation of compressed strings, $\mathcal{X}$ denotes a set of variables. The equation is written as $U = V$, where $U, V \in (\Sigma \cup \mathcal{X})^*$. By $|U|, |V|$ we denote the length of U and V, n denotes the length of the input equation.

A *substitution* is a morphism $S : \mathcal{X} \cup \Sigma \mapsto \Sigma^*$, such that $S(a) = a$ for every $a \in \Sigma$, substitution is naturally extended to $(\mathcal{X} \cup \Sigma)^*$. A *solution* of an equation $U = V$ is a substitution S, such that $S(U) = S(V)$; a solution S is a *length-minimal*, if for every solution S' it holds that $|S(U)| \leq |S'(U)|$.

We want to replace pairs of letters (and blocks) appearing in $S(U)$ and $S(V)$. Since we do not know S, we face the first problem: how to actually list all the pairs and maximal blocks? Clearly some non-deterministic guessing is needed and the below Lemma 3 shows that in fact those guesses can be very local: it is enough to guess the first (say a) and last (say b) letter of $S(X)$ as well as the lengths of the a-prefix and b-suffix of $S(X)$, for each X. Before Lemma 3, though, we need to classify the appearances of pairs and letters:

Definition 1. *Given an equation* $U = V$, *a substitution* S *and a substring* $u \in \Sigma^+$ *of* $S(U)$ *(or* $S(V)$*) we say that this appearance of* u *is* explicit, *if it comes from substring* u *of* U *(or* V, *respectively);* implicit, *if it comes (wholly) from* $S(X)$ *for some variable* X; crossing *otherwise. A string* u *is* crossing *(with respect to* S*) if it has a crossing appearance and* non-crossing *otherwise.*

We say that a pair of ab *is a* crossing pair *(with respect to* S*), if* ab *is crossing. Otherwise, a pair is* non-crossing *(with respect to* S*). Similarly, a letter* $a \in \Sigma$ *has a* crossing block *(with respect to* S*), if there is a block of* as *which has a crossing appearance.*

Unless explicitly stated, we consider crossing/non-crossing pairs ab in which $a \neq b$.

Lemma 3 (cf. [18, Lemma 6]). *Let* S *be a length-minimal solution of* $U = V$.

- *If* ab *is a substring of* $S(U)$, *where* $a \neq b$, *then* ab *is an explicit pair or a crossing pair.*
- *If* a^k *is a maximal block in* $S(U)$ *then there is an appearance of* a^k *which is crossing or explicit.*

Thus, to determine the set of pairs appearing in $S(U)$, $S(V)$ (for a length-minimal S) it is enough to guess the first (a) and last (b) letter of each $S(X)$ and the lengths of the a-prefix, b-suffix of $S(X)$, as those determine all crossing pairs and (lengths of) crossing blocks.

Note that we need to (at least temporarily) store the guessed lengths of the prefixes/suffixes, so it would be unfortunate if they are unbounded. However, it is known that for length-minimal solutions this is not the case, as they can be upper bounded using the well-known exponential bound on exponent of periodicity:

Lemma 4 (Exponent of periodicity bound [8]). *If solution S is length-minimal and $w^\ell \neq \epsilon$ is a substring of $S(U)$, then $\ell \leq 2^{c(|U|+|V|)}$ for some constant $0 < c < 2$.*

The compression of crossing/non-crossing pairs (blocks) are essentially different, as shown in the rest of this section.

Compression of Noncrossing Pairs and Blocks. When ab is non-crossing, each of its appearance in $S(U)$ is either explicit or implicit. Thus, to perform the pair compression of ab on $S(U)$ it is enough to separately replace each explicit pair ab in U and change each ab in $S(X)$ for each variable X. The latter is of course done implicitly (as $S(X)$ is not written down anywhere).

Similarly when none block of a has a crossing appearance, the a's blocks compression consists simply of replacing explicit a blocks.

Algorithm 2. PairCompNCr(a, b)

1: let $c \in \Sigma$ be an unused letter
2: replace each explicit ab in U and V by c

Algorithm 3. BlockCompNCr(a)

1: **for** each explicit a's ℓ-block in U or V **do**
2:     let $a_\ell \in \Sigma$ be an unused letter
3:     replace explicit a's ℓ-blocks in U or V by a_ℓ

We need some notions to formally state that the satisfiability of an equation is preserved by our procedures, especially that there are some nondeterministic choices involved. We say that a nondeterministic procedure *preserves unsatisfiability*, when given a unsatisfiable word equation $U = V$ it cannot transform it to a satisfiable one, regardless of the nondeterministic choices; such a procedure *preserves satisfiability*, if given a satisfiable equation $U = V$ for some nondeterministic choices it returns a satisfiable equation $U' = V'$. Sometimes we explicitly state for which choices the satisfiability is preserved.

Lemma 5. PairCompNCr(a, b) *preserves the unsatisfiability; if ab is a non-crossing pair for some solution S then it preserves satisfiability.*

BlockCompNCr(a) *preserves unsatisfiability; if a has no crossing blocks for some solution S then it preserves satisfiability.*

Crossing Pairs and Blocks Compression. The presented algorithms cannot be directly applied to crossing pairs or to compression of a's blocks that have crossing appearances. However, it is still possible to perform a reduction from the difficult (crossing) to the easy (non-crossing) case, at least for a fixed pair. To do this, we modify the instance: if a pair ab is crossing because there is a variable X such that $S(X) = bw$ for some word w and a is to the left of X, it is enough to *left-pop* b from $S(X)$: replace each X with bX and implicitly change S, so that $S(X) = w$; similar action is applied to variables Y ending with a and with b to the right (*right-popping* a from $S(X)$). Afterwards, ab is non-crossing with respect to S.

Lemma 6. Pop(a, b) *preserves satisfiability and unsatisfiability. If $U = V$ is satisfiable then for some nondeterministic choices the obtained $U' = V'$ has a solution S' such that ab is non-crossing (with respect to S').*
   It introduces at most $2n$ new letters to the equation.

By Lemma 6 the pair ab is non-crossing after Pop(a, b) hence we can compress ab using PairCompNCr.

Lemma 7. PairComp(a, b) *preserves satisfiability and unsatisfiability.*

Algorithm 4. Pop(a, b)

1: **for** $X \in \mathcal{X}$ **do**
2:     **if** b is the first letter of $S(X)$ **then** ▷ Guess
3:        replace each X in U and V by bX
4:     ▷ Implicitly change $S(X) = bw$ to $S(X) = w$
5:        **if** $S(X) = \epsilon$ **then** ▷ Guess
6:           remove X from the equation
7:     ▷ Do a symmetric action for the last letter

Algorithm 5. PairComp(a, b)

1: run Pop(a, b)
2: run PairCompNCr(a, b)

The problems with crossing blocks can be solved in a similar fashion: a has a crossing block, if aa is a crossing pair. So we 'left-pop' a from X until the first letter of $S(X)$ is different than a, we do the same with the ending letter. This can be alternatively seen as removing the whole a-prefix (a-suffix, respectively) from X: suppose that $S(X) = a^\ell w a^r$, where w does not start nor end with a. Then we replace each X by $a^\ell X a^r$, implicitly changing the solution to $S(X) = w$. The obtained equation has a solution for which a has no crossing blocks.

Algorithm 6. CutPrefSuff(a)

1: **for** $X \in \mathcal{X}$ **do**
2:     guess ℓ, r ▷ $S(X) = a^\ell w a^r$
3:     replace each X in U and V by $a^\ell X a^r$
4:                  ▷ implicitly change $S(X) = a^\ell w a^r$ to $S(X) = w$
5:     **if** $S(X) = \epsilon$ **then**   ▷ Guess
6:        remove X from the equation

Lemma 8. CutPrefSuff *preserves unsatisfiability and satisfiability. If* $U = V$ *is satisfiable then for appropriate nondeterministic choices it returns an equation* $U' = V'$ *that has a solution* S' *for which* a *has no crossing blocks.*

After CutPrefSuff we can compresses maximal a blocks.

Algorithm 7. BlockComp(a)

1: run CutPrefSuff(a)
2: BlockCompNCr(a)

Lemma 9. BlockComp *preserves unsatisfiability and satisfiability.*
   It introduces at most $2n$ *letters to the equation.*

The uncrossing procedures (Pop, CutPrefSuff) pop the letters into the equation and so increase equations size. The good news is that the number of crossing pairs and crossing blocks depends solely on the number of variables and not on the size of the actual equation. In particular, it can be bounded in terms of the input equation's size.

Lemma 10. *There are at most* $2n$ *crossing pairs and at most* $2n$ *letters with a crossing blocks.*

The proof is obvious: every crossing pair can be associated with an appearance of a variable in the equation and at most two pairs can be associated with a given appearance, the same applies to blocks.

Main Algorithm. Now, the algorithm for testing satisfiability of word equations can be conveniently stated. We refer to one iteration of the main loop in WordEqSat as one *phase*.

Algorithm 8. WordEqSat Checking the satisfiability of a word equation

---
1: **while** $|U| > 1$ or $|V| > 1$ **do**
2:    $L \leftarrow$ letters in $U = V$ without crossing blocks   ▷ Guess
3:    $L' \leftarrow$ letters in $U = V$ with crossing blocks   ▷ Guess, $\mathcal{O}(n)$ many
4:    **for** $a \in L$ **do** BlockCompNCr(a)
5:    **for** $a \in L'$ **do** BlockComp(a)
6:    $P \leftarrow$ noncrossing pairs in $U = V$ ▷ Guess
7:    $P' \leftarrow$ crossing pairs in $U = V$ ▷ Guess, $\mathcal{O}(n)$ many
8:    **for** $ab \in P$ **do** PairCompNCr(a, b)
9:    **for** $ab \in P'$ **do** PairComp(a, b)
10: Solve the problem naively   ▷ With sides of length 1, the problem is trivial

---

Theorem 2. WordEqSat *nondeterministically verifies the satisfiability of word equations. It can verify an existence of a length-minimal solution of length* N *in* $\mathcal{O}(\text{poly}(n) \log N)$ *time and* $\mathcal{O}(n^2)$ *space.*

As in case of Lemma 2 for SimpleEqualityTesting, the crucial property is that one phase of WordEqSat halves the solution's length, which is formally stated in the below lemma:

Lemma 11. *Let $U = V$ have a solution S. For appropriate nondeterministic choices the equation $U' = V'$ obtained at the end of the phase has a solution S' such that i) at least 2/3 of letters in U or V are compressed in U' or V'; ii) at least 2/3 of letters in $S(U)$ are compressed in $S'(U')$.*

Lemma 11 is enough to show the bound on used memory: on one hand the Pop and CutPrefSuff introduce $\mathcal{O}(n)$ new letters per uncrossed pair or letter and as there are $\mathcal{O}(n)$ such pairs and letters, see Lemma 10, in total there are $\mathcal{O}(n^2)$ letters introduced to the equation in one phase. On the other hand U and V are shortened by a constant factor; together those two yield a quadratic bound on $|U'|$ and $|V'|$. Moreover, Lemma 11 alone yields that for some choices there are $\mathcal{O}(\log N)$ phases.

4   Speeding Up the Recompression

The satisfiability problem for word equations is NP-hard and so 'efficiency' of the implementation of the recompression (running time, exact space consumption) is in general not crucial. However, in some cases (one variable, $\mathcal{O}(1)$ variables) the exact running time and space usage are in fact important, moreover, application of the recompression to SLPs, as described in Section 6, usually yield deterministic polynomial time algorithms, for which the running time is crucial. In this section we present some improvements of the recompression, which are important in some applications, listed in Sections 5 and 6.

4.1   Limiting the Number of Crossing Pairs

The number of crossing pairs can be bounded by $2n_v$, where n_v is the number of appearance sof variables in the equation. Using a simple preprocessing this can be reduced to $2|\mathcal{X}|$, where $|\mathcal{X}|$ is the number of different variables in the equation. The idea is quite simple: for each variable X we left-pop and right-pop a letter from it, in this way each appearance of X is always preceeded (succeeded) by the same letter.

Algorithm 9. Preproc

1: **for** $X \in \mathcal{X}$ **do**
2:     let a be the first letter of $S(X)$ ▷ Guess
3:     replace each X in U and V by aX
4:                          ▷ Implicitly change $S(X) = aw$ to $S(X) = w$
5:     **if** $S(X) = \epsilon$ **then**   ▷ Guess
6:         remove X from the equation
7:                          ▷ Do a symmetric action for the last letter

---

Lemma 12. *After* Preproc *there are at most* $2|\mathcal{X}|$ *crossing pairs.* Preproc *introduces one symbol to the left and one to the right of each appearance of a variable in the equation.*

The proof is fairly obvious. The Preproc is run before the calculation of the crossing pairs. It introduces in total $\mathcal{O}(n_v)$ letters to the equation, so asymptotically does not influence the size of the size of the stored equation. However, for $\mathcal{O}(1)$ variables the improved estimation on the umber of crossing pairs is crucial.

4.2 Parallel Compression of Crossing Pairs

The Pop and PairComp work for a fixed pair and it seems that they cannot uncross and compress arbitrary pairs in parallel. However, they can process in parallel pairs of a specific form: consider a partition of alphabet Σ to Σ_ℓ and Σ_r. Then two pairs from $\Sigma_\ell \Sigma_r$ cannot overlap and so all such pairs can be compressed in parallel. Furthermore, they can all be uncrossed in parallel, using a variant of Pop.

Algorithm 10. PopImp(Σ_ℓ, Σ_r)

1: **for** $X \in \mathcal{X}$ **do**
2:      let $b \leftarrow$ the first letter of $S(X)$ ▷ Guess
3:      **if** $b \in \Sigma_r$ **then**
4:          replace each X in U and V by bX
5:          ▷ Implicitly change $S(X) = bw$ to $S(X) = w$
6:          **if** $S(X) = \epsilon$ **then**   ▷ Guess
7:              remove X from the equation
8:          ▷ Do a symmetric action for the last letter

It can be shown that PopImp uncrosses all pairs from $\Sigma_\ell \Sigma_r$

Lemma 13. Pop(Σ_ℓ, Σ_r) *preserves satisfiability and unsatisfiability. If* $U = V$ *is satisfiable then for some nondeterministic choices the obtained* $U' = V'$ *has a solution* S' *such that no pair from* $\Sigma_\ell \Sigma_r$ *is crossing (with respect to* S'*).*

It introduces at most $2n$ *new letters to the equation.*

Thus after PopImp(Σ_ℓ, Σ_r) we can compress all pairs from $\Sigma_\ell \Sigma_r$.

Still it is left to define a partition (Σ_ℓ, Σ_r). There are basically two ways of doing so: either we generate a set of partitions, such that each crossing pair is in at least one of those partitions, or we can choose a special partition, such that a constant fraction of all appearances of crossing pairs are in this partition.

Lemma 14. *For solution* S *of an equation* $U = V$ *there exists a partition* (Σ_ℓ, Σ_r) *such that at least one fourth of appearances of pairs in* $S(U) = S(V)$ *is in* $\Sigma_\ell \Sigma_r$.

For an equation $U = V$ *there exists a partition* (Σ_ℓ, Σ_r) *such that at least one fourth of appearances of pairs in* $U = V$ *is in* $\Sigma_\ell \Sigma_r$.

For a set of crossing pairs P' *there exists a set of partitions with* $\mathcal{O}(\log |P'|)$ *partitions such that each element of* P' *is in at least one of those partitions.*

Each of the claims follows by the same probabilistic argument: suppose that we partition Σ randomly, with each element going to Σ_ℓ or Σ_r with probability $1/2$. Then a fixed appearance of a pair of different letters is covered by this random partition with probability $1/4$ and so the expected number of pairs covered in $S(U) = S(V)$ (or $U = V$, or in P') is $1/4$, so there is a partition for which this is indeed $1/4$ of all pairs. For the last claim of the lemma note that one partition reduces the size of P' by $3/4$, so we need to iterate it $\mathcal{O}(\log|P'|)$ times.

Thus there are two ways to use the partitions: either we find (guess) the $\mathcal{O}(\log|P'|)$ partitions and compress pairs in each of them, or we find only two partitions: one which covers $1/4$ of pairs in $S(U) = S(V)$ (so that the length of the solution drops by a constant factor) and one that covers $1/4$ of pairs in $U = V$ (so that the length of U and V drop by a constant factor). Both version lead to smaller running time and smaller equation size.

4.3   Uncrossing Blocks in Parallel

It is relatively easy to show that all blocks of letters can be uncrossed in parallel, reducing the running time and the number of letters introduced into the equations during the block compression. To this end it is enough to apply Cut-PrefSuff once, but pop the a-prefix and b-suffix from each variable regardless of what letter a and b are.

Algorithm 11. CutPrefSuffImp

1: **for** $X \in \mathcal{X}$ **do**
2:      guess a ℓ, b and r $\triangleright S(X) = a^\ell w b^r$
3:      replace each X in U and V by $a^\ell X b^r$
4: $\triangleright$ implicitly change $S(X) = a^\ell w b^r$ to $S(X) = w$
5:      **if** $S(X) = \epsilon$ **then** $\triangleright$ Guess
6:          remove X from the equation

Lemma 15. CutPrefSuffImp *preserves unsatisfiability and satisfiability. If $U = V$ is satisfiable then for appropriate nondeterministic choices it returns an equation $U' = V'$ that has a solution S' which has no crossing blocks.*

Using CutPrefSuffImp instead of CutPrefSuff guarantees that only $\mathcal{O}(n_v)$ new letters are introduced into $U = V$.

5   Other Results for Word Equations

Using the approach of recompression we can give (alternative and often simpler) proofs of some (old and new) results on word equations.

Double Exponential Bound on Minimal Solutions. The running time of WordEqSat is polynomial in n and $\log N$ and it is easy to also *lower-bound* it in terms of $\log N$. On the other hand the length of the stored equations is $\mathcal{O}(n)$, which yields that there are exponentially (in n) many different configurations. Comparing those two bounds yields a doubly exponential bound on N.

Exponential Bound on Exponent of Periodicity. For a word w the *exponent of periodicity* $per(w)$ is the maximal k such that u^k is a substring of w, for some $u \in \Sigma^+$; Σ-*exponent of periodicity* $per_\Sigma(w)$ restricts the choice of u to Σ. This notion is naturally transferred to equations: For an equation $U = V$, define the exponent of periodicity as $\max_S[per(S(U))]$, where the maximum is taken over all length-minimal solutions S of $U = V$; define the Σ-*exponent of periodicity* of $U = V$ in a similar way.

One of the important part of Makanin's work was to bound the exponent of periodicity in terms of the equations size. This notion remained important in the following work on word equations and the tight exponential bound was shown by Kościelski and Pacholski [8]. WordEqSat uses only Σ-exponent of periodicity (to bound the lengths of the a-prefixes and suffixes), which can be shown by restricting the proof of Kościelski and Pacholski [8] to algebraic estimations. On the other hand, a more involved analysis of WordEqSat shows that in order to show the upper bound on exponent of periodicity it is enough to show prove bounds on Σ-exponent of periodicity and on running time, both of which we already have.

Linear Space for $\mathcal{O}(1)$ Variables (New). Using the methods described in Section 4 it can be enforced that WordEqSat stores equations of linear length. However, the letters in such an equation can be all different, even if the input equation is over two letters. Hence the (nondeterministic) space usage is $\mathcal{O}(n \log n)$ bits. For $\mathcal{O}(1)$ variables (and unbounded number of appearances) we improve the bit consumption to only constant larger than the input. To this end it is enough to improve the encoding of the words in the equation. Let LinWordEqSat denotes such modified WordEqSat.

Theorem 3. LinWordEqSat *preserves unsatisfiability and satisfiability. For k variables, it runs in (nondeterministic) $\mathcal{O}(mk^{ck})$ space, for some constant c, where m is the size of the input measured in bits.*

One Variable (New). The word equations with one variable are solvable in P (in fact, this is true even for two variables [1,2]). The naive algorithm takes $\mathcal{O}(n^3)$ time. First non-trivial bound was given by Obono, Goralcik and Maksimenko, who devised an $\mathcal{O}(n \log n)$ algorithm [15]. This was improved by Dąbrowksi and Plandowski [3] to $\mathcal{O}(n + \#_X \log n)$, where $\#_X$ is the number of appearances of the variable in the equation.

The WordEqSat determinises in case of one variable and its natural implementation runs in $\mathcal{O}(n + \#_X \log n)$, so the same running time as algorithm of Dąbrowksi and Plandowski [3]. Using a couple of heuristics as well as a better run-time analysis this can be lowered to $\mathcal{O}(n)$, in the RAM model [7].

Representation of All Solutions. Plandowski [17] gave an algorithm that generated a finite, graph-like representation of all solutions of a word equation. It is based on the idea that his algorithm not only preserves satisfiability and unsatisfiability, but it in some sense operates on the solutions: when it transforms

$U = V$ to $U' = V'$ then solutions of $U' = V'$ correspond to solutions of $U = V$. Moreover, each solution of $U = V$ can be represented in this way for some $U' = V'$. Hence, all solutions can be represented as a graph as follows: nodes are labelled with equations of the form $U = V$ and a directed edge leads from $U = V$ to $U' = V'$ if for some nondeterministic choices the former equation is transformed into the latter by the algorithm. Furthermore, the edge describes, how the solutions of $U' = V'$ can be changed into the solutions of $U = V$ (i.e. it says what strings should be substituted for letters in $U' = V'$ to obtain $U = V$).

Nevertheless the necessary changes to the original algorithm were non-trivial and the proof of its correctness involved. On the other hand, the algorithm presented in this paper generalises easily to a generator of all solutions.

Theorem 4 (cf. [17]). *The graph representation of all solutions of an equation $U = V$ can be constructed in* PSPACE. *The size of the constructed graph is at most exponential.*

6   Applications to Straight Line Programmes

A *Straight-Line Programme* (SLP) is a context free grammar G over the alphabet Σ in which every nonterminal generates exactly one string. It is known that SLPs are closely related to practical compression standards (LZW and LZ) in the sense that they are polynomially equivalent to LZ, which is the most powerful among block-based compression standards. The SLPs are widely studied because they offer a very well-structured representation of text, which is suitable for later processing and algorithms, see a recent survey by Lohrey [11].

Since each nonterminal of SLPs defines a unique word, we can imagine them as a collection of word equations $X_i = \alpha_i$, where the SLP contains the rule $X_i \to \alpha_i$. Thus it is clear that the recompression can be applied to SLPs as well. Furthermore, the nondeterminism disappears in this case: it was needed to determine the first/last letter of $S(X)$ and the lengths of the a-prefix/a-suffix of $S(X)$ and those can be calculated in the SLP case by a bottom-up procedure.

Using the recompression, we obtained a couple of results for problems related to SLPs.

Fully Compressed Membership Problem. In the fully compressed membership problem we are given an automaton (NFA or DFA) whose transitions are labelled with compressed words, i.e. SLPs. We are to decide, whether it accepts the input word, which is also supplied as an SLP. It was known that this problem is in PSPACE and NP-hard for NFAs (P-hard for DFAs) and it was conjectured that indeed it is in NP(P, respectively). This was established using the recompression approach [6], and in fact the method was devised do deal with this problem.

Fully Compressed Pattern Matching. In the fully compressed pattern matching we are given two strings, s and p, represented as SLPs, and we are to return all appearances of p in s. The previous-best algorithm for this problem had a cubic running time [10], a recompression-based approach yields a quadratic algorithm [4].

Smallest Grammar Problem. The main disadvantage of the SLP compression is that the problem of outputting the smallest grammar generating a given text is hard: even the size of this grammar is hard to approximate within a constant factor. There are however several algorithms that achieve an $\mathcal{O}(\log(N/n))$ approximation ration, where N is the size of the input text, and n the size of the smallest grammar. Using a recompression we obtain a (yet another) very simple algorithm achieving this bound [5].

Acknowledgements. I would like to thank P. Gawrychowski for initiating my interest in compressed membership problems, which eventually led to this work and for pointing to relevant literature [12,14]; J. Karhumäki, for his question, whether the techniques of local recompression can be applied to the word equations; W. Plandowski for his comments and suggestions and questions concerning the space consumption that led to linear space algorithm for $\mathcal{O}(1)$ variables.

References

1. Charatonik, W., Pacholski, L.: Word equations with two variables. In: Abdulrab, H., Pécuchet, J.P. (eds.) IWWERT 1991. LNCS, vol. 677, pp. 43–56. Springer, Heidelberg (1993)
2. Dąbrowski, R., Plandowski, W.: Solving two-variable word equations. In: Díaz, J., Karhumäki, J., Lepistö, A., Sannella, D. (eds.) ICALP 2004. LNCS, vol. 3142, pp. 408–419. Springer, Heidelberg (2004)
3. Dąbrowski, R., Plandowski, W.: On word equations in one variable. Algorithmica 60(4), 819–828 (2011)
4. Jeż, A.: Faster fully compressed pattern matching by recompression. In: Czumaj, A., Mehlhorn, K., Pitts, A., Wattenhofer, R. (eds.) ICALP 2012, Part I. LNCS, vol. 7391, pp. 533–544. Springer, Heidelberg (2012)
5. Jeż, A.: Approximation of grammar-based compression via recompression. In: Fischer, J., Sanders, P. (eds.) CPM 2013. LNCS, vol. 7922, pp. 165–176. Springer, Heidelberg (2013)
6. Jeż, A.: The complexity of compressed membership problems for finite automata. Theory of Computing Systems, 1–34 (2013), http://dx.doi.org/10.1007/s00224-013-9443-6
7. Jeż, A.: One-variable word equations in linear time. In: Fomin, F.V., Freivalds, R., Kwiatkowska, M., Peleg, D. (eds.) ICALP 2013. LNCS. Springer, Heidelberg (to appear, 2013)
8. Kościelski, A., Pacholski, L.: Complexity of Makanin's algorithm. J. ACM 43(4), 670–684 (1996)
9. Larsson, N.J., Moffat, A.: Offline dictionary-based compression. In: Data Compression Conference, pp. 296–305. IEEE Computer Society (1999)
10. Lifshits, Y.: Processing compressed texts: A tractability border. In: Ma, B., Zhang, K. (eds.) CPM 2007. LNCS, vol. 4580, pp. 228–240. Springer, Heidelberg (2007)

11. Lohrey, M.: Algorithmics on SLP-compressed strings: A survey. Groups Complexity Cryptology 4(2), 241–299 (2012)
12. Lohrey, M., Mathissen, C.: Compressed membership in automata with compressed labels. In: Kulikov, A., Vereshchagin, N. (eds.) CSR 2011. LNCS, vol. 6651, pp. 275–288. Springer, Heidelberg (2011)
13. Makanin, G.S.: The problem of solvability of equations in a free semigroup. Matematicheskii Sbornik 2(103), 147–236 (1977) (in Russian)
14. Mehlhorn, K., Sundar, R., Uhrig, C.: Maintaining dynamic sequences under equality tests in polylogarithmic time. Algorithmica 17(2), 183–198 (1997)
15. Obono, S.E., Goralcik, P., Maksimenko, M.N.: Efficient solving of the word equations in one variable. In: Prívara, I., Rovan, B., Ruzicka, P. (eds.) MFCS 1994. LNCS, vol. 841, pp. 336–341. Springer, Heidelberg (1994)
16. Plandowski, W.: Satisfiability of word equations with constants is in PSPACE. J. ACM 51(3), 483–496 (2004)
17. Plandowski, W.: An efficient algorithm for solving word equations. In: Kleinberg, J.M. (ed.) STOC, pp. 467–476. ACM (2006)
18. Plandowski, W., Rytter, W.: Application of Lempel-Ziv encodings to the solution of words equations. In: Larsen, K.G., Skyum, S., Winskel, G. (eds.) ICALP 1998. LNCS, vol. 1443, pp. 731–742. Springer, Heidelberg (1998)
19. Sakamoto, H.: A fully linear-time approximation algorithm for grammar-based compression. J. Discrete Algorithms 3(2-4), 416–430 (2005)

Joint Spectral Characteristics:
A Tale of Three Disciplines*

Raphaël M. Jungers

ICTEAM Institute, Université catholique de Louvain and FNRS
raphael.jungers@uclouvain.be

Joint spectral characteristics describe the stationary behavior of a discrete time linear switching system. Well, that's what an electrical engineer would say. A mathematician would say that they characterize the asymptotic behavior of a semigroup of matrices, and a computer scientist would perhaps see them as describing languages generated by automata.

Because of their connections with these wide research topics, joint spectral characteristics have been at the center of rich and diverse research efforts in recent years (see [7,11,15] for general surveys). They are notoriously very hard to compute (NP-hardness, Undecidability, etc. are the rule rather than the exception [16,10]), but it turns out that one can often get around these difficulties, and modern optimization techniques seem particularly useful for studying them [14,13].

I will survey and connect several powerful and interesting results, emphasizing the role of optimization methods. I will present applications, ranging from wireless control protocols to viral diseases treatment, malicious agents tracking, etc... I will emphasize the numerous connections with language and automata theory. In particular, I will cover the asymptotics of repetition-free languages [8,4,9,5], the capacity of codes avoiding forbidden differences [12,6,3], and recent results linking automata theory with the generation of Linear Matrix Inequalities for computing the joint spectral radius [1,2]. I will present several open problems and promising research directions.

References

1. Ahmadi, A.A., Jungers, R.M., Parrilo, P., Roozbehani, M.: Analysis of the joint spectral radius via Lyapunov functions on path-complete graphs. In: Proceedings of: Hybrid Systems: Computation and Control (HSCC 2011), Chicago (2011)
2. Ahmadi, A.A., Jungers, R.M., Parrilo, P., Roozbehani, M.: When is a set of LMIs a sufficient condition for stability? In: Proceedings of: ROCOND 2012, Aalborg (2012)
3. Asarin, E., Dima, C.: On the computation of covert channel capacity. RAIRO - Theoretical Informatics and Applications 44(1), 37–58 (2010)
4. Berstel, J.: Growth of repetition-free words–a review. Theoretical Computer Science 340(2), 280–290 (2005)

* Work supported by the Communauté française de Belgique - Actions de Recherche Concertées, and by the Belgian Programme on Interuniversity Attraction Poles initiated by the Belgian Federal Science Policy Office.

5. Blondel, V.D., Cassaigne, J., Jungers, R.M.: On the number of α-power-free words for $2 < \alpha \leq 7/3$. Theoretical Computer Science 410, 2823–2833 (2009)
6. Blondel, V.D., Jungers, R., Protasov, V.Y.: On the complexity of computing the capacity of codes that avoid forbidden difference patterns. IEEE Transactions on Information Theory 52(11), 5122–5127 (2006)
7. Jungers, R.M.: The joint spectral radius, theory and applications. LNCIS, vol. 385. Springer, Heidelberg (2009)
8. Jungers, R.M., Protasov, V.Y., Blondel, V.D.: Overlap-free words and spectra of matrices. Theoretical Computer Science 410, 3670–3684 (2009)
9. Karhumäki, J., Shallit, J.: Polynomial versus exponential growth in repetition-free binary words. Journal of Combinatorial Theory Series A 105(2), 335–347 (2004)
10. Kozyakin, V.A.: Algebraic unsolvability of problem of absolute stability of desynchronized systems. Automation and Remote Control 51, 754–759 (1990)
11. Liberzon, D.: Switching in Systems and Control. Birkhäuser, Boston (2003)
12. Moision, B.E., Orlitsky, A., Siegel, P.H.: On codes that avoid specified differences. IEEE Transactions on Information Theory 47, 433–442 (2001)
13. Parrilo, P.A., Jadbabaie, A.: Approximation of the joint spectral radius of a set of matrices using sum of squares. In: Bemporad, A., Bicchi, A., Buttazzo, G. (eds.) HSCC 2007. LNCS, vol. 4416, pp. 444–458. Springer, Heidelberg (2007)
14. Protasov, V.Y., Jungers, R.M., Blondel, V.D.: Joint spectral characteristics of matrices: a conic programming approach. SIAM Journal on Matrix Analysis and Applications 31(4), 2146–2162 (2010)
15. Shorten, R., Wirth, F., Mason, O., Wulff, K., King, C.: Stability criteria for switched and hybrid systems. SIAM Review 49(4), 545–592 (2007)
16. Tsitsiklis, J.N., Blondel, V.D.: The Lyapunov exponent and joint spectral radius of pairs of matrices are hard- when not impossible- to compute and to approximate. Mathematics of Control, Signals, and Systems 10, 31–40 (1997)

Unambiguous Finite Automata

Christof Löding

RWTH Aachen, Germany
loeding@cs.rwth-aachen.de

In general, a nondeterministic automaton or machine (for example a finite automaton, pushdown automaton or Turing machine) is called unambiguous if each input is accepted by at most one run or computation. Each deterministic automaton is obviously unambiguous. However, in many settings, unambiguous automata are more expressive or admit more succinct automata than deterministic models, while preserving some good algorithmic properties. The aim of this talk is to survey some classical and some more recent results on unambiguous finite automata over different kind of input structures, namely finite words, infinite words, finite trees, and infinite trees.

A typical example is the inclusion problem for automata on finite words, which is solvable in polynomial time for unambiguous automata [9], while it is PSPACE-complete for general nondeterministic automata (see Section 10.6 of [1]). This result can be lifted to unambiguous automata on finite ranked trees [8], and can, for example, be used to derive efficient inclusion tests for certain classes of automata on unranked trees [7].

The method of [9] uses a counting argument for the number of accepting runs for words up to a certain length. This method cannot be used for unambiguous Büchi automata in the setting of infinite input words. As a consequence, the question whether inclusion testing for unambiguous Büchi automata can be done efficiently, is still open. Partial results for a stronger notion of unambiguity taken from [5], and for subclasses of Büchi automata have been obtained in [3] and [6].

Concerning infinite trees, the situation becomes different because unambiguous automata are not expressively equivalent to unrestricted nondeterministic automata anymore. The proof of this result, presented in [4], relies on the fact that it is not possible to define in monadic second-order logic a choice function on the infinite binary tree. Since not all regular languages of infinite trees are unambiguous, a natural decision problem arises: "Given a regular language of an infinite tree, does there exist an unambiguous automaton for it?" It is still unknown whether this problem is decidable, only partial solutions for specific cases have been obtained [2].

References

1. Aho, A.V., Hopcroft, J.E., Ullman, J.D.: The Design and Analysis of Computer Algorithms. Addison-Wesley, New York (1974)
2. Bilkowski, M.: Ambiguity property for complements of deterministic tree languages. Presentation at the Annual Workshop of the ESF Networking Programme on Games for Design and Verification, Oxford (2010)

M.-P. Béal and O. Carton (Eds.): DLT 2013, LNCS 7907, pp. 29–30, 2013.
© Springer-Verlag Berlin Heidelberg 2013

3. Bousquet, N., Löding, C.: Equivalence and inclusion problem for strongly unambiguous Büchi automata. In: Dediu, A.-H., Fernau, H., Martín-Vide, C. (eds.) LATA 2010. LNCS, vol. 6031, pp. 118–129. Springer, Heidelberg (2010)
4. Carayol, A., Löding, C., Niwiński, D., Walukiewicz, I.: Choice functions and well-orderings over the infinite binary tree. Central European Journal of Mathematics 8(4), 662–682 (2010)
5. Carton, O., Michel, M.: Unambiguous Büchi automata. Theor. Comput. Sci. 297(1–3), 37–81 (2003)
6. Isaak, D., Löding, C.: Efficient inclusion testing for simple classes of unambiguous ω-automata. Information Processing Letters 112(14-15), 578–582 (2012)
7. Martens, W., Niehren, J.: On the minimization of xml schemas and tree automata for unranked trees. J. Comput. Syst. Sci. 73(4), 550–583 (2007)
8. Seidl, H.: Deciding equivalence of finite tree automata. SIAM J. Comput. 19(3), 424–437 (1990)
9. Stearns, R.E., Hunt III, H.B.: On the equivalence and containment problems for unambiguous regular expressions, regular grammars and finite automata. SIAM Journal on Computing 14(3), 598–611 (1985)

An Explicit Formula for the Intersection
of Two Polynomials of Regular Languages

Jean-Éric Pin*

LIAFA, University Paris-Diderot and CNRS, France

Abstract. Let $\mathcal{L}$ be a set of regular languages of A^*. An $\mathcal{L}$-polynomial is a finite union of products of the form $L_0 a_1 L_1 \cdots a_n L_n$, where each a_i is a letter of A and each L_i is a language of $\mathcal{L}$. We give an explicit formula for computing the intersection of two $\mathcal{L}$-polynomials. Contrary to Arfi's formula (1991) for the same purpose, our formula does not use complementation and only requires union, intersection and quotients. Our result also implies that if $\mathcal{L}$ is closed under union, intersection and quotient, then its polynomial closure, its unambiguous polynomial closure and its left [right] deterministic polynomial closure are closed under the same operations.

1   Introduction

Let $\mathcal{L}$ be a set of regular languages of A^*. An $\mathcal{L}$-*polynomial* is a finite union of products of the form $L_0 a_1 L_1 \cdots a_n L_n$, where each a_i is a letter of A and each L_i is a language of $\mathcal{L}$. The *polynomial closure* of $\mathcal{L}$, denoted by $\mathrm{Pol}(\mathcal{L})$, is the set of all $\mathcal{L}$-polynomials.

It was proved by Arfi [1] that if $\mathcal{L}$ is closed under Boolean operations and quotient, then $\mathrm{Pol}(\mathcal{L})$ is closed under intersection. This result was obtained by giving an explicit formula for computing the intersection of two polynomials of regular languages.

It follows from the main theorem of [6] that Arfi's result can be extended to the case where $\mathcal{L}$ is only closed under union, intersection and quotient. However, this stronger statement is obtained as a consequence of a sophisticated result involving profinite equations and it is natural to ask for a more elementary proof.

The objective of this paper is to give a new explicit formula for computing the intersection of two $\mathcal{L}$-polynomials. Contrary to the formula given in [1], our formula only requires using union, intersection and quotients of languages of $\mathcal{L}$. Our proof is mainly combinatorial, but relies heavily on the notion of syntactic ordered monoid, a notion first introduced by Schützenberger [14] (see also [10]). The main difficulty lies in finding appropriate notation to state the formula, but then its proof is merely a verification.

---

* Work supported by the project ANR 2010 BLAN 0202 02 FREC.

M.-P. Béal and O. Carton (Eds.): DLT 2013, LNCS 7907, pp. 31–45, 2013.
© Springer-Verlag Berlin Heidelberg 2013

Our result also leads to the following result, that appears to be new: if $\mathcal{L}$ is closed under union, intersection and quotient, then its unambiguous polynomial closure and its left [right] deterministic polynomial closure are closed under the same operations.

Let us mention also that our algorithm can be readily extended to the setting of infinite words by using syntactic ordered ω-semigroups [8].

2   Background and Notation

2.1   Syntactic Order

The *syntactic congruence* of a language L of A^* is the congruence on A^* defined by $u \sim_L v$ if and only if, for every $x, y \in A^*$,

$$xuy \in L \iff xvy \in L$$

The monoid $M = A^*/\!\sim_L$ is the *syntactic monoid* of L and the natural morphism $\eta : A^* \to M$ is called the *syntactic morphism* of L. It is a well-known fact that a language is regular if and only if its syntactic monoid is finite.

The *syntactic preorder*[1] of a language L is the relation $\leqslant_L$ over A^* defined by $u \leqslant_L v$ if and only if, for every $x, y \in A^*$, $xuy \in L$ implies $xvy \in L$. The associated equivalence relation is the syntactic congruence $\sim_L$. Further, $\leqslant_L$ induces a partial order on the syntactic monoid M of L. This partial order $\leqslant$ is compatible with the product and can also be defined directly on M as follows: given $s, t \in M$, one has $s \leqslant t$ if and only if, for all $x, y \in M$, $xsy \in \eta(L)$ implies $xty \in \eta(L)$. The ordered monoid $(M, \leqslant)$ is called the *syntactic ordered monoid* of L.

Let us remind an elementary but useful fact: if $v \in L$ and $\eta(u) \leqslant \eta(v)$, then $u \in L$. This follows immediately form the definition of the syntactic order by taking $x = y = 1$.

2.2   Quotients

Recall that if L is a language of A^* and x is a word, the *left quotient* of L by x is the language $x^{-1}L = \{z \in A^* \mid xz \in L\}$. The *right quotient* Ly^{-1} is defined in a symmetrical way. Right and left quotients commute, and thus $x^{-1}Ly^{-1}$ denotes either $x^{-1}(Ly^{-1})$ or $(x^{-1}L)y^{-1}$. For each word v, let us set

$$[L]_{\uparrow v} = \{u \in A^* \mid \eta(v) \leqslant \eta(u)\}$$
$$[L]_{=v} = \{u \in A^* \mid \eta(u) = \eta(v)\}$$

---

[1] In earlier papers [6,10,13], I used the opposite preorder, but it seems preferable to go back to Schützenberger's original definition.

Proposition 2.1. *The following formulas hold:*

$$[L]_{\uparrow v} = \bigcap_{\{(x,y)\in A^* \times A^* \mid v \in x^{-1}Ly^{-1}\}} x^{-1}Ly^{-1} \tag{1}$$

$$[L]_{=v} = [L]_{\uparrow v} - \bigcup_{\eta(v)<\eta(u)} [L]_{\uparrow u} \tag{2}$$

$$[L]_{\uparrow v} = \bigcup_{\eta(v)\leqslant\eta(u)} [L]_{=u} \tag{3}$$

Proof. A word u belongs to the right hand side of (1) if and only if the condition $v \in x^{-1}Ly^{-1}$ implies $u \in x^{-1}Ly^{-1}$, which is equivalent to stating that $v \leqslant_L u$, or $\eta(v) \leqslant \eta(u)$, or yet $u \in [L]_{\uparrow v}$. This proves (1). Formulas (2) and (3) are obvious. □

Let us make precise a few critical points. First, v always belongs to $[L]_{\uparrow v}$. This is the case even if v cannot be completed into a word of L, that is, if v does not belong to any quotient $x^{-1}Ly^{-1}$. In this case, the intersection on the right hand side of (1) is indexed by the empty set and is therefore equal to A^*.

Secondly, the intersection occurring on the right hand side of (1) and the union occurring on the right hand side of (2) are potentially infinite, but they are finite if L is a regular language, since a regular language has only finitely many quotients.

3 Infiltration Product and Infiltration Maps

The definition below is a special case of a more general definition given in [7]. A word $c_1 \cdots c_r$ belongs to the *infiltration product* of two words $a_1 \cdots a_p$ and $v = b_1 \cdots b_q$, if there are two order preserving maps $\alpha : \{1,\ldots,p\} \to \{1,\ldots,r\}$ and $\beta : \{1,\ldots,q\} \to \{1,\ldots,r\}$ such that

(1) for each $i \in \{1,\ldots,p\}$, $a_i = c_{\alpha(i)}$,

(2) for each $i \in \{1,\ldots,q\}$, $b_i = c_{\beta(i)}$,

(3) the union of the ranges of α and β is $\{1,\ldots,r\}$.

For instance, the set $\{ab, aab, abb, aabb, abab\}$ is the infiltration product of ab and ab and the set $\{aba, bab, abab, abba, baab, baba\}$ is the infiltration product of ab and ba.

A pair of maps (α, β) satisfying Conditions (1)–(3) is called a *pair of infiltration maps*. Note that these conditions imply that $p + q \leqslant r$.

In the example pictured in Figure 1, one has $p = 4$, $q = 2$ and $r = 5$. The infiltration maps α and β are given by $\alpha(1) = 1$, $\alpha(2) = 2$, $\alpha(3) = 3$, $\alpha(4) = 4$ and $\beta(1) = 3$, $\beta(2) = 5$.

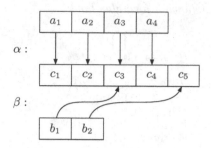

Fig. 1. A pair of infiltration maps

In order to state our main theorem in a precise way, we need to handle the intervals of the form $\{\alpha(i)+1,\ldots,\alpha(i+1)-1\}$, but also the two extremal intervals $\{1,\ldots,\alpha(1)-1\}$ and $\{\alpha(p)+1,\ldots,r\}$. As a means to get a uniform notation, it is convenient to extend α and β to mappings $\alpha:\{0,\ldots,p+1\}\to\{0,\ldots,r+1\}$ and $\beta:\{0,\ldots,q+1\}\to\{0,\ldots,r+1\}$ by setting $\alpha(0)=\beta(0)=0$ and $\alpha(p+1)=\beta(q+1)=r+1$. The two extremal intervals are now of the standard form $\{\alpha(i)+1,\ldots,\alpha(i+1)-1\}$, with $i=0$ and $i=p$, respectively. Further, we introduce the two maps $\bar\alpha:\{0,\ldots,r\}\to\{0,\ldots,p\}$ and $\bar\beta:\{0,\ldots,r\}\to\{0,\ldots,q\}$ defined by

$$\bar\alpha(i)=\max\{k\mid\alpha(k)\leqslant i\}\quad\text{and}\quad\bar\beta(i)=\max\{k\mid\beta(k)\leqslant i\}.$$

For instance, one gets for our example:

$\bar\alpha(0)=0$	$\bar\alpha(1)=1$	$\bar\alpha(2)=2$	$\bar\alpha(3)=3$	$\bar\alpha(4)=4$	$\bar\alpha(5)=4$
$\bar\beta(0)=0$	$\bar\beta(1)=0$	$\bar\beta(2)=0$	$\bar\beta(3)=1$	$\bar\beta(4)=1$	$\bar\beta(5)=2$

These two functions are conveniently represented in Figure 2

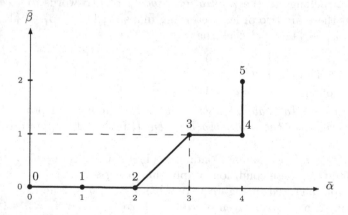

Fig. 2. Graphs of $\bar\alpha$ and $\bar\beta$: for instance, $\bar\alpha(3)=3$ and $\bar\beta(3)=1$

The next lemmas summarize the connections between α and $\bar\alpha$. Of course, similar properties hold for β and $\bar\beta$.

Lemma 3.1. *The following properties hold:*

(1) $\bar{\alpha}(\alpha(k)) = k$, *for* $0 \leqslant k \leqslant p$.

(2) $\bar{\alpha}(s+1) \leqslant \bar{\alpha}(s) + 1$, *for* $0 \leqslant s \leqslant r - 1$.

(3) $k \leqslant \bar{\alpha}(s)$ *if and only if* $\alpha(k) \leqslant s$, *for* $0 \leqslant k \leqslant p$ *and* $0 \leqslant s \leqslant r$.

(4) $k \geqslant \bar{\alpha}(s)$ *if and only if* $\alpha(k+1) \geqslant s+1$, *for* $0 \leqslant k \leqslant p-1$ *and* $0 \leqslant s \leqslant r-1$.

Proof. These properties follow immediately from the definition of $\bar{\alpha}$. □

Lemma 3.2. *For* $0 \leqslant s \leqslant r - 1$, *the conditions* $\bar{\alpha}(s + 1) = \bar{\alpha}(s) + 1$ *and* $\alpha(\bar{\alpha}(s+1)) = s+1$ *are equivalent.*

Proof. Put $k = \bar{\alpha}(s)$ and suppose that $\bar{\alpha}(s+1) = k+1$. Since $k+1 \leqslant \bar{\alpha}(s+1)$, Lemma 3.1 (3) shows that $\alpha(k+1) \leqslant s+1$. Further, since $k \geqslant \bar{\alpha}(s)$, Lemma 3.1 (4) shows that $\alpha(k+1) \geqslant s+1$. Therefore $\alpha(k+1) = s+1$ and finally $\alpha(\bar{\alpha}(s+1)) = s+1$.

Conversely, suppose that $\alpha(\bar{\alpha}(s+1)) = s+1$. Putting $\bar{\alpha}(s+1) = k+1$, one gets $\alpha(k+1) = s+1$ and Lemma 3.1 (4) shows that $k \geqslant \bar{\alpha}(s)$. By Lemma 3.1 (2), one gets $\bar{\alpha}(s+1) \leqslant \bar{\alpha}(s) + 1$ and hence $k \leqslant \bar{\alpha}(s)$. Thus $\bar{\alpha}(s) = k$ and $\bar{\alpha}(s+1) = \bar{\alpha}(s) + 1$. □

Let us denote by $P_\alpha(s)$ the property $\bar{\alpha}(s+1) = \bar{\alpha}(s) + 1$.

Lemma 3.3. *For* $0 \leqslant s \leqslant r-1$, *one of* $P_\alpha(s)$ *or* $P_\beta(s)$ *holds.*

Proof. Since the union of the ranges of α and β is $\{1, \ldots, r\}$, there is an integer $k \geqslant 0$ such that either $\alpha(k+1) = s+1$ or $\beta(k+1) = s+1$. In the first case, one gets $\bar{\alpha}(s+1) = \bar{\alpha}(\alpha(k+1)) = k+1$ and Lemma 3.1 (3) shows that $\bar{\alpha}(s) \leqslant k$. Since $\bar{\alpha}(s+1) \leqslant \bar{\alpha}(s) + 1$ by Lemma 3.1 (2), one also has $k \leqslant \bar{\alpha}(s)$ and finally $\bar{\alpha}(s) = k$, which proves $P_\alpha(s)$. In the latter case, one gets $P_\beta(s)$ by a similar argument. □

4 Main Result

Let $a_1, \ldots, a_p, b_1, \ldots, b_q$ be letters of A and let $K_0, \ldots, K_p, L_0, \ldots, L_q$ be languages of A^*. Let $K = K_0 a_1 K_1 \cdots a_p K_p$ and $L = L_0 b_1 L_1 \cdots b_q L_q$.

A word of $K \cap L$ can be factorized as $u_0 a_1 u_1 \cdots a_p u_p$, with $u_0 \in K_0, \ldots, u_p \in K_p$ and as $v_0 b_1 v_1 \cdots b_q v_q$, with $v_0 \in L_0, \ldots, v_q \in L_q$. These two factorizations can be refined into a single factorization of the form $z_0 c_1 z_1 \cdots c_r z_r$, where $c_1 \cdots c_r$ belongs to the infiltration product of $a_1 \cdots a_p$ and $b_1 \cdots b_q$.

For instance, for $p = 4$ and $q = 2$, one could have $r = 5$, with the relations $c_1 = a_1$, $c_2 = a_2$, $c_3 = a_3 = b_1$, $c_4 = a_4$ and $c_5 = b_2$, leading to the factorization $z_0 c_1 z_1 c_2 z_2 c_3 z_3 c_4 z_4 c_5 z_5$, as pictured in Figure 3.

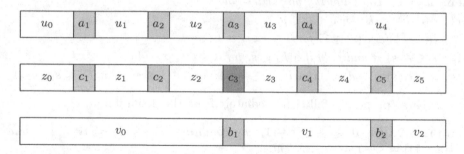

Fig. 3. A word of $K \cap L$ and its factorizations

The associated pair of infiltration maps (α, β) is given by

$$\alpha(1) = 1 \qquad \alpha(2) = 2 \qquad \alpha(3) = 3 \qquad \alpha(4) = 4$$
$$\beta(1) = 3 \qquad \beta(2) = 5$$

Two series of constraints will be imposed on the words z_i:

$$z_0 \in K_0,\ z_1 \in K_1,\ z_2 \in K_2,\ z_3 \in K_3 \text{ and } z_4 c_5 z_5 \in K_4,$$
$$z_0 c_1 z_1 c_2 z_2 \in L_0,\ z_3 c_4 z_4 \in L_1 \text{ and } z_5 \in L_2.$$

We are now ready to state our main result. Let us denote by $I(p, q)$ the set of pairs of infiltration maps (α, β) with domain $\{1, \ldots, p\}$ and $\{1, \ldots, q\}$, respectively. Since $r \leqslant p + q$, the set $I(p, q)$ is finite.

Theorem 4.1. *Let* $K = K_0 a_1 K_1 \cdots a_p K_p$ *and* $L = L_0 b_1 L_1 \cdots b_q L_q$ *be two products of languages. Then their intersection is given by the formulas*

$$K \cap L = \bigcup_{(\alpha, \beta) \in I(p,q)} U(\alpha, \beta) \tag{4}$$

where

$$U(\alpha, \beta) = \bigcup_{(z_0, \ldots, z_r) \in C(\alpha, \beta)} U_0 c_1 U_1 \cdots c_r U_r \tag{5}$$

and, for $0 \leqslant i \leqslant r$,

$$U_i = [K_{\bar{\alpha}(i)}]_{\uparrow z_i} \cap [L_{\bar{\beta}(i)}]_{\uparrow z_i} \tag{6}$$

and $C(\alpha, \beta)$ *is the set of* $(r + 1)$-*tuples* $(z_0, \ldots, z_r)$ *of words such that*

(C$_1$) *for* $0 \leqslant k \leqslant p$, $z_{\alpha(k)} c_{\alpha(k)+1} z_{\alpha(k)+1} \cdots c_{\alpha(k+1)-1} z_{\alpha(k+1)-1} \in K_k$,

(C$_2$) *for* $0 \leqslant k \leqslant q$, $z_{\beta(k)} c_{\beta(k)+1} z_{\beta(k)+1} \cdots c_{\beta(k+1)-1} z_{\beta(k+1)-1} \in L_k$.

For instance, if (α, β) is the pair of infiltration maps of our example, one would have

$$U(\alpha, \beta) = \bigcup_{(z_0, \ldots, z_5) \in C(\alpha, \beta)} ([K_0]_{\uparrow z_0} \cap [L_0]_{\uparrow z_0}) a_1 ([K_1]_{\uparrow z_1} \cap [L_0]_{\uparrow z_1}) a_2$$

$$([K_2]_{\uparrow z_2} \cap [L_0]_{\uparrow z_2}) b_1 ([K_3]_{\uparrow z_3} \cap [L_1]_{\uparrow z_3}) a_4 ([K_4]_{\uparrow z_4} \cap [L_1]_{\uparrow z_4}) b_2 ([K_4]_{\uparrow z_5} \cap [L_2]_{\uparrow z_5})$$

and the conditions (C_1) and (C_2) would be

(C_1) $z_0 \in K_0$, $z_1 \in K_1$, $z_2 \in K_2$, $z_3 \in K_3$, $z_4 c_5 z_5 \in K_4$,

(C_2) $z_0 c_1 z_1 c_2 z_2 \in L_0$, $z_3 c_4 z_4 \in L_1$ and $z_5 \in L_2$.

Before proving the theorem, it is important to note that if the languages K_0, ..., K_p, L_0, ..., L_q are regular, the union indexed by $C(\alpha, \beta)$ is actually a finite union. Indeed, Proposition 2.1 shows that, if R is a regular language, there are only finitely many languages of the form $[R]_z$.

Proof. Let U be the right hand side of (4). We first prove that $K \cap L$ is a subset of U. Let z be a word of $K \cap L$. Then z can be factorized as $u_0 a_1 u_1 \cdots a_p u_p$, with $u_0 \in K_0$, ..., $u_p \in K_p$ and as $v_0 b_1 v_1 \cdots b_q v_q$, with $v_0 \in L_0$, ..., $v_q \in L_q$. The common refinement of these two factorizations leads to a factorization of the form $z_0 c_1 z_1 \cdots c_r z_r$, where each letter c_k is either equal to some a_i or to some b_j or both. This naturally defines a pair of infiltration maps $\alpha : \{1, \ldots, p\} \to \{1, \ldots, r\}$ and $\beta : \{1, \ldots, q\} \to \{1, \ldots, r\}$. Conditions (C_1) and (C_2) just say that the factorization $z_0 c_1 z_1 \cdots c_r z_r$ is a refinement of the two other ones. Now, since, for $0 \leqslant i \leqslant r$, the word z_i belongs to $[K_{\bar{\alpha}(i)}]_{\uparrow z_i} \cap [L_{\bar{\beta}(i)}]_{\uparrow z_i}$, the word z belongs to U. Thus $K \cap L \subseteq U$.

We now prove the opposite inclusion. Let $r \leqslant p + q$ be an integer, let $\alpha : \{1, \ldots, p\} \to \{1, \ldots, r\}$ and $\beta : \{1, \ldots, q\} \to \{1, \ldots, r\}$ be two infiltration maps and let $(z_0, \ldots, z_r) \in C(\alpha, \beta)$ and $c_1, \ldots, c_r$ satisfying (C_1) and (C_2). It suffices to prove that $U_0 c_1 U_1 \cdots c_r U_r$ is a subset of $K \cap L$. We need a stronger version of (C_1) and (C_2).

Lemma 4.2. *The following relations hold:*

(C_3) *for* $0 \leqslant k \leqslant p$, $U_{\alpha(k)} c_{\alpha(k)+1} U_{\alpha(k)+1} \cdots c_{\alpha(k+1)-1} U_{\alpha(k+1)-1} \subseteq K_k$,

(C_4) *for* $0 \leqslant k \leqslant q$, $U_{\beta(k)} c_{\beta(k)+1} U_{\beta(k)+1} \cdots c_{\beta(k+1)-1} U_{\beta(k+1)-1} \subseteq L_k$.

Coming back once again to our main example, these conditions would be

(C_3) $U_0 \subseteq K_0$, $U_1 \subseteq K_1$, $U_2 \subseteq K_2$, $U_3 \subseteq K_3$, $U_4 c_4 U_5 \subseteq K_4$,

(C_4) $U_0 c_1 U_2 c_2 U_2 \subseteq L_0$, $U_3 c_4 U_4 \subseteq L_1$, $U_5 \subseteq L_5$.

Proof. Let η_k be the syntactic morphism of K_k. To simplify notation, let us set $i = \alpha(k) + 1$ and $j = \alpha(k + 1) - 1$. Since $\alpha(k) = i - 1 < i < \cdots < j < \alpha(k + 1)$, one gets $\bar{\alpha}(i - 1) = \bar{\alpha}(i) = \ldots = \bar{\alpha}(j) = k$. Let $u_{i-1} \in U_{i-1}$, $u_i \in U_i$, ..., $u_j \in U_j$. Then $u_{i-1} \in [U_k]_{\uparrow z_{i-1}}$, $u_i \in [U_k]_{\uparrow z_i}$, ..., $u_j \in [U_k]_{\uparrow z_j}$ and by definition, $\eta_k(z_{i-1}) \leqslant \eta_k(u_{i-1})$, $\eta_k(z_i) \leqslant \eta_k(u_i)$, ..., $\eta_k(z_j) \leqslant \eta_k(u_j)$. Therefore we get

$$\eta_k(z_{i-1} c_i z_i \cdots c_j z_j) = \eta_k(z_{i-1}) \eta_k(c_i) \eta_k(z_i) \cdots \eta_k(c_j) \eta_k(z_j)$$
$$\leqslant \eta_k(u_{i-1}) \eta_k(c_i) \eta_k(u_i) \cdots \eta_k(c_j) \eta_k(u_j) = \eta_k(u_{i-1} c_i u_i \cdots c_j u_j)$$

Now, since $z_{i-1} c_i z_i \cdots c_j z_j \in K_k$ by (C_1), we also get $u_{i-1} c_i u_i \cdots c_j u_j \in K_k$, which proves (C_3). The proof of (C_4) is similar. $\square$

Now, since $\bar{\alpha}$ and $\bar{\beta}$ are surjective, Lemma 4.2 shows that $U_0 c_1 U_1 \cdots c_r U_r$ is a subset of $K \cap L$, which concludes the proof of the theorem. $\square$

Example 4.3. Let $K = b^*aA^*ba^*$ and $L = a^*bA^*ab^*$. The algorithm described in Theorem 4.1 gives for $K \cap L$ the expression $aa^*bA^*ba^*a \cup bb^*aA^*ba^*a \cup aa^*bA^*ab^*b \cup bb^*aA^*ab^*b \cup aa^*ba^*a \cup bb^*ab^*b$.

Corollary 4.4. *Let $\mathcal{L}$ be a lattice of regular languages closed under quotient. Then its polynomial closure is also a lattice closed under quotient.*

5   Some Variants of the Product

We consider in this section two variants of the product introduced by Schützenberger in [15]: unambiguous and deterministic products. These products were also studied in [2,3,4,5,9,11,12,13].

5.1   Unambiguous Product

The marked product $L = L_0a_1L_1 \cdots a_nL_n$ of n nonempty languages $L_0, L_1, \ldots, L_n$ of A^* is *unambiguous* if every word u of L admits a unique factorization of the form $u_0a_1u_1 \cdots a_nu_n$ with $u_0 \in L_0$, $u_1 \in L_1$, $\ldots$, $u_n \in L_n$. We require the languages L_i to be nonempty to make sure that subfactorizations remain unambiguous:

Proposition 5.1. *Let $L_0a_1L_1 \cdots a_nL_n$ be an unambiguous product and let i_1, $\ldots$, i_k be a sequence of integers satisfying $0 < i_1 < \ldots < i_k < n$. Finally, let $R_0 = L_0a_1L_1 \cdots a_{i_1-1}L_{i_1-1}$, $R_1 = L_{i_1}a_{i_1+1}L_1 \cdots a_{i_2-1}L_{i_2-1}$, $\ldots$, $R_k = L_{i_k}a_{i_k+1}L_{k+1} \cdots a_nL_n$. Then the product $R_0a_{i_1}R_1 \cdots a_{i_k}R_k$ is unambiguous.*

Proof. Trivial.

The *unambiguous polynomial closure* of a class of languages $\mathcal{L}$ of A^* is the set of languages that are finite unions of unambiguous products of the form $L_0a_1L_1 \cdots a_nL_n$, where the a_i's are letters and the L_i's are elements of $\mathcal{L}$. The term *closure* actually requires a short justification.

Proposition 5.2. *Any unambiguous product of unambiguous products is unambiguous.*

Proof. Let

$$L_0 = L_{0,0}a_{1,0}L_{1,0} \cdots a_{k_0,0}L_{k_0,0}$$
$$L_1 = L_{0,1}a_{1,1}L_{1,1} \cdots a_{k_1,1}L_{k_1,1}$$
$$\vdots \tag{7}$$
$$L_n = L_{0,n}a_{1,n}L_{1,n} \cdots a_{k_n,n}L_{k_n,n}$$

be unambiguous products and let $L = L_0b_1L_1 \cdots b_nL_n$ be an unambiguous product. We claim that the product

$$L_{0,0}a_{1,0}L_{1,0} \cdots a_{k_0,0}L_{k_0,0}b_1L_{0,1}a_{1,1}L_{1,1} \cdots b_nL_{0,n}a_{1,n}L_{1,n} \cdots a_{k_n,n}L_{k_n,n}$$

is unambiguous. Let u be a word of L with two factorizations

$$x_{0,0}a_{1,0}x_{1,0}\cdots a_{k_0,0}x_{k_0,0}b_1x_{0,1}a_{1,1}x_{1,1}\cdots b_nx_{0,n}a_{1,n}x_{1,n}\cdots a_{k_n,n}x_{k_n,n}$$

and

$$y_{0,0}a_{1,0}y_{1,0}\cdots a_{k_0,0}y_{k_0,0}b_1y_{0,1}a_{1,1}x_{1,1}\cdots b_ny_{0,n}a_{1,n}y_{1,n}\cdots a_{k_n,n}y_{k_n,n}$$

with $x_{0,0}, y_{0,0} \in L_{0,0}, \ldots, x_{k_n,n}, y_{k_n,n} \in L_{k_n,n}$. Setting

$$
\begin{aligned}
x_0 &= x_{0,0}a_{1,0}x_{1,0}\cdots a_{k_0,0}x_{k_0,0} & y_0 &= y_{0,0}a_{1,0}y_{1,0}\cdots a_{k,0}y_{k_0,0}\\
x_1 &= x_{0,1}a_{1,1}x_{1,1}\cdots a_{k_1,1}x_{k_1,1} & y_1 &= y_{0,1}a_{1,1}y_{1,1}\cdots a_{k_1,1}y_{k_1,1}\\
&\ \ \vdots & &\ \ \vdots \\
x_n &= x_{0,n}a_{1,n}x_{1,n}\cdots a_{k_n,n}x_{k_n,n} & y_n &= y_{0,n}a_{1,n}y_{1,n}\cdots a_{k_n,n}y_{k_n,n}
\end{aligned}
\tag{8}
$$

we get two factorizations of u: $x_0b_1x_1\cdots b_nx_n$ and $y_0b_1y_1\cdots b_ny_n$. Since the product $L_0b_1L_1\cdots a_nL_n$ is unambiguous, we have $x_0 = y_0, \ldots, x_n = y_n$. Each of these words has now two factorizations given by (8) and since the products of (7) are unambiguous, these factorizations are equal. This proves the claim and the proposition. $\square$

We now consider the intersection of two unambiguous products.

Theorem 5.3. *If the products $K = K_0a_1K_1\cdots a_pK_p$ and $L = L_0b_1L_1\cdots b_qL_q$ are unambiguous, the products occurring in Formula (4) are all unambiguous.*

Proof. Let (α, β) be a pair of infiltration maps, and let $U_i = [K_{\bar{\alpha}(i)}]_{\uparrow z_i} \cap [L_{\bar{\beta}(i)}]_{\uparrow z_i}$, for $0 \leqslant i \leqslant r$. We claim that the product $U = U_0c_1U_1\cdots c_rU_r$ is unambiguous. Let

$$u = u_0c_1u_1\cdots c_ru_r = u_0'c_1u_1'\cdots c_ru_r' \tag{9}$$

be two factorizations of a word u of U such that, for $0 \leqslant i \leqslant r$, $u_i, u_i' \in U_i$. We prove by induction on s that $u_s = u_s'$.

Case $s = 0$. By the properties of α and β, we may assume without loss of generality that $\alpha(1) = 1$, which implies that $c_1 = a_1$. It follows from (C$_3$) that $U_0 \subseteq K_0$. Now the product $K_0a_1(K_1a_2K_2\cdots a_pK_p)$ is unambiguous by Proposition 5.1, and by (C$_3$), $U_1c_2U_2\cdots c_rU_r$ is contained in $K_1a_1K_2\cdots a_pK_p$. Therefore, u admits the two factorizations $u_0a_1(u_1c_2u_2\cdots c_ru_r)$ and $u_0'a_1(u_1'c_2u_2'\cdots c_ru_r')$ in this product. Thus $u_0 = u_0'$.

Induction step. Let $s > 0$ and suppose by induction that $u_i = u_i'$ for $0 \leqslant i \leqslant s-1$. If $s = r$, then necessarily $u_s = u_s'$. If $s < r$, we may assume without loss of generality that s is in the range of α. Thus $\alpha(k) = s$ for some k and $c_s = a_k$. We now consider two cases separately.

If $\alpha(k + 1) = s + 1$ (and $c_{s+1} = a_{k+1}$), it follows from (C$_3$) that u has two factorizations

$$(u_0c_1u_1\cdots c_{s-1}u_{s-1})a_ku_sa_{k+1}(u_{s+1}c_{s+1}u_{s+2}\cdots c_ru_r) \text{ and}$$
$$(u_0c_1u_1\cdots c_{s-1}u_{s-1})a_ku_s'a_{k+1}(u_{s+1}'c_{s+1}u_{s+2}'\cdots c_ru_r')$$

over the product $(K_0 a_1 K_1 \cdots a_{s-1} K_{s-1}) a_k K_s a_{k+1} (K_{s+1} a_{k+2} K_{s+2} \cdots a_p K_p)$. Since this product is unambiguous by Proposition 5.1, we get $u_s = u'_s$.

If $\alpha(k+1) \neq s+1$, then $s+1 = \beta(t+1)$ for some t and $c_{s+1} = b_{t+1}$. Setting $i = \beta(t)$, we get $c_i = b_t$ and it follows from (C$_4$) that u has two factorizations

$$(u_0 c_1 u_1 \cdots c_{i-1} u_{i-1}) b_t (u_i c_{i+1} u_{i+1} \cdots c_s u_s) b_{t+1} (u_{s+1} c_{s+2} u_{s+2} \cdots c_r u_r) \text{ and}$$
$$(u_0 c_1 u_1 \cdots c_{i-1} u_{i-1}) b_t (u'_i c_{i+1} u'_{i+1} \cdots c_s u'_s) b_{t+1} (u'_{s+1} c_{s+2} u'_{s+2} \cdots c_r u'_r)$$

over the product $(L_0 b_1 L_1 \cdots b_{t-1} L_{t-1}) b_t L_t b_{t+1} (L_{t+1} b_{t+1} L_{t+2} \cdots b_p L_p)$. This product is unambiguous by Proposition 5.1, and thus

$$u_i c_{i+1} u_{i+1} \cdots c_s u_s = u'_i c_{i+1} u'_{i+1} \cdots c_s u'_s$$

Now the induction hypothesis gives $u_i = u'_i, \ldots, u_{s-1} = u'_{s-1}$ and one finally gets $u_s = u'_s$.   □

We state separately another interesting property.

Theorem 5.4. *Let* $K = K_0 a_1 K_1 \cdots a_p K_p$ *and* $L = L_0 b_1 L_1 \cdots b_q L_q$ *be two unambiguous products and let* (α, β) *and* (α', β') *be two pairs of infiltration maps of* $I(p, q)$. *If the sets* $U(\alpha, \beta)$ *and* $U(\alpha', \beta')$ *meet, then* $\alpha = \alpha'$ *and* $\beta = \beta'$.

Proof. Suppose that a word u belongs to $U(\alpha, \beta)$ and to $U(\alpha', \beta')$. Then u has two decompositions of the form

$$u = u_0 c_1 u_1 \cdots c_r u_r = u'_0 c'_1 u'_1 \cdots c'_{r'} u'_{r'}$$

Condition (C$_1$) [(C$_2$)] and the unambiguity of the product $K_0 a_1 K_1 \cdots a_p K_p$ [$L_0 b_1 L_1 \cdots b_q L_q$] show that, for $0 \leqslant i \leqslant p$ and for $0 \leqslant j \leqslant q$,

$$u_{\alpha(i)} c_{\alpha(i)+1} u_{\alpha(i)+1} \cdots c_{\alpha(i+1)-1} u_{\alpha(i+1)-1} =$$
$$u'_{\alpha'(i)} c'_{\alpha'(i)+1} u'_{\alpha'(i)+1} \cdots c_{\alpha'(i+1)-1} u_{\alpha'(i+1)-1} \in K_i \tag{10}$$

$$u_{\beta(j)} c_{\beta(j)+1} u_{\beta(j)+1} \cdots c_{\beta(j+1)-1} u_{\beta(j+1)-1} =$$
$$u'_{\beta'(j)} c'_{\beta'(j)+1} u'_{\beta'(j)+1} \cdots c_{\beta'(j+1)-1} u_{\beta'(j+1)-1} \in L_j \tag{11}$$

We prove by induction on s that, for $1 \leqslant s \leqslant \min(r, r')$, the following properties hold:

$\quad$ E$_1(s)$: $u_{s-1} = u'_{s-1}$ and $c_s = c'_s$,
$\quad$ E$_2(s)$: $\bar{\alpha}(s) = \bar{\alpha}'(s)$ and $\bar{\beta}(s) = \bar{\beta}'(s)$,
$\quad$ E$_3(s)$: for $i \leqslant \bar{\alpha}(s)$, $\alpha(i) = \alpha'(i)$ and for $j \leqslant \bar{\beta}(s)$, $\beta(j) = \beta'(j)$.

Case $s = 1$. We know that either $\alpha(1) = 1$ or $\beta(1) = 1$ and that either $\alpha'(1) = 1$ or $\beta'(1) = 1$. Suppose that $\alpha(1) = 1$. We claim that $\alpha'(1) = 1$. Otherwise, one has $\beta'(1) = 1$. Now, Formula (10) applied to $i = 0$ gives

$$u_0 = u'_0 c'_1 u'_1 \cdots c_{\alpha'(1)-1} u'_{\alpha'(1)-1}$$

and Formula (11) applied to $j = 0$ gives

$$u_0 c_1 u_1 \cdots c_{\beta(1)-1} u_{\beta(1)-1} = u_0'.$$

Therefore $u_0 = u_0'$ and $\alpha'(1) = 1$, which proves the claim. It follows also that $a_1 = c_{\alpha(1)} = c_{\alpha'(1)}$ and thus $c_1 = c_1'$. We also have in this case $\bar{\alpha}(1) = \bar{\alpha}'(1) = 1$. A similar argument shows that if $\alpha'(1) = 1$, then $\alpha(1) = 1$. Therefore, the conditions $\alpha(1) = 1$ and $\alpha'(1) = 1$ are equivalent and it follows that $\bar{\alpha}(1) = \bar{\alpha}'(1)$. A dual argument would prove that the conditions $\beta(1) = 1$ and $\beta'(1) = 1$ are equivalent and that $\bar{\beta}(1) = \bar{\beta}'(1)$.

Induction step. Let s be such that $1 \leqslant s+1 \leqslant \min(r, r')$ and suppose by induction that the properties $E_1(i)$, $E_2(i)$, $E_3(i)$ hold for $1 \leqslant i \leqslant s$.

Lemma 5.5. *Suppose that $P_\alpha(s)$ holds and let $k = \bar{\alpha}(s)$. Then*

$$s \leqslant \alpha'(k+1) - 1 \tag{12}$$

and

$$u_s = u_s' c_{s+1}' u_{s+1}' \cdots c_{\alpha'(k+1)-1}' u_{\alpha'(k+1)-1}' \tag{13}$$

Proof. Applying (10) with $i = k$, we get

$$u_{\alpha(k)} c_{\alpha(k)+1} u_{\alpha(k)+1} \cdots c_s u_s =$$
$$u_{\alpha'(k)}' c_{\alpha'(k)+1}' u_{\alpha'(k)+1}' \cdots c_{\alpha'(k+1)-1}' u_{\alpha'(k+1)-1}' \tag{14}$$

Since $\bar{\alpha}(s) = \bar{\alpha}'(s)$ by $E_2(s)$, one has $\bar{\alpha}'(s) = k$ and $\alpha'(k+1) \geqslant s+1$ by Lemma 3.1, which gives (12). Further, since $k = \bar{\alpha}(s)$, it follows from $E_3(s)$ that $\alpha(k) = \alpha'(k)$. Now, for $i \leqslant s$, $E_1(i)$ implies that $u_{i-1} = u_{i-1}'$ and $c_i = c_i'$. It follows that the word $u_{\alpha(k)} c_{\alpha(k)+1} u_{\alpha(k)+1} \cdots c_s$ is a prefix of both sides of (14). Therefore, this prefix can be deleted from both sides of (14), which gives (13). $\square$

We now establish $E_1(s+1)$.

Lemma 5.6. *One has $u_s = u_s'$ and $c_{s+1} = c_{s+1}'$. Further, $P_\alpha(s)$ and $P_{\alpha'}(s)$ are equivalent and $P_\beta(s)$ and $P_{\beta'}(s)$ are equivalent.*

Proof. Let us prove that u_s' is a prefix of u_s. By Lemma 3.3, either $P_\alpha(s)$ or $P_\beta(s)$ holds. Suppose that $P_\alpha(s)$ holds. Then by Lemma 5.5, u_s' is a prefix of u_s. If $P_\beta(s)$ holds, we arrive to the same conclusion by using (11) in place of (10) in the proof of Lemma 5.5.

Now, a symmetrical argument using the pair $(\bar{\alpha}', \bar{\beta}')$ would show that u_s is a prefix of u_s'. Therefore, $u_s = u_s'$. Coming back to (13), we obtain $\alpha'(k+1) = s+1$ and since by $E_2(s)$, $k = \bar{\alpha}(s) = \bar{\alpha}'(s)$, one gets $\alpha'(\bar{\alpha}'(s) + 1) = s + 1$, which, by Lemma 3.2, is equivalent to $P_{\alpha'}(s)$. Thus $P_\alpha(s)$ implies $P_{\alpha'}(s)$ and a dual argument would prove the opposite implication.

We also have $c_{s+1} = c_{\alpha(k+1)} = a_{k+1} = c_{\alpha'(k+1)}' = c_{s+1}'$ and thus $c_{s+1} = c_{s+1}'$. Finally, a similar argument works for β. $\square$

We now come to the proof of $E_2(s+1)$ and $E_3(s+1)$. Since $P_\alpha(s)$ and $P_{\alpha'}(s)$ are equivalent, the next two lemma cover all cases.

Lemma 5.7. *If neither $P_\alpha(s)$ nor $P_{\alpha'}(s)$ hold, then $\bar{\alpha}(s+1) = \bar{\alpha}'(s+1)$ and for $i \leqslant \bar{\alpha}(s+1)$, $\alpha(i) = \alpha'(i)$. Similarly, if neither $P_\beta(s)$ nor $P_{\beta'}(s)$ hold, then $\bar{\beta}(s+1) = \bar{\beta}'(s+1)$ and for $i \leqslant \bar{\beta}(s+1)$, $\beta(i) = \beta'(i)$.*

Proof. We just prove the "α part" of the lemma. If neither $P_\alpha(s)$ nor $P_{\alpha'}(s)$ hold, then $\bar{\alpha}(s+1) = \bar{\alpha}(s)$ and $\bar{\alpha}'(s+1) = \bar{\alpha}'(s)$. Since $\bar{\alpha}(s) = \bar{\alpha}'(s)$ by $E_2(s)$, one gets $\bar{\alpha}(s+1) = \bar{\alpha}'(s+1)$. The second property is an immediate consequence of $E_3(s)$.   □

Lemma 5.8. *If both $P_\alpha(s)$ and $P_{\alpha'}(s)$ hold, then $\bar{\alpha}(s+1) = \bar{\alpha}'(s+1)$ and for $i \leqslant \bar{\alpha}(s+1)$, $\alpha(i) = \alpha'(i)$. Similarly, if both $P_\beta(s)$ and $P_{\beta'}(s)$ hold, then $\bar{\beta}(s+1) = \bar{\beta}'(s+1)$ and for $i \leqslant \bar{\beta}(s+1)$, $\beta(i) = \beta'(i)$.*

Proof. Again, we just prove the "α part" of the lemma. If both $P_\alpha(s)$ and $P_{\alpha'}(s)$ hold, then $\bar{\alpha}(s+1) = \bar{\alpha}(s) + 1$ and $\bar{\alpha}'(s+1) = \bar{\alpha}'(s) + 1$. Since $\bar{\alpha}(s) = \bar{\alpha}'(s)$ by $E_2(s)$, one gets $\bar{\alpha}(s+1) = \bar{\alpha}'(s+1)$. Property $E_3(s)$ shows that for $i \leqslant \bar{\alpha}(s)$, $\alpha(i) = \alpha'(i)$. Since $\bar{\alpha}(s+1) = \bar{\alpha}(s) + 1$, it just remains to prove that

$$\alpha(\bar{\alpha}(s+1)) = \alpha'(\bar{\alpha}(s+1)) \tag{15}$$

But Lemma 3.2 shows that $\alpha(\bar{\alpha}(s+1)) = s+1$ and $\alpha'(\bar{\alpha}'(s+1)) = s+1$, which proves (15) since $\bar{\alpha}(s+1) = \bar{\alpha}'(s+1)$.   □

This concludes the induction step and the proof of Theorem 5.4.   □

Corollary 5.9. *Let $\mathcal{L}$ be a lattice of regular languages closed under quotient. Then its unambiguous polynomial closure is also a lattice closed under quotient.*

If $\mathcal{L}$ is a Boolean algebra, then one can be more precise.

Corollary 5.10. *Let $\mathcal{L}$ be a Boolean algebra of regular languages closed under quotient. Then its unambiguous polynomial closure is also a Boolean algebra closed under quotient.*

Let us conclude with an example which shows that, under the assumptions of Theorem 5.4, the sets $U(\alpha, \beta)$ cannot be further decomposed as a disjoint union of unambiguous products.
Let $K = K_0 a K_1$ and $L = L_0 a L_1$ with $K_0 = L_1 = 1 + b + c + c^2$ and $L_0 = K_1 = a + ab + ba + ac + ca + ac^2 + bab + cac + cac^2$. Then

$$
\begin{aligned}
K \cap L = {}& aa + aab + aba + aac + aca + aac^2 + abab + acac + acac^2 + \\
& baa + baab + baba + baac + baac^2 + babab + caa + \\
& caab + caac + caca + caac^2 + cacac + cacac^2
\end{aligned}
$$

One can write for instance $K \cap L$ as $(1+b+c)aa(1+b+c+c^2) + (1+b)a(1+b)a(1+b) + (1+c)a(1+c)a(1+c+c^2)$ but the three components of this language are not disjoint, since they all contain aa. Note that the words $acab$, $abac$, $baca$ and $caba$ are not in $K \cap L$.

The syntactic ordered monoid of K_0 and L_1 has 4 elements $\{1, a, b, c\}$ and is presented by the relations $a = ba = b^2 = bc = ca = cb = 0$ and $c^2 = b$. Its syntactic order is defined by $a < b < c < 1$.

The syntactic ordered monoid of L_0 and K_1 has 13 elements:

$$\{1, a, b, c, a^2, ab, ac, ba, ca, c^2, ac^2, bab, cac\}$$

and is defined by the relations $cac^2 = bab$ and

$$b^2 = bc = cb = a^2 = aba = aca = bac = cab = c^2a = c^3 = 0.$$

The syntactic order is:

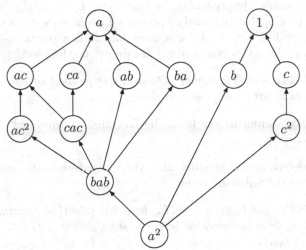

There is only one pair of infiltration maps (α, β) of $I(1, 1)$ that defines a nonempty set $U(\alpha, \beta)$. This pair is defined as follows: $\alpha(1) = 1$ and $\beta(1) = 2$. The triples (z_0, z_1, z_2) of $C(\alpha, \beta)$ are exactly the triples of words such that $z_0 a z_1 a z_2 \in K \cap L$. In particular, $z_0 \in \{1, b, c\}$, $z_1 \in \{1, b, c\}$ and $z_2 \in \{1, b, c, c^2\}$. Now, one has

$[K_0]_{\uparrow 1} = 1$ $[K_0]_{\uparrow b} = 1 + b + c + c^2$ $[K_0]_{\uparrow c} = 1 + c$

$[K_1]_{\uparrow 1} = 1$ $[K_1]_{\uparrow b} = 1 + b$ $[K_1]_{\uparrow c} = 1 + c$ $[K_1]_{\uparrow c^2} = 1 + c + c^2$

$[L_0]_{\uparrow 1} = 1$ $[L_0]_{\uparrow b} = 1 + b$ $[L_0]_{\uparrow c} = 1 + c$

$[L_1]_{\uparrow 1} = 1$ $[L_1]_{\uparrow b} = 1 + b + c + c^2$ $[L_1]_{\uparrow c} = 1 + c$ $[L_1]_{\uparrow c^2} = 1 + b + c + c^2$

which gives the following possibilities for the triples (U_0, U_1, U_2), for the following triples $z = (z_0, z_1, z_2)$:

$$
\begin{array}{llll}
z = (1,1,1) & U_0 = 1 & U_1 = 1 & U_2 = 1 \\
z = (b,b,b) & U_0 = 1+b & U_1 = 1+b & U_2 = 1+b \\
z = (c,c,c) & U_0 = 1+c & U_1 = 1+c & U_2 = 1+c \\
z = (b,c,c^2) & U_0 = 1+b & U_1 = 1+c & U_2 = 1+c+c^2 \\
z = (c,c,c^2) & U_0 = 1+c & U_1 = 1+c & U_2 = 1+c+c^2
\end{array}
$$

5.2 Deterministic Product

The marked product $L = L_0 a_1 L_1 \cdots a_n L_n$ of n nonempty languages L_0, L_1, ..., L_n of A^* is *left deterministic* [*right deterministic*] if, for $1 \leqslant i \leqslant n$, the set $L_0 a_1 L_1 \cdots L_{i-1} a_i$ [$a_i L_i \cdots a_n L_n$] is a prefix [suffix] code. This means that every word of L has a unique prefix [suffix] in $L_0 a_1 L_1 \cdots L_{i-1} a_i$ [$a_i L_i \cdots a_n L_n$]. It is observed in [3, p. 495] that the marked product $L_0 a_1 L_1 \cdots a_n L_n$ is deterministic if and only if, for $1 \leqslant i \leqslant n$, the language $L_{i-1} a_i$ is a prefix code. Since the product of two prefix codes is a prefix code, we get the following proposition.

Proposition 5.11. *Any left [right] deterministic product of left [right] deterministic products is left [right] deterministic.*

Proof. This follows immediately from the fact that the product of two prefix codes is a prefix code. □

Factorizing a deterministic product also gives a deterministic product. More precisely, one has the following result.

Proposition 5.12. *Let $L_0 a_1 L_1 \cdots a_n L_n$ be a left [right] deterministic product and let $i_1, \ldots, i_k$ be a sequence of integers satisfying $0 < i_1 < \ldots < i_k < n$. Finally, let $R_0 = L_0 a_1 L_1 \cdots a_{i_1-1} L_{i_1-1}, \ldots, R_k = L_{i_k} a_{i_k+1} L_{i_k+1} \cdots L_{n-1} a_n L_n$. Then the product $R_0 a_{i_1} R_1 \cdots a_{i_k} R_k$ is left [right] deterministic.*

Proof. Trivial. □

The *left [right] deterministic polynomial closure* of a class of languages $\mathcal{L}$ of A^* is the set of languages that are finite unions of left [right] deterministic products of the form $L_0 a_1 L_1 \cdots a_n L_n$, where the a_i's are letters and the L_i's are elements of $\mathcal{L}$.

We can now state the counterpart of Theorem 5.3 for deterministic products.

Theorem 5.13. *If the products $K = K_0 a_1 K_1 \cdots a_p K_p$ and $L = L_0 b_1 L_1 \cdots b_q L_q$ are deterministic, the products occurring in Formula (4) are all deterministic.*

Proof. Let $i \in \{0, \ldots, r\}$. By construction, there exists $k \geqslant 0$ such that $i + 1 = \alpha(k+1)$ or $i + 1 = \beta(k+1)$. By Lemma 4.2, there exists $j \leqslant i$ such that either $U_j c_{j+1} U_{j+1} \cdots U_i \subseteq K_k$ and $c_{\alpha(k+1)} = a_{k+1}$ or $U_j c_{j+1} U_{j+1} \cdots U_i \subseteq L_k$ and

$c_{\alpha(k+1)} = b_{k+1}$. Suppose we are in the first case and that $U_i c_{i+1}$ is not a prefix code. Then $U_j c_{j+1} U_{j+1} \cdots U_i c_{i+1}$ is not a prefix code and thus $K_k a_{k+1}$ is not a prefix code. This yields a contradiction since the product $K_0 a_1 K_1 \cdots a_p K_p$ is deterministic. $\square$

Corollary 5.14. *Let $\mathcal{L}$ be a lattice of regular languages closed under quotient. Then its deterministic polynomial closure is also closed under quotient.*

Acknowledgements. I would like to thank Mário J. J. Branco for his careful reading of a first version of this article.

References

1. Arfi, M.: Opérations polynomiales et hiérarchies de concaténation. Theoret. Comput. Sci. 91, 71–84 (1991)
2. Branco, M.J.J.: On the Pin-Thérien expansion of idempotent monoids. Semigroup Forum 49(3), 329–334 (1994)
3. Branco, M.J.J.: The kernel category and variants of the concatenation product. Internat. J. Algebra Comput. 7(4), 487–509 (1997)
4. Branco, M.J.J.: Two algebraic approaches to variants of the concatenation product. Theoret. Comput. Sci. 369(1-3), 406–426 (2006)
5. Branco, M.J.J.: Deterministic concatenation product of languages recognized by finite idempotent monoids. Semigroup Forum 74(3), 379–409 (2007)
6. Branco, M.J.J., Pin, J.-É.: Equations defining the polynomial closure of a lattice of regular languages. In: Albers, S., Marchetti-Spaccamela, A., Matias, Y., Nikoletseas, S., Thomas, W. (eds.) ICALP 2009, Part II. LNCS, vol. 5556, pp. 115–126. Springer, Heidelberg (2009)
7. Lothaire, M.: Combinatorics on words, Cambridge Mathematical Library. Cambridge University Press, Cambridge (1997)
8. Perrin, D., Pin, J.-E.: Infinite Words. Pure and Applied Mathematics, vol. 141, Elsevier (2004) ISBN 0-12-532111-2
9. Pin, J.-E.: Propriétés syntactiques du produit non ambigu. In: de Bakker, J.W., van Leeuwen, J. (eds.) ICALP 1980. LNCS, vol. 85, pp. 483–499. Springer, Heidelberg (1980)
10. Pin, J.-E.: A variety theorem without complementation. Russian Mathematics (Iz. VUZ) 39, 80–90 (1995)
11. Pin, J.-E., Straubing, H., Thérien, D.: Locally trivial categories and unambiguous concatenation. J. of Pure and Applied Algebra 52, 297–311 (1988)
12. Pin, J.-E., Thérien, D.: The bideterministic concatenation product. Internat. J. Algebra Comput. 3, 535–555 (1993)
13. Pin, J.-E., Weil, P.: Polynomial closure and unambiguous product. Theory Comput. Systems 30, 1–39 (1997)
14. Schützenberger, M.-P.: Une théorie algébrique du codage, in Séminaire Dubreil-Pisot, année 1955-56, Exposé No. 15, 27 février 1956, 24 pages, Inst. H. Poincaré, Paris (1956)
15. Schützenberger, M.-P.: Sur le produit de concaténation non ambigu. Semigroup Forum 18, 331–340 (1976)

Two Dimensional Prefix Codes of Pictures*

Marcella Anselmo[1], Dora Giammarresi[2], and Maria Madonia[3]

[1] Dipartimento di Informatica, Università di Salerno I-84084 Fisciano (SA) Italy
anselmo@dia.unisa.it
[2] Dipartimento di Matematica, Università di Roma "Tor Vergata",
via della Ricerca Scientifica, 00133 Roma, Italy
giammarr@mat.uniroma2.it
[3] Dipartimento di Matematica e Informatica, Università di Catania,
Viale Andrea Doria 6/a, 95125 Catania, Italy
madonia@dmi.unict.it

Abstract. A two-dimensional code is defined as a set $X \subseteq \Sigma^{**}$ such that any picture over Σ is tilable in at most one way with pictures in X. The codicity problem is undecidable. The subclass of prefix codes is introduced and it is proved that it is decidable whether a finite set of pictures is a prefix code. Further a polynomial time decoding algorithm for finite prefix codes is given. Maximality and completeness of finite prefix codes are studied: differently from the one-dimensional case, they are not equivalent notions. Completeness of finite prefix codes is characterized.

1 Introduction

The theory of word codes is a well established subject of investigation in theoretical computer science. Results are related to combinatorics on words, formal languages, automata theory and semigroup theory. In fact the aim is to find structural properties of codes to be exploited for their construction. We refer to [7] for complete references.

During the last fifty years, many researchers investigated how the formal language theory can be transferred into a two-dimensional (2D) world (e.g. [8,11,12,4,18]). Extensions of classical words to two dimensions bring in general to the definition of polyominoes, labeled polyominoes, directed polyominoes, as well as rectangular labeled polyominoes usually referred to as pictures. Some different attempts were done to generalize the notion of code to those 2D objects. A set C of polyominoes is a code if every polyomino that is tilable with (copies of) elements of C, it is so in a unique way. Most of the results show that in the 2D context we loose important properties. A major result due to D. Beauquier and M. Nivat states that the problem whether a finite set of polyominoes is a code is undecidable, and the same result holds also for dominoes ([6]). Related particular cases were studied in [1]. In [14] codes of directed polyominoes equipped with catenation operations are considered, and some special decidable cases are detected. Codes of labeled polyominoes, called bricks, are studied in [17] and further undecidability results are proved.

* Partially supported by MIUR Project *"Aspetti matematici e applicazioni emergenti degli automi e dei linguaggi formali"*, by 60% Projects of University of Catania, Roma "Tor Vergata", Salerno.

M.-P. Béal and O. Carton (Eds.): DLT 2013, LNCS 7907, pp. 46–57, 2013.
© Springer-Verlag Berlin Heidelberg 2013

As major observation, remark that all mentioned results consider 2D codes independently from a 2D language theory. In this paper we consider codes of pictures, i.e. rectangular arrays of symbols. Two partially operations are generally considered on pictures: the row/column concatenations. Using these operations, in [10] doubly-ranked monoids are introduced and picture codes are studied in order to extend syntactic properties to two dimensions. Unfortunately most of the results are again negative. Even the definition of prefix picture codes in [13] does not lead to any wide enough class.

We study *picture codes* in relation to the family REC of picture languages recognized by tiling systems. REC is defined by naturally extending to two-dimensions a characterization of finite automata by means of local sets and alphabetic projections ([12]). But pictures are much more difficult to deal with than strings and in fact a crucial difference is that REC is intrinsically non-deterministic and the parsing problem is NP-complete ([16]). In [2,3] unambiguous and deterministic tiling systems, together with the corresponding subfamilies DREC and UREC, are defined; it is proved the all the inclusions are proper. Moreover the problem whether a given tiling system is unambiguous is undecidable. Despite these facts, REC has several remarkable properties that let it be the most accredited generalization to 2D of regular string languages.

In this paper the definition of code is given in terms of the operation of *tiling star* as defined in [18]: the tiling star of a set X is the set X^{**} of all pictures that are tilable (in the polyominoes style) by elements of X. Then X is a code if any picture in X^{**} is tilable in one way. Remark that if $X \in$ REC then X^{**} is also in REC. By analogy to the string case, we denote by *flower tiling system* a special tiling system that recognizes X^{**} and show that if X is a picture code then such flower tiling system is unambiguous and therefore X^{**} belongs to UREC. This result sounds like a nice connection to the word code theory, but, unfortunately, again we prove that it is undecidable whether a given set of pictures is a code. This is actually not surprising because it is coherent with the known result of undecidability for UREC.

Inspired by the definition of DREC, we propose *prefix codes*. Pictures are then considered with a preferred scanning direction: from top-left corner to the bottom-right corner. Intuitively, we assume that if X is a prefix code, when decoding a picture p starting from top-left corner, it can be univocally decided which element in X we can start with. The formal definition of prefix codes involves polyominoes. In fact, in the middle of the decoding process, the already decoded part of p is not necessarily rectangular, i.e. it is in general a polyomino. More precisely, we get a special kind of polyominoes that are vertically connected and always contain the whole first row of their minimal bounding box: we refer to them as *comb polyominoes*. Remark that this makes a big difference with the word code theory where a (connected) part of a word is always a word.

We define X to be a *prefix set* of pictures by imposing that any comb polyomino that is tilable by pictures in X cannot "start" with two different pictures of set X. We prove that it can be always verified whether a finite set of picture is a prefix set and that, as in 1D case, every prefix set of picture is a code. Moreover we present a polynomial time decoding algorithm for finite prefix codes. We extend to picture codes also the classical notions of *maximal* and *complete* codes. We present several results regarding the relations between these two notions and give a characterization for maximal complete finite prefix codes.

2 Preliminaries

We introduce some definitions about two-dimensional languages (see [12]).

A *picture* over a finite alphabet Σ is a two-dimensional rectangular array of elements of Σ. Given a picture p, $|p|_{row}$ and $|p|_{col}$ denote the number of rows and columns, respectively; $|p| = (|p|_{row}, |p|_{col})$ denotes the picture *size*. The set of all pictures over Σ of fixed size (m, n) is denoted by $\Sigma^{m,n}$, while Σ^{m*} and Σ^{*n} denote the set of all pictures over Σ with fixed number of rows m and columns n, respectively. The set of all pictures over Σ is denoted by Σ^{**}. A *two-dimensional language* (or *picture language*) over Σ is a subset of Σ^{**}.

The *domain* of a picture p is the set of coordinates $\mathrm{dom}(p) = \{1, 2, \ldots, |p|_{row}\} \times \{1, 2, \ldots, |p|_{col}\}$. We let $p(i, j)$ denote the symbol in p at coordinates (i, j). Positions in $\mathrm{dom}(p)$ are ordered following the lexicographic order: $(i, j) < (i', j')$ if either $i < i'$ or $i = i'$ and $j < j'$. Moreover, to easily detect border positions of pictures, we use initials of words "top", "bottom", "left" and "right": then, for example the *tl-corner* of p refers to position $(1, 1)$. A *subdomain* of $\mathrm{dom}(p)$ is a set d of the form $\{i, i+1, \ldots, i'\} \times \{j, j+1, \ldots, j'\}$, where $1 \leq i \leq i' \leq |p|_{row}$, $1 \leq j \leq j' \leq |p|_{col}$, also specified by the pair $[(i, j), (i', j')]$. The *subpicture of* p associated to $[(i, j), (i', j')]$ is the portion of p corresponding to positions in the subdomain and is denoted by $p[(i, j), (i', j')]$. Given pictures x, p, with $|x|_{row} \leq |p|_{row}$ and $|x|_{col} \leq |p|_{col}$, we say that x is a *prefix* of p if x is a subpicture of p corresponding to its top-left portion, i.e. if $x = p[(1, 1), (|x|_{row}, |x|_{col})]$.

Let $p, q \in \Sigma^{**}$ pictures of size (m, n) and (m', n'), respectively, the *column concatenation* of p and q (denoted by $p \oplus q$) and the *row concatenation* of p and q (denoted by $p \ominus q$) are partial operations, defined only if $m = m'$ and if $n = n'$, respectively, as:

$$p \oplus q = \boxed{\begin{array}{c|c} p & q \end{array}} \qquad p \ominus q = \boxed{\begin{array}{c} p \\ \hline q \end{array}}.$$

These definitions can be extended to define two-dimensional languages row- and column- concatenations and *row-* and *column- stars* ([2,12]).

In this paper we will consider another interesting star operation for picture language introduced by D. Simplot in [18]. The idea is to compose pictures in a way to cover a rectangular area without the restriction that each single concatenation must be a $\ominus$ or $\oplus$ operation. For example, the following figure sketches a possible kind of composition that is not allowed applying only $\ominus$ or a $\oplus$ operations.

Definition 1. *The* tiling star *of X, denoted by X^{**}, is the set of pictures p whose domain can be partitioned in disjoint subdomains $\{d_1, d_2, \ldots, d_k\}$ such that any subpicture p_h of p associated with the subdomain d_h belongs to X, for all $h = 1, \ldots, k$.*

Language X^{**} is called the set of all tilings by X in [18]. In the sequel, if $p \in X^{**}$, the partition $t = \{d_1, d_2, \ldots, d_k\}$ of dom(p), together with the corresponding pictures $\{p_1, p_2, \ldots, p_k\}$, is called a *tiling decomposition* of p in X.

In this paper, while dealing with tiling star of a set X, we will need to manage also non-rectangular "portions" of pictures composed by elements of X: those are actually labeled polyominoes, that we will call polyominoes, for the sake of simplicity.

We extend notations and definitions from pictures to polyominoes by simply defining the *domain of a (labeled) polyomino* as the set of pairs (i, j) corresponding to all positions occupied inside its minimal bounding box, being $(1, 1)$ the tl-corner position of the bounding box. See the examples below.

a	b	a	a
a	a	a	a
b	b	a	a
a	b	a	a

a	b		
	a		
	b	a	a
a	b	a	a

a	b	a	a
a		a	a
b			

(a)   (b)   (c)

Then we can use notion, for a picture x, to be *subpicture* or *prefix of a polyomino c*. Observe that the notion of prefix of polyomino makes sense only if the domain of the polyomino contains $(1, 1)$. If C is a set of polyominoes, *Pref(C)* denotes the (possibly empty) set of all pictures that are prefix of some element in C.

Moreover, we can extend to polyominoes the notion of tiling decomposition in a set of pictures X. We can also define a sort of tiling star that, applied to a set of pictures X, produces the set of all polyominoes that have a tiling decomposition in X. If a polyomino p belongs to the polyomino star of X, we say that p is *tilable* in X.

Let us now recall definitions and properties of recognizable picture languages. Given a finite alphabet Γ, a two-dimensional language $L \subseteq \Gamma^{**}$ is *local* if L coincides with the set of pictures whose sub-blocks of size $(2, 2)$, are all in a given finite set Θ of *allowed tiles* (considering also border positions). Language $X \subseteq \Sigma^{**}$ is *recognizable* if it is the projection of a local language over alphabet Γ (by means of a projection $\pi : \Gamma \to \Sigma$). A *tiling system* for X is a quadruple specifying the four ingredients necessary to recognize X: $(\Sigma, \Gamma, \Theta, \pi)$. The family of all *recognizable* picture languages is denoted by *REC*. A tiling system is *unambiguous* if any recognized picture has a unique pre-image in the local language, and UREC is the class of languages recognized by an unambiguous tiling system [3]. REC family shares several properties with the regular string languages. In particular, REC is closed under row/column concatenations, row/column stars, and tiling star ([12,18]). Note that the notion of locality/recognizability by tiles corresponds to that of finite type/sofic subshifts in symbolic dynamical systems [15].

3 Two-Dimensional Codes

In the literature, many authors afforded the definition of codes in two dimensions. In different contexts, polyomino codes, picture codes, and brick codes were defined. We introduce *two-dimensional codes*, according to the theory of recognizable languages, and in a slight different way from all the mentioned definitions.

Definition 2. $X \subseteq \Sigma^{**}$ *is a code iff any* $p \in \Sigma^{**}$ *has at most one tiling decomposition in* X.

We consider some simple examples. Let $\Sigma = \{a, b\}$ be the alphabet.

Example 1. Let $X = \left\{ \boxed{a\ b}, \ \boxed{\begin{smallmatrix} a \\ b \end{smallmatrix}}, \ \boxed{\begin{smallmatrix} a\ a \\ a\ a \end{smallmatrix}} \right\}$. It is easy to see that X is a code. Any picture $p \in X^{**}$ can be decomposed starting at tl-corner and checking the size $(2, 2)$ subpicture $p[(1, 1), (2, 2)]$: it can be univocally decomposed in X. Then, proceed similarly for the next contiguous size $(2, 2)$ subpictures.

Example 2. Let $X = \left\{ \boxed{a\ b}, \ \boxed{b\ a}, \ \boxed{\begin{smallmatrix} a \\ a \end{smallmatrix}} \right\}$. Set X is not a code. Indeed picture $\begin{smallmatrix} a\ b\ a \\ a\ b\ a \end{smallmatrix}$ has the two following different tiling decompositions in X: $t_1 = \begin{smallmatrix} a\ b\ |a \\ a\ b\ |a \end{smallmatrix}$ and $t_2 = \begin{smallmatrix} a|\ b\ a \\ a|\ b\ a \end{smallmatrix}$.

In the 1D setting, a string language X is a code if and only if a special flower automaton for X^* is unambiguous (cf. [7]). In 2D an analogous result holds for a finite picture language X, by introducing a special tiling system recognizing X^{**}, we call the *flower tiling system of* X, in analogy to the 1D case. The construction (omitted for lack of space) goes similarly to the ones in [12,18], assigning a different "colour" to any different picture in X. In this way, one can show that, a finite language X is a code if and only if the flower tiling system of X is unambiguous. Unfortunately such result cannot be used to decide whether a finite language is a code: the problem whether a tiling system is unambiguous is undecidable [3]. As a positive consequence we obtain the following non-trivial result; recall that in 2D not any recognizable language can be recognized in a unambiguous way.

Proposition 1. *Let* X *be a finite language. If* X *is a code then* X^{**} *is in UREC.*

The unambiguity of a tiling system remains undecidable also for flower tiling systems. In fact, according to the undecidability of similar problems in two dimensions (codicity is undecidable for polyomino codes, picture codes, and brick codes), the codicity problem is undecidable also in our setting.

2D-CODICITY PROBLEM
INPUT: $X \subseteq \Sigma^{**}$, X finite
OUTPUT: TRUE if X is a code, FALSE otherwise.

Proposition 2. *The* 2D-CODICITY PROBLEM *is undecidable.*

Proof. (Sketch) The Thue system word problem reduces to the 2D-CODICITY PROBLEM. The well-known undecidability of the Thue system word problem (see e.g. [9]) will imply the undecidability of the 2D-CODICITY PROBLEM. For a given Thue system (Σ, S) and two words $u, v \in \Sigma^*$ we construct the set $X_{S,u,v}$ of square bricks over an alphabet that extends Σ, in the way defined in [17], Section 3. Each square brick can be regarded to as a picture, and hence $X_{S,u,v}$ can be consider as a picture language $X_{S,u,v} \subseteq \Sigma^{**}$. The authors of [17] show that u and v are equivalent iff $X_{S,u,v}$ is not a brick code. The proof is completed by showing that $X_{S,u,v}$ is a brick code iff it is a picture code. $\qquad\square$

Next step will be to consider subclasses of codes that are decidable. In 1D an important class of codes is that one of *prefix codes*. In the next section we consider a possible extension of the definition to two dimensions.

4   Prefix Codes

In Section 2 we reported the definition of prefix of a picture p, as a subpicture x corresponding to the top-left portion of p. Such definition "translates" to 2D the classic notion of prefix of a string. Starting from this, one could merely define a set of picture X to be prefix whenever it does not contain pictures that are prefixes of other pictures in X. Unfortunately this property would not guarantee the set to be a code. Consider for example the set X introduced in Example 2. No picture in X is prefix of another picture in X; nevertheless X is not a code.

The basic idea in defining a *prefix code* is to prevent the possibility to start decoding a picture in two different ways (as it is for the prefix string codes). One major difference going from 1D to 2D case is that, while any initial part of a decomposition of a string is still a string, the initial part of a decomposition of a picture, at an intermediate step, has not necessarily a rectangular shape. Starting from the tl-corner of a picture, it is possible to reconstruct its tiling decomposition in many different ways, obtaining, as intermediate products, some (labeled) polyominoes whose domain contain always the tl-corner position $(1, 1)$. If we proceed by processing positions (i, j) in lexicographic order, the domains of these polyominoes will have the peculiarity that if they contain a position (h, k) then they contain also all positions above it up to the first row. We name these special polyominoes as follows.

Definition 3. *A corner polyomino is a labeled polyomino whose domain contains position $(1, 1)$. A comb polyomino is a corner polyomino whose domain is the following set of positions for some $n, h_1, h_2, \cdots, h_n \geq 1$:*

$$
\begin{array}{llll}
(1,1) & (1,2) & \cdots\cdots & (1,n) \\
(2,1) & \vdots & & (2,n) \\
\vdots & (h_2, 2) & & \vdots \\
(h_1, 1) & & & (h_n, n)
\end{array}
$$

In other words, a comb polyomino is a column convex corner polyomino whose domain contains all positions in the first row of its minimal bounding box. See the figure in Section 2, where (b) is a corner (but not a comb) polyomino, and (c) is a comb polyomino. In the literature these corner polyominoes correspond to labeled directed polyominoes while comb polyominoes correspond to labeled skyline or Manhattan polyominoes rotated by 180 degrees. Now we can define the comb star, a sort of tiling star that, applied to a set of pictures X, produces a set of comb polyominoes. This set of comb polyominoes tiliable in X will be denoted by X^{**} and will be called the *comb star* of X.

Comb polyominoes and comb star are used to define prefix sets. A set of pictures X is *prefix* if any decomposition of a comb polyomino in X^{**} can "start" (in its tl-corner) in a unique way with a picture of X. To give a formal definition, recall that a picture p is a *prefix of a (comb) polyomino* c if the domain of c includes all positions in $\text{dom}(p) = \{1, 2, \ldots, |p|_{row}\} \times \{1, 2, \ldots, |p|_{col}\}$, and $c(i, j) = p(i, j)$ for all positions $(i, j) \in \text{dom}(p)$.

Definition 4. *A set $X \subseteq \Sigma^{**}$ is* prefix *if any two different pictures in X cannot be both prefix of the same comb polyomino $c \in X^{**}$.*

Example 3. Let $X \subseteq \Sigma^{**}$. One can easily show the following. If $|\Sigma| = 1$ then X is prefix if and only if $|X| = 1$. If $|X| = 2$ and $|\Sigma| \geq 2$, then X is prefix if and only if the two pictures in X are not the power of a same picture (cannot be obtained by column and row concatenation of a same picture). Any set $X \subseteq \Sigma^{m,n}$ containing pictures on Σ of fixed size (m, n), is always prefix.

Example 4. It is easy to verify that the set X of Example 1 is prefix. On the contrary, the set X of Example 2 is not prefix: two different elements of X are prefixes of the comb polyomino $c = \begin{array}{|ccc|} \hline a & b & a \\ a & b & a \\ \hline \end{array}$ that belongs to X^{**}.

Example 5. Let $X = \Big\{\; \boxed{a\ b\ a}\;,\; \boxed{a\ b\ b}\;,\; \boxed{\begin{array}{c} b \\ b \end{array}}\;,\; \boxed{\begin{array}{cc} a & a \\ a & a \end{array}}\;,\; \boxed{\begin{array}{cc} a & a \\ a & b \end{array}}\;,\; \boxed{\begin{array}{cc} a & a \\ b & a \end{array}}\;,\; \boxed{\begin{array}{cc} a & a \\ b & b \end{array}}\;,\; \boxed{\begin{array}{cc} b & a \\ a & a \end{array}}\;,\; \boxed{\begin{array}{cc} b & a \\ a & b \end{array}}\;,\; \boxed{\begin{array}{cc} b & b \\ a & a \end{array}}\;,$ $\boxed{\begin{array}{cc} b & b \\ a & b \end{array}}\Big\}$. Language X is prefix: no two pictures in X can be overlapped in order to be both prefixes of the same comb polyomino.

Let us introduce another notion related to tiling that will be useful for the proofs in the sequel: the notion of *covering*. For example, in the figure below, the picture with thick borders is *(properly) covered* by the others.

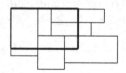

Definition 5. *A picture p is* covered *by a set of pictures X, if there exists a corner polyomino c such that p is prefix of c and the domain of c can be partitioned in rectangular subdomains $\{d_1, ..., d_h\}$ such that each d_i corresponds to a picture in X and the tl-corner of each d_i belongs to the domain of p. Moreover p is* properly covered *if the subdomain containing position $(1, 1)$ corresponds to a picture different from p itself.*

Proposition 3. *A set X is prefix if and only if every $x \in X$ cannot be properly covered by pictures in X.*

Proof. Assume there exists $x \in X$ properly covered by pictures in X, let c be the corresponding corner polyomino, and t its tiling decomposition. Then c can be "completed" to a comb polyomino, by filling all the empty parts of c, that contradict column convexity of c, up to the first row. Indeed it is possible to concatenate some copies of elements of X, that occurr in t, and cover the bottom and right border positions of x

(the tl-corner of each subdomain in t belongs to $\mathrm{dom}(x)$). For example in the figure before Definition 5, one may concatenate a copy of the first two pictures crossing the right border. Conversely, if there exist $x, y \in X$ both prefixes of a comb polyomino c, with a tiling decomposition t in X then, if the subdomain of t containing position $(1, 1)$ does not correspond to x (to y, resp.), one can properly cover x (y, resp.), by taking only the elements of X whose tl-corners belongs to the domain of x (y, resp.).   □

Definition 4 seems to be a right generalization of prefix set of strings since it implies the desired property for the set to be a code. Remark that the following result holds without the hypothesis of finiteness on X, and its converse does not hold.

Proposition 4. *If $X \subseteq \Sigma^{**}$ is prefix then X is a code.*

Proof. (Sketch) Suppose by contradiction that there exists a picture $u \in \Sigma^{**}$ that admits two different tiling decompositions in X, say t_1 and t_2. Now, let (i_0, j_0) the smallest position (in lexicographic order) of u, where t_1 and t_2 differ. Position (i_0, j_0) corresponds in t_1 to position $(1, 1)$ of some $x_1 \in X$, and in t_2 to position $(1, 1)$ of some $x_2 \in X$, with $x_1 \neq x_2$. See the figure below, where a dot indicates position (i_0, j_0) of u in t_1 (on the left) and t_2 (on the right), respectively.

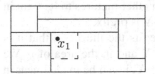 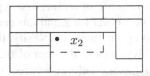

Consider now the size of x_1 and x_2 and suppose, without loss of generality, that $|x_1|_{row} > |x_2|_{row}$ and $|x_1|_{col} < |x_2|_{col}$ (equalities are avoided because of the prefix hypothesis). In this case, x_1 together with other pictures placed to its right in the decomposition t_1 would be a proper cover for x_2 that, by Proposition 3, contradicts the hypothesis of X prefix set.

□

Proposition 4 shows that prefix sets form a class of codes: the *prefix codes*. Contrarily to the case of all other known classes of 2D codes, the family of finite prefix codes has the important property to be decidable. Using results in Proposition 3 together with the hypothesis that X is finite, the following result can be proved.

Proposition 5. *It is decidable whether a finite set of picture $X \subseteq \Sigma^{**}$ is a prefix code.*

We now present a decoding algorithm for a finite prefix picture code.

Proposition 6. *There exists a polynomial algorithm that, given a finite prefix code $X \subseteq \Sigma^{**}$ and a picture $p \in \Sigma^{**}$, finds, if it exists, a tiling decomposition of p in X, otherwise it exits with negative answer.*

Proof. Here is a sketch of the algorithm and its correctness proof. The algorithm scans p using a "current position" (i, j) and a "current partial tiling decomposition" $\mathcal{T}$ that contains some of the positions of p grouped as collections of rectangular p subdomains. Notice that, at each step, the partial decomposition corresponds to a comb polyomino.

1. Sort pictures in X with respect to their size (i.e. by increasing number of rows and, for those of same number of rows, by increasing number of columns).
2. Start from tl-corner position of p ($(i, j) = (1, 1)$) and $\mathcal{T} = \emptyset$.
3. Find the smallest picture $x \in X$ that is a subpicture of p when associated at subdomain with tl-corner in (i, j) and such that positions of this subdomain were not yet put in $\mathcal{T}$. More technically: let (r_x, c_x) indicate the size of x. We find the smallest x such that x is subpicture of p at subdomain $d_h = [(i, j), (i + r_x - 1, j + c_x - 1)]$ and positions in d_h are not in the current $\mathcal{T}$.
4. If x does not exist then exit with negative response, else add the subdomain d_h to $\mathcal{T}$ and set (i, j) as the smallest position not yet added to (subdomains in) $\mathcal{T}$.
5. If (i, j) does not exist then exit returning $\mathcal{T}$, else repeat from step 3.

It is easy to understand that if $p \notin X^{**}$ then at step 3, the algorithm will not find any x, and will exit giving no decomposition in step 4. Suppose now that $p \in X^{**}$ that is there exists a (unique) decomposition $\mathcal{T}$ for p. At each step, the algorithm chooses the right element x to be added to the current decomposition. This can be proved by induction using an argument similar to the one in the proof of Proposition 4. Recall that x is the smallest (as number of rows) picture that is a subpicture of p at current position (i, j). If there is another element y that is subpicture of p at position (i, j), y cannot be part of a decomposition $\mathcal{T}$ of p otherwise y possibly together with other pictures to its right in $\mathcal{T}$ would be a cover for x and this contradicts the fact that X is prefix.

Regarding the time complexity, the most expensive step is the third one that it is clearly polynomial in the sum of the areas of pictures in X and in the area of p. We remark that, using a clever preprocessing on the pictures in X, it can be made linear in the area of set X and picture p. □

To conclude remark that, moving to the context of recognizable picture languages, the problem of finding a decomposition of p, corresponds to check whether picture p belongs to the REC language X^{**} or, equivalently, if p can be recognized by means of the flower tiling system for X^{**}. In the general case the parsing problem in REC is NP-complete ([16]) and there are no known (non deterministic) cases for which the parsing problem is feasible. The presented algorithm shows that, in the case of X finite prefix set, the parsing problem for language X^{**} can be solved in polynomial time.

5   Maximal and Complete Codes

Maximality is a central notion in theory of (word) codes: the subset of any code is a code, and then the investigation may restrict to maximal codes. In 1D maximality coincides with completeness, for thin codes. In this section we investigate on *maximal* and *complete 2D codes*: we show that, differently from the 1D case, complete codes are a proper subset of maximal codes. Moreover we give a full characterization for complete prefix codes. For lack of space all proofs are sketched.

Definition 6. *A code* $X \subseteq \Sigma^{**}$ *is* maximal *over* Σ *if* X *is not properly contained in any other code over* Σ *that is,* $X \subseteq Y \subseteq \Sigma^{**}$ *and* Y *code imply* $X = Y$. *A prefix code* $X \subseteq \Sigma^{**}$ *is* maximal prefix *over* Σ *if it is not properly contained in any other prefix code over* Σ, *that is,* $X \subseteq Y \subseteq \Sigma^{**}$ *and* Y *prefix imply* $X = Y$.

Example 6. The set $X = \Sigma^{m,n}$ of all pictures on Σ of fixed size (m, n), is a prefix code (see Example 3). It is also maximal as a code and as a prefix code.

Let us state the following decidability result.

Proposition 7. *It is decidable whether a finite prefix set X is maximal prefix.*

Proof. The algorithm first tests whether there exists a picture $p \in \Sigma^{**}$ with $|p|_{row} \leq r_X$ and $|p|_{col} \leq c_X$, such that $X \cup \{p\}$ is still prefix. If no such picture exists, then also no picture p' with $|p'|_{row} > r_X$ and $|p'|_{col} > c_X$ exists such that $X \cup \{p'\}$ is prefix (p' belongs to a corneromino that covers a picture in X iff its prefix of size (r_X, c_X) does). Then the algorithm considers pictures p with $|p|_{row} = m < r_X$ and $|p|_{col} = c_X + 1$. Using the fact that the language of pictures with m rows that are covered by X is regular over the alphabet Σ^m, it is possible to find a picture with m rows that can be added to X, or to state that no picture with m rows can be added to X. □

Any prefix code that is maximal, is trivially maximal prefix too. In 1D the converse holds for finite sets. In 2D the converse does not hold. Proposition 8 gives an example of a prefix maximal set that is not a maximal code. Let us show some preliminary result. Informally speaking, a picture p is *unbordered* if it cannot be self overlapped.

Lemma 1. *Let $X \subseteq \Sigma^{**}$, $|\Sigma| \geq 2$. If there exists a picture $p \notin Pref(X^{**})$, then there exists an unbordered picture p', with p prefix of p', and $p' \notin Pref(X^{**})$.*

Proof. Suppose $|p| = (m, n)$, let $p(1, 1) = a$ and let $b \in \Sigma \setminus \{a\}$. The picture p' can be obtained by the following operations: surround p with a row and a column of a; then row concatenate a picture of size $(m, n + 1)$ filled by b, unless for its bl position; finally column concatenate a picture of size $(2m + 1, n + 1)$ filled by b, unless for its bl position. The resulting picture is a picture p' of size $(2m + 1, 2n + 2)$ having p as a prefix. It is possible to claim that p' is unbordered, by focusing on the possible positions of the subpicture $p[(1, n + 1), (m + 1, n + 1)]$ when p eventually self overlaps. □

Proposition 8. *There exists a finite prefix code that is maximal prefix but not maximal code.*

Proof. Let X be the language of Example 5. Applying Proposition 7 one can show that X is a maximal prefix set. On the other hand, X is not a maximal code. Consider

picture $p = \begin{array}{|cccc|} \hline b & b & a & b \\ b & b & b & x \\ a & b & y & z \\ \hline \end{array}$, with $x, y, z \in \Sigma$; $p \notin Pref(X^{**})$. By Lemma 1, there exists

also an unbordered picture p', such that p is a prefix of p', and $p' \notin Pref(X^{**})$. Let us show that $X \cup \{p'\}$ is still a code. Indeed suppose by contradiction, that there is a picture $q \in \Sigma^{**}$ with two different tiling decompositions, say t_1 and t_2, in $X \cup \{p'\}$. By a careful analysis of possible compositions of pictures in X, one can show that if there exists a position (i, j) of q, where t_1 and t_2 differ, while they coincide in all their tl-positions, then this leads to a contradiction if $j = 1$, otherwise it implies that there exists another position to the bottom left of (i, j), where t_1 and t_2 differ, while they coincide in all their tl positions. Repeating the same reasoning for decreasing values of j, this leads again to a contradiction. □

In 1D, the notion of maximality of (prefix) codes is related to the one of (right) completeness. A string language $X \subseteq \Sigma^*$ is defined right complete if the set of all prefixes of X^* is equal to Σ^*. Then a prefix code is maximal if and only if it is right complete. Let us extend the notion of completeness to 2D languages and investigate the relation of completeness with maximality for prefix codes.

Definition 7. *A set $X \subseteq \Sigma^{**}$ is br-complete if $Pref(X^{\overline{**}}) = \Sigma^{**}$.*

Differently from the 1D case, in 2D, the br-completeness of a prefix set implies its prefix maximality, but it is not its characterization.

Proposition 9. *Let $X \subseteq \Sigma^{**}$ be a prefix set. If X is br-complete then X is maximal prefix. Moreover the converse does not hold.*

Proof. It is easy to prove by contradiction that a br-complete prefix language is also maximal prefix. For the converse, let X as in Example 5. Applying Proposition 7, one can show that X is a maximal prefix set. Let us show that X is not br-complete. Consider $p = \begin{array}{|cccc|} \hline b & b & a & b \\ b & b & b & x \\ a & b & y & z \\ \hline \end{array}$ with $x, y, z \in \Sigma$: there is no comb polyomino $c \in X^{\overline{**}}$ that has p as a prefix. Indeed, from a careful analysis of possible compositions of pictures in X, it follows that the symbol b in position $(2, 3)$ of p cannot be tiled by pictures in X.   □

In 2D br-complete languages are less rich than in 1D. Complete picture languages are indeed only languages that are somehow "one-dimensional" complete languages, in the sense specified in the following result. Note that the result shows in particular that completeness of finite prefix sets is decidable.

Proposition 10. *Let $X \subseteq \Sigma^{**}$ be a maximal prefix set. X is br-complete if and only if $X \subseteq \Sigma^{m*}$ or $X \subseteq \Sigma^{*n}$.*

Proof. First observe that when X is a br-complete prefix set, then no two pictures in X can overlap in a way that their tl corners match (otherwise one of them would be covered by X). Suppose that X is br-complete, but $X \not\subseteq \Sigma^{m*}$, for any m. Then there exists $x_1, x_2 \in X$, of size (m_1, n_1) and (m_2, n_2), respectively, with $m_1 > m_2$. We claim that in this case, $n_1 = n_2$. Indeed if $n_1 \neq n_2$, then two cases are possible: $n_1 > n_2$ or $n_1 < n_2$. Define $y = x_1[(1, 1), (m_2, n_1)]$. In the first case, picture p_1, in figure, with $t, z \in \Sigma^{**}$, does not belong to $Pref(X^{\overline{**}})$, against X br-complete. In the second case, picture p_2, in figure, with $t \in \Sigma^{**}$, does not belong to $Pref(X^{\overline{**}})$, against X br-complete. Now, for any picture $x \in X$, the same technique applies either to the pair x, x_1 or x, x_2, concluding that $X \subseteq \Sigma^{*n}$, for some n.

$$p_1 = \begin{array}{|c|c|c|} \hline & x_2 & y \\ x_1 & & \\ \cline{2-3} & x_1 & t \\ \hline z & & \\ \hline \end{array} \qquad p_2 = \begin{array}{|c|c|} \hline x_2 & y \\ \cline{1-2} & x_2 \\ x_1 & \\ \cline{2-2} & t \\ \hline \end{array}$$

Conversely, suppose $X \subseteq \Sigma^{m*}$. Setting $\Gamma = \Sigma^{m,1}$, X can be considered as a set of strings over Γ. Since X is prefix maximal, it is prefix maximal viewed as set of strings over Γ. Therefore, from classical theory of codes (see e.g. [7]), X is right-complete (as a subset of Γ^*), and this easily implies X br-complete (as a subset of Σ^{**}).   □

6   Conclusions

The aim of this paper was to investigate 2D codes in relation with the theory of recognizable picture languages (REC family), starting from the finite case. We determined a meaningful class of 2D codes, the prefix codes, that are decidable and that have very interesting properties to handle with.

Our further researches will follow two main directions. First we will extend the investigation to 2D generalizations of other types of string codes, as for example bifix codes. Secondly, we will try to remove the finiteness hypothesis and consider prefix sets belonging to particular sub-families in REC, such as deterministic ones.

References

1. Aigrain, P., Beauquier, D.: Polyomino tilings, cellular automata and codicity. Theoretical Computer Science 147, 165–180 (1995)
2. Anselmo, M., Giammarresi, D., Madonia, M.: Deterministic and unambiguous families within recognizable two-dimensional languages. Fund. Inform. 98(2-3), 143–166 (2010)
3. Anselmo, M., Giammarresi, D., Madonia, M., Restivo, A.: Unambiguous Recognizable Two-dimensional Languages. RAIRO: Theoretical Informatics and Applications 40(2), 227–294 (2006)
4. Anselmo, M., Jonoska, N., Madonia, M.: Framed Versus Unframed Two-dimensional Languages. In: Nielsen, M., Kučera, A., Miltersen, P.B., Palamidessi, C., Tůma, P., Valencia, F. (eds.) SOFSEM 2009. LNCS, vol. 5404, pp. 79–92. Springer, Heidelberg (2009)
5. Anselmo, M., Madonia, M.: Deterministic and unambiguous two-dimensional languages over one-letter alphabet. Theoretical Computer Science 410-16, 1477–1485 (2009)
6. Beauquier, D., Nivat, M.: A codicity undecidable problem in the plane. Theoret. Comp. Sci. 303, 417–430 (2003)
7. Berstel, J., Perrin, D., Reutenauer, C.: Codes and Automata. Cambridge University Press (2009)
8. Blum, M., Hewitt, C.: Automata on a two-dimensional tape. In: IEEE Symposium on Switching and Automata Theory, pp. 155–160 (1967)
9. Book, R.V., Otto, F.: String-rewriting Systems. Springer (1993)
10. Bozapalidis, S., Grammatikopoulou, A.: Picture codes. ITA 40(4), 537–550 (2006)
11. Giammarresi, D., Restivo, A.: Recognizable picture languages. Int. Journal Pattern Recognition and Artificial Intelligence 6(2 & 3), 241–256 (1992)
12. Giammarresi, D., Restivo, A.: Two-dimensional languages. In: Rozenberg, G., et al. (eds.) Handbook of Formal Languages, vol. III, pp. 215–268. Springer (1997)
13. Grammatikopoulou, A.: Prefix Picture Sets and Picture Codes. In: Procs. CAI 2005, pp. 255–268 (2005)
14. Kolarz, M., Moczurad, W.: Multiset, Set and Numerically Decipherable Codes over Directed Figures. In: Smyth, B. (ed.) IWOCA 2012. LNCS, vol. 7643, pp. 224–235. Springer, Heidelberg (2012)
15. Lind, D., Marcus, B.: An Introduction to Symbolic Dynamics and Coding. Cambridge University Press, Cambridge (1995)
16. Lindgren, K., Moore, C., Nordahl, M.: Complexity of two-dimensional patterns. Journal of Statistical Physics 91(5-6), 909–951 (1998)
17. Moczurad, M., Moczurad, W.: Some Open Problems in Decidability of Brick (Labelled Polyomino) Codes. In: Chwa, K.-Y., Munro, J.I. (eds.) COCOON 2004. LNCS, vol. 3106, pp. 72–81. Springer, Heidelberg (2004)
18. Simplot, D.: A Characterization of Recognizable Picture Languages by Tilings by Finite Sets. Theoretical Computer Science 218-2, 297–323 (1991)

Adjacent Ordered Multi-Pushdown Systems*

Mohamed Faouzi Atig[1], K. Narayan Kumar[2], and Prakash Saivasan[2]

[1] Uppsala University, Sweden
mohamed_faouzi.atig@it.uu.se
[2] Chennai Mathematical Institute, India
{kumar,saivasan}@cmi.ac.in

Abstract. Multi-pushdown systems are formal models of multi-threaded programs. As they are Turing powerful in their full generality, several decidable subclasses, constituting under-approximations of the original system, have been studied in the recent years. Ordered Multi-Pushdown Systems (OMPDSs) impose an order on the stacks and limit pop actions to the lowest non-empty stack. The control state reachability for OMPDSs is 2-ETIME-COMPLETE. We propose a restriction on OMPDSs, called Adjacent OMPDSs (AOMPDS), where values may be pushed only on the lowest non-empty stack or one of its two neighbours. We describe EXPTIME decision procedures for reachability and LTL model-checking, establish matching lower bounds and describe two applications of this model.

1 Introduction

Verification of concurrent recursive programs is an important but difficult problem well studied over the last decade. The theory of pushdown systems has been used very effectively in analysing sequential recursive programs and forms the backbone of a number of verification tools. However, there is no such well established classical theory that underlies concurrent recursive programs. In the case where the number of threads is bounded, it is natural to consider the generalization of pushdown systems to multi-pushdown systems (MPDS), with one pushdown per thread (to model its call stack) and control states to model the contents of the shared (finite) memory.

In their full generality MPDSs are not analyzable as two stacks can simulate the tape of a turing machine. The main focus in this area has hence been to identify decidable subclasses. These subclasses are obtained by placing restrictions on the behaviours of the MPDSs. Qadeer and Rehof [15] show that if the number of times a run switches from accessing one stack to another is bounded a priori then the control state reachability problem is decidable. Such behaviours are called *bounded context* behaviours. Subsequently, this idea of context-bounded analysis has been used and extended to a number of related models very effectively [11,12,6,7,20,18,10].

---

* Partially supported by the Swedish Research Council within UPMARC, CNRS LIA InForMel. The last author was partially funded by Tata Consultancy Services.

M.-P. Béal and O. Carton (Eds.): DLT 2013, LNCS 7907, pp. 58–69, 2013.
© Springer-Verlag Berlin Heidelberg 2013

We may look at such a restriction as identifying a collection of runs of an unrestricted system, constituting an underapproximation of the real system. In the context of linear time model checking, where the aim is to look for a faulty run, verifying such an underapproximation guarantees that there are no faulty runs of the restricted form or identifies the possibility of a faulty run conforming to these restrictions.

Several generalizations of the idea of context bounding have been proposed, notably

- *Bounded Phase MPDSs*[17,19,16]: A *phase* denotes a segment of an execution where all the pop operations are performed on a single stack. An a priori bound is placed on the number of phases.
- *Ordered Multi-pushdowns (OMPDSs)*[2,8]: The stacks are numbered $1, 2, \ldots, n$ and pop moves are permitted only on the lowest non-empty stack.
- *Scope Bounded MPDSs*[21]: A bound is placed on the number of context switches between any push and the corresponding pop (the pop that removes this value from the stack).

The control state reachability problem is decidable for all these models, NP-COMPLETE for bounded context systems [15] and PSPACE-COMPLETE [21] for scope bounded systems. However, for bounded-phase MPDSs and OMPDSs the problem is 2-ETIME-COMPLETE [2,19]. It is interesting to note that these two models allow copying of a stack onto another while the first two do not. OMPDSs can simulate bounded-phase MPDSs.

The run of an OMPDS can be thought as consisting of a sequence of phases, where each phase identifies an active stack, the stack from which values may be popped. Further, during a phase associated with stack i, all stacks j with $j < i$ have to be empty. We impose an additional requirement that values be pushed only on stacks $i - 1$, i or $i + 1$ during such a phase to obtain the subclass of *Adjacent OMPDSs* (AOMPDSs). For this class we show that the state reachability problem can be solved in EXPTIME and prove a matching lower bound. Observe, that this class has the ability to copy a stack onto another, and to our knowledge is the first such class in EXPTIME.

The *repeated reachability problem* has to do with infinite runs and the aim is to determine if there is any run that visits a particular set of states F infinitely often. In formal verification, this problem is of great importance as a solution immediately leads to a model checking algorithm for linear time temporal logic (LTL). We obtain an EXPTIME decision procedure for the repeated reachability problem (and LTL model-checking problem) for AOMPDSs.

We illustrate the power of AOMPDSs via two applications — first, we show that the class of unary OMPDSs, i.e., OMPDSs where the stack alphabet is singleton, (which can be thought of as counter systems with restrictions on decrement operations) can be reduced to AOMPDAs and hence obtain EXP-TIME procedures for reachability, repeated reachability and LTL model-checking for this class. This is likely to be optimal since we also show NP-HARDNESS. Secondly, we show that the control state reachability problem for networks of recursive programs communicating via queues, whose connection topology is a directed forest, can be reduced to reachability in AOMPDS with polynomial blowup, obtaining an EXPTIME solution to this problem. This problem was proposed and solved in [19] where it is also shown that direct forest topology is the only one for which this problem is decidable.

Other Related Work. In [13] Madhusudan and Parlato, show the benefits of studying the structure of runs of MPDSs as graphs. They argue that decidability is closely related to the boundedness of the tree-width of such graphs. The runs arising from all the restrictions of MPDSs described above have bounded tree-width. In [9] a new measure of complexity for multiply nested words called split-width is proposed, which is bounded whenever tree-width is, but seems to yield easier proofs of boundedness.

[18] applies the decidability of bounded-phase MPDSs to the analysis of recursive concurrent programs communicating via queues and obtains a characterization of the decidable topologies. These results have been extended in [10] by placing further restrictions on how messages are consumed from the queue.

In the setting of infinite runs, [1] describes a 2-EXPTIME decision procedure for the repeated reachability and LTL model-checking problems for OMPDSs and bounded-phase MPDSs (also see [5]). Recently, [22] and [4] show that LTL model checking for bounded scope systems can be solved in EXPTIME. In this context, it must be noted that with the bounded context and bounded-phase restrictions every run has to eventually use only a single stack and hence do not form natural restrictions on infinite runs. [3] investigates the existence of ultimately periodic infinite runs, i.e. runs of the form uv^ω, in MPDSs.

Both classes of MPDSs studied in this paper, AOMPDSs and UOMPDSs, are the only known classes of MPDSs in EXPTIME that allow the copying of stack contents and permit infinite runs involving the use multiple stacks infinitely often.

2   Preliminaries

Notation. Let $\mathbb{N}$ denote the set of non-negative integers. For every $i, j \in \mathbb{N}$ such that $i \leq j$, we use $[i..j]$ to denote the set $\{k \in \mathbb{N} \mid i \leq k \leq j\}$. Let Σ be a finite alphabet. We denote by Σ^* (Σ^ω) the set of all finite (resp. infinite) words over Σ, and by ϵ the empty word. Let u be a word over Σ. The length of u is denoted by $|u|$ and $|\epsilon| = 0$. For every $j \in [1..|u|]$, we use $u(j)$ to denote the jth letter of u.

Context-Free Grammars. A *context-free grammar* (CFG) G is a tuple $(\mathcal{X}, \Sigma, P)$ where $\mathcal{X}$ is a finite non-empty set of *variables* (or *nonterminals*), Σ is an alphabet of *terminals*, and $P \subseteq \left(\mathcal{X} \times (\mathcal{X}^2 \cup \Sigma \cup \{\epsilon\})\right)$ a finite set of *productions* and denote the production (X, w) by $X \Rightarrow_G w$. We also write $\Rightarrow$ instead of $\Rightarrow_G$ when the identity of G is clear from the context. Given strings $u, v \in (\Sigma \cup \mathcal{X})^*$ we say $u \Rightarrow_G v$ if there exists a production $(X, w) \in R$ and some words $y, z \in (\Sigma \cup \mathcal{X})^*$ such that $u = yXz$ and $v = ywz$. We use $\Rightarrow_G^*$ for the reflexive transitive closure of $\Rightarrow_G$. For every nonterminal symbol $X \in \mathcal{X}$, we define the context-free language generated from X by $L_G(X) = \{w \in \Sigma^* \mid X \Rightarrow_G^* w\}$.

3   Multi-Pushdown Systems

In this section, we will formally introduce *multi-pushdown systems* and then define two subclasses, namely *ordered multi-pushdown systems* and *adjacent ordered multi-pushdown systems*. A multi-pushdown system has $n \geq 1$ read-write memory tapes (stacks) with a last-in-first-out rewriting policy and optionally a read only input tape. We do not require an input tape as the purpose of this paper is to prove the decidability of the reachability problem.

Definition 1 (Multi-Pushdown Systems). *A* Multi-PushDown System *(MPDS) is a tuple* $M = (n, Q, \Gamma, \Delta, q_0, \gamma_0)$ *where:*

- $n \geq 1$ *is the number of stacks.*
- Q *is the non-empty set of states.*
- Γ *is the finite set of stack symbols containing the special stack symbol* $\perp$. *We use* Γ_ϵ *to denote the set* $\Gamma \cup \{\epsilon\}$.
- $q_0 \in Q$ *is the initial state.*
- $\gamma_0 \in \Gamma \setminus \{\perp\}$ *is the initial stack symbol.*
- $\Delta \subseteq ((Q \times (\Gamma_\epsilon)^n) \times (Q \times (\Gamma^*)^n))$ *is the transition relation such that if* $((q, \gamma_1, \gamma_2, \ldots, \gamma_n), (q', \alpha_1, \alpha_2, \ldots, \alpha_n))$ *is in* Δ, *then for all* $i \in [1..n]$, *we have:* $|\alpha_i| \leq 2$, *if* $\gamma_i \neq \perp$ *then* $\alpha_i \in (\Gamma \setminus \{\perp\})^*$ *and* $\alpha_i \in ((\Gamma \setminus \{\perp\})^* \cdot \{\perp\})$ *otherwise.*

A stack content of M is an element of $Stack(M) = (\Gamma \setminus \{\perp\})^* \{\perp\}$. A configuration of the MPDS M is a (n+1) tuple $(q, w_1, w_2, \cdots, w_n)$ with $q \in Q$, and $w_1, w_2, \ldots, w_n \in Stack(M)$. The set of configurations of the MPDS M is denoted by $\mathcal{C}(M)$. The *initial configuration* c_M^{init} of the MPDS M is $(q_0, \perp, \ldots, \perp, \gamma_0 \perp)$.

If $t = ((q, \gamma_1, \ldots, \gamma_n), (q', \alpha_1, \ldots, \alpha_n))$ is an element of Δ, then $(q, \gamma_1 w_1, \ldots, \gamma_n w_n) \xrightarrow{t}_M (q', \alpha_1 w_1, \ldots, \alpha_n w_n)$ for all $w_1, \ldots, w_n \in \Gamma^*$ such that $\gamma_1 w_1, \ldots, \gamma_n w_n \in Stack(M)$. We define the transition relation $\rightarrow_M$ as $\bigcup_{t \in \Delta} \xrightarrow{t}_M$. Observe that the stack symbol $\perp$ marks the bottom of the stack and our transition relation does not allow this $\perp$ to be popped.

We write $\rightarrow_M^*$ to denote the reflexive and transitive closure of the relation $\rightarrow_M$, representing runs of the system. For every sequence of transitions $\rho = t_1 t_2 \ldots t_m \in \Delta^*$ and two configurations $c, c' \in \mathcal{C}(M)$, we write $c \xrightarrow{\rho}_M^* c'$ to denote that one of the following two cases holds:

1. $\rho = \epsilon$ and $c = c'$.
2. There are configurations $c_0, \cdots, c_m \in \mathcal{C}(M)$ such that $c_0 = c$, $c' = c_m$, and $c_i \xrightarrow{t_{i+1}}_M c_{i+1}$ for all $i \in [0..m-1]$.

An ordered multi-pushdown system is a multi-pushdown system in which one can pop only from the first non-empty stack.

Definition 2 (Ordered Multi-Pushdown Systems). *An Ordered Multi-Pushdown System* *(OMPDS) is a multi-pushdown system* $M = (n, Q, \Gamma, \Delta, q_0, \gamma_0)$ *where for each transition* $((q, \gamma_1, \ldots, \gamma_n), (q', \alpha_1, \ldots, \alpha_n)) \in \Delta$ *there is an index* $i \in [1..n]$ *such that* $\gamma_1 = \cdots = \gamma_{i-1} = \perp$, $\gamma_i \in (\Gamma \setminus \{\perp\})$, *and* $\gamma_{i+1} = \cdots = \gamma_n = \epsilon$ *and further one of the following properties holds*

1. *Operate on the stack i:* $\alpha_j = \perp$ *for all $j < i$ and $\alpha_j = \epsilon$ for all $j > i$.*
2. *Push on the stack $j < i$:* $\alpha_i = \epsilon$, $\alpha_k = \perp$ *if $j \neq k < i$, $\alpha_j \in \Gamma\perp$, and $\alpha_k = \epsilon$ if $k > i$.*
3. *Push on the stack $j > i$:* $\alpha_i = \epsilon$, $\alpha_k = \perp$ *if $k < i$, $\alpha_k = \epsilon$ if $j \neq k > i$ and $|\alpha_j| = 1$.*

If we further restrict the choice of j in item 2 above to be only $i - 1$ and in item 3 to be $i + 1$ we get the subclass of Adjacent OMPDSs *(AOMPDSs). In any AOMPDS, a transition that pops a symbol from stack i is only allowed to push values on one of the stacks $i - 1, i$ and $i + 1$.*

We can extend this definition to allow pushing onto the first stack while popping from the n-th stack without altering the results in the paper.

Definition 3 (Reachability Problem). *Given a MPDS M and a state q of M, the reachability problem is to decide whether $(q_0, \perp, \cdots, \perp, \gamma_0\perp) \rightarrow^*_M (q, \perp, \cdots, \perp)$.*

4   The Reachability Problem for AOMPDS

The purpose of this section is to prove the following theorem:

Theorem 4. *The reachability problem for Adjacent Ordered Multi-Pushdown System is* EXPTIME-COMPLETE.

Upper Bound: Let $M = (n, Q, \Gamma, \Delta, q_0, \gamma_0)$ be an AOMPDS with $n > 1$ (the case where $n = 1$ boils down to the reachability of pushdown systems which is well-known to be in PTIME). The proof of EXPTIME-containment is through an inductive construction that reduces the reachability problem for M to the reachability problem for a pushdown system with only an exponential blow up in size. The key step is to show that we can reduce the reachability problem for M to the reachability problem for an $(n - 1)$-AOMPDS. The key feature of our reduction is that there is no blowup in the state space and the size of the stack alphabet increases quadratically in the number of states. A non-linear blow up in the number of states will result in a complexity higher than EXPTIME.

   We plan to use a single stack to simulate both the first and second stacks of M. It is useful to consider the runs of M to understand how this works. Any run ρ of M starting at the initial configuration naturally breaks up into segments $\sigma_0\rho_1\sigma_1 \ldots \rho_k\sigma_k$ where the segments ρ_i contain configurations where stack 1 is non-empty while in any configuration in the σ_i's stack 1 is empty. Clearly the contents of stack 1 at the beginning of ρ_i contains exactly two symbols, and we assume it to be $a_i\perp$. We further assume that segment ρ_i begins at control state q_i and the segment σ_i in state q'_i. What is the contribution of the segment ρ_i, which is essentially the run of a pushdown automaton starting and ending at the empty stack configuration, to this run?

   Firstly, it transforms the local state from q_i to q'_i. Secondly, a word w_i is pushed on to stack 2 during this segment. It also, consumes the value a_i from stack 1 in this process, but that is not relevant to the rest of the computation. To simulate the effect of ρ_i it would thus suffice to jump from state q_i to q_i' and push the word w_i on stack 2. There are potentially infinitely many possible runs of the form ρ_i that go from q_i to q_i' while removing a_i from stack 1 and thus infinite possibilities for the word that is pushed on stack 2. However, it is easy to see that this set of words $L(q_i, a_i, q_i')$ is a CFL.

If the language $L(q_i, a_i, q_i')$ is a regular language, we could simply *summarize* this run by depositing a word from this language on stack 2 and then proceed with the simulation of stack 2. However, since it is only a CFL this is not possible. Instead, we have to interleave the simulation of stack 2 with the simulation of stack 1, both using stack 2 and there is no a priori bound on the number of switches between the stacks in such a simulation. For general OMPDSs such a simulation would not work as the values pushed by the segments of executions of stack 1 and stack 2 on a third stack would get permuted and this explains why OMPDSs need a different global and more expensive decision procedure.

To simulate the effect of ρ_i, we jump directly to q_i' and push a non-terminal symbol (from the appropriate CFG) that generates the language $L(q_i, a_i, q_i')^R$ (reverse, because stacks are last in first out). Now, when we try to execute σ_i', instead of encountering a terminal symbol on top of stack 2 we might encounter a nonterminal. In this case, we simply rewrite the nonterminal using one of the rules of the CFG applicable to this nonterminal. In effect, we produce a left-most derivation of a word from $L(q_i, a_i, q_i')$ in a lazy manner, interspersed within the execution involving stack 2, generating terminals only when they need to be consumed. This is the main idea in the construction that is formalized below.

For every $i \in [1..n]$, we define the sets of transitions $\Delta_{(i,i-1)}$, $\Delta_{(i,i)}$, and $\Delta_{(i,i+1)}$ of M as follows:

- If $i > 1$ then $\Delta_{(i,i-1)} = \Delta \cap ((Q \times (\{\bot\})^{i-1} \times \Gamma_\epsilon \times (\{\epsilon\})^{n-i}) \times (Q \times (\{\bot\})^{i-2} \times \Gamma^* \times (\{\epsilon\})^{n-i+1}))$. This corresponds to the set of transitions in Δ that pop a symbol from the i-th stack of M while pushing some symbols on the $(i-1)$-th stack of M.
- $\Delta_{(i,i)} = \Delta \cap ((Q \times (\{\bot\})^{i-1} \times \Gamma_\epsilon \times (\{\epsilon\})^{n-i}) \times (Q \times (\{\bot\})^{i-1} \times \Gamma^* \times (\{\epsilon\})^{n-i}))$. This corresponds to the set of transitions in Δ that pop and push exclusively on the i-th stack of M.
- If $i < n$ then $\Delta_{(i,i+1)} = \Delta \cap ((Q \times (\{\bot\})^{i-1} \times \Gamma_\epsilon \times (\{\epsilon\})^{n-i}) \times (Q \times (\{\bot\})^{i-1} \times \{\epsilon\} \times \Gamma^* \times (\{\epsilon\})^{n-i-1}))$. This corresponds to the set of transitions in Δ that pop a symbol from the i-th stack of M while pushing a symbol on the $(i+1)$-th stack of M.

Furthermore, we define

- $\Delta_1 = \Delta_{(1,1)} \cup \Delta_{(1,2)}$
- $\Delta_i = \Delta_{(i,i)} \cup \Delta_{(i,i+1)} \cup \Delta_{(i,i-1)}$ for all $2 \leq i < n$
- $\Delta_n = \Delta_{(n,n)} \cup \Delta_{(n,n-1)}$

We construct a context-free grammar $G_M = (N, (\Gamma \setminus \{\bot\}), P)$ from the AOM-PDA M. The set of non-terminals $N = (Q \times (\Gamma \setminus \{\bot\}) \times Q)$. The set of productions P is defined as the smallest set of rules satisfying:

- For every two states $p, p' \in Q$, and every transition $((q, \gamma, \epsilon, \ldots, \epsilon), (q', \gamma_1 \gamma_2, \epsilon, \ldots, \epsilon))$ in Δ such that $\gamma, \gamma_1, \gamma_2 \in (\Gamma \setminus \{\bot\})$, we have $(q, \gamma, p) \Rightarrow_{G_M} (q', \gamma_1, p')(p', \gamma_2, p)$.

- For every state $p \in Q$, and every transition $((q, \gamma, \epsilon, \ldots, \epsilon), (q', \gamma', \epsilon, \ldots, \epsilon))$ in Δ such that $\gamma, \gamma' \in (\Gamma \setminus \{\bot\})$, we have $(q, \gamma, p) \Rightarrow_{G_M} (q', \gamma', p)$.

- For every transition $((q, \gamma, \epsilon, \ldots, \epsilon), (q', \epsilon, \epsilon, \ldots, \epsilon))$ in Δ such that $\gamma \in (\Gamma \setminus \{\bot\})$, we have $(q, \gamma, q') \Rightarrow_{G_M} \epsilon$.

- For every transition $((q, \gamma, \epsilon, \ldots, \epsilon), (q', \epsilon, \gamma', \epsilon, \ldots, \epsilon))$ in Δ such that $\gamma, \gamma' \in (\Gamma \setminus \{\bot\})$, we have $(q, \gamma, q') \Rightarrow_{G_M} \gamma'$.

Then, it is easy to see that the context-free grammar summarizes the effect of the first stack on the second one. Formally, we have:

Lemma 5. *The context free language $L_{G_M}((q, \gamma, q'))$ is equal to the set of words $\{w^R \in (\Gamma \setminus \{\bot\})^* \mid \exists \rho \in \Delta_1^*. \ (q, \gamma \bot, w_2, \ldots, w_n) \xrightarrow{\rho}_M (q', \bot, w \cdot w_2, \ldots, w_n)\}$ where w^R denotes the reverse of the word w.*

We are now ready to show that reachability problems on M can be reduced to reachability problems on an $(n-1)$-AOMPA N. Further, the number of states of N is linear in $|Q|$, size of the stack alphabet of N is $O(|Q|^2|\Gamma|)$ and the number of transitions is $O(|Q|^3.|\Delta|)$. The upper-bound claimed in Theorem 4 then follows by simple induction.

Let $F \subseteq Q$ be the set of states whose reachability we are interested in, we show how to construct $(n-1)$-AOMPA N such that the reachability question on M can be reduced to reachability question on N. Formally, N is defined by the tuple $(n-1, Q, \Gamma \cup N, \Delta', q_0, \gamma_0)$ where Δ' is defined as the smallest set satisfying the following conditions:

- For any transition $((q, \bot, \gamma_2, \ldots, \gamma_n), (q', \bot, \alpha_2, \ldots, \alpha_n)) \in \Delta$, we have $((q, \gamma_2, \ldots, \gamma_n), (q', \alpha_2, \ldots, \alpha_n)) \in \Delta'$.
- For any transition $((q, \bot, \gamma_2, \epsilon, \ldots, \epsilon), (q', \gamma \bot, \epsilon, \ldots, \epsilon)) \in \Delta_{(2,1)}$, we have $((q, \gamma_2, \epsilon, \ldots, \epsilon), (q'', (q', \gamma, q''), \epsilon, \ldots, \epsilon)) \in \Delta'$ for all $q'' \in Q$.
- For any production rule $X \Rightarrow_{G_M} w$ and state $q \in Q$, we have $((q, X, \epsilon, \ldots, \epsilon), (q, w^R, \epsilon, \ldots, \epsilon)) \in \Delta'$.

The relation between N and M is given by the following lemma:

Lemma 6. *The set of states F is reachable in M iff F is reachable in N.*

The fact that even a single contiguous segment of moves using stack 1 in M may now be interleaved arbitrarily with executions involving other stacks in N, makes proof some what involved. Towards the proof, we define a relation between the configurations of N and M systems. We will denote the set of all configurations of M by $\mathcal{C}^M$ and configurations of N by $\mathcal{C}^N$. For any configuration $c \in \mathcal{C}^M$ and $d \in \mathcal{C}^N$, we say cRd iff one of the following is true.

– d is of the form $(q, \bot, w_3, \cdots, w_n)$ and c is of the form $(q, \bot, \bot, w_3, \cdots, w_n)$.

– d is of the form $(q, \eta_1 v_1 \eta_2 v_2 \cdots \eta_m v_m \bot, w_3, \cdots, w_n)$ and c is of the form $(q, \bot, u_1 v_1 u_2 v_2 \cdots u_m v_m \bot, w_3, \cdots, w_n)$ where $v_1, u_1, v_2, u_2, \ldots, v_m, u_m \in (\Gamma \setminus \{\bot\})^*$, $\eta_1, \eta_2, \ldots, \eta_m \in N^*$ and $\eta_k \Rightarrow^*_{G_M} u_k^R$ for all $k \in [1..m]$.

Thus, cRd verifies that it is possible to replace the nonterminals appearing in stack 2 in d by words they derive (and by tagging an additional empty stack for the missing stack 1) to obtain c. We now show that this abstraction relation faithfully transports runs (to configurations from the initial configuration) in both directions. This is the import of lemmas 7 and 8, which together guarantee that the state reachability in M reduces to state reachability in N.

Lemma 7. *Let $c_1, c_2 \in (Q \times \{\bot\} \times (Stack(M))^{n-1})$ be two configurations such that $c_M^{init} \to^*_M c_1$ and $c_M^{init} \to^*_M c_2$. If $c_1 \xrightarrow{\rho}_M c_2$, with $\rho \in \cup_{i=3}^n \Delta_i \cup (\Delta_{(2,1)} \Delta_1^*) \cup \Delta_{(2,2)} \cup \Delta_{(2,3)}$, then for every configuration $d_1 \in \mathcal{C}^N$ such that $c_1 R d_1$, there is a configuration $d_2 \in \mathcal{C}^N$ such that $c_2 R d_2$ and $d_1 \to^*_N d_2$.*

Lemma 8. *Let $d_1, d_2 \in \mathcal{C}^N$ be two configurations of N such that $c_N^{init} \to^*_N d_1 \xrightarrow{t}_N d_2$ for some $t \in \Delta'$. Then for every configuration $c_2 \in \mathcal{C}_1^M$ such that $c_2 R d_2$, there is a configuration $c_1 \in \mathcal{C}_1^M$ such that $c_1 R d_1$ and $c_1 \to^*_M c_2$.*

Lower Bound: It is known that the following problem is ExpTime-complete [10]: Given a pushdown automaton $\mathcal{P}$ recognizing a context-free language L, and $n - 1$ finite state automata $\mathcal{A}_2, \ldots, \mathcal{A}_n$ recognizing the regular languages $L_2, \ldots, L_n$ respectively, is $L \cap \bigcap_{i=2}^n L_i$ non-empty? We can show that this problem can be reduced, in polynomial time, to the reachability problem for an AOMPDS M with n-stacks. The idea is the following: The first stack is used to simulate $\mathcal{P}$ and write down a word that is accepted to the second stack. Each other stack is then used to check acceptance by one of the finite automata.

5   Repeated Reachability for AOMPDS

In this section, we show that the linear-time model checking problem is ExpTime-complete for AOMPDS. In the following, we assume that the reader is familiar with ω-regular properties expressed in the linear-time temporal logics [14] or the linear time μ-calculus [23]. For more details, the reader is referred to [14,24,23]. Checking whether a MPDS satisfies a property expressed in such a logic reduces to solving the *repeated state reachability* problem, i.e., checking if there is an infinite run that visits control states from a given set F infinitely often.

We use the following theorem to reduce the repeated state reachability problem for OMPDSs to the reachability problem for OMPDSs.

Theorem 9 ([1]). *Let $M = (n, Q, \Gamma, \Delta, q_0, \gamma_0)$ be an OMPDS and q_f be a state of M. There is an infinite run starting from c_M^{init} that visits infinitely often the state q_f if and only if there are $i \in [1..n]$, $q \in Q$, and $\gamma \in \Gamma \setminus \{\bot\}$ such that:*

- $c_M^{init} \rightarrow_M^* (q, \perp^{i-1}, \gamma w, w_{i+1}, \ldots, w_n)$ *for some* $w, w_{i+1}, \ldots, w_n \in \Gamma^*$.
- $(q, \perp^{i-1}, \gamma \perp, \perp^{n-i}) \xrightarrow{\rho_1}_M (q_f, w_1, \ldots, w_n) \xrightarrow{\rho_2}_M (q, \perp^{i-1}, \gamma w_i', w_{i+1}', \ldots, w_n')$
  for some $w_1, \ldots, w_n, w_i', \ldots, w_n' \in \Gamma^*$, $\rho_1 \in \Delta'^*$ *and* $\rho_2 \in \Delta'^+$ *where* Δ' *contains all the transitions of the form* $((q, \perp^{j-1}, \gamma_j, \epsilon, \ldots, \epsilon), (q, \alpha_1, \ldots, \alpha_n)) \in \Delta$ *such that* $1 \leq j \leq i$ *and* $\gamma_j \in (\Gamma \setminus \{\perp\})$.

It is possible to formulate each of the two items listed in the above theorem as simple reachability queries on two AOMPDSs whose sizes are polynomial in the size of M. This gives us the following theorem.

Theorem 10. *Let* $M = (n, Q, \Gamma, \Delta, q_0, \gamma_0)$ *be an AOMPDS and* q_f *be a state of* M. *Then checking whether there is an infinite run starting from* c_M^{init} *that visits the state* q_f *infinitely often can be solved in time* $O(|M|)^{poly(n)}$.

As an immediate consequence of Theorem 10 we have the following corollary.

Theorem 11. *Let* $M = (n, Q, \Gamma, \Delta, q_0, \gamma_0)$ *be an AOMPDS with a labeling function* Λ, *and let* φ *be a linear time* μ-*calculus formula or linear time temporal formula. Then, it is possible to check, in time* $O(|M|)^{poly(n,|\varphi|)}$, *whether there is an infinite run of* M *starting from* c_M^{init} *that does not satisfy* φ.

6 Applications of AOMPDSs

6.1 Unary Ordered Multi-PushDown Systems

The class of Unary Ordered Multi-Pushdown Systems is the subclass of ordered multi-pushdown systems where the stack alphabet contains just one letter other than $\perp$ (i.e., $|\Gamma| = 2$).

Definition 12 (Unary Ordered Multi-Pushdown Systems). *An* Unary Ordered Multi-Pushdown Systems *(UOMPDS) is an ordered multi-pushdown system* $(n, Q, \Gamma, \Delta, q_0, \gamma_0)$ *such that* $| \Gamma | = 2$.

In this section, we prove that the reachability problem for UOMPDS is in EXP-TIME. This is done by reducing the problem to the reachability in an AOMPDS. The key observation is that in a unary OMPDA the order in which elements are pushed on a stack is not important. So, given an UOMPDS M with n-stacks and an alphabet $\Gamma = \{a, \perp\}$ we construct an AOMPDS N with n-stacks and alphabet $\Gamma' = \{(a, 1), (a, 2), \ldots, (a, n)\} \cup \{\perp\}$.

For each i, let π_i denote the function satisfying $\pi_i(a) = (a, i)$, $\pi_i(\perp) = \perp$ and extended homomorphically to all of $a^* \perp + a^*$. The control states of N are precisely the control states of M. The occurrence of the letter (a, i) in any stack in N denotes the occurrence of an a on stack i, so that, counting the number of occurrences of (a, i)'s across all the stacks in a configuration of N gives the contents of stack i in the corresponding configuration in M. If the top element of the left-most non-empty stack i is (a, i) then N simulates a corresponding move of M. If this move involves a pushing α on stack i, it is simulated by pushing

$\pi_i(\alpha)$ on stack i. If it involves a pushing α on stack j, $j > i$ (respectively $j < i$) then $\pi_j(\alpha)$ is pushed on stack $i+1$ (respectively $i-1$). If the top of the left-most nonempty stack i is (a, j) with $j < i$ (respectively $j > i$) then the value is simply copied to stack $i - 1$ (respectively $i + 1$).

Theorem 13. *The reachability, repeated reachability and LTL model-checking for Unary Ordered Multi-PushDown Systems are all solvable in* EXPTIME. *All these problems are also NP-Hard (hence no improvement in the upper bound is likely) even for Adjacent UOMPDS.*

6.2   An Application to Concurrent Recursive Queue Systems

La Torre et al. [18], study the decidability of control state reachability in networks of concurrent processes communicating via queues. Each component process may be recursive, i.e., equipped with a pushdown store, and such systems are called *recursive queuing concurrent programs* (RQCP) in [18]. Further, the state space of the entire system may be global or we may restrict each process to have its own local state space (so that the global state space is the product of the local states). In the terminology of [18] the latter are called RQCPs without shared memory.

An architecture describes the underlying topology of the network, i.e., a graph whose vertices denote the processes and edges correspond to communication channels (queues). One of the main results in [18] is a precise characterization of the architectures for which the reachability problem for RQCP's is decidable. Understandably, given the expressive power of queues and stacks, this class is very restrictive. To obtain any decidability at all, one needs the *well-queuing* assumption, which prohibits any process from dequeuing a message from any of its incoming channels as long as its stack is non-empty. They show that, even under the well-queuing assumption, the only architectures for which the reachability problem is decidable for RQCPs without shared memory are the so called *directed forest* architectures. A directed tree is a tree with a identified root and where all edges are oriented away from the root towards the leaves. A directed forest is a disjoint union of directed trees. They use a reduction to the reachability problem for bounded-phase MPDSs and obtain a double exponential decision procedure.

We now show that this problem can be reduced to the reachability problem for AOMPDS and obtain an EXPTIME upper-bound.[1] The reduction is sketched below. An EXPTIME upper-bound is also obtained via tree-width bounds [13] (Theorem 4.6).

Theorem 14. *The control state reachability problem for RQCPs with a directed forest architecture, without shared memory and under the well-queuing assumption can be solved in* EXPTIME.

Proof. (Sketch) We only consider the directed tree architecture and the result for the directed forest follows quite easily from this. An observation, from [18], is that

---

[1]   The argument in Theorem 4 can also be adapted to show EXPTIME-HARDNESS.

it suffices to only consider executions with the following property: if q is a child of p then p executes all its steps (and hences deposits all its messages for q) before q executes. We fix some topologically sorted order of the tree, say $p_1, p_2, \ldots, p_m$ where p_1 is the root. The AOMPDS we construct only simulates those executions of the RQCP in which all moves of p_i are completed before p_{i+1} begins its execution. We call such a run of the RQCP as a *canonical run*. The number of stacks used is $2m - 1$. The message alphabet is $\Gamma \times \{1, \ldots, m\} \cup \bigcup_{1 \leq i \leq m} \Sigma_i$, where Γ is the communication message alphabet and Σ_i is the stack alphabet of process p_i. We write Γ_i to denote $\Gamma \times \{i\}$ and $w \downarrow \Sigma$ to denote the restriction of a word to the letters in Σ.

We simulate the process in order $p_1, \ldots, p_m$. The invariant we maintain as we simulate a canonical run ρ is that, when we begin simulating process p_i, the contents of stack $2i - 1$ is some α so that $\alpha \downarrow \Gamma_i$ is the contents of the unique input channel to p_i as p_i begins its execution in ρ. Thus we can simulate p_i's contribution to ρ, by popping from stack $2i - 1$ when a value is to be consumed from the input queue. If top of stack $2i-1$ does not belong ot Γ_i, then we transfer it to stack $2i$, as it is not meant for p_i. When p_i sends a message to any other process p_j in ρ (which must be one of its children in the tree) we simulate it by tagging the message with the process identity and pushing it on stack $2i$. Finally, as observed in [18], the stack for p_i can also be simulated on top of stack $2i - 1$ since a value is dequeued only when its local stack is empty (according to the well-queuing assumption). At the end of the simulation of process p_i, we empty any contents left on stack $2i - 1$ (transferring elements of $\Gamma \times \{i+1, \ldots, m\}$ to stack $2i$). Finally, we copy stack $2i$ onto stack $2i + 1$ and simulate process p_{i+1} using stack $2i + 1$ (so that the rear of all queues are on top of the stack.) The state space is linear in the size of the RQCP and hence we conclude that the reachability problem for RQCPs can be solved in EXPTIME using Theorem 4.

References

1. Atig, M.F.: Global model checking of ordered multi-pushdown systems. In: FSTTCS. LIPIcs, vol. 8, pp. 216–227. Schloss Dagstuhl - Leibniz-Zentrum fuer Informatik (2010)
2. Atig, M.F., Bollig, B., Habermehl, P.: Emptiness of multi-pushdown automata is 2ETIME-complete. In: Ito, M., Toyama, M. (eds.) DLT 2008. LNCS, vol. 5257, pp. 121–133. Springer, Heidelberg (2008)
3. Atig, M.F., Bouajjani, A., Emmi, M., Lal, A.: Detecting fair non-termination in multithreaded programs. In: Madhusudan, P., Seshia, S.A. (eds.) CAV 2012. LNCS, vol. 7358, pp. 210–226. Springer, Heidelberg (2012)
4. Atig, M.F., Bouajjani, A., Narayan Kumar, K., Saivasan, P.: Linear-time model-checking for multithreaded programs under scope-bounding. In: Chakraborty, S., Mukund, M. (eds.) ATVA 2012. LNCS, vol. 7561, pp. 152–166. Springer, Heidelberg (2012)
5. Bollig, B., Cyriac, A., Gastin, P., Zeitoun, M.: Temporal logics for concurrent recursive programs: Satisfiability and model checking. In: Murlak, F., Sankowski, P. (eds.) MFCS 2011. LNCS, vol. 6907, pp. 132–144. Springer, Heidelberg (2011)

6. Bouajjani, A., Esparza, J., Schwoon, S., Strejček, J.: Reachability analysis of multithreaded software with asynchronous communication. In: Ramanujam, R., Sen, S. (eds.) FSTTCS 2005. LNCS, vol. 3821, pp. 348–359. Springer, Heidelberg (2005)
7. Bouajjani, A., Fratani, S., Qadeer, S.: Context-bounded analysis of multithreaded programs with dynamic linked structures. In: Damm, W., Hermanns, H. (eds.) CAV 2007. LNCS, vol. 4590, pp. 207–220. Springer, Heidelberg (2007)
8. Breveglieri, L., Cherubini, A., Citrini, C., Crespi-Reghizzi, S.: Multi-push-down languages and grammars. Int. J. Found. Comput. Sci. 7(3), 253–292 (1996)
9. Cyriac, A., Gastin, P., Kumar, K.N.: MSO decidability of multi-pushdown systems via split-width. In: Koutny, M., Ulidowski, I. (eds.) CONCUR 2012. LNCS, vol. 7454, pp. 547–561. Springer, Heidelberg (2012)
10. Heußner, A., Leroux, J., Muscholl, A., Sutre, G.: Reachability analysis of communicating pushdown systems. In: Ong, L. (ed.) FOSSACS 2010. LNCS, vol. 6014, pp. 267–281. Springer, Heidelberg (2010)
11. Lal, A., Reps, T.W.: Reducing concurrent analysis under a context bound to sequential analysis. FMSD 35(1), 73–97 (2009)
12. Lal, A., Touili, T., Kidd, N., Reps, T.W.: Interprocedural analysis of concurrent programs under a context bound. In: Ramakrishnan, C.R., Rehof, J. (eds.) TACAS 2008. LNCS, vol. 4963, pp. 282–298. Springer, Heidelberg (2008)
13. Madhusudan, P., Parlato, G.: The tree width of auxiliary storage. In: POPL, pp. 283–294. ACM (2011)
14. Pnueli, A.: The temporal logic of programs. In: FOCS, pp. 46–57. IEEE (1977)
15. Qadeer, S., Rehof, J.: Context-bounded model checking of concurrent software. In: Halbwachs, N., Zuck, L.D. (eds.) TACAS 2005. LNCS, vol. 3440, pp. 93–107. Springer, Heidelberg (2005)
16. Seth, A.: Global reachability in bounded phase multi-stack pushdown systems. In: Touili, T., Cook, B., Jackson, P. (eds.) CAV 2010. LNCS, vol. 6174, pp. 615–628. Springer, Heidelberg (2010)
17. La Torre, S., Madhusudan, P., Parlato, G.: A robust class of context-sensitive languages. In: LICS, pp. 161–170. IEEE Computer Society (2007)
18. La Torre, S., Madhusudan, P., Parlato, G.: Context-bounded analysis of concurrent queue systems. In: Ramakrishnan, C.R., Rehof, J. (eds.) TACAS 2008. LNCS, vol. 4963, pp. 299–314. Springer, Heidelberg (2008)
19. La Torre, S., Madhusudan, P., Parlato, G.: An infinite automaton characterization of double exponential time. In: Kaminski, M., Martini, S. (eds.) CSL 2008. LNCS, vol. 5213, pp. 33–48. Springer, Heidelberg (2008)
20. La Torre, S., Madhusudan, P., Parlato, G.: Reducing context-bounded concurrent reachability to sequential reachability. In: Bouajjani, A., Maler, O. (eds.) CAV 2009. LNCS, vol. 5643, pp. 477–492. Springer, Heidelberg (2009)
21. La Torre, S., Napoli, M.: Reachability of multistack pushdown systems with scope-bounded matching relations. In: Katoen, J.-P., König, B. (eds.) CONCUR 2011. LNCS, vol. 6901, pp. 203–218. Springer, Heidelberg (2011)
22. La Torre, S., Napoli, M.: A temporal logic for multi-threaded programs. In: Baeten, J.C.M., Ball, T., de Boer, F.S. (eds.) TCS 2012. LNCS, vol. 7604, pp. 225–239. Springer, Heidelberg (2012)
23. Vardi, M.Y.: A temporal fixpoint calculus. In: POPL, pp. 250–259 (1988)
24. Vardi, M.Y., Wolper, P.: An automata-theoretic approach to automatic program verification (preliminary report). In: LICS, pp. 332–344. IEEE Computer Society (1986)

Cuts in Regular Expressions

Martin Berglund[1], Henrik Björklund[1], Frank Drewes[1],
Brink van der Merwe[2], and Bruce Watson[2]

[1] Umeå University, Sweden
{mbe,henrikb,drewes}@cs.umu.se
[2] Stellenbosch University, South Africa
abvdm@cs.sun.ac.za, bruce@fastar.org

Abstract. Most software packages with regular expression matching engines offer operators that extend the classical regular expressions, such as counting, intersection, complementation, and interleaving. Some of the most popular engines, for example those of Java and Perl, also provide operators that are intended to control the nondeterminism inherent in regular expressions. We formalize this notion in the form of the *cut* and *iterated cut* operators. They do not extend the class of languages that can be defined beyond the regular, but they allow for exponentially more succinct representation of some languages. Membership testing remains polynomial, but emptiness testing becomes PSPACE-hard.

1 Introduction

Regular languages are not only a theoretically well-understood class of formal languages. They also appear very frequently in real world programming. In particular, regular expressions are a popular tool for solving text processing problems. For this, the ordinary semantics of regular expressions, according to which an expression simply denotes a language, is extended by an informally defined operational understanding of how a regular expression is "applied" to a string. The usual default in regular expression matching libraries is to search for the leftmost matching substring, and pick the longest such substring [2]. This behavior is often used to repeatedly match different regular expressions against a string (or file contents) using program control flow to decide the next expression to match. Consider the repeatedly matching pseudo-code below, and assume that `match_regex` matches the longest prefix possible:

```
match = match_regex("(a*b)*", s);
if(match != null) then
    if(match_regex("ab*c", match.string_remainder) != null) then
        return match.string_remainder == "";
return false;
```

For the string $s = abac$, this program first matches $R_1 = (a^* \cdot b)^*$ to the substring ab, leaving ac as a remainder, which is matched by $R_2 = a \cdot (b^*) \cdot c$, returning true.

M.-P. Béal and O. Carton (Eds.): DLT 2013, LNCS 7907, pp. 70–81, 2013.
© Springer-Verlag Berlin Heidelberg 2013

The set of strings s for which the program returns "true" is in fact a regular language, but it is *not* the regular language defined by $R_1 \cdot R_2$. Consider for example the string $s = aababcc$, which is matched by $R_1 \cdot R_2$. However, in an execution of the program above, R_1 will match $aabab$, leaving the remainder cc, which is not matched by R_2. The expression $R_1 \cdot R_2$ exhibits non-deterministic behavior which is lost in the case of the earliest-longest-match strategy combined with the explicit *if*-statement. This raises the question, are programs of this type (with arbitrarily many *if*-statements freely nested) always regular, and how can we describe the languages they recognize?

Related Work. Several extensions of regular expressions that are frequently available in software packages, such as counting (or numerical occurrence indicators, not to be confused with counter automata), interleaving, intersection, and complementation, have been investigated from a theoretical point of view. The succinctness of regular expressions that use one or more of these extra operators compared to standard regular expressions and finite automata were investigated, e.g., in [4, 6, 8]. For regular expressions with intersection, the membership problem was studied in, e.g., [10, 14], while the equivalence and emptiness problems were analyzed in [3, 15]. Interleaving was treated in [5, 11] and counting in [9, 12]. To our knowledge, there is no previous theoretical treatment of the cut operator introduced in this paper, or of other versions of possessive quantification.

Paper Outline. In the next section we formalize the control of nondeterminism outlined above by defining the cut and iterated cut operators, which can be included directly into regular expressions, yielding so-called cut expressions. In Section 3, we show that adding the new operators does not change the expressive power of regular expressions, but that it does offer improved succinctness. Section 4 provides a polynomial time algorithm for the uniform membership problem of cut expressions, while Section 5 shows that emptiness is PSPACE-hard. In Section 6, we compare the cut operator to the similar operators found more or less commonly in software packages in the wild (Perl, Java, PCRE, etc.). Finally, Section 7 summarizes some open problems.

2   Cut Expressions

We denote the natural numbers (including zero) by $\mathbb{N}$. The set of all strings over an alphabet Σ is denoted by Σ^*. In particular, Σ^* contains the empty string ε. The set $\Sigma^* \setminus \{\varepsilon\}$ is denoted by Σ^+. We write $pref(u)$ to denote the set of nonempty prefixes of a string u and $pref_\varepsilon(u)$ to denote $pref(u) \cup \{\varepsilon\}$. The canonical extensions of a function $f \colon A \to B$ to a function from A^* to B^* and to a function from 2^A to 2^B are denoted by f as well.

As usual, a regular expression over an alphabet Σ (where $\varepsilon, \emptyset \notin \Sigma$) is either an element of $\Sigma \cup \{\varepsilon, \emptyset\}$ or an expression of one of the forms $(E \mid E')$, $(E \cdot E')$, or (E^*). Parentheses can be dropped using the rule that * (Kleene closure[1])

---

[1] Recall that the Kleene closure of a language L is the smallest language L^* such that $\{\varepsilon\} \cup LL^* \subseteq L^*$.

takes precedence over · (concatenation), which takes precedence over | (union). Moreover, outermost parentheses can be dropped, and $E \cdot E'$ can be written as EE'. The language $\mathcal{L}(E)$ denoted by a regular expression is obtained by evaluating E as usual, where $\emptyset$ stands for the empty language and $a \in \Sigma \cup \{\varepsilon\}$ for $\{a\}$. We denote by $E \equiv E'$ the fact that two regular expressions (or, later on, cut expressions) E and E' are equivalent, i.e., that $\mathcal{L}(E) = \mathcal{L}(E')$. Where the meaning is clear from context we may omit the $\mathcal{L}$ and write E to mean $\mathcal{L}(E)$.

Let us briefly recall finite automata. A nondeterministic finite automaton (NFA) is a tuple $A = (Q, \Sigma, \delta, q_0, F)$ consisting of a finite set Q of states, a initial state $q_0 \in Q$, a set $F \subseteq Q$ of final states, an alphabet Σ, and a transition function $\delta \colon Q \times \Sigma \to 2^Q$. In the usual way, δ extends to a function $\delta \colon \Sigma^* \to 2^Q$, i.e., $\delta(\varepsilon) = \{q_0\}$ and $\delta(wa) = \bigcup_{q \in \delta(w)} \delta(q, a)$. A accepts $w \in \Sigma^*$ if and only if $\delta(w) \cap F \neq \emptyset$, and it recognizes the language $\mathcal{L}(A) = \{w \in \Sigma^* \mid \delta(w) \cap F \neq \emptyset\}$. A deterministic finite automaton (DFA) is the special case where $|\delta(q, a)| \leq 1$ for all $(q, a) \in Q \times \Sigma$. In this case we consider δ to be a function $\delta \colon Q \times \Sigma \to Q$, so that its canonical extension to strings becomes a function $\delta \colon Q \times \Sigma^* \to Q$.

We now introduce cuts, iterated cuts, and cut expressions. Intuitively, $E \mathbin{!} E'$ is the variant of EE' in which E greedily matches as much of a string as it can accommodate, leaving the rest to be matched by E'. The so-called iterated cut $E^{!*}$ first lets E match as much of a string as possible, and seeks to iterate this until the whole string is matched (if possible).

Definition 1 (cut and cut expression). *The* cut *is the binary operation* ! *on languages such that, for languages L, L',*

$$L \mathbin{!} L' = \{uv \mid u \in L, v \in L', uv' \notin L \text{ for all } v' \in pref(v)\}.$$

The iterated cut *of L, denoted by $L^{!*}$, is the smallest language that satisfies*

$$\{\varepsilon\} \cup (L \mathbin{!} (L^{!*})) \subseteq L^{!*}$$

(i.e., $L \mathbin{!} (L \mathbin{!} \cdots (L \mathbin{!} (L \mathbin{!} \{\varepsilon\})) \cdots) \subseteq L^{!*}$ for any number of repetitions of the cut).

Cut expressions *are expressions built using the operators allowed in regular expressions, the cut, and the iterated cut. A cut expression denotes the language obtained by evaluating that expression in the usual manner.*

The precedence rules give $^{!*}$ precedence over ·, which in turn gets precedence over ! which in turn gets precedence over | .

The motivation for the inclusion of the iterated cut is two-fold; (i) it is a natural extension for completeness in that it relates to the cut like the Kleene closure relates to concatenation; and, (ii) in the context of a program like that shown on page 1, the iterated cut permits the modelling of matching regular expressions in loops.

Let us discuss a few examples.

1. The cut expression $ab^* \mathbin{!} b$ yields the empty language. This is because every string in $\mathcal{L}(ab^*b)$ is in $\mathcal{L}(ab^*)$ as well, meaning that the greedy matching of

the first subexpression will never leave a b over for the second. Looking at the definition of the cut, a string in $\mathcal{L}(ab^*\,!\,b)$ would have to be of the form ub, such that $u \in \mathcal{L}(ab^*)$ but $ub \notin \mathcal{L}(ab^*)$. Clearly, such a string does not exist. More generally, if $\varepsilon \notin \mathcal{L}(E')$ then $\mathcal{L}(E\,!\,E') \subseteq \mathcal{L}(EE') \backslash \mathcal{L}(E)$. However, as the next example shows, the converse inclusion does not hold.

2. We have $(a^*\,|\,b^*)\,!\,(ac\,|\,bc) \equiv a^+bc\,|\,b^+ac$.[2] This illustrates that the semantics of the cut cannot be expressed by concatenating subsets of the involved languages. In the example, there are no subsets L_1 and L_2 of $\mathcal{L}(a^*\,|\,b^*)$ and $\mathcal{L}(ac\,|\,bc)$, respectively, such that $L_1 \cdot L_2 = \mathcal{L}(a^*\,|\,b^*)\,!\,\mathcal{L}(ac\,|\,bc)$.

3. Clearly, $((ab)^*\,!\,a)\,!\,b \equiv (ab)^*ab$ whereas $(ab)^*\,!\,(a\,!\,b) \equiv (ab)^*\,!\,ab \equiv \emptyset$ (as in the first example). Thus, the cut is not associative.

4. As an example of an iterated cut, consider $((aa)^*\,!\,a)^*$. We have $(aa)^*\,!\,a \equiv (aa)^*a$ and therefore $((aa)^*\,!\,a)^* \equiv a^*$. This illustrates that matching a string against $(E\,!\,E')^*$ *cannot* be done by greedily matching E, then matching E', and iterating this procedure. Instead, one has to "chop" the string to be matched into substrings and match each of those against $E\,!\,E'$. In particular, $(E\,!\,\varepsilon)^* \equiv E^*$ (since $E\,!\,\varepsilon \equiv E$). This shows that $E^{!*}$ cannot easily be expressed by means of cut and Kleene closure.

5. Let us finally consider the interaction between the Kleene closure and the iterated cut. We have $L^{!*} \subseteq L^*$ and thus $(L^{!*})^* \subseteq (L^*)^* = L^*$. Conversely, $L \subseteq L^{!*}$ yields $L^* \subseteq (L^{!*})^*$. Thus $(L^{!*})^* = L^*$ for all languages L.
   Similarly, we also have $(L^*)^{!*} = L^*$. Indeed, if $w \in L^*$, then it belongs to $(L^*)^{!*}$, since the first iteration of the iterated cut can consume all of w. Conversely, $(L^*)^{!*} \subseteq (L^*)^* = L^*$. Thus, altogether $(L^*)^{!*} = L^* = (L^{!*})^*$

3   Cut Expressions versus Finite Automata

In this section, we compare cut expressions and finite automata. First, we show that the languages described by cut expressions are indeed regular. We do this by showing how to convert cut expressions into equivalent finite automata. Second, we show that cut expressions are succinct: There are cut expressions containing only a single cut (and no iterated cut), such that a minimal equivalent NFA or regular expression is of exponential size.

3.1   Cut Expressions Denote Regular Languages

Let A, A' be DFAs. To prove that the languages denoted by cut expressions are regular, it suffices to show how to construct DFAs recognizing $L(A)\,!\,L(A')$ and $L(A)^{!*}$. We note here that an alternative proof would be obtained by showing how to construct alternating automata (AFAs) recognizing $L(A)\,!\,L(A')$ and $L(A)^{!*}$. Such a construction would be slightly simpler, especially for the iterated cut, but since the conversion of AFAs to DFAs causes a doubly exponential size increase [1], we prefer the construction given below, which (almost) saves one

---

[2] As usual, we abbreviate EE^* by E^+.

level of exponentiality. Moreover, we hope that this construction, though more complex, is more instructive.

We first handle the comparatively simple case $L(A) \,!\, L(A')$. The idea of the construction is to combine A with a kind of product automaton of A and A'. The automaton starts working like A. At the point where A reaches one of its final states, A' starts running in parallel with A. However, in contrast to the ordinary product automaton, the computation of A' is reset to its initial state whenever A reaches one of its final states again. Finally, the string is accepted if and only if A' is in one of its final states.

To make the construction precise, let $A = (Q, \Sigma, \delta, q_0, F)$ and $A' = (Q', \Sigma, \delta', q_0', F')$. In order to disregard a special case, let us assume that $q_0 \notin F$. (The case where $q_0 \in F$ is easier, because it allows us to use only product states in the automaton constructed.) We define a DFA $\overline{A} = (\overline{Q}, \Sigma, \overline{\delta}, q_0, \overline{F})$ as follows:

- $\overline{Q} = Q \cup (Q \times Q')$ and $\overline{F} = Q \times F'$,
- for all $q, r \in Q$, $\overline{q} = (q, q') \in \overline{Q}$, and $a \in \Sigma$ with $\delta(q, a) = r$

$$\overline{\delta}(q, a) = \begin{cases} r & \text{if } r \notin F \\ (r, q_0') & \text{otherwise,} \end{cases} \quad \text{and} \quad \overline{\delta}(\overline{q}, a) = \begin{cases} (r, \delta'(q', a)) & \text{if } r \notin F \\ (r, q_0') & \text{otherwise.} \end{cases}$$

Let $w \in \Sigma^*$. By construction, $\overline{\delta}$ has the following properties:

1. If $u \notin L(A)$ for all $u \in pref_\varepsilon(w)$, then $\overline{\delta}(w) = \delta(w)$.
2. Otherwise, let $w = uv$, where u is the longest prefix of w such that $u \in L(A)$. Then $\overline{\delta}(w) = (\delta(w), \delta'(v))$.

We omit the easy inductive proof of these statements. By the definition of $L(A) \,!\, L(A')$ and the choice of $\overline{F}$, they imply that $L(\overline{A}) = L(A) \,!\, L(A')$. In other words, we have the following lemma.

Lemma 2. *For all regular languages L and L', the language $L \,!\, L'$ is regular.*

Let us now consider the iterated cut. Intuitively, the construction of a DFA recognizing $L(A)^{!*}$ is based on the same idea as above, except that the product construction is iterated. The difficulty is that the straightforward execution of this construction yields an infinite automaton. For the purpose of exposing the idea, let us disregard this difficulty for the moment. Without loss of generality, we assume that $q_0 \notin F$ (which we can do because $L(A)^{!*} = (L(A) \setminus \{\varepsilon\})^{!*}$) and that $\delta(q, a) \neq q_0$ for all $q \in Q$ and $a \in \Sigma$.

We construct an automaton whose states are strings $q_1 \cdots q_k \in Q^+$. The automaton starts in state q_0, initially behaving like A. If it reaches one of the final states of A, say q_1, it continues in state $q_1 q_0$, working essentially like the automaton for $L(A) \,!\, L(A)$. In particular, it "resets" the second copy each time the first copy encounters a final state of A. However, should the second copy reach a final state q_2 of A (while $q_1 \notin F$), a third copy is spawned, thus resulting in a state of the form $q_1 q_2 q_0$, and so on.

Formally, let $\delta_a : Q \to Q$ be given by $\delta_a(q) = \delta(q, a)$ for all $a \in \Sigma$ and $q \in Q$. Recall that functions to extend to sequences, so $\delta_a : Q^* \to Q^*$ operates element-wise. We construct the (infinite) automaton $\widehat{A} = (\widehat{Q}, \Sigma, \widehat{\delta}, q_0, \widehat{F})$ as follows:

– $\widehat{Q} = (Q \setminus \{q_0\})^* Q.$

– For all $s = q_1 \cdots q_k \in \widehat{Q}$ and $a \in \Sigma$ with $\delta_a(s) = q'_1 \cdots q'_k$

$$\widehat{\delta}(s, a) = \begin{cases} q'_1 \cdots q'_k & \text{if } q'_1, \ldots, q'_k \notin F \\ q'_1 \cdots q'_l q_0 & \text{if } l = \min\{i \in \{1, \ldots, k\} \mid q'_i \in F\}. \end{cases} \quad (1)$$

– $\widehat{F} = \{q_1 \cdots q_k \in \widehat{Q} \mid q_k = q_0\}.$

Note that $\widehat{\delta}(s, a) \in \widehat{Q}$ since we assume that $\delta(q, a) \neq q_0$ for all $q \in Q$ and $a \in \Sigma$.

Similar to the properties of $\overline{A}$ above, we have the following:

Claim 1. Let $w = v_1 \cdots v_k \in \Sigma^*$, where $v_1 \cdots v_k$ is the unique decomposition of w such that (a) for all $i \in \{1, \ldots, k-1\}$, v_i is the longest prefix of $v_i \cdots v_k$ which is in $L(A)$ and (b) $pref_\varepsilon(v_k) \cap L(A) = \emptyset$.[3] Then $\widehat{\delta}(w) = \delta(v_1 \cdots v_k)\delta(v_2 \cdots v_k) \cdots \delta(v_k)$. In particular, $\widehat{A}$ accepts w if and only if $w \in L(A)^{!*}$.

Again, we omit the straightforward inductive proof.

It remains to be shown how to turn the set of states of $\widehat{A}$ into a finite set. We do this by verifying that repetitions of states of A can be deleted. To be precise, let $\pi(s)$ be defined as follows for all $s = q_1 \cdots q_k \in \widehat{Q}$. If $k = 1$ then $\pi(s) = s$. If $k > 1$ then

$$\pi(s) = \begin{cases} \pi(q_1 \cdots q_{k-1}) & \text{if } q_k \in \{q_1, \ldots, q_{k-1}\} \\ \pi(q_1 \cdots q_{k-1})q_k & \text{otherwise.} \end{cases}$$

Let $\pi(\widehat{A})$ be the NFA obtained from $\widehat{A}$ by taking the quotient with respect to π, i.e., by identifying all states $s, s' \in \widehat{Q}$ such that $\pi(s) = \pi(s')$. The set of final states of $\pi(\widehat{A})$ is the set $\pi(\widehat{F})$.

This completes the construction. The following lemmas prove its correctness.

Lemma 3. *For all $s \in \widehat{Q}$ and $a \in \Sigma$ it holds that $\pi(\widehat{\delta}(s, a)) = \pi(\widehat{\delta}(\pi(s), a))$.*

Proof. By the very definition of π, for every function $f: Q \to Q$ and all $s \in \widehat{Q}$ we have $\pi(f(s)) = \pi(f(\pi(s)))$. In particular, this holds for $f = \delta_a$. Now, let $s = q_1 \cdots q_k$ be as in the definition of $\widehat{\delta}$. Since the same set of symbols occurs in $\delta_a(s)$ and $\delta_a(\pi(s))$, the same case of Equation 1 applies for the construction of $\widehat{\delta}(s, a)$ and $\widehat{\delta}(\pi(s), a)$. In the first case $\pi(\widehat{\delta}(s, a)) = \pi(\delta_a(s)) = \pi(\delta_a(\pi(s))) = \pi(\widehat{\delta}(\pi(s), a))$. In the second case

$$\begin{aligned} \pi(\widehat{\delta}(s, a)) &= \pi(\delta_a(q_1 \cdots q_l)q_0) \\ &= \pi(\delta_a(q_1 \cdots q_l))q_0 \\ &= \pi(\delta_a(\pi(q_1 \cdots q_l)))q_0 \\ &= \pi(\delta_a(\pi(q_1 \cdots q_l))q_0) \\ &= \pi(\widehat{\delta}(\pi(s), a)). \end{aligned}$$

Note that the second and the fourth equality make use of the fact that $q_0 \notin \{q_1, \ldots, q_{k-1}\}$, which prevents π from deleting the trailing q_0. □

---

[3] The strings $v_1, \ldots, v_k$ are well defined because $\varepsilon \notin L(A)$.

Lemma 4. *The automaton $\pi(\widehat{A})$ is a DFA such that $L(\pi(\widehat{A})) = L(A)^{!*}$. In particular, $L^{!*}$ is regular for all regular languages L.*

Proof. To see that $\pi(\widehat{A})$ is a DFA, let $a \in \Sigma$. By the definition of $\pi(\widehat{A})$, its transition function $\widehat{\delta}_\pi$ is given by

$$\widehat{\delta}_\pi(t, a) = \{\pi(\widehat{\delta}(s, a)) \mid s \in \widehat{Q}, \; t = \pi(s)\}$$

for all $t \in \pi(\widehat{Q})$. However, by Lemma 3, $\pi(\widehat{\delta}(s, a)) = \pi(\widehat{\delta}(t, a))$ is independent of the choice of s. In other words, $\widehat{A}$ is a DFA. Furthermore, by induction on the length of $w \in \Sigma^*$, Lemma 3 yields $\widehat{\delta}_\pi(w) = \pi(\widehat{\delta}(w))$. Thus, by Claim 1, $L(\pi(\widehat{A})) = L(A)^{!*}$. In particular, for a regular language L, this shows that $L^{!*}$ is regular, by picking A such that $\mathcal{L}(A) = L$.   □

We note here that, despite the detour via an infinite automaton, the construction given above can effectively be implemented. Unfortunately, it results in a DFA of size $\mathcal{O}(n!)$, where n is the number of states of the original DFA.

Theorem 5. *For every cut expression E, $\mathcal{L}(E)$ is regular.*

Proof. Follows from combining Lemmas 2 and 4.   □

3.2 Succinctness of Cut Expressions

In this section we show that for some languages, cut expressions provide an exponentially more compact representation than regular expressions and NFAs.

Theorem 6. *For every $k \in \mathbb{N}_+$, there exists a cut expression E_k of size $\mathcal{O}(k)$ such that every NFA and every regular expression for $\mathcal{L}(E_k)$ is of size $2^{\Omega(k)}$. Furthermore, E_k does not contain the iterated cut and it contains only one occurrence of the cut.*

Proof. We use the alphabets $\Sigma = \{0, 1\}$ and $\Gamma = \Sigma \cup \{], [\}$. For $k \in \mathbb{N}_+$, let

$$E_k = (\varepsilon \mid [\Sigma^* 0 \Sigma^{k-1} 1 \Sigma^*] \mid [\Sigma^* 1 \Sigma^{k-1} 0 \Sigma^*]) \, ! \, [\Sigma^{2k}].$$

Each string in the language $\mathcal{L}(E_k)$ consists of one or two bitstrings enclosed in square brackets. If there are two, the first has at least two different bits at a distance of exactly k positions and the second is an arbitrary string in Σ^{2k}. However, when there is only a single pair of brackets the bitstring enclosed is of length $2k$ and its second half will be an exact copy of the first.

We argue that any NFA that recognizes $\mathcal{L}(E_k)$ must have at least 2^k states. Assume, towards a contradiction, that there is an NFA A with fewer than 2^k states that recognizes $\mathcal{L}(E_k)$.

Since $|\Sigma^k| = 2^k$ there must exist two distinct bitstrings w_1 and w_2 of length k such that the following holds. There exist a state q of A and accepting runs ρ_1 and ρ_2 of A on $[w_1 w_1]$ and $[w_2 w_2]$, resp., such that ρ_1 reaches q after reading $[w_1$

and ρ_2 reaches q after reading $[w_2$. This, in turn, means that there are accepting runs ρ_1' and ρ_2' of A_q on $w_1]$ and $w_2]$, respectively, where A_q is the automaton obtained from A by making q the sole initial state. Combining the first half of ρ_1 with ρ_2' gives an accepting run of A on $[w_1w_2]$. This is a contradiction and we conclude that there is no NFA for E_k with fewer than 2^k states.

The above conclusion also implies that every regular expression for $\mathcal{L}(E_k)$ has size $2^{\Omega(k)}$. If there was a smaller regular expression, the Glushkov construction [7] would also yield a smaller NFA. □

Remark 7. The only current upper bound is the one implied by Section 3.1, from which automata of non-elementary size cannot be ruled out as it yields automata whose sizes are bounded by powers of twos.

A natural restriction on cut expressions is to only allow cuts to occur at the topmost level of the expression. This gives a tight bound on automata size.

Lemma 8. *Let E be a cut expression, without iterated cuts, such that no subexpression of the form C^* or $C \cdot C'$ contains cuts. Then the minimal equivalent DFA has $2^{\mathcal{O}(|E|)}$ states, and this bound is tight.*

Proof (sketch). Given any DFAs A, A', using product constructions we get DFAs for $L(A) \mid L(A')$ and $L(A)\,!\,L(A')$ whose number of states is proportional to the product of the number of states in A and A'. (See Lemma 2 for the case $L(A)\,!\,L(A')$.) Thus, one can construct an exponential-sized DFA in a bottom-up manner. Theorem 6 shows that this bound is tight. □

4   Uniform Membership Testing

We now present an easy membership test for cut expressions that uses a dynamic programming approach (or, equivalently, memoization). Similarly to the Cocke-Younger-Kasami algorithm, the idea is to check which substrings of the input string belong to the languages denoted by the subexpressions of the given cut expression. The pseudocode of the algorithm is shown in Algorithm 1. Here, the string $u = a_1 \cdots a_n$ to be matched against a cut expression E is a global variable. For $1 \leq i \leq j \leq n+1$, $Match(E, i, j)$ will check whether $a_i \cdots a_{j-1} \in \mathcal{L}(E)$. We assume that an implicit table is used in order to memoize computed values for a given input triple. Thus, recursive calls with argument triples that have been encountered before will immediately return the memoized value rather than executing the body of the algorithm.

Theorem 9. *The uniform membership problem for cut expressions can be decided in time $\mathcal{O}(m \cdot n^3)$, where m is the size of the cut expression and n is the length of the input string.*

Proof. Consider a cut expression E_0 of size m and a string $u = a_1 \cdots a_n$. It is straightforward to show by induction on $m+n$ that $Match(E, i, j) = true$ if and only if $a_i \cdots a_{j-1} \in \mathcal{L}(E)$, where $1 \leq i \leq j \leq n+1$ and E is a subexpression of

---

Algorithm 1. $Match(E, i, j)$

---

if $E = \emptyset$ then return *false*
else if $E = \varepsilon$ then return $i = j$
else if $E \in \Sigma$ then return $j = i + 1 \wedge E = a_i$
else if $E = E_1 \mid E_2$ then return $Match(E_1, i, j) \vee Match(E_2, i, j)$
else if $E = E_1 \cdot E_2$ then
    for $k = 0, \ldots, j - i$ do
        if $Match(E_1, i, i + k) \wedge Match(E_2, i + k, j)$ then return *true*
    return *false*
else if $E = E_1^*$ then
    for $k = 1, \ldots, j - i$ do
        if $Match(E_1, i, i + k) \wedge Match(E, i + k, j)$ then return *true*
    return $i = j$
else if $E = E_1 \,!\, E_2$ then
    for $k = j - i, \ldots, 0$ do
        if $Match(E_1, i, i + k)$ then return $Match(E_2, i + k, j)$
    return *false*
else if $E = E_1^{!*}$ then
    for $k = j - i, \ldots, 1$ do
        if $Match(E_1, i, i + k)$ then return $Match(E, i + k, j)$
    return $i = j$

E_0. For $E = E_1 \,!\, E_2$, this is because of the fact that $v \in \mathcal{L}(E)$ if and only if v has a longest prefix $v_1 \in \mathcal{L}(E_1)$, and the corresponding suffix v_2 of v (i.e., such that $v = v_1 v_2$) is in $\mathcal{L}(E_2)$. Furthermore, it follows from this and the definition of the iterated cut that, for $E = E_1^{!*}$, $v \in \mathcal{L}(E)$ if either $v = \varepsilon$ or v has a longest prefix $v_1 \in \mathcal{L}(E_1)$ such that the corresponding suffix v_2 is in $\mathcal{L}(E)$.

Regarding the running time of $Match(E, 1, n + 1)$, by memoization the body of $Match$ is executed at most once for every subexpression of E and all i, j, $1 \leq i \leq j \leq n + 1$. This yields $\mathcal{O}(m \cdot n^2)$ executions of the loop body. Moreover, a single execution of the loop body involves at most $\mathcal{O}(n)$ steps (counting each recursive call as one step), namely if $E = E_1^*$, $E = E_1 \,!\, E_2$ or $E = E^{!*}$.   □

5 Emptiness Testing of Cut Expressions

Theorem 10. *Given a cut expression E, it is PSPACE-hard to decide whether $\mathcal{L}(E) = \emptyset$. This remains true if $E = E_1 \,!\, E_2$, where E_1 and E_2 are regular expressions.*

Proof. We prove the theorem by reduction from regular expression universality, i.e. deciding for a regular expression R and an alphabet Σ whether $\mathcal{L}(R) = \Sigma^*$. This problem is well known to be PSPACE-complete [12]. Given R, we construct a cut expression E such that $\mathcal{L}(E) = \emptyset$ if and only if $\mathcal{L}(R) = \Sigma^*$.

We begin by testing if $\varepsilon \in \mathcal{L}(R)$. This can be done in polynomial time. If $\varepsilon \notin \mathcal{L}(R)$, then we set $E = \varepsilon$, satisfying $\mathcal{L}(E) \neq \emptyset$. Otherwise, we set $E = R \,!\, \Sigma$. If R is universal, there is no string ua such that $u \in \mathcal{L}(R)$ but $ua \notin \mathcal{L}(R)$. Thus $\mathcal{L}(E)$ is empty. If R is not universal, since $\varepsilon \in \mathcal{L}(R)$ there are $u \in \Sigma^*$ and $a \in \Sigma$ such that $u \in \mathcal{L}(R)$ and $ua \notin \mathcal{L}(R)$, which means that $ua \in \mathcal{L}(E) \neq \emptyset$.   □

Lemma 11. *For cut expressions E the problems whether $\mathcal{L}(E) = \emptyset$ and $\mathcal{L}(E) = \Sigma^*$ are LOGSPACE-equivalent.*

Proof. Assume that $\# \notin \Sigma$, and let $\Sigma' = \Sigma \cup \{\#\}$. The lemma then follows from these two equivalences: (i) $E \equiv \emptyset$ if and only if $((\varepsilon \,|\, E\Sigma^*)\,!\,\Sigma^+)\,|\,\varepsilon \equiv \Sigma^*$; and; (ii) $E \equiv \Sigma^*$ if and only if $(\varepsilon \,|\, E\#(\Sigma')^*)\,!\,\Sigma^*\# \equiv \emptyset$. □

6 Related Concepts in Programming Languages

Modern regular expression matching engines have numerous highly useful features, some of which improve succinctness (short-hand operators) and some of which enable expressions that specify non-regular languages. Of interest here is that most regular expression engines in practical use feature at least *some* operation intended to control nondeterminism in a way that resembles the cut. They are however only loosely specified in terms of *backtracking*, the specific evaluation technique used by many regular expression engines. This, combined with the highly complex code involved, makes formal analysis difficult.

All these operations appear to trace their ancestry to the first edition of "Mastering Regular Expressions" [2], which contains the following statement:

> *"A feature I think would be useful, but that no regex flavor that I know of has, is what I would call possessive quantifiers. They would act like normal quantifiers except that once they made a decision that met with local success, they would never backtrack to try the other option. The text they match could be unmatched if their enclosing subexpression was unmatched, but they would never give up matched text of their own volition, even in deference to the overall match."[2]*

The cut operator certainly fits this somewhat imprecise description, but as we shall see implementations have favored different interpretations. Next we give a brief overview of three different operations implemented in several major regular expression engines, that exhibit some control over nondeterminism. All of these operators are of great practical value and are in use. Still, they feature some idiosyncrasies that should be investigated, in the interest of bringing proper regular behavior to as large a set of regular expression functionality as possible.

Possessive Quantifiers. Not long after the proposal for the possessive quantifier, implementations started showing up. It is available in software such as Java, PCRE, Perl, etc. For a regular expression R the operation is denoted R^{*+}, and behaves like R^* except it never backtracks. This is already troublesome, since "backtracking" is poorly defined at best, and, in fact, by itself $\mathcal{L}(R^{*+}) = \mathcal{L}(R^*)$, but $\mathcal{L}(R^{*+} \cdot R') = \mathcal{L}(R^* \,!\, R')$ for all R'. That is, extending regular expressions with possessive quantifiers makes it possible to write expressions such that $\mathcal{L}(E \cdot E') \neq \mathcal{L}(E) \cdot \mathcal{L}(E')$, an example being given by $E = a^{*+}$ and $E' = a$. This violates the compositional spirit of regular expressions.

Next, consider Table 1. The expression on the first row, call it R, is tested in each of the given implementations, and the language recognized is shown. The results on the first row are easy to accept from every perspective. The second

Table 1. Some examples of possessive quantifier use

Expression	Perl 5.16.2	Java 1.6.0u18	PCRE 8.32
$(aa)^{*+}a$	$\{a, aaa, aaaaa, \dots\}$	$\{a, aaa, aaaaa, \dots\}$	$\{a, aaa, aaaaa, \dots\}$
$((aa)^{*+}a)^*$	$\{\varepsilon, a, aaa, aaaaa, \dots\}$	$\{\varepsilon, a, aaa, aaaaa, \dots\}$	$\{a, aaa, aaaaa, \dots\}$
$((aa)^{*+}a)^*a$	$\{a\}$	$\{a\}$	$\{a\}$

Table 2. Comparison between Perl and PCRE when using the (*PRUNE) operator

Expression	Perl 5.10.1	Perl 5.16.2	PCRE 8.32
$(aa)^*(\text{*PRUNE})a$	$\{a, aaa, aaaaa, \dots\}$	$\{a, aaa, aaaaa, \dots\}$	$\{a, aaa, aaaaa, \dots\}$
$((aa)^*(\text{*PRUNE})a)^*$	$\{\varepsilon, a, aa, aaa, \dots\}$	$\emptyset$	$\emptyset$
$a^*(\text{*PRUNE})a$	$\{a, aa, aaa, \dots\}$	$\emptyset$	$\emptyset$

row however has the expression R^*, and despite $a \in \mathcal{L}(R)$ no implementation gives $aa \in \mathcal{L}(R^*)$, which violates the classical compositional meaning of the Kleene closure (in addition, in PCRE we have $\varepsilon \notin \mathcal{L}(R^*)$). The third row further illustrates how the compositional view of regular expressions breaks down when using possessive quantifiers.

Independent Groups or Atomic Subgroups. A practical shortcoming of the possessive quantifiers is that the "cut"-like operation cannot be separated from the quantifier. For this reason most modern regular expression engines have also introduced atomic subgroups ("independent groups" in Java). An atomic subgroup containing the expression R is denoted $(?\!>\!R)$, and described as "preventing backtracking". Any subexpression $(?\!>\!R^*)$ is equivalent to R^{*+}, but subexpressions of the form $(?\!>\!R)$ where the topmost operation in R is not a Kleene closure may be hard to translate into an equivalent expression using possessive quantifiers.

Due to the direct translation, atomic subgroups suffer from all the same idiosyncrasies as possessive quantifiers, such as $\mathcal{L}(((?\!>\!(aa)^*)a)^*a) = \{a\}$.

*Commit Operators and (*PRUNE).* In Perl 6 several interesting "commit operators" relating to nondeterminism control were introduced. As Perl 5 remains popular they were back-ported to Perl 5 in version 5.10.0 with different syntax. The one closest to the pure cut is (*PRUNE), called a "zero-width pattern", an expression that matches ε (and therefore always succeeds) but has some engine side-effect. As with the previous operators the documentation depends on the internals of the implementation. *"[(*PRUNE)] prunes the backtracking tree at the current point when backtracked into on failure"*[13].

These operations are available both in Perl and PCRE, but interestingly their semantics in Perl 5.10 and Perl 5.16 differ in subtle ways; see Table 2. Looking at the first two rows we see that Perl 5.10 matches our compositional understanding of the Kleene closure (i.e., row two has the same behavior as $((aa)*!a)*$). On the other hand Perl 5.10 appears to give the wrong answer in the third row example.

7 Discussion

We have introduced cut operators and demonstrated several of their properties. Many open questions and details remain to be worked out however:

- There is a great distance between the upper and lower bounds on minimal automata size presented in Section 3.2, with an exponential lower bound for both DFA and NFA, and a non-elementary upper bound in general.
- The complexity of uniform membership testing can probably be improved as the approach followed by Algorithm 1 is very general. (It can do complementation, for example.)
- The precise semantics of the operators discussed in Section 6 should be studied further, to ensure that all interesting properties can be captured.

Acknowledgments. We thank Yves Orton who provided valuable information about the implementation and semantics of (*PRUNE) in Perl.

References

[1] Chandra, A.K., Kozen, D.C., Stockmeyer, L.J.: Alternation. J. ACM 28(1), 114–133 (1981)
[2] Friedl, J.E.F.: Mastering Regular Expressions. Reilly & Associates, Inc., Sebastopol (1997)
[3] Fürer, M.: The complexity of the inequivalence problem for regular expressions with intersection. In: de Bakker, J.W., van Leeuwen, J. (eds.) ICALP 1980. LNCS, vol. 85, pp. 234–245. Springer, Heidelberg (1980)
[4] Gelade, W.: Succinctness of regular expressions with interleaving, intersection and counting. Theor. Comput. Sci. 411(31-33), 2987–2998 (2011)
[5] Gelade, W., Martens, W., Neven, F.: Optimizing schema languages for XML: Numerical constraints and interleaving. SIAM J. Comput. 38(5), 2021–2043 (2009)
[6] Gelade, W., Neven, F.: Succinctness of the complement and intersection of regular expressions. ACM Trans. Comput. Logic 13(1), Article 4 (2012)
[7] Glushkov, V.M.: The abstract theory of automata. Russian Mathematical Surveys 16, 1–53 (1961)
[8] Gruber, H., Holzer, M.: Tight bounds on the descriptional complexity of regular expressions. In: Diekert, V., Nowotka, D. (eds.) DLT 2009. LNCS, vol. 5583, pp. 276–287. Springer, Heidelberg (2009)
[9] Kilpeläinen, P., Tuhkanen, R.: Regular expressions with numerical occurrence indicators - preliminary results. In: SPLST, pp. 163–173 (2003)
[10] Kupferman, O., Zuhovitzky, S.: An improved algorithm for the membership problem for extended regular expressions. In: Diks, K., Rytter, W. (eds.) MFCS 2002. LNCS, vol. 2420, pp. 446–458. Springer, Heidelberg (2002)
[11] Mayer, A.J., Stockmeyer, L.J.: Word problems - this time with interleaving. Inform. and Comput. 115(2), 293–311 (1994)
[12] Meyer, A.R., Stockmeyer, L.J.: The equivalence problem for regular expressions with squaring requires exponential time. In: SWAT (FOCS), pp. 125–129 (1972)
[13] Perl 5 Porters. perlre (2012), http://perldoc.perl.org/perlre.html (accessed January 16, 2013)
[14] Petersen, H.: The membership problem for regular expressions with intersection is complete in LOGCFL. In: Alt, H., Ferreira, A. (eds.) STACS 2002. LNCS, vol. 2285, pp. 513–522. Springer, Heidelberg (2002)
[15] Robson, J.M.: The emptiness of complement problem for semi extended regular expressions requires c^n space. Inform. Processing Letters 9(5), 220–222 (1979)

Quantum Finite Automata and Linear Context-Free Languages: A Decidable Problem

Alberto Bertoni[1], Christian Choffrut[2], and Flavio D'Alessandro[3]

[1] Dipartimento di Scienze dell'Informazione, Università degli Studi di Milano
Via Comelico 39, 20135 Milano, Italy
[2] Laboratoire LIAFA, Université de Paris 7
2, pl. Jussieu, 75251 Paris Cedex 05
[3] Dipartimento di Matematica, Sapienza Università di Roma
Piazzale Aldo Moro 2, 00185 Roma, Italy

Abstract. We consider the so-called measure once finite quantum automata model introduced by Moore and Crutchfield in 2000. We show that given a language recognized by such a device and a linear context-free language, it is recursively decidable whether or not they have a nonempty intersection. This extends a result of Blondel et al. which can be interpreted as solving the problem with the free monoid in place of the family of linear context-free languages.

Keywords: Quantum automata, Context-free languages, Algebraic groups, Decidability.

1 Introduction

Quantum finite automata or simply quantum automata were introduced at the beginning of the previous decade in [9] as a new model of language recognizer. Numerous publications have ever since compared their decision properties to those of the older model of probabilistic finite automata. Some undecidable problems for probabilistic finite automata turn out to be decidable for quantum finite automata. The result in [4] which triggered our investigation can be viewed as asserting that the intersection emptiness problem of a language recognized by a finite quantum automaton with the free monoid is recursively decidable. The present result concerns the same problem where instead of the free monoid, more generally a language belonging to some classical families of languages such as the context-free languages and the bounded semilinear languages is considered.

An ingredient of the proof in [4] consists of expressing the emptiness problem in the first order theory of the reals and then to apply Tarski-Seidenberg quantifier elimination. This is possible because an algebraic subset, i.e., a closed subset in the Zariski topology $\mathcal{A} \subseteq \mathbb{R}^n$, is naturally associated to this intersection and even more miraculously because this subset can be effectively computed (cf. also [5]).

Here we show that the (actually semi-)algebraicity of $\mathcal{A}$ still holds when considering not only the free monoid but more generally arbitrary context-free

M.-P. Béal and O. Carton (Eds.): DLT 2013, LNCS 7907, pp. 82–93, 2013.
© Springer-Verlag Berlin Heidelberg 2013

languages and bounded semilinear languages. Unfortunately, its effective construction is only guaranteed under stricter conditions such as the fact that the language is context-free and linear or is bounded semilinear. In particular, in the case of context-free languages, we are not able to settle the nonlinear case yet.

We now give a more formal presentation of our work. The free monoid generated by the finite alphabet Σ is denoted by Σ^*. The elements of Σ^* are *words*. We consider all finite dimensional vector spaces as provided with the Euclidian norm $\|\cdot\|$. A *quantum automaton* is a quadruple $Q = (s, \varphi, P, \lambda)$ where $s \in \mathbb{R}^n$ is a row-vector of unit norm, P is a projection of $\mathbb{R}^n$, φ is a representation of the free monoid Σ^* into the group of *orthogonal* $n \times n$-matrices in $\mathbb{R}^{n \times n}$ and the *threshold* λ has value in $\mathbb{R}$. We recall that a real matrix M is orthogonal if its inverse equals its transpose: $M^{-1} = M^T$. We denote by O_n the group of $n \times n$-orthogonal matrices. We are mainly interested in effective properties which require the quantum automaton to be effectively given. We say that the quantum automaton is *rational* if all the coefficients of the components of the automaton are rational numbers, i.e., φ maps Σ^* into $\mathbb{Q}^{n \times n}$ and $\lambda \in \mathbb{Q}$. This hypothesis is not a restriction since all we use for the proofs is the fact that the arithmetic operations and the comparison are effective in the field of rational numbers. This is the "measure once" model introduced by Moore and Crutchfield in 2000 [9]. For a real threshold λ, the languages recognized by Q with strict and nonstrict threshold λ are

$$|Q_>| = \{w \in \Sigma^* \mid \|s\varphi(w)P\| > \lambda\}, \quad |Q_\geq| = \{w \in \Sigma^* \mid \|s\varphi(w)P\| \geq \lambda\}$$

Blondel et al. in [4] proved that the emptiness problem of $|Q_>|$ is decidable and the emptiness problem of $|Q_\geq|$ is undecidable. It is worth to remark that these results are in contrast with the corresponding ones for probabilistic finite automata. Indeed for this class of automata the above mentioned problems are both undecidable (see [11], Thm 6.17). It is also proven in [7] that the problems remain both undecidable for the "measure many" model of quantum automata, a model not computationally equivalent to the "measure once", introduced by Kondacs and Watrous in [8].

The result of decidability proved in [4] for quantum automata can be interpreted as saying that the emptiness problem of the intersection of a language accepted by a quantum automaton and the specific language Σ^* is decidable. In other word, it falls into the category of issues asking for the decision status of the intersection of two languages. It is known that such a problem is already undecidable at a very low level of the complexity hierarchy of recursive languages, namely for linear context-free languages to which Post Correspondence Problem can be easily reduced.

A few words on the technique used in the above paper. Observe that, with the natural meaning of the notation $|Q_\leq|$, the emptiness problem for languages $|Q_>|$ is equivalent to the inclusion

$$\Sigma^* \subseteq |Q_\leq| \tag{1}$$

Since the function $M \to \|sMP\|$ is continuous, it is sufficient to prove that for all matrices M in the topological closure of $\varphi(\Sigma^*)$ the condition $\|sMP\| \leq \lambda$

holds. The nonemptiness is clearly semidecidable. In order to prove that the emptiness is semidecidable the authors resort to two ingredients. They observe that the topological closure of the monoid of matrices $\varphi(\Sigma^*)$ is algebraic, i.e., when considering the $n \times n$-entries of a matrix M in the topological closure of $\varphi(\Sigma^*)$ as as many unknowns in the field of reals, they are precisely the zeros of a polynomial in $\mathbb{R}[x_{1,1}, \ldots, x_{n,n}]$. This allows them to express the property (1) in first-order logic of the field of reals. The second ingredient consists of applying Tarski-Seidenberg quantifier elimination and Hilbert basis results, which yields decidability.

We generalize the problem by considering families of languages $\mathcal{L}$ instead of the fixed language Σ^*. The question we tackle is thus the following

(L, Q) INTERSECTION

INPUT: a language L in a family of languages $\mathcal{L}$ and a finite quantum automaton $\mathcal{Q}$.

QUESTION: does $L \cap |\mathcal{Q}_>| = \emptyset$ hold?

Our main result shows that whenever $\mathcal{L}$ is the family of linear context-free languages or is the family of bounded semilinear languages, and whenever the automaton is rational, the problem is decidable. It can be achieved, not only because the orthogonal matrices associated with L are semialgebraic (a more general property than algebraic, which is defined by more general first-order formulas), but also because these formulas can be computed "in the limit".

We can prove the semialgebraicity of more general families of languages: arbitrary subsemigroups which is a trivial case and context-free languages which is less immediate.

A version of this paper completed with the proofs of all the results can be found in [3].

2   Preliminaries

A quantum automaton $\mathcal{Q}$ is a quadruple (s, φ, P, λ) where, as mentioned in the Introduction, $s \in \mathbb{R}^n$ is a vector of unit norm, P is a projection of $\mathbb{R}^n$, φ is a representation of the free monoid Σ^* into the group O_n of orthogonal $n \times n$-matrices in $\mathbb{R}^{n \times n}$. The behaviour of $\mathcal{Q}$ heavily depends on the topological properties of the semigroup of matrices $\varphi(\Sigma^*)$. This is why, before returning to quantum automata, we first focus our attention on these matrices for their own sake.

2.1   Topology

The following result is needed in the proof of the main theorem. Though valid under weaker conditions, it will be considered in the particular case of orthogonal matrices. Given a subset E of a finite dimensional vector space, we denote by $\mathbf{Cl}(E)$ the topological closure for the topology induced by the Euclidian norm. Given a k-tuple of matrices $(M_1, \ldots, M_k)$, denote by f the k-ary product

$f(M_1, \ldots, M_k) = M_1 \cdots M_k$ and extend the notation to subsets ρ of k-tuples of matrices by posing $f(\rho) = \{f(M_1, \ldots, M_k) \mid (M_1, \ldots, M_k) \in \rho\}$. The following result will be applied in several instances of this paper. It says that because we are dealing with compact subsets, the two operators of matrix multiplication and the topological closure commute.

Theorem 1. *Let $\mathcal{C}$ be a compact subset of matrices and let $\rho \subseteq \mathcal{C}^k$ be a k-ary relation. Then we have*

$$Cl(f(\rho)) = f(Cl(\rho))$$

Consequently, if ρ is a binary relation which is a direct product $\rho_1 \times \rho_2$, we have $Cl(\rho_1 \rho_2) = f(Cl(\rho_1 \times \rho_2))$. It is an elementary result of topology that $Cl(\rho_1 \times \rho_2) = Cl(\rho_1) \times Cl(\rho_2)$ holds. Because of $Cl(\rho_1 \rho_2) = f(Cl(\rho_1 \times \rho_2)) = f(Cl(\rho_1) \times Cl(\rho_2)) = Cl(\rho_1) \, Cl(\rho_2)$ we have

Corollary 1. *The topological closure of the product of two sets of matrices included in a compact subspace is equal to the product of the topological closures of the two sets.*

2.2 Algebraic and Semialgebraic Sets

Let us give first the definition of algebraic set over the field of real numbers (cf. [2,10]).

Definition 1. *A subset $\mathcal{A} \subseteq \mathbb{R}^n$ is algebraic (over the field of real numbers), if it satisfies one of the following equivalent conditions:*

(i) $\mathcal{A}$ is the zero set of a polynomial $p \in \mathbb{R}[x_1, \ldots, x_n]$, i.e.,

$$v \in \mathcal{A} \iff p(v) = 0. \tag{2}$$

(ii) $\mathcal{A}$ is the zero set of an arbitrary set of polynomials $\mathcal{P}$ with coefficients in $\mathbb{R}[x_1, \ldots, x_n]$, i.e., for every vector $v \in \mathbb{R}^n$,

$$v \in \mathcal{A} \iff \forall\, p \in \mathcal{P} : p(v) = 0. \tag{3}$$

The equivalence of the two statements is a consequence of Hilbert finite basis Theorem. Indeed, the theorem claims that given a family $\mathcal{P}$ there exists a finite subfamily $p_1, \ldots, p_r$ generating the same ideal which implies in particular that for all $p \in \mathcal{P}$ there exist $q_1, \ldots, q_r$ with

$$p = q_1 p_1 + \cdots + q_r p_r$$

Then $p_j(v) = 0$ for $j = 1, \ldots, r$ implies $p(v) = 0$. Now this finite set of equations can be reduced to the single equation

$$\sum_{i=1}^{n} p_j(x)^2 = 0$$

As a trivial example, a singleton $\{v\}$ is algebraic since it is the unique solution of the equation

$$\sum_{i=1}^{n}(x_i - v_i)^2 = 0$$

where v_i, with $1 \le i \le n$, is the i-th component of the vector v.

It is routine to check that the family of algebraic sets is closed under finite unions and intersections. However, it is not closed under complement and projection. The following more general class of subsets enjoys extra closure properties and is therefore more robust. The equivalence of the two definitions below is guaranteed by Tarski-Seidenberg quantifier elimination result.

Definition 2. *A subset $A \subseteq \mathbb{R}^n$ is* semialgebraic (over the field of real numbers) *if it satisfies one of the two equivalent conditions*

(i) A is the set of vectors satisfying a finite Boolean combination of predicates of the form $p(x_1, \ldots, x_n) > 0$ where $p \in \mathbb{R}[x_1, \ldots, x_n]$.

(ii) A is first-order definable in the theory of the structure whose domain are the reals and whose predicates are of the form $p(x_1, \ldots, x_n) > 0$ and $p(x_1, \ldots, x_n) = 0$ with $p \in \mathbb{R}[x_1, \ldots, x_n]$.

We now specify these definitions to square matrices.

Definition 3. *A set $A \subseteq \mathbb{R}^{n \times n}$ of matrices is algebraic, resp. semialgebraic, if considered as a set of vectors of dimension n^2, it is algebraic, resp. semialgebraic.*

We now combine the notions of zero sets and of topology. In the following two results we rephrase Theorem 3.1 of [4] by emphasizing the main features that serve our purpose (see also [4,10]). Given a subset E of a group, we denote by $\langle E \rangle$ and by E^* the subgroup and the submonoid it generates, respectively.

Theorem 2. *Let $S \subseteq \mathbb{R}^{n \times n}$ be a set of orthogonal matrices and let E be any subset of S satisfying $\langle S \rangle = \langle E \rangle$. Then we have $\mathbf{Cl}(S^*) = \mathbf{Cl}(\langle E \rangle)$. In particular $\mathbf{Cl}(S^*)$ is a group.*

The main consequence of the next theorem is that the topological closure of a monoid of orthogonal matrices is algebraic.

Theorem 3. *Let E be a set of orthogonal matrices. Then $\mathbf{Cl}(\langle E \rangle)$ is a subgroup of orthogonal matrices and it is the zero set of all polynomials $p[x_{1,1}, \ldots, x_{n,n}]$ satisfying the conditions*

$$p(I) = 0 \quad and \quad p(eX) = p(X) \quad for \ all \ e \in E$$

Furthermore, if the matrices in E have rational coefficients, the above condition may be restricted to polynomials with coefficients in $\mathbb{Q}$.

Combining the previous two theorems, we get the general result

Corollary 2. *Let $L \subseteq \Sigma^*$. Then $\mathbf{Cl}(\varphi(L)^*)$ is algebraic.*

2.3   Effectiveness Issues

We now return to the (L, Q) INTERSECTION problem as defined in the Introduction. We want to prove the implication

$$\forall X : X \in \varphi(L) \Rightarrow \|sXP\| \leq \lambda$$

We observed that due to the fact that the function $X \to \|sXP\|$ is continuous the implication is equivalent to the implication

$$\forall X : X \in \mathbf{Cl}(\varphi(L)) \Rightarrow \|sXP\| \leq \lambda$$

It just happens that under certain hypotheses, $\mathbf{Cl}(\varphi(L))$ is semialgebraic, i.e., it is defined by a first-order formula which turns the above statement into a first order formula. In the simplest examples, the closure is defined by an infinite conjunction of equations which by Hilbert finite basis result reduces to a unique equation. Thus Theorem 3 guarantees the existence of the formula but does not give an upper bound on the finite number of equations which must be tested. Therefore the following definition is instrumental for the rest of the paper. It conveys the idea that given a subset $\mathcal{A}$ of matrices there exists a sequence of formulas defining a non-increasing sequence of matrices which eventually coincide with $\mathcal{A}$. Each formula of the sequence can thus be considered as an approximation of the ultimate formula.

Definition 4. *A subset $\mathcal{A}$ of matrices is* effectively eventually definable *if there exists a constructible sequence of first-order formulas ϕ_i satisfying the conditions*

1) for all $i \geq 0$ $\phi_{i+1} \Rightarrow \phi_i$

2) for all $i \geq 0$ $\mathcal{A} \models \phi_i$

3) there exists $n \geq 0$ $\mathcal{B} \models \phi_n \Rightarrow \mathcal{B} \subseteq \mathcal{A}$

The following is a first application of the notion and illustrates the discussion before the definition.

Proposition 1. *Let $\mathcal{Q}$ be a rational quantum automaton. Let $L \subseteq \Sigma^*$ be such that the set $\mathbf{Cl}(\varphi(L))$ is effectively eventually definable. It is recursively decidable whether or not $L \cap |\mathcal{Q}_>| = \emptyset$ holds.*

We state a sufficient condition for a subset of matrices to be effectively eventually definable.

Let $S \subseteq \mathbb{R}^{n \times n}$ be a set of orthogonal matrices and let E be any subset satisfying $\langle S \rangle = \langle E \rangle$.

Proposition 2. *Let $L \subseteq \Sigma^*$ and let $E \subseteq \mathbb{Q}^{n \times n}$ be a finite subset of orthogonal matrices satisfying $\langle \varphi(L) \rangle = \langle E \rangle$. Then $\mathbf{Cl}(\varphi(L)^*)$ is effectively eventually definable.*

2.4   Closure Properties

In this paragraph we investigate some closure properties of the three different classes of matrices: algebraic, semialgebraic and effectively eventually definable, under the main usual operations as well as new operations.

We define the *sandwich* operation denoted by $\diamond$ whose first operand is a set of pairs of matrices $\mathcal{A} \subseteq \mathbb{R}^{n \times n} \times \mathbb{R}^{n \times n}$ and the second operand a set of matrices $\mathcal{B} \subseteq \mathbb{R}^{n \times n}$ by setting

$$\mathcal{A} \diamond \mathcal{B} = \{XYZ \mid (X, Z) \in \mathcal{A} \text{ and } Y \in \mathcal{B}\}$$

The next operation will be used. Given a bijection

$$\pi : \{(i,j) \mid i,j \in \{1,\ldots,n\}\} \to \{(i,j) \mid i,j \in \{1,\ldots,n\}\} \tag{4}$$

and a matrix $M \in \mathbb{R}^{n \times n}$ denote by $\pi(M)$ the matrix $\pi(M)_{i,j} = M_{\pi(i,j)}$. Extend this operation to subsets of matrices $\mathcal{A}$. Write $\pi(\mathcal{A})$ to denote the set of matrices $\pi(M)$ for all $M \in \mathcal{A}$.

The last operation is the *sum* of square matrices $M_1, \ldots, M_k$ whose result is the square block matrix

$$M_1 \oplus \cdots \oplus M_k = \begin{pmatrix} M_1 & 0 & \cdots & 0 \\ 0 & M_2 & \cdots & 0 \\ \vdots & \vdots & \vdots & \vdots \\ 0 & 0 & 0 & M_k \end{pmatrix} \tag{5}$$

These notations extend to subsets of matrices in the natural way. Here we assume that all k matrices have the same dimension $n \times n$. Observe that if the matrices are orthogonal, so is their sum. Such matrices form a subgroup of orthogonal matrices of dimension $kn \times kn$.

Logic provides an elegant way to formulate properties in the present context. Some conventions are used throughout this work. E.g., we write $\exists^n X$ when we mean that X is a vector of n bound variables. Furthermore, a vector of $n \times n$ variables can be interpreted as an $n \times n$ matrix of variables. As a consequence of Tarski-Seidenberg result, consider two semialgebraic subsets of matrices, say $\mathcal{A}_1$ and $\mathcal{A}_2$, defined by two first-order formulas $\phi_1(X_1)$ and $\phi_2(X_2)$ where X_1 and X_2 are two families of n^2 free variables viewed as two $n \times n$ matrices of variables. Then the product

$$\mathcal{A}_1 \mathcal{A}_2 = \{M_1 M_2 \mid M_1 \in \mathcal{A}_1, M_2 \in \mathcal{A}_2\}$$

is defined by the following formula where X is a family of n^2 free variables viewed as an $n \times n$ matrix

$$\exists^{n \times n} X_1 \exists^{n \times n} X_2 : X = X_1 X_2 \wedge \phi_1(X_1) \wedge \phi_2(X_2)$$

where $X = X_1 X_2$ is an abbreviation for the predicate defining X as the matrix product of X_1 and X_2. This proves that the product of two semialgebraic sets of matrices is semialgebraic. Similarly we have the following closure properties whose verification is routine.

Proposition 3. *Let $\mathcal{A}_1, \mathcal{A}_2 \subseteq \mathbb{R}^{n \times n}$ be two sets of matrices and let π be a one-to-one mapping as in (4).*

1) If $\mathcal{A}_1$ and $\mathcal{A}_2$ are algebraic so are $\mathcal{A}_1 \cup \mathcal{A}_2$ and $\pi(\mathcal{A}_1)$.

2) If $\mathcal{A}_1$ and $\mathcal{A}_2$ are semialgebraic, resp. effectively eventually definable, so are $\mathcal{A}_1 \cup \mathcal{A}_2$, $\mathcal{A}_1 \mathcal{A}_2$ and $\pi(\mathcal{A}_1)$.

Proposition 4. *Let $\mathcal{A}_1 \subseteq \mathbb{R}^{n \times n} \times \mathbb{R}^{n \times n}$ and $\mathcal{A}_2 \subseteq \mathbb{R}^{n \times n}$ be semialgebraic, resp. effectively eventually definable. Then $\mathcal{A}_1 \diamond \mathcal{A}_2$ is semialgebraic, resp. effectively eventually definable.*

Proposition 5. *Let $\mathcal{A}$ be a semialgebraic, resp. effectively eventually definable, set of $kn \times kn$ matrices of the form (5). The set*

$$\{X_1 \cdots X_k \mid X_1 \oplus \cdots \oplus X_k \in \mathcal{A}\}$$

is semialgebraic, resp. effectively eventually definable.

Proposition 6. *If $\mathcal{A}_1, \ldots, \mathcal{A}_k \subseteq \mathbb{R}^{n \times n}$ are semialgebraic, resp. effectively eventually definable sets of matrices then so is the set $\mathcal{A}_1 \oplus \cdots \oplus \mathcal{A}_k$.*

3   Context-Free languages

For the sake of self-containment and in order to fix notation, we recall the basic properties and notions concerning the family of context-free languages which can be found in all introductory textbooks on theoretical computer science (see, for instance, [6]).

A *context-free grammar* G is a quadruple $\langle V, \Sigma, P, S \rangle$ where Σ is the alphabet of *terminal symbols*, V is the set of *nonterminal symbols*, P is the set of *rules*, and S is the *axiom* of the grammar. A word over the alphabet Σ is called *terminal*. As usual, the nonterminal symbols are denoted by uppercase letters A, B, ... A typical rule of the grammar is written as $A \to \alpha$. The *derivation* relation of G is denoted by $\overset{*}{\Rightarrow}$.

A grammar is *linear* if every right hand side α contains at most one occurrence of nonterminal symbols, i.e., if it belongs to $\Sigma^* \cup \Sigma^* V \Sigma^*$.

The idea of the following notation is to consider the set of all pairs of left and right contexts in the terminal alphabet of a self-embedding nonterminal symbol. In the next definition, the initial "C" is meant to suggest the term "context" as justified by the following.

Definition 5. *With each nonterminal symbol $A \in V$ associate its terminal contexts defined as*

$$C_A = \{(\alpha, \beta) \in \Sigma^* \times \Sigma^* : A \overset{*}{\Rightarrow} \alpha A \beta\}.$$

It is convenient to define the sandwich operation also for languages in the following way. With C_A as above and L' an arbitrary language, we define

$$C_A \diamond L' = \{uwv \mid (u, v) \in C_A \text{ and } w \in L'\}$$

As the proof of the main theorem proceeds by induction on the number of nonterminal symbols, we need to show how to recombine a grammar from simpler ones obtained by choosing an arbitrary non-axiom symbol as the new axiom and by canceling all the rules involving S. This is the reason for introducing the next notation

Definition 6. *Let $G = \langle V, \Sigma, P, S \rangle$ be a context-free grammar. Set $V' = V \backslash \{S\}$.*

For every $A \in V'$, define the context-free grammar $G_A = \langle V', \Sigma, P_A, A \rangle$ where the set P_A consists of all the rules $B \to \gamma$ of G of the form

$$B \in V', \quad \gamma \in (V' \cup \Sigma)^*$$

and denote by L_A the language of all terminal words generated by the grammar G_A.

The next definition introduces the language of terminal words obtained in a derivation where S occurs at the start only.

Definition 7. *Let $L'(G)$ denote the set of all the words of Σ^* which admit a derivation*

$$S \Rightarrow \gamma_1 \Rightarrow \cdots \Rightarrow \gamma_\ell \Rightarrow w \tag{6}$$

where, for every $i = 1, \ldots, \ell$, $\gamma_i \in (V' \cup \Sigma)^*$.

The language $L'(G)$ can be easily expressed in terms of the languages L_A for all $A \in V'$. Indeed, consider the set of all rules of the grammar G of the form

$$S \to \beta, \quad \beta \in (V' \cup \Sigma)^* \tag{7}$$

Factorize every such β as

$$\beta = w_1 A_1 w_2 A_2 \cdots w_\ell A_\ell w_{j_{\ell+1}} \tag{8}$$

where $w_1, \ldots, w_{\ell+1} \in \Sigma^*$ and $A_1, A_2, \ldots A_\ell \in V'$. The following is a standard exercise.

Lemma 1. *With the notation of (8), the language $L'(G)$ is the (finite) union of the languages*

$$w_1 L_{A_1} w_2 L_{A_2} \cdots w_\ell L_{A_\ell} w_{j_{\ell+1}}$$

when β ranges over all rules (7).

Proposition 7. *With the previous notation L is a finite union of languages of the form $C_S \diamond L''$ where*

$$L'' = w_1 L_{A_1} w_2 L_{A_2} \cdots w_\ell L_{A_\ell} w_{\ell+1}$$

Proof. In order to prove the inclusion of the right- into left- hand side, it suffices to consider $w = \alpha u \beta$, with $u \in L'(G)$ and $(\alpha, \beta) \in C_S$. One has $S \overset{*}{\Rightarrow} u$ and $S \overset{*}{\Rightarrow} \alpha S \beta$ and thus $S \overset{*}{\Rightarrow} \alpha S \beta \overset{*}{\Rightarrow} \alpha u \beta$.

Let us prove the opposite inclusion. A word $w \in L$ admits a derivation $S \overset{*}{\Rightarrow} w$. If the symbol S does not occur in the derivation except at the start of the derivation, then $w \in L'(G)$. Otherwise factor this derivation into $S \overset{*}{\Rightarrow} \alpha S \beta \overset{*}{\Rightarrow} w$ such that S does not occur in the second part of the derivation except in the sentential form $\alpha S \beta$. Reorder the derivation $\alpha S \beta \overset{*}{\Rightarrow} w$ into $\alpha S \beta \overset{*}{\Rightarrow} \gamma S \delta \overset{*}{\Rightarrow} w$ so that $\gamma, \delta \in \Sigma^*$. This implies $w = \gamma u \delta$ for some word $u \in L'(G)$, completing the proof.

4 The Main Results

Here we prove that the problem is decidable for two families of languages, namely the linear context-free languages and the linear bounded languages.

4.1 The Bounded Semilinear Languages

We solve the easier case. We recall that a *bounded semilinear* language is a finite union of *linear* languages which are languages of the form

$$L = \{w_1^{n_1} \cdots w_k^{n_k} \mid (n_1, \ldots, n_k) \in R\} \tag{9}$$

for some fixed words $w_i \in \Sigma^*$ for $i = 1, \ldots, k$ and $R \subseteq \mathbb{N}^k$ is a linear set, i.e., there exists $v_0, v_1, \ldots, v_p \in \mathbb{N}^k$ such that

$$R = \{v_0 + \lambda_1 v_1 + \cdots + \lambda_p v_p \mid \lambda_1, \ldots, \lambda_p \in \mathbb{N}\}$$

Proposition 8. *If L is bounded semilinear then its closure $Cl(\varphi(L))$ is semialgebraic. Furthermore, if the quantum automaton Q is rational, the (L, Q) intersection is decidable.*

4.2 The Case of Context-Free Languages

Here we show that $Cl(\varphi(L))$ is effectively eventually definable for languages generated by linear grammars and rational quantum automata.

We adopt the notation from Section 3 for context-free grammars. We recall the following notion that will be used in the proof of the next result (see [12]). A subset of a monoid M is *regular* if it is recognized by some finite M-automaton which differs from an ordinary finite nondeterministic automaton over the free monoid by the fact the transitions are labeled by elements in M.

Proposition 9. *If L is generated by a context-free grammar, then $Cl(\varphi(L))$ is semialgebraic. Furthermore, if the grammar is linear and if the quantum automaton is rational then $Cl(\varphi(L))$ is effectively eventually definable and the (L, Q) intersection is decidable.*

Proof. With the notation of Section 3 the language L is a finite union of languages of the form $C_S \diamond L''$ with

$$L'' = w_1 L_{A_1} w_2 L_{A_2} \cdots w_\ell L_{A_\ell} w_{\ell+1} \tag{10}$$

where, for every $1 \leq i \leq \ell + 1$, $w_i \in \Sigma^*$ and $A_i \in V'$. It suffices to show by induction on the number of nonterminal symbols that, with the previous notation, the subsets

$$\mathbf{Cl}(\varphi(C_S \diamond L'')) \tag{11}$$

are semialgebraic in all cases and effectively eventually definable when the quantum automaton is rational and the grammar of the language is linear. As a preliminary remark let us show this property for $\mathbf{Cl}(\varphi(C_S))$. Define $\varphi^T : \Sigma^* \to \mathbb{R}^{n \times n}$ as $\varphi^T(u) = \varphi(u)^T$ and set

$$M = \{\varphi(a) \oplus \varphi^T(b) \mid (a, b) \in C_S\}.$$

Observe that M is a monoid since if $\varphi(a) \oplus \varphi^T(b)$ and $\varphi(c) \oplus \varphi^T(d)$ are in M then we have

$$\varphi^T(b)\varphi^T(d) = \varphi(b)^T \varphi(d)^T = (\varphi(d)\varphi(b))^T = \varphi(db)^T = \varphi^T(db)$$

which yields

$$(\varphi(a) \oplus \varphi^T(b))(\varphi(c) \oplus \varphi^T(d)) = \varphi(ac) \oplus \varphi^T(db).$$

As a first consequence, by Corollary 2, $\mathbf{Cl}(M)$ is algebraic. Furthermore we can show that $\mathbf{Cl}(M)$ is effectively eventually definable. Indeed M is a regular submonoid of the group of orthogonal matrices $O_n \oplus O_n$ if the grammar is linear. Precisely, it is recognized by the finite O_{2n}-automaton whose states are the nonterminal symbols, the transitions are of the form $A \xrightarrow{\varphi(a) \oplus \varphi^T(b)} B$ where $A \to aBb$ is a rule of the grammar and where the initial and final states coincide with S. Now, the subgroup generated by a regular subset of a monoid has an effective finite generating set [1] (see also [12]) and thus by Proposition 2 $\mathbf{Cl}(M)$ is effectively eventually definable if $\varphi(\Sigma^*) \subseteq \mathbb{Q}^{n \times n}$.

We now proceed with the proof by induction on the number of nonterminal symbols. If the set of nonterminal symbols is reduced to S then L is reduced to $C_S \diamond L'(G)$ and $L'(G)$ is finite. We may further assume that there is a unique terminal rule $S \to w$. By Theorem 1 we have

$$\mathbf{Cl}(\varphi(L)) = \{X\varphi(w)Y^T \mid X \oplus Y \oplus \{\varphi(w)\} \in \mathbf{Cl}(M \oplus \varphi(w))\}$$

By Corollary 1 we have

$$\mathbf{Cl}(M \oplus \varphi(w)) = \mathbf{Cl}(M) \oplus \mathbf{Cl}(\varphi(w)) = \mathbf{Cl}(M) \oplus \varphi(w)$$

which, by Proposition 6, is semialgebraic, resp. effectively eventually definable. In that latter case the (L, Q) intersection is decidable.

Now assume V contains more than one nonterminal symbol. We first prove that for each nonterminal symbol A, $\mathbf{Cl}(\varphi(C_S \diamond L_A))$ is semialgebraic in the general case and effectively eventually definable when the grammar is linear and the quantum automaton is rational. By Theorem 1 and Corollary 1, $\mathbf{Cl}(\varphi(C_S \diamond L''))$ is the subset

$$\mathbf{Cl}(\varphi(C_S \diamond L'')) = \{XZY^T \mid X \oplus Y \oplus Z \in \mathbf{Cl}(M) \oplus \mathbf{Cl}(\varphi(L''))\}$$

with L'' as in (10), i.e.,

$$\{XZY^T \mid X \oplus Y \oplus Z \in \mathbf{Cl}(M) \oplus \mathbf{Cl}(\varphi(w_1)\varphi(L_{A_1}) \cdots \varphi(w_\ell)\varphi(L_{A_\ell})\varphi(w_{\ell+1}))\}$$

By Cororally 1 we have

$$\mathbf{Cl}(\varphi(w_1)\varphi(L_{A_1}) \cdots \varphi(w_\ell)\varphi(L_{A_\ell})\varphi(w_{\ell+1}))$$
$$= \varphi(w_1)\mathbf{Cl}(\varphi(L_{A_1})) \cdots \varphi(w_\ell)\mathbf{Cl}(\varphi(L_{A_\ell}))\varphi(w_{\ell+1})$$

which shows, via Proposition 3 and by induction hypothesis that this subset is semialgebraic, resp. effectively, eventually definable. Then its direct sum with $\mathbf{Cl}(M)$ is semialgebraic and effectively, eventually definable if the grammar is linear and the quantum automaton is rational. We conclude by applying Proposition 5.

References

1. Anisimov, A.V., Seifert, F.D.: Zur algebraischen Charakteristik der durch Kontext-freie Sprachen definierten Gruppen. Elektron. Inform. Verarb. u. Kybern. 11, 695–702 (1975)
2. Basu, S., Pollack, R., Roy, M.-F.: Algorithms in Real Algebraic Geometry. Springer, Berlin (2003)
3. Bertoni, A., Choffrut, C., D'Alessandro, F.: Quantum finite automata and linear context-free languages, Preprint arXiv:1303.2967 (2013)
4. Blondel, V.D., Jeandel, E., Koiran, P., Portier, N.: Decidable and Undecidable Problems about Quantum Automata. SIAM J. Comput. 34, 1464–1473 (2005)
5. Derksen, H., Jeandel, E., Koiran, P.: Quantum automata and algebraic groups. J. Symb. Comput. 39, 357–371 (2005)
6. Hopcroft, J., Ullman, J.D.: Introduction to Automata Theory, Languages and Computation. Addison-Wesley (1979)
7. Jeandel, E.: Indécidabilité sur les automates quantiques. Master's thesis. ENS Lyon (2002)
8. Kondacs, A., Watrous, J.: On the power of quantum finite state automata. In: Proceedings of the 38th Annual Symposium on Foundations of Computer Science, pp. 66–75 (1997)
9. Moore, C., Crutchfield, J.: Quantum automata and quantum grammars. Theoret. Comput. Sci. 237, 275–306 (2000)
10. Onishchik, A., Vinberg, E.: Lie Groups and Algebraic Groups. Springer, Berlin (1990)
11. Paz, A.: Introduction to Probabilistic Automata. Academic Press, New York (1971)
12. Sakarovitch, J.: Elements of Automata Theory. Cambridge University Press, Cambridge (2009)

On the Asymptotic Abelian Complexity
of Morphic Words*

Francine Blanchet-Sadri[1] and Nathan Fox[2]

[1] Department of Computer Science, University of North Carolina,
P.O. Box 26170, Greensboro, NC 27402–6170, USA
blanchet@uncg.edu
[2] Department of Mathematics, Rutgers University,
Hill Center for the Mathematical Sciences,
110 Frelinghuysen Rd., Piscataway, NJ 08854–8019, USA
fox@math.rutgers.edu

Abstract. The subword complexity of an infinite word counts the number of subwords of a given length, while the abelian complexity counts this number up to letter permutation. Although a lot of research has been done on the subword complexity of morphic words, i.e., words obtained as fixed points of iterated morphisms, little is known on their abelian complexity. In this paper, we undertake the classification of the asymptotic growths of the abelian complexities of fixed points of binary morphisms. Some general results we obtain stem from the concept of factorization of morphisms. We give an algorithm that yields all canonical factorizations of a given morphism, describe how to use it to check quickly whether a binary morphism is Sturmian, discuss how to fully factorize the Parry morphisms, and finally derive a complete classification of the abelian complexities of fixed points of uniform binary morphisms.

1 Introduction

The *subword complexity* of an infinite word w, denoted ρ_w, is the function mapping each positive integer n to the number of distinct subwords of w of length n. On the other hand, the *abelian complexity* of w, denoted ρ_w^{ab}, is the function mapping each positive integer n to the number of distinct Parikh vectors of subwords of w of length n. Here, we assume the standard alphabet $A_k = \{0, \ldots, k-1\}$, and the *Parikh vector* of a finite word over A_k is the vector whose ith entry is the number of occurrences of letter $i-1$ in the word.

An infinite word is a *morphic word* if it is the fixed point of some morphism at some letter. For compactness of notation, we frequently denote a morphism φ over A_k, $\varphi : A_k^* \to A_k^*$, as an ordered k-tuple $\varphi = (w_0, \ldots, w_{k-1})$, where

---

* This material is based upon work supported by the National Science Foundation under Grant No. DMS–1060775. A World Wide Web server interface has been established at www.uncg.edu/cmp/research/abeliancomplexity2 for automated use of the programs.

M.-P. Béal and O. Carton (Eds.): DLT 2013, LNCS 7907, pp. 94–105, 2013.
© Springer-Verlag Berlin Heidelberg 2013

$\varphi(a) = w_a$ for each $a \in A_k$. The *fixed point of φ at a*, denoted $\varphi^\omega(a)$, is the limit as $n \to \infty$ of $\varphi^n(a)$. The fixed point exists precisely when the limit exists.

A lot of research has been done on the subword complexity of morphic words, e.g., Ehrenfeucht and Rozenberg [9] showed that the fixed points of *uniform morphisms*, i.e., morphisms φ over A_k satisfying $|\varphi(a)| = |\varphi(b)|$ for all $a, b \in A_k$, have at most linear subword complexity, Berstel and Séébold [3] gave a characterization of *Sturmian morphisms*, i.e., morphisms φ over the binary alphabet $A_2 = \{0, 1\}$ such that $\varphi^\omega(0)$ exists and is Sturmian, in other words, $\rho_{\varphi^\omega(0)}(n) = n+1$ for all n, and Frid [11] obtained a formula for the subword complexity of the fixed points of binary uniform morphisms. On the other hand, abelian complexity is a relatively new research topic. Balková, Břinda, and Turek [2] studied the abelian complexity of infinite words associated with quadratic Parry numbers, Currie and Rampersad [6] studied recurrent words with constant abelian complexity, and Richomme, Saari, and Zamboni [14] investigated abelian complexity of minimal subshifts.

In this paper, we are interested in classifying the asymptotic growths of the abelian complexities of words obtained as fixed points of iterated morphisms over A_2 at 0. This classification has already been done for the subword complexity of morphisms over A_k [7–10, 12] (see also [4, Section 9.2]). Pansiot's classification of the asymptotic growths of the subword complexity of morphic words not only depends on the type of morphisms but also on the distribution of so-called bounded letters [12]. We assume without loss of generality that the first letter in the image of 0 is 0 and we assume that all of our morphisms are nonerasing. Also for conciseness, we frequently use the term "abelian complexity of a morphism" (when referring to its fixed point at 0).

As mentioned above, we are mainly concerned with the asymptotic behaviors of the abelian complexities rather than their specific values. Some general results we obtain stem from the concept of *factorization of morphisms*. We mainly examine the monoid of binary morphisms under composition, but we also consider factorization in a more general setting.

The binary morphism types whose asymptotic abelian complexities we classify are the following: morphisms with ultimately periodic fixed points (Proposition 1(10)), Sturmian morphisms (Proposition 2(2)), morphisms with equal ratios of zeroes to ones in both images (Theorem 3), Parry morphisms, i.e., morphisms of the type $(0^p 1, 0^q)$ with $p \geq q \geq 1$ or of the type $(0^p 1, 0^q 1)$ with $p > q \geq 1$ studied in [2] and cyclic shifts of their factorizations (Proposition 2(4) along with Corollaries 2 and 3), most morphisms where the image of 1 contains only ones (Theorem 4), and most uniform morphisms and cyclic shifts of their factorizations (Theorem 6).

All of the asymptotic abelian complexity classes we obtain, where $f(n) = 1$ if $\log_2 n \in \mathbb{Z}$, and $f(n) = \log n$ otherwise, are the following (listed in increasing order of growth): $\Theta(1)$, e.g., $(01, 10)$, $\tilde{\Theta}(f(n))$, e.g., $(01, 00)$, $\Theta(\log n)$, e.g., $(001, 110)$, $\Theta(n^{\log_a b})$ for $a > b > 1$, e.g., $(0001, 0111)$, $\Theta\left(\frac{n}{\log n}\right)$, e.g., $(001, 11)$, and $\Theta(n)$, e.g., $(0001, 11)$.

The contents of our paper are as follows. In Section 2, we discuss some preliminaries and simple results. In Section 3, we study morphism factorizations. We give an algorithm that yields all canonical factorizations of a morphism φ over A_k into two morphisms each over an alphabet of at most k letters and we describe how to use it for checking quickly whether a binary morphism is Sturmian. In Section 4, we obtain our main results. Among other things, we show how to fully factorize the Parry morphisms and we also derive a complete classification of the abelian complexities of fixed points of uniform binary morphisms. Finally in Section 5, we conclude with suggestions for future work. Some proofs have been omitted and others have only been outlined due to the 12-page space constraint.

2 Preliminary Definitions and Results

Given $C \geq 0$, a (finite or infinite) word w over A_k is called C-balanced if for all letters a in w and for all integers $0 < n \leq |w|$ (or just $0 < n$ if w is infinite), the difference between the maximum and the minimum possible counts of letter a in a length-n subword of w is less than or equal to C.

Given an infinite word w over A_2, $z_M(n)$ (resp., $z_m(n)$) denotes the maximum (resp., minimum) number of zeroes in a length-n subword of w. For ease of notation, $z(v)$ denotes the number of zeroes in a binary word v (as opposed to the standard $|v|_0$). The number of ones in v is then $|v| - z(v)$.

Here are some facts about abelian complexity and zero counts.

Proposition 1. *If w is an infinite word over A_k, then the following hold:*

1. $\rho_w^{ab}(n) = \Theta(1)$ *if and only if w is C-balanced for some C [14, Lemma 2.2];*
2. *If w is Sturmian, then $\rho_w^{ab}(n) = 2$ for all n [5];*
3. $\rho_w^{ab}(n) = O(n^{k-1})$ *[14, Theorem 2.4];*
4. $\rho_w^{ab}(n) \leq \rho_w(n)$;
5. *If $k = 2$, then $\rho_w^{ab}(n) = z_M(n) - z_m(n) + 1$;*
6. *If $k = 2$, then $z_M(m + n) \leq z_M(m) + z_M(n)$;*
7. *If $k = 2$, then $z_m(m + n) \geq z_m(m) + z_m(n)$;*
8. *If $k = 2$, then $z_M(n + 1) - z_M(n) \in \{0, 1\}$ and $z_m(n + 1) - z_m(n) \in \{0, 1\}$;*
9. *If $k = 2$, then $\left| \rho_w^{ab}(n + 1) - \rho_w^{ab}(n) \right| \leq 1$ for all positive integers n;*
10. *If w is ultimately periodic, then $\rho_w^{ab}(n) = \Theta(1)$.*

Here are some morphisms that are classified based on prior results. For 3, any such word is ultimately periodic.

Proposition 2. *The fixed points at 0 of the following morphisms over A_2 have $\Theta(1)$ abelian complexity:*

1. *The Thue-Morse morphism $(01, 10)$ [14, Theorem 3.1];*
2. *Any Sturmian morphism (this includes $(01, 0)$) [14, Theorem 1.2];*
3. *Any morphism whose fixed point contains finitely many zeroes or ones (this includes $(01, 11)$);*

4. *Any morphism of the form* $(0^p 1, 0^q)$ *with* $p \geq q \geq 1$ *or of the form* $(0^p 1, 0^q 1)$ *with* $p > q \geq 1$ *[2, Corollary 3.1].*

Let $f(x)$ be a real function, and define

$$f_m(x) := \inf_{a \geq x} f(a), f_M(x) := \sup_{a \leq x} f(a).$$

Now, let $g(x)$ be also a real function. We write $f(x) = \tilde{\Omega}(g(x))$ if $f_m(x) = \Omega(g_m(x))$ and $f_M(x) = \Omega(g_M(x))$, $f(x) = \tilde{O}(g(x))$ if $f_m(x) = O(g_m(x))$ and $f_M(x) = O(g_M(x))$, and $f(x) = \tilde{\Theta}(g(x))$ if $f_m(x) = \Theta(g_m(x))$ and $f_M(x) = \Theta(g_M(x))$.

3   Morphism Factorizations

Some more general results we obtain stem from the concept of factorization of morphisms. We mainly examine the monoid of binary morphisms under composition, but we also consider factorization in a more general setting.

Let φ be a morphism over A_k. If φ cannot be written as $\phi \circ \zeta$ for two morphisms $\phi : A_{k'}^* \to A_k^*$, $\zeta : A_k^* \to A_{k'}^*$ where neither is a permutation of $A_{k'}$, then φ is *irreducible* over $A_{k'}$. Otherwise, φ is *reducible*. (We use the convention that a permutation of the alphabet is not irreducible.) If $\varphi = \phi \circ \zeta$ for some morphisms $\phi : A_{k'}^* \to A_k^*$, $\zeta : A_k^* \to A_{k'}^*$, we say that φ *factors as* $\phi \circ \zeta$ over $A_{k'}$.

Here are two propositions that together lead to a factorization algorithm.

Proposition 3. *Let* $\varphi = (w_0, w_1, \ldots, w_{k-1})$ *be a morphism over* A_k. *If there exist* $v_0, v_1, \ldots, v_{k'-1} \in A_k^+$ *such that* $w_0, w_1, \ldots, w_k \in \{v_0, v_1, \ldots, v_{k'-1}\}^*$, *then there exists a morphism* $\zeta : A_k^* \to A_{k'}^*$ *such that* $\varphi = \phi \circ \zeta$, *where* $\phi = (v_0, v_1, \ldots, v_{k'-1})$. *Conversely every factorization* $\varphi = \phi \circ \zeta$, *where* $\phi : A_{k'}^* \to A_k^*$ *and* $\zeta : A_k^* \to A_{k'}^*$, *corresponds to* $v_i = \phi(i)$ *for* $i = 0, 1, \ldots, k'-1$, *where* $w_0, w_1, \ldots, w_{k-1} \in \{v_0, v_1, \ldots, v_{k'-1}\}^*$.

Proposition 4. *Let* σ *be a function that permutes elements of a* k-*tuple and let* ψ_σ *be the morphism corresponding to the* k-*tuple obtained by applying* σ *to* $(0, 1, \ldots, k-1)$. *Let* φ *be a morphism over* A_k *that factors as* $\varphi = \phi \circ \zeta$ *for some morphisms* ϕ *and* ζ. *Then,* $\varphi = \sigma(\phi) \circ \psi_\sigma(\zeta)$.

Proposition 4 allows us to define the notion of a *canonical factorization*. Let φ be a morphism over A_k that factors as $\varphi = \phi \circ \zeta$ for some morphisms ϕ and ζ. Let $v = \zeta(0)\zeta(1)\cdots\zeta(k-1)$. We say that the factorization $\varphi = \phi \circ \zeta$ is *canonical* if $v[0] = 0$ and the first occurrence of each letter a, $a \neq 0$, in v is after the first occurrence of letter $a - 1$ in v. It is clear from Proposition 4 that given a factorization $\phi \circ \zeta$ of φ we can put it in canonical form by applying some permutation σ to the letters in the images in ζ and to the order of the images in ϕ. Hence, every factorization of φ corresponds to one in canonical form.

Before we give our factorization algorithm, here is an important note: the monoid of binary morphisms does not permit unique factorization into irreducible morphisms, even if the factorizations are canonical. Indeed, letting

$\varphi = (00, 11)$, we have $\varphi = (0, 11) \circ (00, 1) = (00, 1) \circ (0, 11)$. These are distinct canonical factorizations of φ into irreducibles.

We now give an algorithm, Algorithm 1, that yields all canonical factorizations over $A_{k'}$ of a given morphism φ into two morphisms. The basis of this algorithm is the subroutine it calls, Algorithm 2, which creates a factorization by recursively finding $v_0, v_1, \ldots, v_{k'-1}$, as specified in Proposition 3, and then backtracking to find more factorizations. It always attempts to match or create a v_i at the beginning of the shortest image in the morphism, as that has the fewest possibilities to consider. Algorithm 1 works as follows:

- Call Algorithm 2 with φ, k', and an empty list. Given morphism φ', integer k'', and a list of words $v_0, \ldots, v_m$, Algorithm 2 does the following:
  1. If φ' has no letters, return $\{(v_0, \ldots, v_m) \circ \varphi'\}$;
  2. If $k'' > 0$, try each prefix of a minimal-length image in φ' as v_{m+1}. Call this same subroutine each time with $k'' - 1$, pruning that prefix off that image. Consolidate the results of the recursive call and add to the set of factorization pairs with appropriate right factor;
  3. Try matching each v_i to a prefix of a minimal-length image in φ'. If there is a match, call this same subroutine with k'', pruning that prefix off that image. Consolidate the results of the recursive call and add to the set of factorization pairs with appropriate right factor;
  4. Return the set of factorization pairs.
- Put the resulting factorizations into canonical form.

Theorem 1. *Algorithm 1 can be applied recursively (and optionally along with a lookup table to avoid recomputing things) to obtain complete (canonical) factorizations of a given morphism into irreducible morphisms.*

Proof. Algorithm 1's correctness follows from Propositions 3 and 4. To obtain all factorizations (not just canonical ones), run Algorithm 1 and then apply all possible permutations to the resulting factorizations.   □

Given as input $\varphi = (01, 00)$ and $k' = 2$, Algorithm 1 outputs the canonical factorizations $(0, 1) \circ (01, 00)$, $(01, 0) \circ (0, 11)$, and $(01, 00) \circ (0, 1)$:

φ'	k''	v_0	v_1	φ'	k''	v_0	v_1	φ'	k''	v_0	v_1
$(01, 00)$	2			$(01, 00)$	2			$(01, 00)$	2		
$(1, 00)$	1	0		$(\varepsilon, 00)$	1	01		$(\varepsilon, 00)$	1	01	
$(\varepsilon, 00)$	0	0	1	$(\varepsilon, 0)$	1	01	0	$(\varepsilon, \varepsilon)$	1	01	00
$(\varepsilon, 0)$	0	0	1	$(\varepsilon, \varepsilon)$	0	01	0				
$(\varepsilon, \varepsilon)$	0	0	1								

We conclude this section with a discussion on checking whether a binary morphism φ is Sturmian. Berstel and Séébold in [3] prove that φ is Sturmian if and only if $\varphi(10010010100101)$ is 1-balanced and primitive (not a power of a shorter word). This leads to an algorithm for deciding whether a given morphism is Sturmian. While the resulting algorithm is typically fast to give a negative answer, a positive answer requires computing (essentially) $|\varphi\,(10010010100101)|$ balance values and checking for primitivity. (Also, a check that our morphism's fixed point does not contain finitely many zeroes is needed.)

Richomme in [13] gives a note that leads to an alternative approach. The note says that a binary morphism is Sturmian if and only if it can be written as compositions of the morphisms $(1, 0)$, $(0, 01)$, $(10, 1)$, $(0, 10)$, and $(01, 1)$. No canonical factorization of a Sturmian morphism ever contain $(1, 0)$ or $(10, 1)$, but these two can be combined to form $(01, 0)$, which we must add to our list. Hence, we have the criterion that a binary morphism is Sturmian if and only if it has a canonical factorization that is a composite of $(0, 01)$, $(01, 0)$, $(0, 10)$, and $(01, 1)$. (We also disallow composites that are powers of the morphisms with ultimately periodic or finite fixed points so we can keep our fixed points aperiodic.) The task of factoring a Sturmian morphism is well suited to repeated application of Algorithm 1. In fact, we can speed up the algorithm specifically in this case by pre-seeding v_0 and v_1 with each of the possible factors for a Sturmian morphism (as in, directly calling Algorithm 2 with φ, 2, and $[v_0, v_1]$, where (v_0, v_1) is equal to each of the four possible morphisms). This algorithm is fast in both cases where the given morphism is or is not Sturmian.

4   Main Results

We have the following theorem which we can prove using the following lemma, commonly known as Fekete's Lemma.

Lemma 1. *Let* $\{a_n\}_{n\geq 1}$ *be a sequence such that* $a_{m+n} \geq a_m + a_n$ *(resp.,* $a_{m+n} \leq a_m + a_n$). *Then,* $\lim_{n\to\infty} \frac{a_n}{n}$ *exists and equals* $\sup \frac{a_n}{n}$ *(resp.,* $\inf \frac{a_n}{n}$).

Theorem 2. *Let* w *be an infinite binary word and* ψ *be a binary morphism. Then,* $\rho^{ab}_{\psi(w)}(n) = \tilde{O}\left(\rho^{ab}_w(n)\right)$.

This leads to the following corollary.

Corollary 1. *Let* ϕ *and* ψ *be binary morphisms. Then,*

$$\rho^{ab}_{(\psi\circ\phi)^\omega(0)}(n) = \tilde{\Theta}\left(\rho^{ab}_{(\phi\circ\psi)^\omega(0)}(n)\right).$$

The following result is a generalization of one direction of [14, Theorem 3.3]. Note that it holds for alphabets of any size.

Theorem 3. *Let* ψ *be a morphism over* A_k *such that there exist positive integers* $n_0, n_1, \ldots, n_{k-1}$ *such that for all* $a, b \in A_k$, $\psi(a)^{n_a}$ *and* $\psi(b)^{n_b}$ *are abelian equivalent (have the same Parikh vector). Then, for any infinite word* w *over* A_k, $\rho^{ab}_{\psi(w)}(n) = \Theta(1)$. *In particular, if the fixed point of* ψ *at* 0 *exists,* $\rho^{ab}_{\psi^\omega(0)}(n) = \Theta(1)$.

The following criterion allows the classification of more morphisms.

Theorem 4. *Let φ be a binary morphism such that $\varphi(1) = 1^m$ for some $m \geq 1$. Let c be the number of zeroes in $\varphi(0)$. Assume that $c + m > 2$ (so that the fixed point at 0 can exist), and, if $m = 1$, then assume $\varphi(0)$ ends in 1. Then, one of the following cases holds: $\rho^{ab}_{\varphi^\omega(0)}(n) = \Theta(n)$ if $c > m$, $\rho^{ab}_{\varphi^\omega(0)}(n) = \Theta\left(\frac{n}{\log n}\right)$ if $c = m$, $\rho^{ab}_{\varphi^\omega(0)}(n) = \Theta\left(n^{\log_m c}\right)$ if $1 < c < m$, and $\rho^{ab}_{\varphi^\omega(0)}(n) = \Theta(1)$ if $c = 1$.*

Proof. First, the case where $\Theta(1)$ if $c = 1$ follows from the fact that $\varphi^\omega(0) = 01^\omega$. Next, all other cases use the fact that $\rho^{ab}_{\varphi^\omega(0)}$ is monotone increasing ($\varphi^\omega(0)$ contains arbitrarily long blocks of ones). Finally, in each case, we consider limits of ratios of the maximal number of zeroes in a subword and the target complexity and can show that they exist and are between 0 and ∞.   □

4.1   Factorization of Parry Morphisms

When we say Parry morphisms, we mean those studied in [2] that we stated have fixed points with bounded abelian complexity in Proposition 2(4). We describe all canonical factorizations of such morphisms, which allow us to construct additional morphisms with bounded abelian complexity, due to Corollary 1.

The following theorem states how to fully factor the two types of Parry morphisms.

Theorem 5.
 – *If $\varphi = (0^p 1, 0^q 1)$ with $1 \leq q < p$, then all factorizations of φ are of the form $(\prod_{i=1}^m \phi_i) \circ (01, 1)$, where $\phi_i = (0^{p_i}, 1)$ for some prime p_i or $\phi_i = (0, 01)$.*
 – *If $\varphi = (0^p 1, 0^q)$ with $1 \leq q \leq p$, then for all choices of a nonnegative odd integer N, of a sequence of nonnegative integers $a_0, a_1, \ldots, a_N$ with all but possibly the last positive, and of integers q_0, q' with $q_0 \geq 0$ and $q' > 0$ where*

$$\prod_{i=0}^{\frac{N-1}{2}} a_{2i} + q_0 + \sum_{i=0}^{\frac{N-1}{2}} \left(a_{2i+1} \prod_{j=0}^{i} a_{2j} \right) = p, \qquad q' \prod_{i=0}^{\frac{N-1}{2}} a_{2i} = q,$$

there exists a complete canonical factorization:

$$\varphi = (0, 01)^{q_0} \circ \left(\prod_{j=0}^{\frac{N-1}{2}} \left(\left(\prod_{i=1}^{m_{2j}} (0^{p_{i,2j}}, 1) \right) \circ (0, 01)^{a_{2j+1}} \right) \right)$$

$$\circ (01, 0) \circ \left(\prod_{i=1}^{m'} \left(0, 1^{q'_i} \right) \right),$$

where each of the m_{2j}'s is a positive integer, all of the $p_{i,2j}$'s are prime, $\prod_{i=1}^{m_{2j}} p_{i,2j} = a_{2j}$, all of the q'_i's are prime, and $\prod_{i=1}^{m'} q'_i = q'$.
In both cases, any composites of the necessary forms yield a Parry morphism of the proper type (where for the complicated case, all we require is that the complicated p value exceed the complicated q value).

Proof. To prove this result, we need to show how to completely canonically factor various types of morphisms:

1. Every complete canonical factorization of the morphism $(0, 1^q)$ has the form $(0, 1^{p_1}) \circ (0, 1^{p_2}) \circ \cdots \circ (0, 1^{p_m})$ for $p_1, \ldots, p_m$ primes such that $\prod_{i=1}^{m} p_i = q$. Also, if $\prod_{i=1}^{m} p_i = q$, then $(0, 1^{p_1}) \circ (0, 1^{p_2}) \circ \cdots \circ (0, 1^{p_m}) = (0, 1^q)$.

2. Every complete canonical factorization of $(0^p, 1)$ has the form $(0^{p_1}, 1) \circ (0^{p_2}, 1) \circ \cdots \circ (0^{p_m}, 1)$ for $p_1, \ldots, p_m$ primes such that $\prod_{i=1}^{m} p_i = p$, and if $\prod_{i=1}^{m} p_i = p$, then $(0^{p_1}, 1) \circ (0^{p_2}, 1) \circ \cdots \circ (0^{p_m}, 1) = (0^p, 1)$.

3. Every complete canonical factorization of $(0^p, 0^q 1)$ has the form $\prod_{i=1}^{m} \phi_i$, where $\phi_i = (0^{p_i}, 1)$ for some prime p_i, or $\phi_i = (0, 01)$. Also, any composite of the form $\prod_{i=1}^{m} \phi_i$, where $\phi_i = (0^{p_i}, 1)$ for some prime p_i or $\phi_i = (0, 01)$, yields a morphism of the form $(0^p, 0^q 1)$.

4. Every complete canonical factorization of $(0^p 1, 1)$ has the form $(0^{p_1}, 1) \circ (0^{p_2}, 1) \circ \cdots \circ (0^{p_m}, 1) \circ (01, 1)$ for $p_1, \ldots, p_m$ primes such that $\prod_{i=1}^{m} p_i = p$, and if $\prod_{i=1}^{m} p_i = p$, then $(0^{p_1}, 1) \circ (0^{p_2}, 1) \circ \cdots \circ (0^{p_m}, 1) \circ (01, 1) = (0^p 1, 1)$.

5. Every complete canonical factorization of $(0^p 1, 0^q)$ has the form

$$\left(\prod_{i=1}^{m} \phi_i \right) \circ (01, 0) \circ \left(\prod_{j=1}^{m'} (0, 1^{q_j}) \right),$$

where $\phi_i = (0^{p_i}, 1)$ for some prime p_i or $\phi_i = (0, 01)$, and each of the q_j's is prime (we allow the second product to be empty, in which case $m' = 0$). Also, any composite of the form

$$\left(\prod_{i=1}^{m} \phi_i \right) \circ (01, 0) \circ \left(\prod_{j=1}^{m'} (0, 1^{q_j}) \right),$$

where $\phi_i = (0^{p_i}, 1)$ for some prime p_i or $\phi_i = (0, 01)$, and each of the q_j's is prime (and $m' = 0$ is allowed), yields a morphism of the form $(0^p 1, 0^q)$.

6. Every complete canonical factorization of $(0^p 1, 0^q 1)$ with $p > q$ has the form $(\prod_{i=1}^{m} \phi_i) \circ (01, 1)$, where $\phi_i = (0^{p_i}, 1)$ for some prime p_i or $\phi_i = (0, 01)$. Also, any composite of the form $(\prod_{i=1}^{m} \phi_i) \circ (01, 1)$, where $\phi_i = (0^{p_i}, 1)$ for some prime p_i or $\phi_i = (0, 01)$ yields a morphism of the form $(0^p 1, 0^q 1)$ with $p > q$.

The result for the first type of Parry morphism follows directly from item 6. We now prove the result for the second type of Parry morphism. By item 5, all complete canonical factorizations of $\varphi = (0^p 1, 0^q)$ are of the form

$$\varphi = \left(\prod_{i=1}^{m} \phi_i \right) \circ (01, 0) \circ \left(\prod_{j=1}^{m'} (0, 1^{q_j}) \right),$$

where $\phi_i = (0^{p_i}, 1)$ for some prime p_i or $\phi_i = (0, 01)$, and each of the q_j's is prime. We can use item 2, item 1, and the fact that $(0, 01)^m = (0, 0^m 1)$ to assert

$$\varphi = (0, 0^{q_0} 1) \circ \left(\prod_{j=0}^{\frac{N-1}{2}} ((0^{a_{2j}}, 1) \circ (0, 0^{a_{2j+1}} 1)) \right) \circ (01, 0) \circ \left(0, 1^{q'} \right), \qquad (1)$$

for some odd $N \geq 0$, some $q_0 \geq 0$, some $q' > 0$, and some sequence $a_1, a_2, \ldots, a_N$ all positive except for possibly the last, which is nonnegative (all of these are integers). This factorization is (probably) not complete. Item 2, item 1, and the fact that $(0, 01)^m = (0, 0^m 1)$ combine to give the complete form, which is precisely what the theorem requires (and is not restated here).

We begin by defining two sequences: $p_i = 1$ if $i = 0$, $a_{i-1} p_{i-1}$ if i is odd, and p_{i-1} otherwise, and $q_i = q_0$ if $i = 0$, $a_{i-1} p_{i-1} + q_{i-1}$ if i is even, and q_{i-1} otherwise. We can prove by induction on m that

$$(0, 0^{q_0} 1) \circ \left(\prod_{j=0}^{m} ((0^{a_{2j}}, 1) \circ (0, 0^{a_{2j+1}} 1)) \right) = (0^{p_{2m+2}}, 0^{q_{2m+2}} 1).$$

As this is the beginning of the factorization in Eq. (1), this implies that $\varphi = (0^{p_{N+1}}, 0^{q_{N+1}} 1) \circ (01, 0) \circ \left(0, 1^{q'} \right) = (0^{p_{N+1}}, 0^{q_{N+1}} 1) \circ \left(01, 0^{q'} \right)$, which is equal to $\left(0^{p_{N+1} + q_{N+1}} 1, 0^{q' p_{N+1}} \right)$. So, we have $p = p_{N+1} + q_{N+1}$ and $q = q' p_{N+1}$. We can then prove by induction on m that

$$p_{2m} = \prod_{i=0}^{m-1} a_{2i}, \qquad q_{2m} = q_0 + \sum_{i=0}^{m-1} \left(a_{2i+1} \prod_{j=0}^{i} a_{2j} \right).$$

Substituting $\frac{N+1}{2}$ for m proves the desired formulas:

$$\prod_{i=0}^{\frac{N-1}{2}} a_{2i} + q_0 + \sum_{i=0}^{\frac{N-1}{2}} \left(a_{2i+1} \prod_{j=0}^{i} a_{2j} \right) = p_{N+1} + q_{N+1} = p,$$

$$q' \prod_{i=0}^{\frac{N-1}{2}} a_{2i} = q' p_{N+1} = q,$$

thereby completing this direction of the proof.

The converses follow from the various preceding items.   □

Theorem 5, when combined with Corollary 1, yields the following corollaries.

Corollary 2. *Let φ be a morphism with a complete canonical factorization of the form*

$$\left(\prod_{i=1}^{m_0} \phi_{0,i} \right) \circ (01, 1) \circ \left(\prod_{j=1}^{m_1} \phi_{1,j} \right),$$

for some integers $m_0, m_1 \geq 0$, where $\phi_{m,n} = (0^{p_n}, 1)$ for some prime p_n or $\phi_{m,n} = (0, 01)$. Then $\rho^{ab}_{\varphi^\omega(0)}(n) = \Theta(1)$.

Corollary 3. *Let φ be a morphism with a (probably not complete) canonical factorization that is a cyclic shift of the following composite:*

$$(0, 0^{q_0}1) \circ \left(\prod_{j=0}^{\frac{N-1}{2}} ((0^{a_{2j}}, 1) \circ (0, 0^{a_{2j+1}}1)) \right) \circ (01, 0) \circ \left(0, 1^{q'} \right)$$

for some odd $N \geq 0$ (in case $N = -1$, the product term is absent), some $q_0 \geq 0$, some $q' > 0$, and some sequence $a_0, a_1, \ldots, a_N$ all nonnegative. If

$$\prod_{i=0}^{\frac{N-1}{2}} a_{2i} + q_0 + \sum_{i=0}^{\frac{N-3}{2}} \left(a_{2i+1} \prod_{j=0}^{i} a_{2j} \right) \geq q' \prod_{i=0}^{\frac{N-1}{2}} a_{2i},$$

then $\rho^{ab}_{\varphi^\omega(0)}(n) = \Theta(1)$.

An example of a morphism classifiable by Corollary 2 is $(001001, 00101)$, which has a complete canonical factorization $(0, 01) \circ (01, 1) \circ (0, 01) \circ (00, 1)$, which satisfies the conditions of Corollary 2. An example of a morphism classifiable by Corollary 3 is $(0011, 0)$, which has a complete canonical factorization $(0, 11) \circ (0, 01) \circ (01, 0)$. This is a cyclic shift of $(0, 01) \circ (01, 0) \circ (0, 11)$, so we have $q_0 = 1$, $q' = 2$, and $N = -1$.

4.2 Classification of Uniform Morphisms

We now derive a complete classification of the abelian complexities of fixed points of uniform binary morphisms.

Let φ be a uniform binary morphism with fixed point at 0. The *length* of φ (denoted $\ell(\varphi)$ or just ℓ if φ is unambiguous) is equal to $|\varphi(0)|$ (which equals $|\varphi(1)|$). The *difference* of φ (denoted $d(\varphi)$ or just d if φ is unambiguous) equals $|z(\varphi(0)) - z(\varphi(1))|$. The *delta* of φ (denoted $\Delta(\varphi)$ or just Δ if φ is unambiguous) equals $z_M(\ell(\varphi)) - \max\{z(\varphi(0)), z(\varphi(1))\}$, where $z_M(\ell(\varphi))$ denotes the maximum number of zeroes in a subword of length $\ell(\varphi)$ of $\varphi^\omega(0)$. Also, if φ is unambiguous, we denote $z(\varphi(0))$ by z_0, $z(\varphi(1))$ by z_1, and $\rho^{ab}_{\varphi^\omega(0)}(n)$ by $\rho^{ab}(n)$.

Theorem 6. *Let φ be a uniform binary morphism, and define $f(n) = 1$ if $\log_2 n \in \mathbb{Z}$, and $f(n) = \log n$ otherwise. Then the following hold:*

$$\rho^{ab}(n) = \begin{cases} \Theta(1), & d = 0; \\ \Theta(1), & \varphi = ((01)^{\lfloor \frac{\ell}{2} \rfloor} 0, (10)^{\lfloor \frac{\ell}{2} \rfloor} 1) \text{ if } \ell \text{ is odd}; \\ \Theta(1), & \varphi = (01^{\ell-1}, 1^\ell); \\ \tilde{\Theta}(f(n)), & d = 1, \Delta = 0, \text{ and not earlier cases}; \\ O(\log n), & d = 1, \Delta > 0; \\ \Theta(n^{\log_\ell d}), & d > 1. \end{cases}$$

Proof. We prove the fourth case. Assume that $d = 1$, $\Delta = 0$, and we are not in the second or third case. We can show that there exist integers i and j such that $\varphi^2(0)[i] = \varphi^2(1)[i] = 0$ and $\varphi^2(0)[j] = \varphi^2(1)[j] = 1$. We can also show that $d(\varphi^2) = 1$ and $\Delta(\varphi^2) = 0$, so $\rho^{ab}(\ell^2 n - 1) = \rho^{ab}(\ell^2 n + 1) = \rho^{ab}(\ell^2 n) + 1$. Consider the sequence defined by $a_0 = 1$ and $a_i = \ell^2 a_{i-1} + 1$. It can be easily shown by induction that $a_i = \sum_{j=0}^{i} \ell^{2j}$. Also, we know that $\rho^{ab}(a_i) = i$. Hence, we have $\rho^{ab}\left(\sum_{j=0}^{i} \ell^{2j}\right) = i$, so the a_i's give a subsequence of $\rho^{ab}(n)$ with logarithmic growth. We now show that $\rho^{ab}(n)$ is $O(\log n)$ completing our proof that it is $\tilde{\Theta}(f(n))$. Let c be the maximum value of $\rho^{ab}(n)$ for $n < \ell$. For $r \in \{0, \ldots, \ell - 1\}$, we can show that that

$$\rho^{ab}(\ell n + r) \leq d\rho^{ab}(n) - d + 2 + 2\Delta + \rho^{ab}(r) \leq \rho^{ab}(n) + 1 + c.$$

As we increase by (approximately) a factor of ℓ in the argument, we can increase by at most a constant in value. This is $\log n$ behavior, as required.

In the fifth case, $d = 1$ and $\Delta > 0$, and the inequality $\rho^{ab}(\ell n + r) \leq d\rho^{ab}(n) - d + 2 + 2\Delta + \rho^{ab}(r)$ leads similarly to $\rho^{ab}(n) = O(\log n)$. The same inequality leads to the $O(n^{\log_\ell d})$ bound in the sixth case, and a similar inequality yields the $\Omega(n^{\log_\ell d})$ bound.   □

Note that some uniform morphisms have nontrivial factorizations. Hence, Theorem 6 gives a classification of the abelian complexities of some nonuniform morphisms as well via Corollary 1. For example, $(01, 00) = (01, 0) \circ (0, 11)$ and $(0, 11) \circ (01, 0) = (011, 0)$. Let $\varphi = (01, 00)$ and $\psi = (011, 0)$. Since $\rho^{ab}_{\varphi^\omega(0)}(n) = \tilde{\Theta}(f(n))$, $\rho^{ab}_{\psi^\omega(0)}(n) = \tilde{\Theta}(f(n))$ as well, though ψ is not uniform.

Referring to the fifth case of Theorem 6, we conjecture an $\Omega(\log n)$ bound abelian complexity for all uniform binary morphisms with $d = 1$ and $\Delta > 0$.

Conjecture 1. Let φ be a uniform binary morphism with $d = 1$ and $\Delta > 0$. For all $h \geq 1$ and $n \geq \ell^h$, $\rho^{ab}(n) \geq h + 2$.

5   Future Work

Problems to be considered in the future include: prove (or disprove) the conjectured $\Omega(\log n)$ bound for uniform binary morphisms with $d = 1$ and $\Delta > 0$, carry out worst and average case running time analyses on the factorization algorithm, examine additional classes of morphisms, and attempt to extend some results to $k > 2$. Most of our results about abelian complexity are about binary words. A notable exception is Theorem 3.

In general, if the alphabet size k is greater than 2, we lose the property that for an infinite word w, $\left|\rho^{ab}_w(n+1) - \rho^{ab}_w(n)\right| \leq 1$. We also can no longer reduce questions about abelian complexity to simply counting zeroes. In general, if w is an infinite word over a k-letter alphabet, Proposition 1(3) says that $\rho^{ab}_w(n) = O(n^{k-1})$. If w is required to be the fixed point of a morphism, we can give a better bound. Corollary 10.4.9 in [1] says that if w is the fixed point of a morphism, then $\rho_w(n) = O(n^2)$. Hence, by Proposition 1(4), if w is the fixed point of a morphism, then $\rho^{ab}_w(n) = O(n^2)$, no matter how large k is.

In many cases, we can give an even better upper bound. Allouche and Shallit [1] define a *primitive morphism* as a morphism φ for which there exists an integer $n \geq 1$ such that given any letters a and b in the alphabet, a occurs in $\varphi^n (b)$. Then [1, Theorem 10.4.12] states that if w is the fixed point of a primitive morphism, then $\rho_w (n) = O(n)$. Hence if w is the fixed point of a primitive morphism, then $\rho_w^{ab} (n) = O(n)$, no matter how large k is.

Finally, we note that the truth value of Corollary 1 has not been examined in depth for alphabets of size greater than 2. Our proof of Theorem 2 certainly depends on the alphabet size, but we have not yet seen a counterexample to it for a larger alphabet. Since binary morphisms can be factorized over larger alphabets, the truth of Corollary 1 would allow us to classify the abelian complexities of the fixed points of many morphisms with $k > 2$ simply based on the results we have here for binary morphisms.

References

1. Allouche, J.P., Shallit, J.: Automatic Sequences: Theory, Applications, Generalizations. Cambridge University Press (2003)
2. Balková, L., Břinda, K., Turek, O.: Abelian complexity of infinite words associated with quadratic Parry numbers. Theoret. Comput. Sci. 412, 6252–6260 (2011)
3. Berstel, J., Séébold, P.: A characterization of Sturmian morphisms. In: Borzyszkowski, A.M., Sokolowski, S. (eds.) MFCS 1993. LNCS, vol. 711, pp. 281–290. Springer, Heidelberg (1993)
4. Choffrut, C., Karhumäki, J.: Combinatorics of Words. In: Rozenberg, G., Salomaa, A. (eds.) Handbook of Formal Languages, vol. 1, ch. 6, pp. 329–438. Springer, Berlin (1997)
5. Coven, E.M., Hedlund, G.A.: Sequences with minimal block growth. Math. Systems Theory 7, 138–153 (1973)
6. Currie, J., Rampersad, N.: Recurrent words with constant abelian complexity. Adv. Appl. Math. 47, 116–124 (2011)
7. Ehrenfeucht, A., Lee, K.P., Rozenberg, G.: Subword complexities of various classes of deterministic developmental languages without interactions. Theoret. Comput. Sci. 1, 59–75 (1975)
8. Ehrenfeucht, A., Rozenberg, G.: On the subword complexity of D0L languages with a constant distribution. Inform. Process. Lett. 13, 108–113 (1981)
9. Ehrenfeucht, A., Rozenberg, G.: On the subword complexity of square-free D0L languages. Theoret. Comput. Sci. 16, 25–32 (1981)
10. Ehrenfeucht, A., Rozenberg, G.: On the subword complexity of locally catenative D0L languages. Inform. Process. Lett. 16, 121–124 (1983)
11. Frid, A.E.: The subword complexity of fixed points of binary uniform morphisms. In: Chlebus, B.S., Czaja, L. (eds.) FCT 1997. LNCS, vol. 1279, pp. 179–187. Springer, Heidelberg (1997)
12. Pansiot, J.J.: Complexité des facteurs des mots infinis engendrés par morphismes itérés. In: Paredaens, J. (ed.) ICALP 1984. LNCS, vol. 172, pp. 380–389. Springer, Heidelberg (1984)
13. Richomme, G.: Conjugacy of morphisms and Lyndon decomposition of standard Sturmian words. Theoret. Comput. Sci. 380, 393–400 (2007)
14. Richomme, G., Saari, K., Zamboni, L.Q.: Abelian complexity in minimal subshifts. J. London Math. Soc. 83, 79–95 (2011)

Strict Bounds for Pattern Avoidance*

Francine Blanchet-Sadri[1] and Brent Woodhouse[2]

[1] Department of Computer Science, University of North Carolina,
P.O. Box 26170, Greensboro, NC 27402–6170, USA
blanchet@uncg.edu
[2] Department of Mathematics, Purdue University,
150 N. University Street, West Lafayette, IN 47907–2067
bwoodhou@purdue.edu

Abstract. Cassaigne conjectured in 1994 that any pattern with m distinct variables of length at least $3(2^{m-1})$ is avoidable over a binary alphabet, and any pattern with m distinct variables of length at least 2^m is avoidable over a ternary alphabet. Building upon the work of Rampersad and the power series techniques of Bell and Goh, we obtain both of these suggested strict bounds. Similar bounds are also obtained for pattern avoidance in partial words, sequences where some characters are unknown.

1 Introduction

Let Σ be an alphabet of letters, denoted by $a, b, c, \ldots$, and Δ be an alphabet of variables, denoted by $A, B, C, \ldots$. A *pattern* p is a word over Δ. A word w over Σ is an *instance* of p if there exists a non-erasing morphism $\varphi : \Delta^* \to \Sigma^*$ such that $\varphi(p) = w$. A word w is said to *avoid* p if no factor of w is an instance of p. For example, $aa\ b\ aa\ c$ contains an instance of ABA while $abaca$ avoids AA.

A pattern p is *avoidable* if there exist infinitely many words w over a finite alphabet such that w avoids p, or equivalently, if there exists an infinite word that avoids p. Otherwise p is *unavoidable*. If p is avoided by infinitely many words over a k-letter alphabet, p is said to be *k-avoidable*. Otherwise, p is *k-unavoidable*. If p is avoidable, the minimum k such that p is k-avoidable is called the *avoidability index* of p. If p is unavoidable, the avoidability index is defined as ∞. For example, ABA is unavoidable while AA has avoidability index 3.

If a pattern p occurs in a pattern q, we say p *divides* q. For example, $p = ABA$ divides $q = \underline{ABC}\ \underline{BB}\ \underline{ABC}\ A$, since we can map A to ABC and B to BB and this maps p to a factor of q. If p divides q and p is k-avoidable, there exists an infinite word w over a k-letter alphabet that avoids p; w must also avoid q, thus q is necessarily k-avoidable. It follows that the avoidability index of q is less than or equal to the avoidability index of p. Chapter 3 of Lothaire [5] is a nice summary of background results in pattern avoidance.

* This material is based upon work supported by the National Science Foundation under Grant No. DMS–1060775. We thank the referees of preliminary versions of this paper for their very valuable comments and suggestions.

M.-P. Béal and O. Carton (Eds.): DLT 2013, LNCS 7907, pp. 106–117, 2013.
© Springer-Verlag Berlin Heidelberg 2013

It is not known if it is generally decidable, given a pattern p and integer k, whether p is k-avoidable. Thus various authors compute avoidability indices and try to find bounds on them. Cassaigne [4] listed avoidability indices for unary, binary, and most ternary patterns (Ochem [7] determined the remaining few avoidability indices for ternary patterns). Based on this data, Cassaigne conjectured in his 1994 Ph.D. thesis [4, Conjecture 4.1] that any pattern with m distinct variables of length at least $3(2^{m-1})$ is avoidable over a binary alphabet, and any pattern with m distinct variables of length at least 2^m is avoidable over a ternary alphabet. This is also [5, Problem 3.3.2].

The contents of our paper are as follows. In Section 2, we establish that both bounds suggested by Cassaigne are strict by exhibiting well-known sequences of patterns that meet the bounds. Note that the results of Section 2 were proved by Cassaigne in his Ph.D. thesis with the same patterns (see [4, Proposition 4.3]). We recall them here for sake of completeness. In Section 3, we provide foundational results for the power series approach to this problem taken by Bell and Goh [1] and Rampersad [8], then proceed to prove the strict bounds in Section 4. In Section 5, we apply the power series approach to obtain similar bounds for avoidability in partial words, sequences that may contain some do-not-know characters, or holes, which are *compatible* or match any letter in the alphabet. The modifications include that now we must avoid all partial words compatible with instances of the pattern. Lots of additional work with inequalities is necessary. Finally in Section 6, we conclude with various remarks and conjectures. Due to the 12-page space constraint, we have decided to put all the proofs of our main result, which is the affirmative answer to the long-standing conjecture of Cassaigne, within the allowed 12 pages, and to omit the proofs of the lemmas in Section 5.

2   Two Sequences of Unavoidable Patterns

The following proposition allows the construction of sequences of unavoidable patterns.

Proposition 1. ([5, Proposition 3.1.3]) Let p be a k-unavoidable pattern over Δ and $A \in \Delta$ be a variable that does not occur in p. Then the pattern pAp is k-unavoidable.

Let $A_1, A_2, \ldots$ be distinct variables in Δ. Define $Z_0 = \varepsilon$, the empty word, and for all integers $m \geq 0$, $Z_{m+1} = Z_m A_{m+1} Z_m$. The patterns Z_m are called Zimin words. Since ε is k-unavoidable for every positive integer k, Proposition 1 implies Z_m is k-unavoidable for all $m \in \mathbb{N}$ by induction on m. Thus all the Zimin words are unavoidable. Note that Z_m is over m variables and $|Z_m| = 2^m - 1$. Thus there exists a 3-unavoidable pattern over m variables with length $2^m - 1$ for all $m \in \mathbb{N}$.

Likewise, define $R_1 = A_1 A_1$ and for all integers $m \geq 1$, $R_{m+1} = R_m A_{m+1} R_m$. Since $A_1 A_1$ is 2-unavoidable, Proposition 1 implies R_m is 2-unavoidable for all $m \in \mathbb{N}$ by induction on m. Note that R_m is over m variables; induction also yields $|R_m| = 3(2^{m-1}) - 1$. Thus there exists a 2-unavoidable pattern over m variables with length $3(2^{m-1}) - 1$ for all $m \in \mathbb{N}$.

3 The Power Series Approach

The following theorem was originally presented by Golod (see [10, Lemma 6.2.7]) and rewritten and proven with combinatorial terminology by Rampersad.

Theorem 1. ([8, Theorem 2]) Let S be a set of words over a k-letter alphabet with each word of length at least two. Suppose that for each $i \geq 2$, the set S contains at most c_i words of length i. If the power series expansion of

$$B(x) := \left(1 - kx + \sum_{i \geq 2} c_i x^i\right)^{-1}$$

has non-negative coefficients, then there are at least $[x^n]B(x)$ words of length n over a k-letter alphabet that have no factors in S.

To count the number of words of length n avoiding a pattern p, we let S consist of all instances of p. To use Theorem 1, we require an upper bound c_i on the number of words of length i in S. The following lemma due to Bell and Goh provides a useful upper bound.

Lemma 1. ([1, Lemma 7]) Let $m \geq 1$ be an integer and p be a pattern over an alphabet $\Delta = \{A_1, \ldots, A_m\}$. Suppose that for $1 \leq i \leq m$, the variable A_i occurs $d_i \geq 1$ times in p. Let $k \geq 2$ be an integer and let Σ be a k-letter alphabet. Then for $n \geq 1$, the number of words of length n over Σ that are instances of the pattern p is no more than $[x^n]C(x)$, where

$$C(x) := \sum_{i_1 \geq 1} \cdots \sum_{i_m \geq 1} k^{i_1 + \cdots + i_m} x^{d_1 i_1 + \cdots + d_m i_m}.$$

Note that this approach for counting instances of a pattern is based on the frequencies of each variable in the pattern, so it will not distinguish $AABB$ and $ABAB$, for example.

4 Derivation of the Strict Bounds

First we prove a technical inequality.

Lemma 2. *Suppose $k \geq 2$ and $m \geq 1$ are integers and $\lambda > \sqrt{k}$. For any integer P and integers d_j for $1 \leq j \leq m$ such that $d_j \geq 2$ and $P = d_1 + \cdots + d_m$,*

$$\prod_{i=1}^{m} \frac{1}{\lambda^{d_i} - k} \leq \left(\frac{1}{\lambda^2 - k}\right)^{m-1} \left(\frac{1}{\lambda^{P-2(m-1)} - k}\right). \qquad (1)$$

Proof. The proof is by induction on m. For $m = 1$, $d_1 = P$ and the inequality is trivially satisfied. Suppose Eq. (1) holds for m and $d_1 + d_2 + \cdots + d_{m+1} = P$ with $d_j \geq 2$ for $1 \leq j \leq m+1$. Note that $P \geq 4$.

Letting $P' = P - d_{m+1} = d_1 + \cdots + d_m$, the inductive hypothesis implies

$$\prod_{i=1}^{m} \frac{1}{\lambda^{d_i} - k} \leq \left(\frac{1}{\lambda^2 - k} \right)^{m-1} \left(\frac{1}{\lambda^{P'-2(m-1)} - k} \right). \tag{2}$$

If $d_{m+1} = 2$, multiplying both sides by

$$\frac{1}{\lambda^{d_{m+1}} - k} = \frac{1}{\lambda^2 - k}$$

yields the desired inequality.

Otherwise, $d_{m+1} > 2$. If $P' - 2(m-1) = 2$, multiplying both sides of Eq. (2) by

$$\frac{1}{\lambda^{d_{m+1}} - k} = \frac{1}{\lambda^{P-2m} - k}$$

yields the desired inequality. In the remaining case, $P' - 2(m - 1) > 2$. Let $c_1 = P' - 2(m - 1)$ and $c_2 = d_{m+1}$. Since $\lambda > \sqrt{k}$ and $c_1, c_2 > 2$,

$$(\lambda^{c_1-1} - \lambda)(\lambda^{c_2-1} - \lambda) \geq 0,$$

$$\lambda^{c_1+c_2-2} - \lambda^{c_1} - \lambda^{c_2} + \lambda^2 \geq 0,$$

$$\lambda^{c_1+c_2-2} + \lambda^2 \geq \lambda^{c_1} + \lambda^{c_2},$$

$$-k(\lambda^{c_1+c_2-2} + \lambda^2) \leq -k(\lambda^{c_1} + \lambda^{c_2}),$$

$$(\lambda^{c_1} - k)(\lambda^{c_2} - k) \geq (\lambda^{c_1+c_2-2} - k)(\lambda^2 - k),$$

$$\frac{1}{(\lambda^{c_1} - k)(\lambda^{c_2} - k)} \leq \frac{1}{(\lambda^{c_1+c_2-2} - k)(\lambda^2 - k)}.$$

Substituting the c_i,

$$\frac{1}{(\lambda^{P'-2(m-1)} - k)(\lambda^{d_{m+1}} - k)} \leq \frac{1}{(\lambda^{P'-2m+d_{m+1}} - k)(\lambda^2 - k)}. \tag{3}$$

Multiplying Eq. (2) by $\frac{1}{\lambda^{d_{m+1}-k}}$,

$$\prod_{i=1}^{m+1} \frac{1}{\lambda^{d_i} - k} \leq \left(\frac{1}{\lambda^2 - k} \right)^{m-1} \left(\frac{1}{\lambda^{P'-2(m-1)} - k} \right) \frac{1}{\lambda^{d_{m+1}} - k}.$$

Substituting Eq. (3),

$$\prod_{i=1}^{m+1} \frac{1}{\lambda^{d_i} - k} \leq \left(\frac{1}{\lambda^2 - k} \right)^{m} \left(\frac{1}{\lambda^{P'+d_{m+1}-2m} - k} \right)$$

$$= \left(\frac{1}{\lambda^2 - k} \right)^{(m+1)-1} \left(\frac{1}{\lambda^{P-2((m+1)-1)} - k} \right),$$

as desired. $\qquad\square$

Remark 1. We have written Lemma 2 in terms of partitions of P with parts of size at least 2. However, as it will be used with $P = |p|$ for some pattern p containing d_j occurrences of variable A_j, its statement and its proof could also be written in terms of patterns defining p' to be p without its d_{m+1} instances of the $(m + 1)$th variable. Then using the inductive hypothesis on p', the proof would follow as it is.

The remaining arguments in this section are based on those of [8], but add additional analysis to obtain the optimal bound.

Lemma 3. *Let m be an integer and p be a pattern over an alphabet $\Delta = \{A_1, \ldots, A_m\}$. Suppose that for $1 \leq i \leq m$, A_i occurs $d_i \geq 2$ times in p.*

1. *If $m \geq 3$ and $|p| \geq 4m$, then for $n \geq 0$, there are at least $(1.92)^n$ words of length n over a binary alphabet that avoid p.*
2. *If $m \geq 2$ and $|p| \geq 12$, then for $n \geq 0$, there are at least $(2.92)^n$ words of length n over a ternary alphabet that avoid p (for $m \geq 6$, this implies that every pattern with each variable occurring at least twice is 3-avoidable).*

Proof. Let Σ be an alphabet of size $k \in \{2, 3\}$. Define S to be the set of all words in Σ^* that are instances of the pattern p. By Lemma 1, the number of words of length n in S is at most $[x^n]C(x)$, where

$$C(x) := \sum_{i_1 \geq 1} \cdots \sum_{i_m \geq 1} k^{i_1 + \cdots + i_m} x^{d_1 i_1 + \cdots + d_m i_m}.$$

By hypothesis, $d_i \geq 2$ for $1 \leq i \leq m$. In order to use Theorem 1 on Σ, define

$$B(x) := \sum_{i \geq 0} b_i x^i = (1 - kx + C(x))^{-1},$$

and set the constant $\lambda = k - 0.08$. Clearly $b_0 = 1$ and $b_1 = k$. We show that $b_n \geq \lambda b_{n-1}$ for all $n \geq 1$, hence $b_n \geq \lambda^n$ for all $n \geq 0$. Then all coefficients of B are non-negative, thus Theorem 1 implies there are at least $b_n \geq \lambda^n$ words of length n avoiding S. By construction of S, these words all avoid p.

We show by induction on n that $b_n \geq \lambda b_{n-1}$ for all $n \geq 1$. We can easily verify $b_1 \geq (k - 0.08)(1) = \lambda b_0$. Now suppose that for all $1 \leq j < n$, we have $b_j \geq \lambda b_{j-1}$. By definition of B, $B(x)(1 - kx + C(x)) = 1$, hence for $n \geq 1$, $[x^n]B(1 - kx + C) = 0$. Expanding the left hand side,

$$B(1 - kx + C) = \left(\sum_{i \geq 0} b_i x^i\right)\left(1 - kx + \sum_{i_1 \geq 1} \cdots \sum_{i_m \geq 1} k^{i_1 + \cdots + i_m} x^{d_1 i_1 + \cdots + d_m i_m}\right),$$

thus

$$[x^n]B(1 - kx + C) = b_n - kb_{n-1} + \sum_{i_1 \geq 1} \cdots \sum_{i_m \geq 1} k^{i_1 + \cdots + i_m} b_{n-(d_1 i_1 + \cdots + d_m i_m)} = 0.$$

Rearranging and adding and subtracting λb_{n-1},

$$b_n = \lambda b_{n-1} + (k-\lambda)b_{n-1} - \sum_{i_1 \geq 1} \cdots \sum_{i_m \geq 1} k^{i_1 + \cdots + i_m} b_{n-(d_1 i_1 + \cdots + d_m i_m)}.$$

To complete the induction, it thus suffices to show

$$(k-\lambda)b_{n-1} - \sum_{i_1 \geq 1} \cdots \sum_{i_m \geq 1} k^{i_1 + \cdots + i_m} b_{n-(d_1 i_1 + \cdots + d_m i_m)} \geq 0. \tag{4}$$

Because $b_j \geq \lambda b_{j-1}$ for $1 \leq j < n$, $b_{n-i} \leq b_{n-1}/\lambda^{i-1}$ for $1 \leq i \leq n$. Note that the bound on b_{n-i} is stated for $1 \leq i \leq n$, but actually it is used also for $i > n$, with the implicit convention that $b_{n-i} = 0$ in this case. Therefore,

$$\sum_{i_1 \geq 1} \cdots \sum_{i_m \geq 1} k^{i_1 + \cdots + i_m} b_{n-(d_1 i_1 + \cdots + d_m i_m)}$$

$$\leq \sum_{i_1 \geq 1} \cdots \sum_{i_m \geq 1} \frac{k^{i_1 + \cdots + i_m}}{\lambda^{d_1 i_1 + \cdots + d_m i_m}} \lambda b_{n-1} = \lambda b_{n-1} \sum_{i_1 \geq 1} \frac{k^{i_1}}{\lambda^{d_1 i_1}} \cdots \sum_{i_m \geq 1} \frac{k^{i_m}}{\lambda^{d_m i_m}}.$$

Since $d_j \geq 2$ for $1 \leq j \leq m$, $k \leq 3$, and $\lambda > \sqrt{3}$,

$$\frac{k}{\lambda^{d_j}} \leq \frac{3}{\lambda^2} < 1,$$

thus all the geometric series converge. Computing the result, for $1 \leq j \leq m$,

$$\sum_{i_j \geq 1} \frac{k^{i_j}}{\lambda^{d_j i_j}} = \frac{k/\lambda^{d_j}}{1 - k/\lambda^{d_j}} = \frac{k}{\lambda^{d_j} - k}.$$

Thus

$$\sum_{i_1 \geq 1} \cdots \sum_{i_m \geq 1} k^{i_1 + \cdots + i_m} b_{n-(d_1 i_1 + \cdots + d_m i_m)} \leq k^m \lambda b_{n-1} \prod_{i=1}^{m} \frac{1}{\lambda^{d_i} - k}.$$

Applying Lemma 2 to $P = |p|$,

$$\sum_{i_1 \geq 1} \cdots \sum_{i_m \geq 1} k^{i_1 + \cdots + i_m} b_{n-(d_1 i_1 + \cdots + d_m i_m)}$$

$$\leq k^m \lambda b_{n-1} \left(\frac{1}{\lambda^2 - k} \right)^{m-1} \left(\frac{1}{\lambda^{|p|-2(m-1)} - k} \right). \tag{5}$$

It thus suffices to show

$$(k-\lambda) \geq \lambda k^m \left(\frac{1}{\lambda^2 - k} \right)^{m-1} \left(\frac{1}{\lambda^{|p|-2(m-1)} - k} \right), \tag{6}$$

since multiplying this by b_{n-1} and using Eq. (5) derives Eq. (4).

To show Statement 1, let $k = 2$ and recall we restricted $m \geq 3$ and $|p| \geq 4m$. Note that the right hand side of Eq. (6) decreases as $|p|$ increases, thus it suffices to verify the case $|p| = 4m$. Taking $m = 3$, $|p| = 12$ and

$$k - \lambda = 0.08 \geq 0.02956 \cdots = 1.92 \frac{2^3}{((1.92)^2 - 2)^2 (1.92^{12-2(3-1)} - 2)}$$

$$= \lambda k^m \left(\frac{1}{\lambda^2 - k} \right)^{m-1} \left(\frac{1}{\lambda^{|p|-2(m-1)} - k} \right).$$

Now consider an arbitrary $m' \geq 3$ and p' with $|p'| = 4m'$. Substituting $\lambda = 1.92$ and $k = 2$, it follows that

$$c := \left(\frac{k}{\lambda^2 - k} \right)^{m'-m} \left(\frac{\lambda^{|p|-2(m-1)} - k}{\lambda^{|p'|-2(m'-1)} - k} \right)$$

$$\leq (1.19)^{m'-m} \left(\frac{1}{\lambda^{|p'|-2(m'-1)-(|p|-2(m-1))}} \right) = (1.19)^{m'-m} \left(\frac{1}{\lambda^{2(m'-m)}} \right) < 1.$$

Thus we conclude

$$k - \lambda \geq c\lambda k^m \left(\frac{1}{\lambda^2 - k} \right)^{m-1} \left(\frac{1}{\lambda^{|p|-2(m-1)} - k} \right)$$

$$= \lambda k^{m'} \left(\frac{1}{\lambda^2 - k} \right)^{m'-1} \left(\frac{1}{\lambda^{|p'|-2(m'-1)} - k} \right).$$

Likewise for Statement 2, for any $m \geq 2$, it suffices to verify Eq. (6) for $|p| = \max\{12, 2m\}$ (clearly every pattern in which each variable occurs at least twice satisfies $|p| \geq 2m$). For $m = 2$ through $m = 5$ and $|p| = 12$, the equation is easily verified. For $m \geq 6$, $|p| = 2m$ and

$$\lambda k^m \left(\frac{1}{\lambda^2 - k} \right)^{m-1} \left(\frac{1}{\lambda^{|p|-2(m-1)} - k} \right) = 2.92 \left(\frac{3}{(2.92)^2 - 3} \right)^m$$

$$\leq 2.92(0.5429)^m \leq 2.92(0.5429)^6 = 0.07476 \cdots < 0.08 = k - \lambda.$$

This completes the induction and the proof of the lemma.   □

Remark 2. Referring to Statement 2 of Lemma 3 "for $m \geq 6$, every pattern with each variable occurring at least twice is 3-avoidable" is mentioned by Bell and Goh (not as a theorem, but as a remark at the end of [1, Section 4]). They provide a slightly better constant 2.9293298 for the exponential growth in this case. As a consequence, Statement 2 is new only for $m \in \{2, 3, 4, 5\}$. For $m \in \{2, 3\}$, patterns of length 12 where known to be avoidable [9,4] but without an exponential lower bound.

Here are the main results. As discussed in Section 2, both bounds below are strict in the sense that for every positive integer m, there exists a 2-unavoidable pattern with m variables and length $3(2^{m-1}) - 1$ as well as a 3-unavoidable pattern with m variables and length $2^m - 1$.

Theorem 2. *Let p be a pattern with m distinct variables.*

1. *If $|p| \geq 3(2^{m-1})$, then p is 2-avoidable.*
2. *If $|p| \geq 2^m$, then p is 3-avoidable.*

Proof. We prove Statement 1 (the proof for Statement 2 is similar). We show by induction on m that if p is 2-unavoidable, $|p| < 3(2^{m-1})$. For $m = 1$, note that A^3 is 2-avoidable [5], hence A^ℓ is 2-avoidable for all $\ell \geq 3$. Thus if a unary pattern p is 2-unavoidable, $|p| < 3 = 3(2^{1-1})$. For $m = 2$, it is known that all binary patterns of length 6 are 2-avoidable [9], hence all binary patterns of length at least 6 are also 2-avoidable. Thus if a binary pattern p is 2-unavoidable, $|p| < 6 = 3(2^{2-1})$. Now assume the statement holds for $m \geq 2$ and suppose p is a 2-unavoidable pattern with $m + 1$ variables. For the sake of contradiction, assume that $|p| \geq 3(2^m)$. There are two cases to consider.

First, if p has a variable A that occurs exactly once, let $p = p_1 A p_2$, where p_1 and p_2 are patterns with at most m variables. Without loss of generality, suppose $|p_1| \geq |p_2|$. Since $|p| \geq 3(2^m)$,

$$|p_1| \geq \left\lceil \frac{|p| - 1}{2} \right\rceil \geq \left\lceil \frac{3(2^m) - 1}{2} \right\rceil = 3(2^{m-1}).$$

By the contrapositive of the inductive hypothesis, p_1 is 2-avoidable. But p_1 divides p, hence p is 2-avoidable, a contradiction.

Alternatively, suppose every variable in p occurs at least twice. Since $|p| \geq 3(2^m) \geq 4(m+1)$ for $m \geq 2$, Lemma 3 indicates there are infinitely many words over a binary alphabet that avoid p, thus p is 2-avoidable, a contradiction. These contradictions imply $|p| < 3(2^{(m+1)-1})$, which completes the induction. □

5   Extension to Partial Words

A partial word over an alphabet Σ is a concatenation of characters from the extended alphabet $\Sigma_\diamond = \Sigma \cup \{\diamond\}$, where $\diamond$ is called the hole character and represents any unknown letter. If u and v are two partial words of equal length, we say u is *compatible* with v, denoted $u \uparrow v$, if $u[i] = v[i]$ whenever $u[i], v[i] \in \Sigma$. A partial word w over Σ is an instance of a pattern p over Δ if there exists a non-erasing morphism $\varphi : \Delta^* \to \Sigma^*$ such that $\varphi(p) \uparrow w$; the partial word w avoids p if none of its factors is an instance of p. For example, $\underline{aa}\,\underline{b}\,\underline{a}\diamond\,c$ contains an instance of ABA while it avoids AAA.

A pattern p is called *k-avoidable* in partial words if for every $h \in \mathbb{N}$ there is a partial word with h holes over a k-letter alphabet avoiding p, or, equivalently, if there is a partial word over a k-letter alphabet with infinitely many holes which avoids p. The *avoidability index* for partial words is defined analogously to that of full words. For example, AA is unavoidable in partial words since a factor of the form $a\diamond$ or $\diamond a$ must occur, where $a \in \Sigma_\diamond$, while the pattern $AABB$ has avoidability index 3 in partial words. Classification of avoidability indices for unary and binary patterns is complete and the ternary classification is nearly complete [2,3].

The power series method previously used for full words can also count partial words avoiding patterns, and similar results are obtained. Before we can use the power series approach to develop bounds for partial words, we must obtain an upper bound for the number of partial words over Σ that are compatible with instances of the pattern. This result is comparable with Lemma 1 for full words.

Lemma 4. *Let $m \geq 1$ be an integer and p be a pattern over an alphabet $\Delta = \{A_1, \ldots, A_m\}$. Suppose that for $1 \leq i \leq m$, the variable A_i occurs $d_i \geq 1$ times in p. Let $k \geq 2$ be an integer and let Σ be a k-letter alphabet. Then for $n \geq 1$, the number of partial words of length n over Σ that are compatible with instances of the pattern p is no more than $[x^n]C(x)$, where*

$$C(x) := \sum_{i_1 \geq 1} \cdots \sum_{i_m \geq 1} \left(\prod_{j=1}^{m} \left(k(2^{d_j} - 1) + 1 \right)^{i_j} \right) x^{d_1 i_1 + \cdots + d_m i_m}.$$

Once again we require a technical inequality.

Lemma 5. *Suppose $(k, \lambda) \in \{(2, 2.97), (3, 3.88)\}$ and $m \geq 1$ is an integer. For any integer P and integers d_j for $1 \leq j \leq m$ such that $d_j \geq 2$ and $P = d_1 + \cdots + d_m$,*

$$\prod_{i=1}^{m} \frac{k(2^{d_i} - 1) + 1}{\lambda^{d_i} - (k(2^{d_i} - 1) + 1)} \leq \left(\frac{3k + 1}{\lambda^2 - (3k + 1)} \right)^{m-1} \left(\frac{k}{(\frac{\lambda}{2})^{P - 2(m-1)} - k} \right). \quad (7)$$

When all variables in the pattern occur at least twice, we obtain the following exponential lower bounds.

Lemma 6. *Let $m \geq 4$ be an integer and p be a pattern over an alphabet $\Delta = \{A_1, \ldots, A_m\}$. Suppose that for $1 \leq i \leq m$, A_i occurs $d_i \geq 2$ times in p.*

1. *If $|p| \geq 15(2^{m-3})$, then for $n \geq 0$, there are at least $(2.97)^n$ partial words of length n over a binary alphabet that avoid p.*
2. *If $|p| \geq 2^m$, then for $n \geq 0$, there are at least $(3.88)^n$ partial words of length n over a ternary alphabet that avoid p.*

Thus for certain patterns, there exist λ^n partial words of length n that avoid the pattern, for some λ. It is not immediately clear that this is enough to prove the patterns are avoidable in partial words. The next lemma asserts this count is so large that it must include partial words with arbitrarily many holes, thus the patterns are 2-avoidable or 3-avoidable in partial words.

Lemma 7. *Suppose $k \geq 2$ is an integer, $k < \lambda < k + 1$, Σ is an alphabet of size k, and S is a set of partial words over Σ with at least λ^n words of length n for each $n > 0$. For all integers $h \geq 0$, S contains a partial word with at least h holes.*

Unfortunately, the pattern $A^2BA^2CA^2$ of length $8 = 2^3$ is unavoidable in partial words (since some $a\diamond$ must occur infinitely often), thus to obtain the 2^m bound for avoidability as in the full word case, we require information about quaternary patterns of length $16 = 2^4$. Fortunately, for certain patterns, constructions can be made from full words avoiding a pattern to partial words avoiding a pattern that provide upper bounds on avoidability indices. We obtain the following bounds.

Theorem 3. *Let p be a pattern with m distinct variables.*

1. *If $m \geq 3$ and $|p| \geq 15(2^{m-3})$, then p is 2-avoidable in partial words.*
2. *If $m \geq 3$ and $|p| \geq 5(2^{m-2})$, then p is 3-avoidable in partial words.*
3. *If $m \geq 4$ and $|p| \geq 2^m$, then p is 4-avoidable in partial words.*

Proof. For Statement 1, we prove by induction on m that if p is 2-unavoidable, $|p| < 15(2^{m-3})$. The base case of ternary patterns ($m = 3$) is handled by a list of over 800 patterns in the appendix of [2]. The maximum length 2-unavoidable ternary pattern in partial words is $A^2BA^2CA^2BA^2$, length $11 < 15 = 15(2^{3-3})$.

Now suppose the result holds for m and let p be a pattern with $m + 1 \geq 4$ distinct variables. If every variable in p is repeated at least twice, Statement 1 of Lemma 6 implies there exists a set S of partial words with at least $(2.97)^n$ binary words of length n that avoid p for each $n \geq 0$. Applying Lemma 7 to S, we find that for each $h \geq 0$, there exists a partial word with at least h holes that avoids p. Thus p is 2-avoidable. If p has a variable that occurs exactly once, we reason as in the proof of Theorem 2 to complete the induction.

For Statement 2, we prove by induction on m that if p is 3-unavoidable, $|p| < 5(2^{m-2})$. For $m = 3$, all patterns of length $10 = 5(2^{3-2})$ are shown to be 3-avoidable in [2]. For $m \geq 4$, Statement 2 of Lemma 6 and Lemma 7 imply that every pattern of length at least 2^m in which each variable appears at least twice is 3-avoidable. If p has a variable that occurs exactly once, we reason as in the proof of Theorem 2 to complete the induction.

For Statement 3, we show by induction on m that if p is 4-unavoidable, $|p| < 2^m$. We first establish the base case $m = 4$ by showing that every pattern p of length $16 = 2^4$ is 4-avoidable. Using the data in [2], the ternary patterns which have avoidability index greater than 4 are $AABCABA$, $ABACAAB$, $ABACBAA$, and $ABBCBAB$ of length 7 (up to reversal and renaming of variables).

Consider any p with $|p| = 16$. If every variable in p occurs at least twice, Statement 2 of Lemma 6 implies there exists a set S with at least $(3.88)^n$ ternary partial words of length n that avoid p for each $n \geq 0$. Applying Lemma 7 to S, we find that for each $h \geq 0$, there exists a ternary partial word with at least h holes that avoids p. Thus p is 3-avoidable. Otherwise, p contains a variable α that occurs exactly once and $p = p_1 \alpha p_2$ for patterns p_1 and p_2 with at most 3 distinct variables. Note that $|p_1| + |p_2| = 15$. If p_1 has length at least 9, then p_1 is 4-avoidable, hence p is 4-avoidable by divisibility (likewise for p_2).

Thus the only remaining cases are when $|p_1| = 8$ and $|p_2| = 7$ or vice versa. Suppose $|p_1| = 8$ and $|p_2| = 7$ (the other case is similar). If p_1 or p_2 is not in the list of ternary patterns above, it is 4-avoidable, hence p is 4-avoidable. Otherwise $p_1 = A^2BA^2CA^2$ up to a renaming of the variables. Note that p_1 contains a factor of the form A^2BA, which fits the form of [2, Theorem 6(2)] for $q_1 = B$. All of the possible values of p_2 are on three variables, so they must contain B. Thus setting $q_2 = B$, [2, Theorem 6(2)] implies p is 4-avoidable.

For $m \geq 5$, Lemma 6 and Lemma 7 imply that every pattern with length at least 2^m in which each variable appears at least twice is 3-avoidable. If p has a variable that occurs exactly once, we reason as in the proof of Theorem 2 to complete the induction.   □

6   Concluding Remarks and Conjectures

Overall, the power series method is a useful way to show existence of infinitely many words avoiding patterns in full words and partial words. It is mainly helpful to obtain upper bounds as derived here, since it utilizes the frequencies of each variable in the pattern and not their placement relative to one another. Only patterns where each variable occurs at least twice can be investigated in this way, but induction arguments as in Theorem 2 then imply bounds for all patterns. For patterns with a variable that appears exactly once, the counts used in Lemma 1 and Lemma 4 grow too quickly, thus the power series method is not applicable.

It would be nice to attain strict bounds for 2-avoidability and 3-avoidability in partial words. Statement 1 of the following conjecture appears in [2], and we add Statement 2.

Conjecture 1. Let p be a pattern with m distinct variables.

1. If $|p| \geq 3(2^{m-1})$, then p is 2-avoidable in partial words.
2. If $m \geq 4$ and $|p| \geq 2^m$, then p is 3-avoidable in partial words.

Both bounds would then be strict, using the same sequences of patterns given for full words in Section 2.

To show Statement 1 using the power series method, we require either an improvement of the bound $15(2^{m-3})$ to $3(2^{m-1})$ in Statement 1 of Lemma 6 or some additional data about avoidability indices of patterns over 4 variables. It may be possible to improve the count used in Lemma 4 to improve this bound. To show Statement 2 using the power series method, we require additional data about avoidability indices of patterns over 4 variables. Unfortunately, finding avoidability indices using HD0L systems as in [2] is likely infeasible for patterns over 4 variables. Perhaps some constructions can be made from words avoiding long enough 2-avoidable or 3-avoidable patterns in full words to prove there exist infinitely many partial words that avoid the pattern over 2 or 3 letters.

Finally, it may be possible to make better approximations than Theorem 1 and Lemma 1 based on the Goulden-Jackson method for avoiding a finite number of words [6]. The method works better when the growth rate of words avoiding a k-avoidable pattern is close to k, whereas it is known that for the pattern $AABBCABBA$, where $k = 2$, the growth rate is close to 1. There is no hope for the pattern $ABWACXBCYBAZCA$, where $k = 4$, since only polynomially many words over 4 letters avoid it (here the growth rate is 1). Perhaps the method could handle the cases where each variable of the pattern occurs at least twice, but even the case of the pattern AA, where $k = 3$, seems to be challenging with a 1.31 growth rate.

Note that our paper was submitted to DLT 2013 on January 2, 2013. Some referees made us aware that Theorem 2 has also been found, completely independently and almost simultaneouly, by Pascal Ochem and Alexandre Pinlou (P. Ochem and A. Pinlou, Application of entropy compression in pattern avoidance, arXiv:1301.1873, January 9, 2013). Their proof of Statement 1 uses Bell and Goh's method, while their proof of Statement 2 uses the entropy compression method.

References

1. Bell, J., Goh, T.L.: Exponential lower bounds for the number of words of uniform length avoiding a pattern. Information and Computation 205, 1295–1306 (2007)
2. Blanchet-Sadri, F., Lohr, A., Scott, S.: Computing the partial word avoidability indices of ternary patterns. In: Arumugam, S., Smyth, B. (eds.) IWOCA 2012. LNCS, vol. 7643, pp. 206–218. Springer, Heidelberg (2012)
3. Blanchet-Sadri, F., Mercaş, R., Simmons, S., Weissenstein, E.: Avoidable binary patterns in partial words. Acta Informatica 48, 25–41 (2011)
4. Cassaigne, J.: Motifs évitables et régularités dans les mots. Ph.D. thesis, Paris VI (1994)
5. Lothaire, M.: Algebraic Combinatorics on Words. Cambridge University Press, Cambridge (2002)
6. Noonan, J., Zeilberger, D.: The Goulden-Jackson cluster method: Extensions, applications, and implementations. Journal of Differential Equations and Applications 5, 355–377 (1999)
7. Ochem, P.: A generator of morphisms for infinite words. RAIRO-Theoretical Informatics and Applications 40, 427–441 (2006)
8. Rampersad, N.: Further applications of a power series method for pattern avoidance. The Electronic Journal of Combinatorics 18, P134 (2011)
9. Roth, P.: Every binary pattern of length six is avoidable on the two-letter alphabet. Acta Informatica 29, 95–106 (1992)
10. Rowen, L.: Ring Theory. Pure and Applied Mathematics 128, vol. II. Academic Press, Boston (1988)

A Fresh Approach to Learning Register Automata*

Benedikt Bollig[1], Peter Habermehl[2],
Martin Leucker[3], and Benjamin Monmege[1]

[1] LSV, ENS Cachan, CNRS & Inria, France
[2] Univ Paris Diderot, Sorbonne Paris Cité, LIAFA, CNRS, France
[3] ISP, University of Lübeck, Germany

Abstract. This paper provides an Angluin-style learning algorithm for
a class of register automata supporting the notion of *fresh* data values.
More specifically, we introduce *session automata* which are well suited for
modeling protocols in which sessions using fresh values are of major interest, like in security protocols or ad-hoc networks. We show that session
automata (i) have an expressiveness partly extending, partly reducing
that of register automata, (ii) admit a symbolic regular representation,
and (iii) have a decidable equivalence and model-checking problem (unlike register automata). Using these results, we establish a learning algorithm to infer session automata through membership and equivalence
queries. Finally, we strengthen the robustness of our automaton by its
characterization in monadic second-order logic.

1 Introduction

Learning automata deals with the inference of automata based on some partial
information, for example samples, which are words that either belong to their
accepted language or not. A popular framework is that of active learning defined
by Angluin [2] in which a learner may consult a teacher for so-called membership
and equivalence queries to eventually infer the automaton in question. Learning
automata has a lot of applications in computer science. Notable examples are
the use in model checking [12] and testing [3]. See [18] for an overview.

While active learning of regular languages is meanwhile well understood and
is supported by freely available libraries such as learnlib [19] and libalf [8], extensions beyond plain regular languages are still an area of active research. Recently,
automata dealing with potentially infinite data as first class citizens have been
studied. Seminal works in this area are that of [1,15] and [14]. While the first
two use abstraction and refinement techniques to cope with infinite data, the
second approach learns a sub-class of register automata.

In this paper, we follow the work on learning register automata. However,
we study a different model than [14], having the ability to require that input
data is *fresh* in the sense that it has not been seen so far. This feature has been

---

* This work is partially supported by EGIDE/DAAD-Procope (LeMon).

M.-P. Béal and O. Carton (Eds.): DLT 2013, LNCS 7907, pp. 118–130, 2013.
© Springer-Verlag Berlin Heidelberg 2013

proposed in [24] in the context of semantics of programming languages, as, for example, fresh names are needed to model object creation in object-oriented languages. Moreover, fresh data values are important ingredients in modeling security protocols which often make use of so-called fresh nonces to achieve their security assertions [17]. Finally, fresh names are also important in the field of network protocols and are one of the key ingredients of the π-calculus [20].

In general, the equivalence problem of register automata is undecidable (even without freshness). This limits their applicability in active learning, as equivalence queries cannot be implemented (correctly and completely). Therefore, we restrict the studied automaton model to either store fresh data values or read data values from registers. In the terminology of [24], we retain global freshness, while local freshness is discarded. We call our model *session automata*. They are well-suited whenever fresh values are important for a finite period, for which they will be stored in one of the registers. Session automata correspond to the model from [7] without stacks. They are incomparable with the model from [14].

Session automata accept data words, i.e., words over an alphabet $\Sigma \times D$, where Σ is a finite set of labels and D an infinite set of data values. A data word can be mapped to a so-called symbolic word where we record for each different data value the register in which it was stored (when appearing for the first time) or from which it was read later. To each symbolic word we define a symbolic word in unique normal form representing the same data words by fixing a canonical way of storing data values in registers. Then, we show how to transform a session automaton into a unique canonical automaton that accepts the same data language. This canonical automaton can be seen as a classical finite-state automaton and, therefore, we can define an active learning algorithm for session automata in a natural way. In terms of the size of the canonical automaton, the number of membership and equivalence queries needed is polynomial (both in the number of states and in the number of registers). When the reference model are arbitrary (data) deterministic automata, the complexity is polynomial in the number of states and exponential in the number of registers.

Applicability of our framework in verification (e.g., compositional verification [10] and infinite state regular model checking [13]) is underpinned by the fact that session automata form a robust language class: While inclusion is undecidable for register automata [21], we show that it is decidable for session automata. In [7], model checking session automata was shown decidable wrt. a powerful monadic second-order logic with data-equality predicate (dMSO). Here, we also provide a natural fragment of dMSO that precisely captures session automata.

To summarize, we show that session automata (i) have a unique canonical form, (ii) have a decidable inclusion problem, (iii) enjoy a logical characterization, and (iv) can be learned via an active learning algorithm. Altogether, this provides a versatile learning framework for languages over infinite alphabets.

Outline. In Section 2 we introduce session automata. Section 3 presents an active learning algorithm for them and in Section 4 we give some language-theoretic properties of our model and a logical characterization. Missing proofs can be found in the long version: `http://hal.archives-ouvertes.fr/hal-00743240`

2   Data Words and Session Automata

We let $\mathbb{N}$ (respectively, $\mathbb{N}_{>0}$) be the set of natural numbers (respectively, non-zero natural numbers). For $n \in \mathbb{N}$, we let $[n]$ denote the set $\{1, \ldots, n\}$. In the following, we fix a non-empty finite alphabet Σ of *labels* and an infinite set D of *data values*. In examples, we usually use $D = \mathbb{N}$. A *data word* is a sequence of elements of $\Sigma \times D$, i.e., an element from $(\Sigma \times D)^*$. An example data word over $\Sigma = \{a, b\}$ and $D = \mathbb{N}$ is $(a, 4)(b, 2)(b, 4)$.

Our automata will not be able to distinguish between data words that are equivalent up to permutation of data values. Intuitively, this corresponds to saying that data values can only be compared wrt. equality. When two data words w_1 and w_2 are equivalent in that sense, we write $w_1 \approx w_2$, e.g. $(a, 4)(b, 2)(b, 4) \approx (a, 2)(b, 5)(b, 2)$. The equivalence class of a data word w wrt. $\approx$ is written $[w]_\approx$.

We can view a data word as being composed of (not necessarily disjoint) sessions, each session determining the scope in which a given data value is used. Let $w = (a_1, d_1) \cdots (a_n, d_n) \in (\Sigma \times D)^*$ be a data word. We let $Fresh(w) \overset{\text{def}}{=} \{i \in [n] \mid d_i \neq d_j \text{ for all } j \in \{1, \ldots, i - 1\}\}$ be the set of positions of w where a data value occurs for the first time. Accordingly, we let $Last(w) \overset{\text{def}}{=} \{i \in [n] \mid d_i \neq d_j \text{ for all } j \in \{i+1, \ldots, n\}\}$. A set $S \subseteq [n]$ is a *session* of w if there are $i \in Fresh(w)$ and $j \in Last(w)$ such that $S = \{i, \ldots, j\}$ and $d_i = d_j$. For $i \in [n]$, let $Session(i)$ denote the unique session S with $d_{\min(S)} = d_i$. Thus $Session(i)$ is the scope in which d_i is used. Note that $Fresh(w) = \{\min(Session(i)) \mid i \in [n]\}$. For $k \geq 1$, we say that w is k-*bounded* if every position of w belongs to at most k sessions. A language L is k-bounded if every word in L is so. The set of all data words is not k-bounded, for any k. Fig. 1 illustrates a data word w with four sessions. It is 2-bounded, as no position shares more than 2 sessions. We have $Session(7) = \{4, \ldots, 9\}$ and $Fresh(w) = \{1, 2, 4, 6\}$.

Intuitively, k is the number of resources that will be needed to execute a k-bounded word. Speaking in terms of automata, a resource is a register that can store a data value. Our automata will be able to write a fresh data value into some register r, denoted $f(r)$, or reuse a data value that has already been stored in r, denoted $r(r)$. In other words, automata will work over (a finite subset of) the alphabet $\Sigma \times \Gamma$ where $\Gamma \overset{\text{def}}{=} \{f(r), r(r) \mid r \in \mathbb{N}_{>0}\}$. A word over $\Sigma \times \Gamma$ is called a *symbolic word*. Given a symbolic word $u = (a_1, t_1) \cdots (a_n, t_n)$ and a position $i \in [n]$, we let $reg(i)$ denote the register r that is used at i, i.e., such that $t_i \in \{f(r), r(r)\}$. Similarly, we define the type $type(i) \in \{f, r\}$ of i.

Naturally, a register has to be initialized before it can be used. So, we call u *well formed* if, for all $j \in [n]$ with $type(j) = r$, there is $i \leq j$ such that

$$
\begin{array}{cccccccccc}
\text{1} & \text{2} & \text{3} & \text{4} & \text{5} & \text{6} & \text{7} & \text{8} & \text{9} \\
a & b & a & a & c & c & b & c & c \\
4 & 2 & 4 & 3 & 2 & 1 & 3 & 1 & 3
\end{array}
\qquad
\begin{array}{ccccccccc}
\text{1} & \text{2} & \text{3} & \text{4} & \text{5} & \text{6} & \text{7} & \text{8} & \text{9} \\
a & b & a & a & c & c & b & c & c \\
f(1) & f(2) & r(1) & f(1) & r(2) & f(2) & r(1) & r(2) & r(1)
\end{array}
$$

Fig. 1. A data word and its sessions   **Fig. 2.** A symbolic word

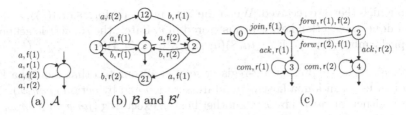

Fig. 3. (a) Session automaton, (b) Client-server system, (c) P2P protocol

$t_i = f(reg(j))$. Let WF denote the set of well formed words. A well formed symbolic word is illustrated in Fig. 2. We have $type(5) = r$ and $reg(5) = 2$.

A symbolic word $u = (a_1, t_1) \cdots (a_n, t_n) \in$ WF generates a set of data words. Intuitively, a position i with $t_i = f(r)$ opens a new session, writing a fresh data value in register r. The same data value is reused at positions $j > i$ with $t_j = r(r)$, unless r is reinitialized at some position i' with $i < i' < j$. Formally, $w \in (\Sigma \times D)^*$ is a *concretization* of u if it is of the form $(a_1, d_1) \cdots (a_n, d_n)$ such that, for all $i, j \in [n]$ with $i \leq j$, (i) $i \in Fresh(w)$ iff $type(i) = f$, and (ii) $d_i = d_j$ iff both $reg(i) = reg(j)$ and there is no position i' with $i < i' \leq j$ such that $t_{i'} = f(reg(i))$. For example, the data word from Fig. 1 is a concretization of the symbolic word from Fig. 2. By $\gamma(u)$, we denote the set of concretizations of a well formed word u. We extend γ to sets $L \subseteq (\Sigma \times \Gamma)^*$ and let $\gamma(L) \stackrel{\text{def}}{=} \{\gamma(u) \mid u \in L \cap \text{WF}\}$.

Remark 1. Let us state some simple properties of γ. It is easily seen that $w \in \gamma(u)$ implies $\gamma(u) = [w]_\approx$. Let $k \geq 1$. If $u \in \text{WF} \cap (\Sigma \times \Gamma_k)^*$ where $\Gamma_k \stackrel{\text{def}}{=} \{f(r), r(r) \mid r \in [k]\}$, then all data words in $\gamma(u)$ are k-bounded. Moreover, $\gamma((\Sigma \times \Gamma_k)^*)$ is the set of all k-bounded data words.

Session Automata. As suggested, we consider automata over the alphabet $\Sigma \times \Gamma$ to process data words. Actually, they are equipped with a *finite* number $k \geq 1$ of registers so that we rather deal with finite automata over $\Sigma \times \Gamma_k$.

Definition 1. *Let $k \geq 1$. A k-register session automaton (or just session automaton) over Σ and D is a finite automaton over $\Sigma \times \Gamma_k$, i.e., a tuple $\mathcal{A} = (Q, q_0, F, \delta)$ where Q is the finite set of states, $q_0 \in Q$ the initial state, $F \subseteq Q$ the set of accepting states, and $\delta : Q \times (\Sigma \times \Gamma_k) \rightarrow 2^Q$ the transition function.*

The *symbolic language* $L_{symb}(\mathcal{A}) \subseteq (\Sigma \times \Gamma_k)^*$ of $\mathcal{A}$ is defined in the usual way, considering $\mathcal{A}$ as a finite automaton. Its *(data) language* is $L_{data}(\mathcal{A}) \stackrel{\text{def}}{=} \gamma(L_{symb}(\mathcal{A}))$. By Remark 1, $L_{data}(\mathcal{A})$ is closed under $\approx$. Moreover, it is k-bounded, which motivates the naming of our automata.

Example 1. Consider the 2-register session automaton $\mathcal{A}$ from Fig. 3(a). It recognizes the set of all 2-bounded data words over $\Sigma = \{a\}$.

Example 2. The 2-register session automaton $\mathcal{B}$ over $\Sigma = \{a, b\}$ from Fig. 3(b) represents a client-server system. A server can receive requests on two channels of capacity 1, represented by the two registers. Requests are acknowledged in the order in which they are received. When the automaton performs $(a, \mathsf{f}(r))$, a client gets a unique transaction key, which is stored in r. Later, the request is acknowledged performing $(b, \mathsf{r}(r))$. E.g., $(a, 8)(a, 4)(b, 8)(a, 3)(b, 4)(b, 3) \in L_{data}(\mathcal{B})$.

Example 3. Next, we present a 2-register session automaton that models a P2P protocol. A user can join a host with address x, denoted by action $(join, x)$. The request is either forwarded by x to another host y, executing $(forw_1, x)(forw_2, y)$, or acknowledged by (ack, x). In the latter case, a connection between the user and x is established so that they can communicate, indicated by action (com, x). Note that the sequence of actions $(forw_1, x)(forw_2, y)$ should be considered as an encoding of a single action $(forw, x, y)$ and is a way of dealing with actions that actually take two or more data values. An example execution of our protocol is $(join, 145)(forw, 145, 978)(forw, 978, 14)(ack, 14)(com, 14)(com, 14)(com, 14)$. In Fig. 3(c), we show the 2-register session automaton for the P2P protocol.

Session automata come with two natural notions of determinism. We call $\mathcal{A} = (Q, q_0, F, \delta)$ *symbolically deterministic* if $|\delta(q, (a, t))| \leq 1$ for all $q \in Q$ and $(a, t) \in \Sigma \times \Gamma_k$. Then, δ can be seen as a partial function $Q \times (\Sigma \times \Gamma_k) \to Q$. We call $\mathcal{A}$ *data deterministic* if it is symbolically deterministic and, for all $q \in Q$, $a \in \Sigma$, and $r_1, r_2 \in [k]$ with $r_1 \neq r_2$, we have that $\delta(q, (a, \mathsf{f}(r_1))) \neq \emptyset$ implies $\delta(q, (a, \mathsf{f}(r_2))) = \emptyset$. Intuitively, given a data word as input, the automaton is data deterministic if, in each state, given a letter and a data value, there is at most one fireable transition. While "data deterministic" implies "symbolically deterministic", the converse is not true. E.g., the session automata from Fig. 3(a) and 3(b) are symbolically deterministic but not data deterministic. However, the automaton of Fig. 3(b) with the dashed transition removed is data deterministic.

Theorem 1. *Session automata are strictly more expressive than data deterministic session automata.*

The proof can be found in the long version of the paper. Intuitively, data deterministic automata cannot guess if a data value in a register will be reused later.

Session automata are expressively incomparable with the various register automata models considered in [16,21,23,9,14]. In particular, due to freshness, the languages from Ex. 1, 2, and 3 are not recognizable by the models for which a learning algorithm exists [9,14]. On the other hand, our model cannot recognize "the set of all data words" or "every two consecutive data values are distinct". Our automata are subsumed by fresh-register automata [24], class memory automata [5], and data automata [6]. However, no algorithm for the inference of the latter is known. Note that, for ease of presentation, we consider one-dimensional data words, unlike [14] where labels have an arity and can carry several data values. Following [7], our automata can be easily extended to multi-dimensional data words (cf. Ex. 3). This also holds for the learning algorithm.

Canonical Session Automata. Our goal will be to infer the data language of a session automaton $\mathcal{A}$ in terms of a canonical session automaton $\mathcal{A}^C$.

As a first step, we associate with a data word $w = (a_1, d_1) \cdots (a_n, d_n) \in (\Sigma \times D)^*$ a *symbolic normal form* $snf(w) \in \text{WF}$ such that $w \in \gamma(snf(w))$, based on the idea that data values are always stored in the first register whose data value is not needed anymore. To do so, we will determine $t_1, \ldots, t_n \in \Gamma$ and set $snf(w) = (a_1, t_1) \cdots (a_n, t_n)$. We define $\tau : Fresh(w) \to \mathbb{N}_{>0}$ inductively by $\tau(i) = \min(FreeReg(i))$ where $FreeReg(i) \stackrel{\text{def}}{=} \mathbb{N}_{>0} \setminus \{\tau(i') \mid i' \in Fresh(w) \text{ such that } i' < i \text{ and } i \in Session(i')\}$. With this, we set, for all $i \in [n]$, $t_i = \mathsf{f}(\tau(i))$ if $i \in Fresh(w)$ and $t_i = \mathsf{r}(\tau(\min(Session(i))))$ otherwise. One readily verifies that $snf(w) = (a_1, t_1) \cdots (a_n, t_n)$ is well formed and that properties (i) and (ii) in the definition of a concretization hold. This proves $w \in \gamma(snf(w))$. E.g., Fig. 2 shows the symbolic normal form of the data word from Fig. 1. The mapping snf carries over to languages in the expected manner.

We consider again $\mathcal{B}$ of Fig. 3(b). Let $\mathcal{B}'$ be the automaton that we obtain from $\mathcal{B}$ when we remove the dashed transition. We have $L_{data}(\mathcal{B}) = L_{data}(\mathcal{B}')$, but $snf(L_{data}(\mathcal{B})) = L_{symb}(\mathcal{B}') \subsetneq L_{symb}(\mathcal{B})$.

Lemma 1. *Let L be a regular language over $\Sigma \times \Gamma_k$. Then, $snf(\gamma(L))$ is a regular language over $\Sigma \times \Gamma_k$.*

In other words, for every k-register session automaton $\mathcal{A}$, there is a k-register session automaton $\mathcal{A}'$ such that $L_{symb}(\mathcal{A}') = snf(L_{data}(\mathcal{A}))$ and, therefore, $L_{data}(\mathcal{A}') = L_{data}(\mathcal{A})$. We denote by $\mathcal{A}^C$ the minimal symbolically deterministic automaton $\mathcal{A}'$ satisfying $L_{symb}(\mathcal{A}') = snf(L_{data}(\mathcal{A}))$. Note that the number k' of registers effectively used in $\mathcal{A}^C$ may be smaller than k, and we actually consider $\mathcal{A}^C$ to be a k'-register session automaton.

Theorem 2. *Let $\mathcal{A} = (Q, q_0, F, \delta)$ be a k-register session automaton. Then, $\mathcal{A}^C$ has at most $2^{O(|Q| \times (k+1)! \times 2^k)}$ states. If $\mathcal{A}$ is data deterministic, then $\mathcal{A}^C$ has at most $O(|Q| \times (k+1)! \times 2^k)$ states. Finally, $\mathcal{A}^C$ uses at most k registers.*

3   Learning Session Automata

In this section, we introduce an active learning algorithm for session automata. In the usual active learning setting (as introduced by Angluin [2]), a *learner* interacts with a so-called minimally adequate *teacher* (MAT), an oracle which can answer *membership* and *equivalence* queries. In our case, the learner is given the task to infer the data language $L_{data}(\mathcal{A})$ defined by a given session automaton $\mathcal{A}$. We suppose here that the teacher knows the session automaton or any other device accepting $L_{data}(\mathcal{A})$. In practice, this might not be the case — $\mathcal{A}$ could be a black box — and equivalence queries could be (approximately) answered, for example, by extensive testing. The learner can ask if a *data* word is accepted by $\mathcal{A}$ or not. Furthermore it can ask equivalence queries which consist in giving an *hypothesis* session automaton to the teacher who either answers yes, if the hypothesis is equivalent to $\mathcal{A}$ (i.e., both data languages are the same), or gives a data word which is a counterexample, i.e., a data word that is either accepted by the hypothesis automaton but should not, or vice versa.

Given the data language $L_{data}(\mathcal{A})$ accepted by a session automaton $\mathcal{A}$ over Σ and D, our algorithm will learn the canonical k-register session automaton $\mathcal{A}^C$, i.e., the minimal symbolically deterministic automaton recognizing the data language $L_{data}(\mathcal{A})$ and the regular language $L_{symb}(\mathcal{A}^C)$ over $\Sigma \times \Gamma_k$. Therefore one can consider that the learning target is $L_{symb}(\mathcal{A}^C)$ and use any active learning algorithm for regular languages. However, as the teacher answers only questions over data words, queries have to be adapted. Since $\mathcal{A}^C$ only accepts symbolic words which are in normal form, a membership query for a given symbolic word u not in normal form will be answered negatively (without consulting the teacher); otherwise, the teacher will be given one data word included in $\gamma(u)$ (all the answers on words of $\gamma(u)$ are the same). Likewise, before submitting an equivalence query to the teacher, the learning algorithm checks if the current hypothesis automaton accepts symbolic words not in normal form[1]. If yes, one of those is taken as a counterexample, else an equivalence query is submitted to the teacher. Since the number of registers needed to accept a data language is a priori not known, the learning algorithm starts by trying to learn a 1-register session automaton and increases the number of registers as necessary.

Any active learning algorithm for regular languages may be adapted to our setting. Here we describe a variant of Rivest and Schapire's [22] algorithm which is itself a variant of Angluin's L^* algorithm [2]. An overview of learning algorithms for deterministic finite state automata can be found, for example, in [4].

The algorithm is based on the notion of *observation table* which contains the information accumulated by the learner during the learning process. An observation table over a given alphabet $\Sigma \times \Gamma_k$ is a triple $\mathcal{O} = (T, U, V)$ with U, V two sets of words over $\Sigma \times \Gamma_k$ such that $\varepsilon \in U \cap V$ and T is a mapping $(U \cup U \cdot (\Sigma \times \Gamma_k)) \times V \to \{+, -\}$. A table is partitioned into an upper part U and a lower part $U \cdot (\Sigma \times \Gamma_k)$. We define for each $u \in U \cup U \cdot (\Sigma \times \Gamma_k)$ a mapping $row(u): V \to \{+, -\}$ where $row(u)(v) = T(u, v)$. An observation table must satisfy the following property: for all $u, u' \in U$ such that $u \neq u'$ we have $row(u) \neq row(u')$, i.e., there exists $v \in V$ such that $T(u, v) \neq T(u', v)$. This means that the rows of the upper part of the table are pairwise distinct. A table is *closed* if, for all u' in $U \cdot (\Sigma \times \Gamma_k)$, there exists $u \in U$ such that $row(u) = row(u')$. From a closed table we can construct a symbolically deterministic session automaton whose states correspond to the rows of the upper part of the table:

Definition 2. *For a closed table $\mathcal{O} = (T, U, V)$ over a finite alphabet $\Sigma \times \Gamma_k$, we define a symbolically deterministic k-register session automaton $A_{\mathcal{O}} = (Q, q_0, F, \delta)$ over $\Sigma \times \Gamma_k$ by $Q = U$, $q_0 = \varepsilon$, $F = \{u \in Q \mid T(u, \epsilon) = +\}$, and for all $u \in Q$ and $(a, t) \in \Sigma \times \Gamma_k$, $\delta(u, (a, t)) = u'$ if $row(u(a, t)) = row(u')$. This is well defined as the table is closed.*

We now describe in detail our active learning algorithm for a given session automaton $\mathcal{A}$ given in Table 1. It is based on a loop which repeatedly constructs a

---

[1] This can be checked in polynomial time over the trimmed hypothesis automaton with a fixed point computation labelling the states with the registers that should be used again before overwriting them.

Table 1. The learning algorithm for a session automaton $\mathcal{A}$

---

initialize $k := 1$ and
$\mathcal{O} := (T, U, V)$ by $U = V = \{\varepsilon\}$ and $T(u, \varepsilon)$ for all $u \in U \cup U \cdot (\Sigma \times \Gamma_k)$ with membership queries
repeat
        while $\mathcal{O}$ is not closed
        do
                find $u \in U$ and $(a, t) \in \Sigma \times \Gamma_k$ such that for all $u \in U$: $row(u(a,t)) \neq row(u)$
                extend table to $\mathcal{O} := (T', U \cup \{u(a,t)\}, V)$ by membership queries
        from $\mathcal{O}$ construct the hypothesized automaton $\mathcal{A}_\mathcal{O}$ (cf. Definition 2)
        if $\mathcal{A}_\mathcal{O}$ accepts symbolic words not in normal form
            then let z be one of those
            else **if** $L_{data}(\mathcal{A}) = L_{data}(\mathcal{A}_\mathcal{O})$
                then equivalence test succeeds
                else get counterexample $w \in (L_{data}(\mathcal{A}) \setminus L_{data}(\mathcal{A}_\mathcal{O})) \cup (L_{data}(\mathcal{A}_\mathcal{O}) \setminus L_{data}(\mathcal{A}))$
                        set $z := snf(w)$; find minimal k' such that $z \in \Sigma \times \Gamma_{k'}$
                        if $k' > k$
                        then set $k := k'$
                              extend table to $\mathcal{O} := (T', U, V)$ over $\Sigma \times \Gamma_k$ by membership queries
        if $\mathcal{O}$ is closed /* is true if $k' \leq k$ */
            then find a breakpoint for z where v is the distinguishing word
                extend table to $\mathcal{O} := (T', U, V \cup \{v\})$ by membership queries
    until equivalence test succeeds
return $\mathcal{A}_\mathcal{O}$

---

closed table using membership queries, builds the corresponding automaton and
then asks an equivalence query. This is repeated until $\mathcal{A}$ is learned. An important part of any active learning algorithm is the treatment of counterexamples
provided by the teacher as an answer to an equivalence query. Suppose that for
a given $\mathcal{A}_\mathcal{O}$ constructed from a closed table $\mathcal{O} = (T, U, V)$ the teacher answers
by a counterexample data word w. Let $z = snf(w)$. If z uses more registers
than available in the current alphabet, we extend the alphabet and then the
table. If the obtained table is not closed, we restart from the beginning of the
loop. Otherwise – and also if z does not use more registers – we use Rivest and
Schapire's [22] technique to extend the table by adding a suitable v to V making it non-closed. The technique is based on the notion of breakpoint. As z is
a counterexample, (1) $z \in L_{symb}(\mathcal{A}_\mathcal{O}) \iff z \notin L_{symb}(\mathcal{A}^C)$. Let $z = z_1 \cdots z_m$.
Then, for any i with $1 \leq i \leq m + 1$, let z be decomposed as $z = u_i v_i$, where
$u_1 = v_{m+1} = \varepsilon$, $v_1 = u_{m+1} = z$ and the length of u_i is equal to $i - 1$ (we have
also $z = u_i z_i v_{i+1}$ for all i such that $1 \leq i \leq m$). Let s_i be the state visited by
z just before reading the ith letter, along the computation of z on $\mathcal{A}_\mathcal{O}$: i is a
breakpoint if $s_i z_i v_{i+1} \in L_{symb}(\mathcal{A}_\mathcal{O}) \iff s_{i+1} v_{i+1} \notin L_{symb}(\mathcal{A}^C)$. Because of
(1) such a break-point must exist and can be obtained with $O(\log(m))$ member-
ship queries by a dichotomous search. The word v_{i+1} is called the distinguishing
word. If V is extended by v_{i+1} the table is not closed anymore ($row(s_i)$ and
$row(s_i z_i)$ are different). Now, the algorithm closes the table again, then asks another equivalence query and so forth until termination. At each iteration of the
loop the number of rows (each of those correspond to a state in the automaton
$\mathcal{A}^C$) is increased by at least one. Notice that the same counterexample might be
given several times. The treatment of the counterexample only guarantees that
the table will contain one more row in its upper part. We obtain the following:

Theorem 3. *Let A be a k'-register session automaton over Σ and D. Let A^C be the corresponding canonical k-register session automaton. Let N be its number of states, K be the size of $\Sigma \times \Gamma_k$ and M the length of the longest counterexample returned by an equivalence query. Then, the learning algorithm for A terminates with at most $O(KN^2 + N\log(M))$ membership and $O(N)$ equivalence queries.*

Proof: This follows directly from the proof of correctness and complexity of Rivest and Schapire's algorithm [4,22]. Notice that the equivalence query cannot return a counterexample whose normal form uses more than k registers, as such a word is rejected by both A^C (by definition) and by A_O, (by construction). ■

Let us discuss the complexity of our algorithm. In terms of the canonical session automaton, the number of required membership and equivalence queries is polynomial. When we consider data deterministic session automata, the complexity is still polynomial in the number of states, but exponential in k (with constant base). As usual, we have to add one exponent wrt. (data) non-deterministic automata. In [14], the number of equivalence queries is polynomial in the size of the underlying automaton. In contrast, the number of membership queries contains a factor n^k where n is the number of states and k the number of registers. This may be seen as a drawback, as n is typically large. Note that [14] restrict to deterministic automata, since classical register automata are not determinizable.

Table 2. The successive observation tables

O_1

	ε
ε	+
$(b,r(1))$	−
$(a,f(1))$	+
$(b,r(1))\text{-}$	−

$\Rightarrow$

O_2

	ε	$(b,r(1))$
ε	+	−
$(b,r(1))$	−	−
$(a,f(1))$	+	+
$(b,r(1))\text{-}$	−	−
$(a,f(1))(a,f(1))$	+	+
$(a,f(1))(b,r(1))$	+	+

$\Rightarrow$

O_3

	ε	$(b,r(1))$
ε	+	−
$(b,r(1))$	−	−
$(a,f(1))$	+	+
$(a,f(2))$	−	−
$(b,r(2))$	−	−
$(b,r(1))\text{-}$	−	−
$(a,f(1))(a,f(1))$	+	+
$(a,f(1))(b,r(1))$	+	+
$(a,f(1))(a,f(2))$	−	+
$(a,f(1))(b,r(2))$	−	−

$\Rightarrow$

O_4

	ε	$(b,r(1))$
ε	+	−
$(b,r(1))$	−	−
$(a,f(1))$	+	+
$(a,f(1))(a,f(2))$	−	+
$(a,f(2))$	−	−
$(b,r(2))$	−	−
$(b,r(1))\text{-}$	−	−
$(a,f(1))(a,f(1))$	+	+
$(a,f(1))(b,r(1))$	+	+
$(a,f(1))(b,r(2))$	−	−
$(a,f(1))(a,f(2))(a,f(1))$	−	−
$(a,f(1))(a,f(2))(b,r(1))$	+	+
$(a,f(1))(a,f(2))(a,f(2))$	−	+
$(a,f(1))(a,f(2))(b,r(2))$	−	+

$\Rightarrow$

O_5

	ε	$(b,r(1))$	$(b,r(2))$
ε	+	−	−
$(b,r(1))$	−	−	−
$(a,f(1))$	+	+	−
$(a,f(1))(a,f(2))$	−	+	−
$(a,f(1))(a,f(2))(b,r(1))$	+	+	+
$(a,f(2))$	−	−	−
$(b,r(2))$	−	−	−
$(b,r(1))\text{-}$	−	−	−
$(a,f(1))(a,f(1))$	+	+	−
$(a,f(1))(b,r(1))$	+	+	−
$(a,f(1))(b,r(2))$	−	−	−
$(a,f(1))(a,f(2))(a,f(1))$	−	−	−
$(a,f(1))(a,f(2))(a,f(2))$	−	+	−
$(a,f(1))(a,f(2))(b,r(2))$	−	+	−
$(a,f(1))(a,f(2))(b,r(1))(a,f(1))$	+	+	+
$(a,f(1))(a,f(2))(b,r(1))(b,r(1))$	+	+	+
$(a,f(1))(a,f(2))(b,r(1))(a,f(2))$	−	+	−
$(a,f(1))(a,f(2))(b,r(1))(b,r(2))$	+	+	+

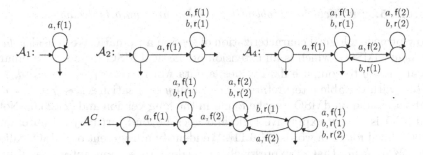

Fig. 4. The successive hypothesis automata

Example 4. We apply our learning algorithm on the data language generated by a single state automaton with loops labelled by $(a, \mathsf{f}(1))$, $(b, \mathsf{r}(1))$, $(a, \mathsf{f}(2))$ and $(b, \mathsf{r}(2))$. Table 2 shows the successive observation tables constructed by the algorithm[2], and Fig. 4 the successive automata constructed from the closed observation tables. For sake of clarity we omit the sink states. We start with the alphabet $\Sigma \times \Gamma_1 = \{(a, \mathsf{f}(1)), (a, \mathsf{r}(1)), (b, \mathsf{f}(1)), (b, \mathsf{r}(1))\}$. Table $\mathcal{O}_1$ is obtained after initialization and closing by adding $(b, \mathsf{r}(1))$ to the top: hypothesis automaton $\mathcal{A}_1$ is constructed. Suppose that the equivalence query gives back as counterexample the data word $(a, 3)(b, 3)$ whose normal form is $(a, \mathsf{f}(1))(b, \mathsf{r}(1))$. Here the breakpoint yields the distinguishing word $(b, \mathsf{r}(1))$. Adding it to V and closing the table by adding $(a, \mathsf{f}(1))$ to the top, we get table $\mathcal{O}_2$ yielding hypothesis automaton $\mathcal{A}_2$. Notice that $L_{symb}(\mathcal{A}_2) = L_{symb}(\mathcal{A}^C) \cap (\Sigma \times \Gamma_1)^*$: the equivalence query must now give back a data word whose normal form is using at least 2 registers (here $(a, 7)(a, 4)(b, 7)$ with normal form $(a, \mathsf{f}(1))(a, \mathsf{f}(2))(b, \mathsf{r}(1)))$. Then we must extend the alphabet to $\Sigma \times \Gamma_2$ and obtain table $\mathcal{O}_3$. We close the table and get $\mathcal{O}_4$. After the equivalence query with the hypothesis automaton $\mathcal{A}_4$ we get $(a, \mathsf{f}(1))(a, \mathsf{f}(2))(b, \mathsf{r}(1))(b, \mathsf{r}(2))$ as normal form of the data word counterexample $(a, 9)(a, 3)(b, 9)(b, 3)$. After adding $(b, \mathsf{r}(2))$ to V and closing the table by moving $(a, \mathsf{f}(1))(a, \mathsf{f}(2))(b, \mathsf{r}(1))$ to the top, we get the table $\mathcal{O}_5$ from which the canonical automaton $\mathcal{A}^C$ is obtained and the equivalence query succeeds.

4   Language Theoretical Results

In this section, we establish some language theoretical properties of session automata, which they inherit from classical regular languages. These results demonstrate a certain robustness as required in verification tasks such as compositional verification [10] and infinite-state regular model checking [13].

Theorem 4. *Data languages recognized by session automata are closed under intersection and union. They are also closed under complementation in the following sense: given a k-register session automaton $\mathcal{A}$, the language $\gamma((\Sigma \times \Gamma_k)^*) \setminus L_{data}(\mathcal{A})$ is recognized by a k-register session automaton.*

---

[2] To save space some letters whose rows contain only $-$'s are omitted. Moreover, we use $\_$ to indicate that all letters will lead to the same row.

Theorem 5. *The inclusion problem for session automata is decidable.*

We now provide a logical characterization of session automata. We consider *data MSO logic* (dMSO), which is an extension of classical MSO logic by the binary predicate $x \sim y$ to compare data values: a data word $w = (a_1, d_1) \cdots (a_n, d_n) \in (\Sigma \times D)^*$ with variable interpretation $x \mapsto i$ and $y \mapsto j$ satisfies $x \sim y$ if $d_i = d_j$. More background on dMSO may be found in the long version and [21,23,6]. Note that dMSO is a very expressive logic and goes beyond virtually all automata models defined for data words [21,6,11]. We identify a fragment of dMSO, called *session MSO logic*, that is expressively equivalent to session automata. While register automata also enjoy a logical characterization [11], we are not aware of logics capturing the automata model considered in [14].

Definition 3. *A session MSO (sMSO) formula is a dMSO sentence of the form* $\varphi = \exists X_1 \cdots \exists X_m \, (\alpha \wedge \forall x \forall y \, (x \sim y \leftrightarrow \beta))$ *such that α and β are classical MSO formulas (not containing the predicate $\sim$).*

Example 5. For instance, $\varphi_1 = \forall x \forall y \, (x \sim y \leftrightarrow x = y)$ is an sMSO formula. Its semantics $L_{data}(\varphi_1)$ is the set of data words in which every data value occurs at most once. Moreover, $\varphi_2 = \forall x \forall y \, (x \sim y \leftrightarrow true)$ is an sMSO formula, and $L_{data}(\varphi_2)$ is the set of data words where all data values coincide. As a last example, let $\varphi_3 = \exists X \, \forall x \forall y \, (x \sim y \leftrightarrow (\neg \exists z \in X \, (x < z \leq y \vee y < z \leq x)))$. Then, $L_{data}(\varphi_3)$ is the set of 1-bounded data words. Intuitively, the second-order variable X represents the set of positions where a fresh data value is introduced.

Theorem 6. *A data language is recognized by a session automaton iff it is definable by an sMSO formula.*

In [7], it was already shown (for a more powerful model with pushdown stacks) that model checking for the *full* dMSO logic is decidable:

Theorem 7 ([7]). *Given a session automaton $\mathcal{A}$ and a dMSO sentence φ, one can decide whether $L_{data}(\mathcal{A}) \subseteq L_{data}(\varphi)$.*

5   Conclusion

In this paper, we provided a complete framework for algorithmic learning of session automata, a special class of register automata to process data words. As a key ingredient, we associated with every session automaton a canonical one, which revealed close connections with classical regular languages. This also allowed us to show that session automata form a robust language class with good closure and decidability properties as well as a characterization in MSO logic. As a next step, we plan to employ our setting for various verification tasks.

Acknowledgment. We are grateful to Thomas Schwentick for suggesting the symbolic normal form of data words.

References

1. Aarts, F., Heidarian, F., Kuppens, H., Olsen, P., Vaandrager, F.W.: Automata learning through counterexample guided abstraction refinement. In: Giannakopoulou, D., Méry, D. (eds.) FM 2012. LNCS, vol. 7436, pp. 10–27. Springer, Heidelberg (2012)
2. Angluin, D.: Learning regular sets from queries and counterexamples. Information and Computation 75(2), 87–106 (1987)
3. Berg, T., Grinchtein, O., Jonsson, B., Leucker, M., Raffelt, H., Steffen, B.: On the correspondence between conformance testing and regular inference. In: Cerioli, M. (ed.) FASE 2005. LNCS, vol. 3442, pp. 175–189. Springer, Heidelberg (2005)
4. Berg, T., Raffelt, H.: Model Checking. In: Broy, M., Jonsson, B., Katoen, J.-P., Leucker, M., Pretschner, A. (eds.) Model-Based Testing of Reactive Systems. LNCS, vol. 3472, pp. 557–603. Springer, Heidelberg (2005)
5. Björklund, H., Schwentick, T.: On notions of regularity for data languages. Theoretical Computer Science 411(4-5), 702–715 (2010)
6. Bojanczyk, M., David, C., Muscholl, A., Schwentick, T., Segoufin, L.: Two-variable logic on data words. ACM Trans. Comput. Log. 12(4), 27 (2011)
7. Bollig, B., Cyriac, A., Gastin, P., Narayan Kumar, K.: Model checking languages of data words. In: Birkedal, L. (ed.) FOSSACS 2012. LNCS, vol. 7213, pp. 391–405. Springer, Heidelberg (2012)
8. Bollig, B., Katoen, J.-P., Kern, C., Leucker, M., Neider, D., Piegdon, D.R.: libalf: The automata learning framework. In: Touili, T., Cook, B., Jackson, P. (eds.) CAV 2010. LNCS, vol. 6174, pp. 360–364. Springer, Heidelberg (2010)
9. Cassel, S., Howar, F., Jonsson, B., Merten, M., Steffen, B.: A succinct canonical register automaton model. In: Bultan, T., Hsiung, P.-A. (eds.) ATVA 2011. LNCS, vol. 6996, pp. 366–380. Springer, Heidelberg (2011)
10. Cobleigh, J.M., Giannakopoulou, D., Păsăreanu, C.S.: Learning assumptions for compositional verification. In: Garavel, H., Hatcliff, J. (eds.) TACAS 2003. LNCS, vol. 2619, pp. 331–346. Springer, Heidelberg (2003)
11. Colcombet, T., Ley, C., Puppis, G.: On the use of guards for logics with data. In: Murlak, F., Sankowski, P. (eds.) MFCS 2011. LNCS, vol. 6907, pp. 243–255. Springer, Heidelberg (2011)
12. Giannakopoulou, D., Magee, J.: Fluent model checking for event-based systems. In: ESEC / SIGSOFT FSE, pp. 257–266. ACM (2003)
13. Habermehl, P., Vojnar, T.: Regular Model Checking Using Inference of Regular Languages. In: INFINITY 2004. ENTCS, vol. 138, pp. 21–36. Elsevier (2005)
14. Howar, F., Steffen, B., Jonsson, B., Cassel, S.: Inferring canonical register automata. In: Kuncak, V., Rybalchenko, A. (eds.) VMCAI 2012. LNCS, vol. 7148, pp. 251–266. Springer, Heidelberg (2012)
15. Jonsson, B.: Learning of automata models extended with data. In: Bernardo, M., Issarny, V. (eds.) SFM 2011. LNCS, vol. 6659, pp. 327–349. Springer, Heidelberg (2011)
16. Kaminski, M., Francez, N.: Finite-memory automata. Theoretical Computer Science 134(2), 329–363 (1994)
17. Kürtz, K.O., Küsters, R., Wilke, T.: Selecting theories and nonce generation for recursive protocols. In: FMSE, pp. 61–70. ACM (2007)
18. Leucker, M.: Learning meets verification. In: de Boer, F.S., Bonsangue, M.M., Graf, S., de Roever, W.-P. (eds.) FMCO 2006. LNCS, vol. 4709, pp. 127–151. Springer, Heidelberg (2007)

19. Margaria, T., Raffelt, H., Steffen, B., Leucker, M.: The LearnLib in FMICS-jETI. In: ICECCS, pp. 340–352. IEEE Computer Society Press (2007)
20. Milner, R., Parrow, J., Walker, D.: A calculus of mobile processes, Parts I and II. Information and Computation 100, 1–77 (1992)
21. Neven, F., Schwentick, T., Vianu, V.: Finite state machines for strings over infinite alphabets. ACM Transactions on Computational Logic 5(3), 403–435 (2004)
22. Rivest, R., Schapire, R.: Inference of finite automata using homing sequences. Information and Computation 103, 299–347 (1993)
23. Segoufin, L.: Automata and logics for words and trees over an infinite alphabet. In: Ésik, Z. (ed.) CSL 2006. LNCS, vol. 4207, pp. 41–57. Springer, Heidelberg (2006)
24. Tzevelekos, N.: Fresh-register automata. In: POPL, pp. 295–306. ACM (2011)

Suffixes, Conjugates and Lyndon Words

Silvia Bonomo, Sabrina Mantaci, Antonio Restivo,
Giovanna Rosone, and Marinella Sciortino

University of Palermo, Dipartimento di Matematica e Informatica, Italy
{sabrina,restivo,giovanna,mari}@math.unipa.it,
bonomosilvia@gmail.com

Abstract. In this paper we are interested in the study of the combinatorial aspects connecting three important constructions in the field of string algorithms: the suffix array, the Burrows-Wheeler transform (BWT) and the extended Burrows-Wheeler transform (EBWT). Such constructions involve the notions of suffixes and conjugates of words and are based on two different order relations, denoted by $<_{lex}$ and $\prec_\omega$, that, even if strictly connected, are quite different from the computational point of view. In this study an important role is played by Lyndon words. In particular, we improve the upper bound on the number of symbol comparisons needed to establish the $\prec_\omega$ order between two primitive words by using a preliminary knowledge of the $<_{lex}$ order of the corresponding Lyndon conjugates. Moreover, we propose an algorithm that efficiently sorts, according to the $\prec_\omega$ order, the list of conjugates of a multiset of Lyndon words. Finally, we show that the Lyndon factorization of a word helps the construction of its suffix array, allowing a reduction of the number of symbol comparisons needed to lexicographically sort the suffixes of the word.

Keywords: Lyndon words, Lyndon factorization, BWT, Suffix array, EBWT, Circular words, Conjugacy.

1   Introduction

In the field of String Algorithms, there are three constructions that have received an increasing attention during the last decades: the suffix array (SA) [25], the Burrows-Wheeler Transform (BWT) [5] and a more recent extension of BWT to a multiset of words [22].

The SA of a word w is defined as the permutation of integers giving the starting positions of the suffixes of w in lexicographical order (denoted by $<_{lex}$). It is a space-saving data structure alternative to suffix tree and it is used efficiently in string processing problems where suffix tree methodology is applicable.

The BWT of a word w is a word $\mathtt{bwt}(w)$ obtained by a letters permutation of w induced by the sorted list of the conjugates of w. It is an important preprocessing for several text compressors for its property of compression booster (cf. [26,10]).

In [7], Crochemore, Désarménien and Perrin pointed out the very interesting fact that the BWT coincides with a particular case of a bijection defined in [13]

M.-P. Béal and O. Carton (Eds.): DLT 2013, LNCS 7907, pp. 131–142, 2013.
© Springer-Verlag Berlin Heidelberg 2013

by Gessel and Reutenauer. This important remark suggested to define in [22] an extension of the BWT (called EBWT) to a multiset S of primitive words, that is somehow an algorithmic presentation of the bijection introduced in [13]. The EBWT of a multiset S is a word obtained by letter permutation of the words in S induced by the sorted list of the conjugates of words in S according with an order relation (denoted by $\prec_\omega$) defined by using lexicographic order between infinite words.

The EBWT has been used for circular words comparison [23], for circular pattern matching [17], for the preprocessing of a compressor (cf. [19,15,21]). In all the applications related to the EBWT, the bottleneck is the sorting process, i.e. the $\prec_\omega$ order computation.

The three constructions are connected to each other, with the difference that in the suffix array we sort the *suffixes* of a word according with the $<_{lex}$ order, whereas in the BWT we sort the *conjugates* of a word according to the $<_{lex}$ order and in the EBWT we sort the *conjugates* of the words in the multiset S according with $\prec_\omega$ order.

If we consider the number of symbol comparisons needed to establish the order between two words, there exists a hierarchy among the complexity of the three constructions: the computation of the SA is the fastest in practice because we only compare the suffixes of a word, whereas the computation of the EBWT appears to be the most expensive because the number of symbol comparisons to establish the $\prec_\omega$ order between two words exceeds, in general, the lengths of the words.

Actually, in several implementations of the BWT (or the EBWT) in order to improve the efficiency of the algorithm, end-marker symbols are appended to the end of the words. In this way, the BWT computation can be reduced to a SA construction. However, it is important to observe that as consequence of this alteration, BWT produces a different output. Note that the strict connection between BWT and SA has an important applicative impact (cf. [1]). Moreover, many combinatorial properties of SA by using the connection with BWT are proved in [18]. In the case of EBWT, the idea of appending the same end-marker to each word of the multiset fails, because we lose the reversibility of the transformation. Such property is essential in several applications in the field of data compression. It would be useful to append a different special symbol to each word in S (cf. [2]), but besides producing different outputs, it destroys the circular nature of the transformation. So we don't consider here this variant.

In this paper we study the combinatorics underlying the above three constructions. In this study an important role is played by the notion of Lyndon word, which is the least word, with respect the lexicographic order, in a conjugacy class. The general idea at the base of our results is that, in order to establish the $<_{lex}$ order of two suffixes (or the $\prec_\omega$ order of two conjugates) of two Lyndon words, one can take advantage of a preliminary knowledge of the $<_{lex}$ order of the Lyndon words.

In particular, in Section 3 we show that the sorting of the conjugates of two Lyndon words and the lexicographic order between their suffixes are related. From this fact, we deduce an improvement of the upper bound on the number of symbol comparisons needed to establish the $\prec_\omega$ order between two primitive words when the lexicographic order between the corresponding Lyndon conjugates is known.

Section 4 is devoted to the construction of an algorithm that by using the results of Section 3 produces the $\prec_\omega$ order among all the conjugates of the Lyndon words in a lexicographically sorted multiset by a suitable comparison among the suffixes of such words. Such an algorithm can be used for an efficient computation of the EBWT.

Lyndon words also play a leading role in the computation of the suffix array of a word w. Indeed, in Section 5, by using the Lyndon factorization of a word w, we prove that the lexicographical order between two suffixes of w can be established by a number of symbol comparisons that in general is smaller than the usual one that involves the longest common prefix between the suffixes. The results of this section could suggest new strategies for the suffix array computation.

2   Preliminaries

Let $\Sigma = \{c_1, c_2, \ldots, c_\sigma\}$ be a finite alphabet with $c_1 < c_2 < \ldots < c_\sigma$. Given a finite word $w = a_1 a_2 \cdots a_n$, $a_i \in \Sigma$ for $i = 1, \ldots, n$, a *factor* of w is written as $w[i, j] = a_i \cdots a_j$. A factor $w[1, j]$ is called a *prefix*, while a factor $w[i, n]$ is called a *suffix*. In this paper, we also denote by $suf_k(w)$ (resp. $pref_k(w)$) the suffix (resp. prefix) of w that has length k.

We say that $x, y \in \Sigma^*$ are *conjugate* or y *is a conjugate of* x if $x = uv$ and $y = vu$ for some $u, v \in \Sigma^*$. Recall that conjugacy is an equivalent relation. We denote by $conj_k(w)$ the conjugate of w starting from the position $|w| - k + 1$, i.e. if $w = xy$ and $|y| = k$ then $yx = conj_k(w)$. A word $v \in \Sigma^*$ is *primitive* if $v = u^n$ implies $v = u$ and $n = 1$. In what follows we only deal with primitive words.

If $u \in \Sigma^*$, we denote by u^ω the infinite word obtained by infinitely iterating u, i.e. $u^\omega = uuuuu\ldots$. The usual lexicographic order $<_{lex}$ between finite words can be naturally extended on infinite words, that is, taken two infinite words $x = x_1 x_2 \cdots$ and $y = y_1 y_2 \cdots$, with $x_i, y_i \in \Sigma$, we say that $x <_{lex} y$ if there exists an index $j \in \mathbb{N}$ such that $x_i = y_i$ for $i = 1, 2, \ldots, j - 1$ and $x_j < y_j$. Note that $x \leq_{lex} y$ means that either $x = y$ or $x <_{lex} y$.

Lexicographic order on infinite words allows to define the following order relation on finite words: given two primitive words u and v, we say that $u \preceq_\omega v \Leftrightarrow u^\omega <_{lex} v^\omega$. Note that this order can also be defined for non-primitive words (cf. [22]), but this case is not considered in this paper.

Although the $\prec_\omega$ order of u and v is defined by using infinite words, the following theorem (cf. [22]), that is a consequence of Fine and Wilf theorem in [12], shows that this order can be established with a bounded number of symbol comparison.

Theorem 1. *Let u, v be two primitive words over a finite alphabet Σ. If $k = |u| + |v| - \gcd(|u|, |v|)$ we have that:*

$$u \prec_\omega v \Leftrightarrow pref_k(u^\omega) <_{lex} pref_k(v^\omega).$$

The following example shows that the $<_{lex}$ order can be different from the $\prec_\omega$ order when one word is a prefix of the other. Moreover it shows that the bound given in Theorem 1 is tight.

Example 1. Consider $u = abaab$ and $v = abaababa$. Although $u <_{lex} v$, we have $v \prec_\omega u$. Moreover u^ω and v^ω differ for the character in position $k = 12 = 5+8-1$. Remark that, for any $h < k$, $pref_h(u^\omega) = pref_k(v^\omega)$, i.e. k is tight.

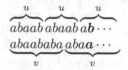

A *Lyndon* word is a primitive word which is also the minimum in its conjugacy class, with respect to the lexicographic order relation. In [20,8], one can find a linear algorithm that for any word $w \in \Sigma^*$ computes the Lyndon word of its conjugacy class. We call it the Lyndon word of w and we denote it by T_w. Lyndon words are involved in a nice and important factorization property of words.

Theorem 2. *[6] Every word $w \in \Sigma^+$ has a unique factorization $w = w_1 \cdots w_s$ such that $w_1 \geq_{lex} \cdots \geq_{lex} w_s$ is a non-increasing sequence of Lyndon words.*

The Lyndon factorization of a given word can be computed in linear time (see for instance [8,20]).

The Burrows-Wheeler Transform (BWT) [5] is intuitively described as follows: given a word $w \in \Sigma^*$, $\mathtt{bwt}(w)$ is a word obtained by sorting the list of the conjugates of w and by concatenating the last symbol of each element in the sorted list. If w and v are conjugate words, it is easy to see that $\mathtt{bwt}(w) = \mathtt{bwt}(v)$. With the additional information of the position of w in the sorted list, the BWT becomes an invertible transform, i.e., we can recover w from $\mathtt{bwt}(w)$.

The Extended Burrows-Wheeler transform (EBWT) [22] is defined as follows: given a multiset of words $\mathcal{S} = \{S_1, S_2, \ldots, S_k\}$, $\mathtt{ebwt}(\mathcal{S})$, is obtained by sorting the list of conjugates of $\mathcal{S}$ according to the $\prec_\omega$ order relation and by concatenating the last symbol of each element in the sorted list. Due to results in [13], EBWT is a reversible transformation. As one can easily see, the hardest computational step of EBWT consists in sorting the conjugates of a set of words according to the $\prec_\omega$ order relation.

3   Comparing Conjugates and Suffixes of Lyndon Words

In this section we consider two Lyndon words T_1 and T_2 and we show that there exists a relation between the $<_{lex}$ order of their suffixes and the $\prec_\omega$ order of their conjugates.

Let us first consider the special case of the suffixes and the conjugates of a single Lyndon word T. The following theorem can be deduced as trivial consequence of the fact that, for words of the same length, the $<_{lex}$ order coincides with the $\prec_\omega$ order and as consequence of the properties of Lyndon words (cf. [14, Lemma 12]).

Theorem 3. *Let T be a Lyndon word. For any integers h, k with $1 \leq h, k \leq |T|$, the following statements are equivalent:*

i) $conj_h(T) <_{lex} conj_k(T)$;
ii) $conj_h(T) \prec_\omega conj_k(T)$;
iii) $suf_h(T) <_{lex} suf_k(T)$.

Consider now two distinct Lyndon words T_1 and T_2. We first show that the $<_{lex}$ order and the $\prec_\omega$ order coincide for Lyndon words.

Theorem 4. *Let T_1 and T_2 be two Lyndon words, then $T_1 \leq_{lex} T_2$ if and only if $T_1 \prec_\omega T_2$.*

The following lemmas take into consideration the generic conjugates $conj_h(T_1)$ and $conj_k(T_2)$ of two distinct Lyndon words.

Lemma 1. *Let T_1 and T_2 be two distinct Lyndon words. If $T_1 <_{lex} T_2$ and $h \leq k$ then*

$$conj_h(T_1) \prec_\omega conj_k(T_2) \Leftrightarrow suf_h(T_1) \leq_{lex} suf_k(T_2).$$

The following example shows that if $k < h$ and $suf_k(T_2)$ is a prefix of $suf_h(T_1)$, the relative order of the suffixes is not sufficient to establish the $\prec_\omega$ order between the conjugates.

Example 2. Let $\Sigma = \{a, b, c\}$, $T_1 = aacab$ and $T_2 = acbcc$, where $T_1 \prec_\omega T_2$. We consider the following conjugates: $conj_3(T_1) = cabaa$, $conj_1(T_2) = cacbc$. In this case we have $T_1 <_{lex} T_2$ and $conj_3(T_1) \prec_\omega conj_1(T_2)$.
Consider now the Lyndon words $T_1 = aacab$ and $T_2 = aacbc$, where $T_1 \prec_\omega T_2$. We consider the following conjugates: $conj_3(T_1) = cabaa$, $conj_1(T_2) = caacb$. In this case we have $T_1 <_{lex} T_2$ and $conj_1(T_2) \prec_\omega conj_3(T_1)$.

Lemma 2. *Let T_1 and T_2 be two distinct Lyndon words. If $T_1 <_{lex} T_2$ and $h > k$, then*

$$conj_h(T_1) \prec_\omega conj_k(T_2) \Leftrightarrow suf_h(T_1) \leq_{lex} suf_k(T_2)pref_{h-k}(T_2^\omega).$$

Remark that, both in Lemma 1 and in Lemma 2 we need at most h symbol comparisons in order to establish the $\prec_\omega$ order between $conj_h(T_1)$ and $conj_k(T_2)$. Hence, from previous lemmas one can derive the following theorem that relates the $\prec_\omega$ order between the conjugates of two arbitrary primitive words and the $<_{lex}$ order between some prefixes of their infinite iterations.

Theorem 5. *Let u and v be primitive words, let T_u and T_v be their corresponding Lyndon words. Let suppose $T_u <_{lex} T_v$ and let r be the integer such that $u = conj_r(T_u)$. Then*

$$u \prec_\omega v \Leftrightarrow pref_r(u^\omega) \leq_{lex} pref_r(v^\omega).$$

The theorem states that we can determine the $\prec_\omega$ order of two primitive words u and v only by looking at the first r characters of the infinite words u^ω and v^ω: if there is a mismatch within the first r characters, then the $\prec_\omega$ order is determined by the order of the letters corresponding to such a mismatch; otherwise, the $\prec_\omega$ order is decided by the order of the corresponding Lyndon words T_u and T_v.

Remark that even if r can be much smaller than $|u|+|v|-gcd(|u|,|v|)$, Theorem 5 does not contradict the tightness of the bound given in Theorem 1, since here we have the supplementary information on the order of the corresponding Lyndon words.

Consider, for instance, the words in Example 1, $u = abaab$ and $v = abaababa$. The corresponding Lyndon words are $T_u = aabab$ and $T_v = aabaabab$. We have that $T_v <_{lex} T_u$, $u = conj_2(T_u)$ and $v = conj_7(T_v)$. Consider the infinite words

$$u^\omega = abaababaabab \cdots$$
$$v^\omega = abaababaabaa \cdots$$

The first mismatch between u^ω and v^ω is in the position 12. Nevertheless since $pref_7(v^\omega) = pref_7(u^\omega)$, by Theorem 5 we can conclude that $v \prec_\omega u$.

Theorem 5 suggests that, in order to establish the $\prec_\omega$ order of two primitive words u and v, one could use the following procedure: first determine the $<_{lex}$ order of T_u and T_v, and then decide the $\prec_\omega$ order by looking only the first r characters of u^ω and v^ω.

However the above example shows that the total number of comparisons with this procedure is the same as the one given by the bound in Theorem 1. In the above example we need 5 comparisons in order to state that $T_v <_{lex} T_u$ and 7 comparisons to state that $v \prec_\omega u$, i.e. 12 comparisons, which is the same number of comparisons that we need in order to find a mismatch between u^ω and v^ω.

However, such a strategy takes advantage with respect the usual comparison between conjugates when more than two pairs of conjugates have to be compared. In fact in this case the cost of comparing the Lyndon words is amortized on the total number of comparisons among conjugates. Such a consideration is enforced when multisets of words are considered. This is a starting point for the algorithm proposed in the next section.

4   Sorting the Conjugates of a Multiset of Lyndon Words

In this section we assume that $\mathcal{T} = \{T_1, T_2, \ldots, T_m\}$ is a lexicographically sorted multiset of Lyndon words. Such an hypothesis, although strong, is not restrictive because if $\mathcal{S}$ is a generic multiset of primitive words we can obtain $\mathcal{T}$ in linear time by computing for each word the corresponding Lyndon conjugate (cf. [20]), and then by sorting this multiset.

Nevertheless we can notice that when the Lyndon factorization of any word w is performed (cf. [8]), this sorted list of Lyndon words is naturally obtained "for free". As shown in previous section, the hypothesis that the elements of $\mathcal{T}$ are Lyndon words suggests to connect the problem to the $\prec_\omega$ sorting of the conjugates to the lexicographic sorting of the suffixes of the elements of $\mathcal{T}$. In this connection, the results of the previous section also show an asymmetry in the roles played by the Lyndon words involved into the sorting. An immediate application of the considerations of the previous section could lead to an algorithm that is more efficient in terms of comparisons with the bound imposed by Theorem 1 among all conjugates. A more careful use of the results has allowed us to develop the algorithm described in this section that produces the $\prec_\omega$ sorting of the conjugates of a multiset $\mathcal{T}$ of Lyndon words by a suitable comparison among the suffixes of such words.

Our strategy analyzes all the conjugates of the words in $\mathcal{T}$ from all the conjugates of the greatest Lyndon word to all the conjugates of the smallest one.

Due the circular nature of the conjugacy relation, we assume that both $conj_0(T_i)$ and $conj_{|T_i|}(T_i)$ denote the word T_i. We denote by $CA(\mathcal{T})$ the conjugate array of $\mathcal{T}$, i.e. the list of the positions of all the conjugates of the words of $\mathcal{T}$ sorted according to the $\prec_\omega$ relation. In particular, $CA(\mathcal{T})[\gamma] = (i, h)$ if the conjugate $conj_h(T_i)$ is the γ-th smallest conjugate in the $\prec_\omega$ sorted list of conjugates. We denote by $CA_i(\mathcal{T})$ the partial conjugate array containing the positions of the $\prec_\omega$ sorted list of all the conjugates of T_j, $j = i, \ldots, m$. Moreover, $CA_i^h(\mathcal{T})$ denotes the array containing the positions of the $\prec_\omega$ sorted list of all the conjugates of T_j, $j = i+1, \ldots, m$ together with conjugates $conj_k(T_i)$ with $0 \le k \le h$. Note that $CA_i(\mathcal{T}) = CA_i^{|T_i|-1}(\mathcal{T})$.

Finally, we denote by $B_i(\mathcal{T})$ (resp. $B_i^k(\mathcal{T})$) the array of characters such that, if $CA_i(\mathcal{T})[\gamma] = (j, h)$ (resp. $CA_i^k(\mathcal{T})[\gamma] = (j, h)$), then $B_i(\mathcal{T})[\gamma]$ (resp. $B_i^k(\mathcal{T})[\gamma]$) contains the last character of $conj_h(T_j)$, or in other words, it is the symbol in position $|T_j| - h$ in the word T_j. Note that $CA(\mathcal{T}) = CA_1(\mathcal{T})$. Remark that the concatenations of the symbols in $B_1(\mathcal{T})$ coincides with the $\mathtt{ebwt}(\mathcal{S}) = \mathtt{ebwt}(\mathcal{T})$ defined in [22] (see also [17]) and the conjugate array is related to the generalized suffix array defined in [4].

Example 3. Let $\mathcal{T} = \{aaacab, acbcc\}$ an ordered multiset of Lyndon words. $\mathcal{T}$ can be represented by a right justified table in which the rows are the Lyndon words and the columns represent the starting points of the conjugates of the words. Such a table is depicted in Figure 1. One can see that
$CA(\mathcal{T}) = [(1,6), (1,5), (1,2), (1,4), (2,5), (1,1), (2,3), (1,3), (2,1), (2,4), (2,2)]$
and $B_1(\mathcal{T}) = bacacacacab$.

Our algorithm consists of m steps and at each step i (from m down to 1) the partial conjugate array $CA_i(\mathcal{T})$ and the array B_i are computed, i.e. the $\prec_\omega$ sorted list of all conjugates of $T_i, T_{i+1}, \ldots, T_m$ is created. During each step i the sorted list is incrementally built by considering all the conjugates of the word T_i from the rightmost to the leftmost one, and by adding such a conjugate according with the $\prec_\omega$ order to the partial sorted list computed in the previous

Fig. 1. Table representing the multiset $\mathcal{T} = \{aaacab, acbcc\}$

steps. Such insertion does not affect the relative order of the conjugates already inserted in the partial sorted list. In this way the arrays $CA_i^h(\mathcal{T})$ are built.

More formally, at each step i from m down to 1 we consider the word T_i and we distinguish two different phases: 1. determining the position of the word $T_i = conj_0(T_i) = conj_{|T_i|}(T_i)$ into the sorted partial list; 2. determining the position of the conjugate $conj_r(T_i)$, with $r = 1, \ldots, |T_i| - 1$ into the sorted partial list.

In the algorithm we also use the following notations. Let w be a string. For any character $x \in \Sigma$, let $C(w, x)$ denote the number of characters in w that are smaller than x, and let $rank(w, x, t)$ denote the number of occurrences of x in $pref_t(w)$. Such functions have been introduced in [11] for the FM-index.

Figure 2 gives a sketch of the algorithm BUILD_CA($\mathcal{T}$).

Algorithm BUILD_CA($\mathcal{T}$)

1   $CA_{m+1} = NULL$;
2   $B_{m+1} = NULL$;
3   $i = m$;
4   **for** *each Lyndon word $T_i \in \mathcal{T}$, for $i = m, \ldots, 1$* **do**
5      $CA_i = CA_{i+1}$;
6      $B_i = B_{i+1}$;
7   **Insert** $(i, |T_i|)$ **in position 1 of** CA_i;
8   **Insert** $T_i[|T_i|]$ **in position 1 of** B_i;
9      $prevPos = 1$;
10     $prevSymb = B_i[1]$;
11   **for** *each conjugate $conj_h(T_i)$ for $h = 1, \ldots, |T_i| - 1$* **do**
12        $\gamma = C[prevSymb] + rank(B_i, prevPos, prevSymb) + 1$;
13   **Insert** (i, h) **in position** γ **of** CA_i;
14   **Insert** $conj_h[|T_i|]$ **in position** γ **of** B_i;
15        $prevPos = \gamma$;
16        $prevSymb = B_i[\gamma]$;

Fig. 2. Construction of the conjugate array of a multiset of Lyndon words

Theorem 6. *Given the multiset $\mathcal{T}$ of lexicographically sorted Lyndon words, the algorithm* BUILD_CA($\mathcal{T}$) *correctly constructs* $CA(\mathcal{T})$.

The proof of the theorem can be deduced by the following lemmas. Such lemmas show, for each step, the correctness of the two phases above described.

Lemma 3. *For each step i from m to 1, the position of the pair $(i, |T_i|)$ in the partial conjugate array $CA_i(\mathcal{T})$ is 1.*

In the phase 2 of each step i we have to establish the position in the list where we have to insert the conjugate $conj_h(T_i)$, with $h \geq 1$. Clearly, it must be inserted after all conjugates that starting with a symbol y smaller than the initial of $conj_h(T_i)$. The order between two conjugates $conj_h(T_i)$ and $conj_k(T_j)$ starting with the same symbol $a \in \Sigma$ is established by using the (already known) order between the conjugates $conj_{h-1}(T_i)$ and $conj_{k-1}(T_j)$.

Lemma 4. *For each step i from m to 1 and for $h = 1, \ldots, |T_i| - 1$ the position of $conj_h(T_i)$ in the array $CA_i^h(\mathcal{T})$ is*

$$\gamma = C(B_i^{h-1}(\mathcal{T}), x) + rank(B_i^{h-1}(\mathcal{T}), x, t) + 1. \tag{1}$$

where x is the first character of $conj_h(T_i)$ and t is the position in $CA_i^{h-1}(\mathcal{T})$ of the pair corresponding to the conjugate $conj_{h-1}(T_i)$.

With reference to the computational complexity of the algorithm we can formulate the following theorem.

Theorem 7. *Let $\mathcal{T} = \{T_1, \ldots, T_m\}$ a lexicographically sorted multiset of Lyndon words and let n be the size of $\mathcal{T}$ (i.e. $n = \sum_{i=1}^m |T_i|$). The algorithm* BUILD_CA($\mathcal{T}$) *runs in $O(n(t_1 + t_2 + t_3))$ where t_1, t_2, t_3 are the cost of the operations insertion in the array $CA_i(\mathcal{T})$, insertion and rank in the array $B_i(\mathcal{T})$, respectively.*

One can deduce from the previous theorem that the complexity of the algorithm depends on the suitable data structures used for the rank operations and for the insertion operations. Navarro and Nekrich's recent result [24] on optimal representations of dynamic sequences shows that one can insert symbols at arbitrary positions and compute the rank function in the optimal time $O(\frac{\log n}{\log \log n})$ within essentially $nH_0(s) + O(n)$ bits of space, for a sequence s of length n. Moreover, it is possible to give also an external memory implementation of the algorithm BUILD_CA($\mathcal{T}$) by adapting the strategy used in [2,3]. In this case the used memory is negligible and the time complexity depends on the time of I/O operations.

5   Suffix Array of a Word through Its Lyndon Factorization

Recall that the suffix array of a word w, that is here denoted by SA_w, is defined as the permutation of integers giving the indexes of the starting positions of the suffixes of w, lexicographically ordered. We refer interested readers to [25] for further reading on suffix arrays.

In this section we show that arguments similar to those described in Section 3 can be also used for designing a new strategy for computing the order of the suffixes of a given word (i.e. its suffix array). Note that one can verify that the lexicographic order among the suffixes of a text is different from the order induced by the $\prec_\omega$ sorting of the conjugates of such factors as used in [15,19]. For sake of simplicity, here we use $<$ rather than $<_{lex}$.

The basic idea described in the following theorem states that, if the Lyndon factorization of a word w is given, in order to establish the mutual order of two suffixes of w starting at positions p and q, we just need to perform a number of symbol comparisons depending on the Lyndon factors that contain p and q, respectively.

Let $w = a_1 \cdots a_n \in \Sigma^*$ and let $w_1 w_2 \cdots w_s$ be its Lyndon factorization. Let $j_1, j_2, \ldots, j_s$ be the positions of the last characters in factors $w_1, w_2, \ldots, w_s$, respectively. Obviously $j_s = n$. Let p be a position in w, we define

$$L(p) = \min\{j_k \mid 1 \leq j_k \leq s \text{ and } p \leq j_k\},$$

and

$$l(p) = L(p) - p + 1.$$

Theorem 8. *Let $w \in \Sigma^*$ and let p and q be two positions in w, $p < q$. Then*

$$w[q, n] < w[p, n] \Leftrightarrow pref_{l(q)}(w[q, n]) \leq pref_{l(q)}(w[p, n]).$$

Remark 1. If p and q are in the same Lyndon factor, i.e. $L(p) = L(q)$, then the order of the two suffixes is the same as the order of the suffixes inside their common Lyndon factor.

Remark 2. Let p and q be two positions in the word w. If $p < q$, let us denote by $lcp(p, q)$ the length of the longest common prefix between the suffixes $w[p, n]$ and $w[q, n]$. The previous theorem states that in order to get the mutual order between $w[p, n]$ and $w[q, n]$ one needs $\min(lcp(p, q)+1, l(q))$ symbol comparisons. The following example shows that $l(q)$ can be much smaller than $lcp(p, q) + 1$.

Example 4. Let $w = abaaaabaaaaabaaaabaaaaaab$. The Lyndon factorization of w is $ab|aaaab|aaaaabaaaab|aaaaaab$.
Consider the suffixes $w[2, 25] = b|aaaab|aaaaabaaaab|aaaaaab$ and $w[13, 25] = baaaab|aaaaaab$. We have $lcp(2, 13) = 11$.

$$
\begin{array}{cc}
l(13) & lcp(2,13)+1 \\
\downarrow & \downarrow
\end{array}
$$
$$
\begin{aligned}
w[2, 25] &= baaaa\ b\ aaaaa\ b\ aaaabaaaaaab \\
w[13, 25] &= baaaa\ b\ aaaaa\ a\ b
\end{aligned}
$$

By Theorem 8 we just need to perform $l(13) = 6 < 12 = lcp(2, 13) + 1$ symbol comparisons. So, even if $w[2, 7] = w[13, 18]$, the mutual order is established by the Lyndon properties.

Remark 3. Since the underlying arguments are similar to the ones described in the previous section, a possible implementation for the construction of suffix array can be obtained by adapting the strategies in the previous section.

The *suffix permutation* (cf. [9]) of a word $w = a_1 \cdots a_n$ is the permutation π_w over $\{1, \ldots, n\}$, where $\pi_w(i)$ is the *rank* of the suffix $w[i, n]$ in the set of the lexicographically sorted suffixes of w. In other words the suffix permutation π_w is the inverse permutation defined by the suffix array SA_w. Given a permutation π over $\{1, \ldots, n\}$, an integer i $(1 \leq i \leq n)$ is a *left-to-right minimum* of π if $\pi(j) > \pi(i)$, for all $j < i$.

Similar considerations used to prove Theorem 8 can be used to deduce the following result given for the first time in [16, Corollary 3.1]. Such a result further strengthens and highlights the close connection between the Lyndon factorization of a text and the lexicographic order of its suffixes.

Theorem 9. *Let w be a word, let $i_1 = 1, i_2, \ldots, i_k$ be the start positions of the factors in its Lyndon factorization and let π_w be its suffix permutation. Then the values $i_1, i_2, \ldots i_k$ correspond to the positions of the left to right minima of π_w.*

Example 5. Given the word $w = abaaaabaaaaabaaaabaaaaaab$, its suffix array is

$$SA_w = [19, 20, 8, 21, 14, 3, 9, 22, 15, 4, 10, 23, 16, 5, 11, 24, 17, 6, 12, 1, 25, 18, 7, 13, 2].$$

and its suffix permutation is

$$\pi_w = \begin{pmatrix} 1 & 2 & 3 & 4 & 5 & 6 & 7 & 8 & 9 & 10 & 11 & 12 & 13 & 14 & 15 & 16 & 17 & 18 & 19 & 20 & 21 & 22 & 23 & 24 & 25 \\ 20 & 25 & 6 & 10 & 14 & 18 & 23 & 3 & 7 & 11 & 15 & 19 & 24 & 5 & 9 & 13 & 17 & 22 & 1 & 2 & 4 & 8 & 12 & 16 & 21 \end{pmatrix}$$

where the left-to-right minima are marked by the arrows. The Lyndon factorization of w is $ab|aaaab|aaaaabaaaab|aaaaaab$.

References

1. Adjeroh, D., Bell, T., Mukherjee, A.: The Burrows-Wheeler Transform: Data Compression, Suffix Arrays, and Pattern Matching, 1st edn. Springer Publishing Company, Incorporated (2008)
2. Bauer, M.J., Cox, A.J., Rosone, G.: Lightweight BWT construction for very large string collections. In: Giancarlo, R., Manzini, G. (eds.) CPM 2011. LNCS, vol. 6661, pp. 219–231. Springer, Heidelberg (2011)
3. Bauer, M.J., Cox, A.J., Rosone, G.: Lightweight algorithms for constructing and inverting the BWT of string collections. Theoret. Comput. Sci. 483, 134–148 (2013)
4. Bauer, M.J., Cox, A.J., Rosone, G., Sciortino, M.: Lightweight LCP construction for next-generation sequencing datasets. In: Raphael, B., Tang, J. (eds.) WABI 2012. LNCS, vol. 7534, pp. 326–337. Springer, Heidelberg (2012)
5. Burrows, M., Wheeler, D.J.: A block sorting data compression algorithm. Technical report, DIGITAL System Research Center (1994)
6. Chen, K.T., Fox, R.H., Lyndon, R.C.: Free differential calculus. IV. The quotient groups of the lower central series. Ann. of Math. 68(2), 81–95 (1958)

7. Crochemore, M., Désarménien, J., Perrin, D.: A note on the Burrows-Wheeler transformation. Theoret. Comput. Sci. 332, 567–572 (2005)
8. Duval, J.-P.: Factorizing words over an ordered alphabet. Journal of Algorithms 4(4), 363–381 (1983)
9. Duval, J.-P., Lefebvre, A.: Words over an ordered alphabet and suffix permutations. RAIRO Theor. Inform. Appl. 36(3), 249–259 (2002)
10. Ferragina, P., Giancarlo, R., Manzini, G., Sciortino, M.: Boosting textual compression in optimal linear time. J. ACM 52(4), 688–713 (2005)
11. Ferragina, P., Manzini, G.: Opportunistic data structures with applications. In: Proceedings of the 41st Annual Symposium on Foundations of Computer Science, pp. 390–398. IEEE Computer Society (2000)
12. Fine, N.J., Wilf, H.S.: Uniqueness theorem for periodic functions. Proc. Am. Mathematical Society (16), 109–114 (1965)
13. Gessel, I.M., Reutenauer, C.: Counting permutations with given cycle structure and descent set. J. Combin. Theory Ser. A 64(2), 189–215 (1993)
14. Giancarlo, R., Restivo, A., Sciortino, M.: From first principles to the Burrows and Wheeler transform and beyond, via combinatorial optimization. Theoret. Comput. Sci. 387(3), 236–248 (2007)
15. Gil, J.Y., Scott, D.A.: A bijective string sorting transform. CoRR, abs/1201.3077 (2012)
16. Hohlweg, C., Reutenauer, C.: Lyndon words, permutations and trees. Theoret. Comput. Sci. 307(1), 173–178 (2003)
17. Hon, W.-K., Ku, T.-H., Lu, C.-H., Shah, R., Thankachan, S.V.: Efficient algorithm for circular Burrows-Wheeler transform. In: Kärkkäinen, J., Stoye, J. (eds.) CPM 2012. LNCS, vol. 7354, pp. 257–268. Springer, Heidelberg (2012)
18. Kucherov, G., Tóthmérész, L., Vialette, S.: On the combinatorics of suffix arrays. CoRR, abs/1206.3877 (2012)
19. Kufleitner, M.: On bijective variants of the Burrows-Wheeler transform. In: Proceedings of the Prague Stringology Conference 2009, pp. 65–79 (2009)
20. Lothaire, M.: Applied Combinatorics on Words (Encyclopedia of Mathematics and its Applications). Cambridge University Press, New York (2005)
21. Mantaci, S., Restivo, A., Rosone, G., Sciortino, M.: An Extension of the Burrows Wheeler Transform and Applications to Sequence Comparison and Data Compression. In: Apostolico, A., Crochemore, M., Park, K. (eds.) CPM 2005. LNCS, vol. 3537, pp. 178–189. Springer, Heidelberg (2005)
22. Mantaci, S., Restivo, A., Rosone, G., Sciortino, M.: An extension of the Burrows-Wheeler Transform. Theoret. Comput. Sci. 387(3), 298–312 (2007)
23. Mantaci, S., Restivo, A., Rosone, G., Sciortino, M.: A new combinatorial approach to sequence comparison. Theory Comput. Syst. 42(3), 411–429 (2008)
24. Navarro, G., Nekrich, Y.: Optimal dynamic sequence representations. In: Proc. 24th Annual ACM-SIAM Symposium on Discrete Algorithms (SODA), pp. 865–876 (2013)
25. Puglisi, S.J., Smyth, W.F., Turpin, A.H.: A taxonomy of suffix array construction algorithms. ACM Comput. Surv. 39 (2007)
26. Seward, J.: The bzip2 home page, http://www.bzip.org

Extremal Words in the Shift Orbit Closure
of a Morphic Sequence

James D. Currie, Narad Rampersad, and Kalle Saari

Department of Mathematics and Statistics
University of Winnipeg
515 Portage Avenue
Winnipeg, MB R3B 2E9
Canada
j.currie@uwinnipeg.ca, {narad.rampersad,kasaar2}@gmail.com

Abstract. Given an infinite word $\mathbf{x}$ over an alphabet A, a letter b occurring in $\mathbf{x}$, and a total order σ on A, we call the smallest word with respect to σ starting with b in the shift orbit closure $S_\mathbf{x}$ of $\mathbf{x}$ an *extremal word* of $\mathbf{x}$. In this paper we consider the extremal words of morphic words. If $\mathbf{x} = g(f^\omega(a))$ for some morphisms f and g, we give a simple condition on f and g that guarantees that all extremal words are morphic. An application of this condition shows that all extremal words of binary pure morphic words are morphic. Our technique also yields easy characterizations of extremal words of the Period-doubling and Chacon words and a new proof of the form of the lexicographically least word in the shift orbit closure of the Rudin-Shapiro word.

Keywords: Lexicographic order, morphic sequence, extremal word, Period-doubling word, Chacon word, Rudin-Shapiro word.

1 Introduction

Given an infinite word $\mathbf{x} \in A^\mathbb{N}$, it is natural to inquire about the nature of the lexicographically extremal words in its shift orbit closure. We get different extremal words depending on the choice of the total order on the alphabet A and the initial letter of the extremal word. For example, if $A = \{0, 1\}$ and $\mathbf{x}$ a Sturmian word, it is well-known that the extremal words with respect to $0 < 1$ are $0\mathbf{c}$ and $10\mathbf{c}$, where $\mathbf{c}$ is the characteristic word whose slope equals that of $\mathbf{x}$, and if we order $1 < 0$, then the extremal words are $1\mathbf{c}$ and $01\mathbf{c}$, see for example [15]. As another example, if $\mathbf{x}$ is k-automatic, then its extremal words are k-automatic as well [4]. For related results, see also [1–3, 10]

The motivation for this paper comes from the following question: given a morphic sequence $\mathbf{x}$, when are the corresponding extremal words also morphic? While we are not able to solve the question in full generality, we give a fairly general condition (1) on the morphisms that generate $\mathbf{x}$ guaranteeing that the extremal words are morphic as well (Theorem 2). Using this condition we show that the extremal words of all binary pure morphic words are morphic (Theorem 4). Then we move on to find characterizations of the extremal words of the

M.-P. Béal and O. Carton (Eds.): DLT 2013, LNCS 7907, pp. 143–154, 2013.
© Springer-Verlag Berlin Heidelberg 2013

Period-doubling (Theorem 5) and Chacon (Theorem 6) words. On our way to proving the main results, we show that if $\mathbf{x}$ is a pure morphic word generated by a morphism f and $\mathbf{t}$ is in the shift orbit closure of $\mathbf{x}$ such that $f(\mathbf{t}) = \mathbf{t}$, then $\mathbf{t}$ is morphic (Theorem 1).

2   Preliminaries

We will follow the standard terminology and notation of combinatorics on words as established, for example, in [5, 13].

If A is an alphabet, then $A^{\mathbb{N}}$ denotes the set of all infinite words over A. If $X \subset A^*$, then X^{ω} denotes the set of all infinite words obtained by concatenating elements of X.

If $f \colon A^* \to A^*$ is a morphism such that $f(a) = ax$ for some letter $a \in A$ and a word $x \in A^+$ such that $f^n(x) \neq \varepsilon$, the empty word, for all $n \geq 0$, then there exists an infinite word $f^{\omega}(a) := \lim_{n \to \infty} f^n(a)$ such that $f^n(a)$ is a prefix of $f^{\omega}(a)$ for all $n \geq 0$, and it is called a *pure morphic word generated by f*. Notice that $f^{\omega}(a)$ is a fixed point of f, that is $f\big(f^{\omega}(a)\big) = f^{\omega}(a)$, but in general a fixed point of a morphism is not necessarily generated by the morphism (however, see Theorem 1).

If $c \colon A^* \to B^*$ is a coding, that is a letter-to-letter morphism, then $c(f^{\omega}(a))$ is called a *morphic sequence*. It is clear that all ultimately periodic sequences are morphic. The following result on morphic sequences is well-known, see Theorems 7.6.1 and 7.6.3 and Corollary 7.7.5 in [5].

Lemma 1. *Let $\mathbf{x} \in A^{\mathbb{N}}$ be a morphic sequence, $w \in A^*$, and $g \colon A^* \to B^*$ a non-erasing morphism. Then the words $w\mathbf{x}$, $w^{-1}\mathbf{x}$, and $g(\mathbf{x})$ are morphic.*

Let $\mathbf{x} \in A^{\mathbb{N}}$ be an infinite word. The set of factors of $\mathbf{x}$ is denoted by $F(\mathbf{x})$. We denote by $\mathcal{S}_{\mathbf{x}}$ the set of all infinite words $\mathbf{y} \in A^{\mathbb{N}}$ such that $F(\mathbf{y}) \subseteq F(\mathbf{x})$. Thus $\mathcal{S}_{\mathbf{x}}$ is the *shift orbit closure* of $\mathbf{x}$.

Now we are ready for the key definition of this paper. Let $f \colon A^* \to B^*$ be a morphism and $\mathbf{x} \in A^{\mathbb{N}}$. We will write

$$f \in \mathcal{M}_{\mathbf{x}} \tag{1}$$

if the following condition holds: for each letter $b \in A$, there exists a finite word $p_b \in B^+$ such that if $\mathbf{y} \in \mathcal{S}_{\mathbf{x}}$ begins with b, then $f(\mathbf{y})$ begins with p_b, and if $a \in A$ with $a \neq b$, then neither of p_a and p_b is a prefix of the other. Notice that then f is necessarily non-erasing.

Example 1. Let us illustrate the above definition with a morphism appearing in [12]. Let f be given by $0 \mapsto 02$, $1 \mapsto 02$, and $2 \mapsto 1$, and let $\mathbf{x}$ be the unique fixed point of f. It is easy to see that if $0\mathbf{y} \in \mathcal{S}_{\mathbf{x}}$, then $\mathbf{y}$ must begin with 2; hence $f(0\mathbf{y})$ begins with 021. Similarly, if $1\mathbf{y} \in \mathcal{S}_{\mathbf{x}}$, then $\mathbf{y}$ must begin with 0; hence $f(1\mathbf{y})$ begins with 020. Finally, $f(2\mathbf{y})$ begins with 1 regardless of $\mathbf{y}$. Therefore we may let $p_0 = 021$, $p_1 = 020$, and $p_2 = 1$, and consequently $f \in \mathcal{M}_{\mathbf{x}}$.

Example 2. Let f be the morphism $0 \mapsto 010$, $1 \mapsto 21$, $2 \mapsto 211$, and let $\mathbf{x} = f^\omega(0)$. Now we have $f \notin \mathcal{M}_{\mathbf{x}}$ because $f(10\cdots) = 210\cdots$ and $f(12\cdots) = 212\cdots$, so that if p_1 existed, it would have to be a prefix of 21, which is a prefix of $f(2)$. Therefore no matter how p_2 is chosen, one of p_1 and p_2 is necessarily a prefix of the other.

Let $\sigma = \sigma_A$ be a total order on an alphabet A, that is, a transitive and anti-symmetric relation for which either (a, b) or (b, a) is in σ for all distinct letters $a, b \in A$. If $(a, b) \in \sigma$, we denote $a <_\sigma b$. The order σ extends to a lexicographic order on finite and infinite words over A in the usual way. Let $a \in A$ be a letter and $\mathbf{x} \in A^{\mathbb{N}}$ an infinite word in which a occurs. Then there exists a unique lexicographically smallest word in $\mathcal{S}_{\mathbf{x}}$ with respect to σ that begins with the letter a, and we will denote it by

$$\mathbf{l}_{a,\sigma,\mathbf{x}}.$$

Words of this form are collectively called the *extremal words* of $\mathbf{x}$ or $\mathcal{S}_{\mathbf{x}}$. We also denote by $\mathbf{s}_{a,\sigma,\mathbf{x}}$ the infinite word obtained from $\mathbf{l}_{a,\sigma,\mathbf{x}}$ by erasing the first letter, that is,

$$\mathbf{l}_{a,\sigma,\mathbf{x}} = a\, \mathbf{s}_{a,\sigma,\mathbf{x}}.$$

For the remainder of this section, let us fix a morphism $f \colon A^* \to A^*$. A word $u \in A^*$ is called *bounded under* f if there exists a constant $k > 0$ such that $|f^n(u)| < k$ for all $n \geq 0$. It is clear that every letter occurring in a bounded word is bounded. Let $B_f \subset A$ denote the set of bounded letters; the letters in $C_f := A \setminus B_f$ are said to be *growing under* f.

The following result is proved in [6, Prop. 4.7.62].

Lemma 2. *Suppose that* $\mathbf{x} \in A^{\mathbb{N}}$ *is a pure morphic sequence generated by* f. *There exists a finite subset* Q *of* $C_f \times B_f^* \times B_f^* \times B_f^* \times B_f^* \times B_f^* \times C_f$ *such that* $F(\mathbf{x}) \cap C_f B_f^* C_f$ *equals the set of all words of the form* $c_1 y_1 z_1^k x z_2^k y_2 c_2$ *with* $(c_1, y_1, z_1, x, z_2, y_2, c_2) \in Q$ *and* $k \in \mathbb{N}$.

Lemma 3. *Suppose that* $\mathbf{x} \in A^{\mathbb{N}}$ *is a pure morphic sequence generated by* f. *If* $\mathbf{z} \in \mathcal{S}_{\mathbf{x}} \cap B_f^{\mathbb{N}}$, *then* $\mathbf{z}$ *is ultimately periodic.*

Proof. Suppose that $\mathbf{z} \in \mathcal{S}_{\mathbf{x}} \cap B_f^{\mathbb{N}}$. If $\mathbf{x}$ has a suffix that is in $B_f^{\mathbb{N}}$, then it is ultimately periodic, which is proved in [6, Lemma 4.7.65], and then so is $\mathbf{z}$. Therefore we may assume that there are infinitely many occurrences of growing letters in $\mathbf{x}$. Let u_n be a sequence of factors of $\mathbf{x}$ such that u_n is a prefix of u_{n+1} for all $n \geq 1$ and $\mathbf{z} = \lim_{n \to \infty} u_n$. Since the first letter of $\mathbf{x}$ is necessarily growing and $\mathbf{x}$ has infinitely many occurrences of growing letters, it follows that each u_n is a factor of a word w_n such that $w_n \in C_f B_f^+ C_f \cap F(\mathbf{x})$. Since the set Q in Lemma 2 is finite, there exist letters $c_1, c_2 \in C_f$ and words $y_1, y_2, z_1, z_2, x \in B_f^*$ such that $w_{n_k} = c_1 y_1 z_1^{i_k} x z_2^{i_k} y_2 c_2$ for some subsequence n_k. By chopping off a prefix of length $|y_1|$ and a suffix of length $|y_2|$ from u_{n_k} if necessary, we may assume that each sufficiently long u_{n_k} is a factor of the biinfinite word $\mathbf{q} := {}^\omega z_1 . x z_2^\omega$, where the the word x occurs in position 0. Now we have two possibilities: If there exists an integer $j \in \mathbb{Z}$ such that infinitely many u_{n_k} occurs in $\mathbf{q}$ in a position $\geq j$, then $\lim_{k\to\infty} u_{n_k}$ has suffix z_2^ω. If no such j exists, then $\lim_{k\to\infty} u_{n_k}$ has suffix z_1^ω. In the first case $\mathbf{z}$ has suffix z_2^ω and in the second case it has suffix z_1^ω.

Let $M_f \subset A$ denote the set of letters b such that $f^i(b) = \varepsilon$ for some integer $i \geq 1$, and let $t \geq 1$ be the smallest integer such that $f^t(b) = \varepsilon$ for all $b \in M_f$. Let

$$G_f = \{\, f^t(a) \mid a \in A \text{ such that } f(a) = xay \text{ for some } x, y \in M_f^* \,\}$$

Notice that each word $f^t(a)$ in G_f is a finite fixed point of f because

$$f^t(a) = f^{t-1}(x) \cdots f(x) xay f(y) \cdots f^{t-1}(y).$$

In particular, all words in G_f are bounded. The following result is by Head and Lando [11], see also [5, Theorem 7.3.1].

Lemma 4. *Let $\mathbf{t} \in A^\mathbb{N}$ be an infinite word. We have $f(\mathbf{t}) = \mathbf{t}$ if and only if at least one of the following two conditions holds:*

(a) $\mathbf{t} \in G_f^\omega$; *or*
(b) $\mathbf{t} = w f^{t-1}(x) \cdots f(x) xay f(y) f^2(y) \cdots$ *for some $w \in G_f^*$ and $a \in A$ such that $f(a) = xay$ with $x \in M_f^*$ and $y \notin M_f^*$.*

Lemma 5. *Let $f\colon A^* \to A^*$ be a morphism. If $\mathbf{t} \in A^\mathbb{N}$ can be written in the form $\mathbf{t} = w x f(x) f^2(x) f^3(x) \cdots$, where $w \in A^*$ and $x \notin M_f^*$, then $\mathbf{t}$ is morphic.*

Proof. Let b be a new letter that does not occur in A. Then the infinite word $b x f(x) f^2(x) \cdots$ is morphic as it is generated by a morphism $g\colon (A \cup \{b\})^* \to (A \cup \{b\})^*$ for which $g(b) = bx$ and $g(a) = f(a)$ for all $a \in A$. Thus it follows from Lemma 1 that $\mathbf{t}$ is morphic.

Theorem 1. *Let $f\colon A^* \to A^*$ be a morphism, and suppose that $\mathbf{x} \in A^\mathbb{N}$ is a pure morphic word generated by f. If $\mathbf{t} \in \mathcal{S}_\mathbf{x}$ satisfies $f(\mathbf{t}) = \mathbf{t}$, then $\mathbf{t}$ is morphic.*

Proof. According to Lemma 4, either $\mathbf{t}$ is in G_f^ω or it is of the form

$$\mathbf{t} = w f^{t-1}(x) \cdots f(x) xay f(y) f^2(y) \cdots .$$

In the former case $\mathbf{t} \in B_f^\mathbb{N}$, so $\mathbf{t}$ is ultimately periodic by Lemma 3, and thus morphic. In the latter case $\mathbf{t}$ is morphic by Lemma 5.

3   Main Theorem

Lemma 6. *Let $\mathbf{x} \in A^\mathbb{N}$ be an infinite word and $f\colon A^* \to B^*$ a morphism. If $\mathbf{y} \in B^\mathbb{N}$ is in $\mathcal{S}_{f(\mathbf{x})}$, then there exist a letter $a \in A$ and an infinite word $\mathbf{z}$ such that $a\mathbf{z} \in \mathcal{S}_\mathbf{x}$ and $\mathbf{y} = u f(\mathbf{z})$, where u is a nonempty suffix of $f(a)$.*

Proof. Let L_n denote the length-n prefix of $\mathbf{y}$; then L_n is a factor of $f(\mathbf{x})$ by the definition of $\mathcal{S}_{f(\mathbf{x})}$. Consequently, if $n \geq \max_{a \in A} |f(a)|$, there exist letters $a_n, b_n \in A$ and a word $v_n \in A^*$ such that $a_n v_n b_n$ occurs in $\mathbf{x}$ and we have $L_n = s_n f(v_n) p_n$, where s_n is a nonempty suffix of $f(a_n)$ and p_n is a possibly empty prefix of $f(b_n)$. Since there are only finitely many different possibilities for a_n and s_n, there exists a letter $a \in A$ and a word u such that $a_{n_i} = a$ and $s_{n_i} = u$ for infinitely many n_i. The set of words $\{v_{n_i}\}$ being infinite, König's Lemma implies that there exists an infinite word $\mathbf{z}$ such that every prefix of $\mathbf{z}$ is a prefix of some v_{n_i}. Since each of av_{n_i} is a factor of $\mathbf{x}$, we have $a\mathbf{z} \in \mathcal{S}_\mathbf{x}$. Furthermore, since each prefix z of $\mathbf{z}$ is a prefix of some v_{n_i}, the word $uf(z)$ is a prefix of $\mathbf{y}$, and consequently $\mathbf{y} = uf(\mathbf{z})$.

Lemma 7. *Let $f \colon A^* \to B^*$ be a morphism and $\mathbf{x} \in A^{\mathbb{N}}$ such that $f \in \mathcal{M}_\mathbf{x}$. Let $b \in B$ be a letter that occurs in $f(\mathbf{x})$ and let ρ be a total order on B. Then there exist a total order σ on A, a letter $a \in A$, and a possibly empty proper suffix v of $f(a)$ such that*

$$\mathbf{s}_{b,\rho,f(\mathbf{x})} = vf(\mathbf{s}_{a,\sigma,\mathbf{x}}). \tag{2}$$

Proof. By Lemma 6, we can write $\mathbf{l}_{b,\rho,f(\mathbf{x})} = b\mathbf{s}_{b,\rho,f(\mathbf{x})} = uf(\mathbf{z})$, where u is a nonempty suffix of $f(a)$ for some $a \in A$ and $a\mathbf{z} \in \mathcal{S}_\mathbf{x}$. Since $f \in \mathcal{M}_\mathbf{x}$, there exist words $p_x \in B^+$ for every $x \in A$ such that $p_x \neq p_y$ whenever $x \neq y$. Thus we can define a total order σ on A such that, for all letters $x, y \in A$, we have $x <_\sigma y$ if and only if $p_x <_\rho p_y$.

We claim that $\mathbf{z} = \mathbf{s}_{a,\sigma,\mathbf{x}}$. If this is not the case, then $\mathbf{z} >_\sigma \mathbf{s}_{a,\sigma,\mathbf{x}}$ because both $a\mathbf{z}$ and $a\mathbf{s}_{a,\sigma,\mathbf{x}} = \mathbf{l}_{a,\sigma,\mathbf{x}}$ are in $\mathcal{S}_\mathbf{x}$ and $\mathbf{l}_{a,\sigma,\mathbf{x}}$ is the smallest word in $\mathcal{S}_\mathbf{x}$ starting with the letter a. Therefore $\mathbf{z} = wy\mathbf{t}$ and $\mathbf{s}_{a,\sigma,\mathbf{x}} = wx\mathbf{t}'$ with $x, y \in A$ satisfying $y >_\sigma x$. Since $f(x\mathbf{t})$ begins with p_x and $f(y\mathbf{t}')$ begins with p_y and neither of p_x and p_y is a prefix of the other, we have $f(y\mathbf{t}) >_\rho f(x\mathbf{t}')$ by the definition of σ, and this gives

$$\mathbf{l}_{b,\rho,f(\mathbf{x})} = uf(\mathbf{z}) = uf(w)f(y\mathbf{t}) >_\rho uf(w)f(x\mathbf{t}') = uf(\mathbf{s}_{a,\sigma,\mathbf{x}}).$$

But this contradicts the definition of $\mathbf{l}_{b,\rho,f(\mathbf{x})}$ because $uf(\mathbf{s}_{a,\sigma,\mathbf{x}})$ starts with the letter b and is in $\mathcal{S}_{f(\mathbf{x})}$. Therefore we have shown that $\mathbf{z} = \mathbf{s}_{a,\sigma,\mathbf{x}}$, and so (2) holds with $v = b^{-1}u$.

Lemma 8. *Let $f \colon A^* \to A^*$ be a morphism and $\mathbf{x} \in A^{\mathbb{N}}$ such that $f \in \mathcal{M}_\mathbf{x}$ and $f(\mathbf{x}) = \mathbf{x}$. Then for any total order ρ on A and any letter $b \in A$ occurring in $\mathbf{x}$, there exist a total order σ on A, a letter $a \in A$, words $u, v \in A^*$, and integers $k, m \geq 1$ such that*

$$\mathbf{s}_{b,\rho,\mathbf{x}} = uf^k(\mathbf{s}_{a,\sigma,\mathbf{x}}) \quad and \quad \mathbf{s}_{a,\sigma,\mathbf{x}} = vf^m(\mathbf{s}_{a,\sigma,\mathbf{x}}). \tag{3}$$

Proof. Since $f(\mathbf{x}) = \mathbf{x}$, Lemma 7 implies that $\mathbf{s}_{b,\rho,\mathbf{x}} = v_0 f(\mathbf{s}_{a_1,\sigma_1,\mathbf{x}})$ for some total order σ_1 on A, a letter $a_1 \in A$, and a possibly empty suffix v_0 of $f(a_1)$. By applying Lemma 7 next on $\mathbf{s}_{a_1,\sigma_1,\mathbf{x}}$ and further, we get a sequence of identities

$$\mathbf{s}_{a_k,\sigma_k,\mathbf{x}} = v_k f(\mathbf{s}_{a_{k+1},\sigma_{k+1},\mathbf{x}}) \quad (k \geq 0),$$

where we denote $a_0 = b$ and $\sigma_0 = \rho$. Therefore,

$$\mathbf{s}_{a_k, \sigma_k, \mathbf{x}} = v_k f(v_{k+1}) \cdots f^{m-1}(v_{k+m-1}) f^m(\mathbf{s}_{a_{k+m}, \sigma_{k+m}, \mathbf{x}}),$$

for all integers $k \geq 0$ and $m \geq 1$. Since there are only finitely many different letters and total orders on A, there is a choice of k and m such that $a_k = a_{k+m}$ and $\sigma_k = \sigma_{k+m}$. Thus by denoting $a = a_k$, $\sigma = \sigma_k$,

$$u = v_0 f(v_1) \cdots f^{k-1}(v_{k-1}), \quad \text{and} \quad v = v_k f(v_{k+1}) \cdots f^{m-1}(v_{k+m-1}),$$

we have the identities in (3).

Lemma 9. *Let $f \colon A^* \to A^*$ be a morphism and $\mathbf{x} \in A^{\mathbb{N}}$ such that $f \in \mathcal{M}_{\mathbf{x}}$ and $f(\mathbf{x}) = \mathbf{x}$. Then for any total order ρ on A and any letter $b \in A$ occurring in $\mathbf{x}$, there exist a finite word $w \in A^+$, an infinite word $\mathbf{t} \in \mathcal{S}_{\mathbf{x}}$, and an integer $m \geq 1$ such that*

$$\mathbf{l}_{b, \rho, \mathbf{x}} = w\mathbf{t} \tag{4}$$

and either

$$\mathbf{t} = f^m(\mathbf{t}) \tag{5}$$

or

$$\mathbf{t} = \lim_{n \to \infty} x f^m(x) f^{2m}(x) \cdots f^{nm}(x) \tag{6}$$

for some finite word $x \in A^+$.

Proof. According to Lemma 8, there exist a total order σ on A, a letter $a \in A$, words $u, v \in A^*$, and integers $k, m \geq 1$ such that

$$\mathbf{s}_{b, \rho, \mathbf{x}} = u f^k(\mathbf{s}_{a, \sigma, \mathbf{x}}) \quad \text{and} \quad \mathbf{s}_{a, \sigma, \mathbf{x}} = v f^m(\mathbf{s}_{a, \sigma, \mathbf{x}}).$$

Denote $w = bu$ and $\mathbf{t} = f^k(\mathbf{s}_{a, \sigma, \mathbf{x}})$. Then $\mathbf{t} \in \mathcal{S}_{\mathbf{x}}$ and Eq. (4) holds. By denoting $x = f^k(v)$, we get $\mathbf{t} = x f^m(\mathbf{t})$. If $x = \varepsilon$, then we have $\mathbf{t} = f^m(\mathbf{t})$, and Eq. (5) holds. If $x \neq \varepsilon$, then

$$\mathbf{t} = x f^m(\mathbf{t}) = x f^m(x) f^{2m}(\mathbf{t}) = \cdots = x f^m(x) f^{2m}(x) \cdots f^{nm}(x) f^{(n+1)m}(\mathbf{t}),$$

for all integers $n \geq 0$. The morphism f is non-erasing because $f \in \mathcal{M}_{\mathbf{x}}$, and therefore the words $x f^m(x) f^{2m}(x) \cdots f^{nm}(x)$ get longer and longer as n grows. Thus Eq. (6) holds.

Theorem 2. *Let $f \colon A^* \to A^*$ be a morphism. If $\mathbf{x} \in A^{\mathbb{N}}$ is a pure morphic word generated by f and $f \in \mathcal{M}_{\mathbf{x}}$, then all extremal words in $\mathcal{S}_{\mathbf{x}}$ are morphic.*

Proof. Let ρ be a total order on A and $b \in A$ a letter occurring in $\mathbf{x}$. We will show that $l_{b,\rho,\mathbf{x}}$ is morphic. Lemma 9 says that there exist a finite word $w \in A^+$, an infinite word $\mathbf{t} \in \mathcal{S}_{\mathbf{x}}$, and an integer $m \geq 1$ such that $l_{b,\rho,\mathbf{x}} = w\mathbf{t}$ and either $\mathbf{t} = f^m(\mathbf{t})$ or $\mathbf{t} = \lim_{n\to\infty} x f^m(x) f^{2m}(x) \cdots f^{nm}(x)$ for some finite word $x \in A^+$. Since f^m generates $\mathbf{x}$, the claim that $\mathbf{t}$ is morphic follows in the former case from Theorem 1 and in the latter case from Lemma 5.

Theorem 3. *Let $f\colon A^* \to A^*$ and $g\colon A^* \to B^*$ be morphisms and $\mathbf{x} \in A^{\mathbb{N}}$ such that $f, g \in \mathcal{M}_{\mathbf{x}}$. If $\mathbf{x}$ is a pure morphic word generated by f, then all extremal words in $\mathcal{S}_{g(\mathbf{x})}$ are morphic.*

Proof. Let ρ be a total order ρ on B and $b \in B$. According to Lemma 7, there exists a total order σ on A, a letter $a \in A$, and a word $v \in B^*$ such that

$$\mathbf{s}_{b,\rho,g(\mathbf{x})} = v g(\mathbf{s}_{a,\sigma,\mathbf{x}})$$

Thus it follows from Theorem 2 and Lemma 1 that $l_{b,\rho,g(\mathbf{x})}$ is morphic.

4   Extremal Words of Binary Pure Morphic Words

In this section we show that the extremal words of binary pure morphic words are morphic.

Lemma 10. *Let $f\colon \{0,1\}^* \to \{0,1\}^*$ be a morphism such that $f(01) \neq f(10)$. Then $f \in \mathcal{M}_{\mathbf{x}}$ for every $\mathbf{x} \in \{0,1\}^{\mathbb{N}}$.*

Proof. Let us denote $u = f(0)$ and $v = f(1)$. We have two possibilities:

Case 1. The word u is not a prefix of v^ω. Then there exists an integer $n \geq 0$ such that $u = v^n p a s$ and $v = p b t$, where $p, s, t \in \{0,1\}^*$ and $a, b \in \{0,1\}$ with $a \neq b$. Now it is easy to see that, for every $\mathbf{y} \in \{0,1\}^{\mathbb{N}}$, the word $f(1\mathbf{y})$ begins with $v^n p b$ and $f(0\mathbf{y})$ begins with $v^n p a$. Therefore $f \in \mathcal{M}_{\mathbf{x}}$ because we may choose $p_1 = v^n p b$ and $p_0 = v^n p a$.

Case 2. The word u is a prefix of v^ω. Then $v = xy$ and $u = v^n x$ for some integer $n \geq 0$ and words x, y. Now it is easy to see that, for every $\mathbf{y} \in \{0,1\}^{\mathbb{N}}$, the word $f(0\mathbf{y})$ begins with $(xy)^n xxy$ and $f(1\mathbf{y})$ begins with $(xy)^n xyx$. Since $f(01) \neq f(10)$, it follows that $xy \neq yx$. Denote $xy = pas$ and $yx = pbt$ with a, b distinct letters. Then $f \in \mathcal{M}_{\mathbf{x}}$ because we may let $p_0 = (xy)^n xpa$ and $p_1 = (xy)^n xpb$.

Theorem 4. *If $\mathbf{x} \in \{0,1\}^{\mathbb{N}}$ is a binary pure morphic sequence, then all extremal words of $\mathbf{x}$ are morphic.*

Proof. Let f be a binary morphism that generates $\mathbf{x}$. If $f(01) = f(10)$, then $\mathbf{x}$ is purely periodic, and the claim holds. If $f(01) \neq f(10)$, then $f \in \mathcal{M}_{\mathbf{x}}$ by Lemma 10, so that $\mathbf{x}$ is morphic by Theorem 2.

There are exactly two total orders on the binary alphabet $\{0,1\}$; let ρ denote the natural order $0 <_\rho 1$ and $\bar{\rho}$ the other order $1 <_{\bar{\rho}} 0$. The following lemma simplifies the search for the extremal words of a binary pure morphic word, and we will use it later.

Lemma 11. *If $\mathbf{x} \in \{0,1\}^{\mathbb{N}}$ is a recurrent word in which both 0 and 1 occur, then*

$$l_{1,\rho,\mathbf{x}} = 1l_{0,\rho,\mathbf{x}} \qquad\qquad l_{0,\overline{\rho},\mathbf{x}} = 0l_{1,\overline{\rho},\mathbf{x}}. \qquad (7)$$

Therefore also

$$s_{1,\rho,\mathbf{x}} = 0s_{0,\rho,\mathbf{x}} \qquad\qquad s_{0,\overline{\rho},\mathbf{x}} = 1s_{1,\overline{\rho},\mathbf{x}}. \qquad (8)$$

Proof. Consider the first equation in (7). On the one hand, $1l_{0,\rho,\mathbf{x}}$ is in $\mathcal{S}_{\mathbf{x}}$ because the recurrence of $\mathbf{x}$ implies that $al_{0,\rho,\mathbf{x}}$ is in $\mathcal{S}_{\mathbf{x}}$ for some $a \in \{0,1\}$ and if a equaled 0, then the inequality $0l_{0,\rho,\mathbf{x}} < l_{0,\rho,\mathbf{x}}$ would contradict the definition of $l_{0,\rho,\mathbf{x}}$. On the other hand, $1l_{0,\rho,\mathbf{x}}$ must equal $l_{1,\rho,\mathbf{x}}$ because otherwise $l_{1,\rho,\mathbf{x}} < 1l_{0,\rho,\mathbf{x}}$, which implies $1^{-1}l_{1,\rho,\mathbf{x}} < l_{0,\rho,\mathbf{x}}$, and this contradicts the definition of $l_{0,\rho,\mathbf{x}}$. The second equation in (7) is proved similarly. The identities (8) follow immediately from (7). $\qquad\blacksquare$

5   Extremal Words of the Period-Doubling Word

Let f denote the morphism $0 \mapsto 01$, $1 \mapsto 00$ and let $\mathbf{d} = f^{\omega}(0)$ denote the *period-doubling word* [5, 8, 14]. According to Lemma 10, we have $f \in \mathcal{M}_{\mathbf{d}}$.

Let ρ denote the natural order $0 <_{\rho} 1$ and $\overline{\rho}$ the reversed order $1 <_{\overline{\rho}} 0$. Using the observation that neither 0000 nor 11 occur in $\mathbf{d}$ and Lemma 11, the reader has no trouble verifying that the following words start as shown.

$$s_{0,\rho,\mathbf{d}} = 00100\cdots \qquad\qquad s_{1,\overline{\rho},\mathbf{d}} = 010100\cdots \qquad (9)$$

$$s_{1,\rho,\mathbf{d}} = 0001\cdots \qquad\qquad s_{0,\overline{\rho},\mathbf{d}} = 1010100\cdots. \qquad (10)$$

Lemma 7 implies that $s_{0,\rho,\mathbf{d}} = vf(s_{a,\sigma,\mathbf{d}})$ for some $a \in \{0,1\}$, proper suffix v of $f(a)$, and $\sigma \in \{\rho,\overline{\rho}\}$. The only possible such factorization has to be of the form $s_{0,\rho,\mathbf{d}} = 0f(01\cdots)$, so from (9) and (10) we see that $s_{a,\sigma,\mathbf{d}} = s_{1,\overline{\rho},\mathbf{d}}$. Thus

$$s_{0,\rho,\mathbf{d}} = 0f(s_{1,\overline{\rho},\mathbf{d}}).$$

We can deduce similarly that

$$s_{1,\overline{\rho},\mathbf{d}} = f(001\cdots) = f(s_{0,\rho,\mathbf{d}}). \qquad (11)$$

Therefore $s_{0,\rho,\mathbf{d}} = 0f^2(s_{0,\rho,\mathbf{d}})$, which implies

$$f^2(l_{0,\rho,\mathbf{d}}) = 01l_{0,\rho,\mathbf{d}}. \qquad (12)$$

We claim that $l_{0,\rho,\mathbf{d}}$ is the fixed point of the morphism $g \colon 0 \mapsto 0001$ and $1 \mapsto 0101$. Let us denote the unique fixed point of g by $\mathbf{z}$, that is $\mathbf{z} = g^{\omega}(0)$. An easy induction proof shows that $01g(w) = f^2(w)01$ for all $w \in \{0,1\}^*$. Therefore

$$01\mathbf{z} = 01g(\mathbf{z}) = f^2(\mathbf{z}).$$

Thus by (12), both $\mathbf{z}$ and $l_{0,\rho,\mathbf{d}}$ satisfy the same relation $01\mathbf{x} = f^2(\mathbf{x})$, which is easily seen to admit a unique solution; thus $\mathbf{z} = l_{0,\rho,\mathbf{d}}$. Hence, using (11) and Lemma 11, the following result is obtained.

Theorem 5. *Let* d *denote the period-doubling word and let* z *denote the unique fixed point of the morphism* $0 \mapsto 0001$, $1 \mapsto 0101$. *Then we have*

$$l_{0,\rho,d} = z \qquad\qquad l_{1,\rho,d} = 1z$$
$$l_{1,\bar{\rho},d} = 0^{-1}f(z) \qquad\qquad l_{0,\bar{\rho},d} = f(z).$$

6 Extremal Words of the Chacon Word

The Chacon word [9, 16] is the fixed point $c = f^{\omega}(0)$, where f is the morphism $0 \mapsto 0010$, $1 \mapsto 1$. Lemma 10 guarantees that $f \in \mathcal{M}_x$. Let ρ denote the natural order $0 <_{\rho} 1$ and $\bar{\rho}$ the reversed order $1 <_{\bar{\rho}} 0$ as before.

As in Section 5, we use the observation that neither 0000 nor 11 occur in c and Lemma 11, to deduce that the following words start as shown.

$$s_{0,\rho,c} = 001000101\cdots \qquad\qquad s_{1,\bar{\rho},c} = 010010\cdots$$
$$s_{1,\rho,c} = 0001000101\cdots \qquad\qquad s_{0,\bar{\rho},c} = 1010010\cdots.$$

Applying Lemma 7 as in the previous section, we find

$$s_{0,\rho,c} = f(001\cdots) = f(s_{0,\rho,c}).$$

Since $s_{0,\rho,c}$ begins with 0, we thus have $s_{0,\rho,c} = c$ and $l_{0,\rho,c} = 0c$.

Similarly, recalling that $s_{0,\bar{\rho},c} = 1s_{1,\bar{\rho},c}$ by Lemma 11, we deduce using Lemma 7 that

$$s_{1,\bar{\rho},c} = 0f(10\cdots) = 0f(s_{0,\bar{\rho},c}) = 01f(s_{1,\bar{\rho},c}).$$

Therefore $l_{1,\bar{\rho},c}$ can be expressed as $l_{1,\bar{\rho},c} = \tau g^{\omega}(b)$, where b is a new symbol, g is a morphism for which $g(b) = b01$ and $g(a) = f(a)$ for $a \in \{0,1\}$, and $\tau(b) = 1$ and $\tau(a) = a$ for $a \in \{0,1\}$. Thus a final application of Lemma 11 allows us to wrap up the results of this section as follows.

Theorem 6. *Let* c *denote the Chacon word. Then we have*

$$l_{0,\rho,c} = 0c \qquad\qquad l_{1,\bar{\rho},c} = \tau g^{\omega}(b)$$
$$l_{1,\rho,c} = 10c \qquad\qquad l_{0,\bar{\rho},c} = 0\tau g^{\omega}(b),$$

where g *and* τ *are the morphisms given above.*

7 The Least Word in the Shift Orbit Closure of the Rudin-Shapiro Word

In this section, we give a new proof for the form of the lexicographically smallest word in the shift orbit closure of the Rudin-Shapiro word. This result was first derived in [7]. Considerations in this section are more involved than the ones in the previous sections because a coding is needed in the definition of the Rudin-Shapiro word. In what follows, we denote the natural order on letters $0, 1, 2, 3$ by ρ. Thus we have $0 <_\rho 1 <_\rho 2 <_\rho 3$.

Let f and g be the morphisms

$$f: \begin{cases} 0 \mapsto 01 \\ 1 \mapsto 02 \\ 2 \mapsto 31 \\ 3 \mapsto 32 \end{cases} \qquad \text{and} \qquad g: \begin{cases} 0 \mapsto 0 \\ 1 \mapsto 0 \\ 2 \mapsto 1 \\ 3 \mapsto 1 \end{cases}$$

Denote

$$\mathbf{u} = f^\omega(0) = 0102013101023202010201313231013101020131\cdots$$

and

$$\mathbf{w} = g(\mathbf{u}) = 0001001000011101000100101110001000010010\cdots.$$

Then $\mathbf{w}$ is the Rudin-Shapiro word, and our goal is to prove the identity $\mathbf{l}_{0,\rho,\mathbf{w}} = 0\mathbf{w}$. To that end, we need the next two lemmas. Let us denote $\Sigma_4 = \{0, 1, 2, 3\}$.

Lemma 12. *Let σ and σ' be two total orders on Σ_4. If σ and σ' order the pairs $(0, 3)$ and $(1, 2)$ in the same way, i.e., $0 <_\sigma 3$ iff $0 <_{\sigma'} 3$ and $1 <_\sigma 2$ iff $1 <_{\sigma'} 2$, then $\mathbf{l}_{d,\sigma,\mathbf{u}} = \mathbf{l}_{d,\sigma',\mathbf{u}}$ for all $d \in \Sigma_4$.*

Proof. Suppose that $\mathbf{l}_{d,\sigma,\mathbf{u}} = uat$ and $\mathbf{l}_{d,\sigma',\mathbf{u}} = ubt'$ with distinct letters $a, b \in \Sigma_4$. Since σ and σ' agree on $(0, 3)$ and $(1, 2)$, it follows that either $a \in \{0, 3\}$ and $b \in \{1, 2\}$, or vice versa. Furthermore, if c denotes the last letter of u, then both ca and cb occur in $\mathbf{u}$. This contradicts the fact that none of the words 00, 03, 11, 12, 21, 22, 30, 33 occur in $\mathbf{u}$.

The next lemma is interesting in its own right. It was also proved in [7].

Lemma 13. *We have $\mathbf{l}_{0,\rho,\mathbf{u}} = \mathbf{u}$.*

Proof. Since clearly $f \in \mathcal{M}_\mathbf{u}$, Lemma 7 implies that there exist a letter $a \in \Sigma_4$, a proper suffix v of $f(a)$, and a total order σ on Σ_4 such that $\mathbf{s}_{0,\rho,\mathbf{u}} = vf(\mathbf{s}_{a,\sigma,\mathbf{u}})$. An easy case analysis based on the observation that 00 does not occur in $\mathbf{u}$ yields $\mathbf{s}_{0,\rho,\mathbf{u}} = 10201\cdots = 1f(10\cdots)$, and hence

$$v = 1 \qquad \text{and} \qquad \mathbf{s}_{a,\sigma,\mathbf{u}} = 10\cdots.$$

Since $v = 1$ is a suffix of $f(a)$, we have $a = 0$ or $a = 2$. Furthermore since $\mathbf{l}_{a,\sigma,\mathbf{u}}$ starts with $a1$, we must have $a = 0$ because 21 does not occur in $\mathbf{u}$. Thus $\mathbf{s}_{0,\rho,\mathbf{u}} = 1f(\mathbf{s}_{0,\sigma,\mathbf{u}})$.

Next we claim $\mathbf{s}_{0,\sigma,\mathbf{u}} = \mathbf{s}_{0,\rho,\mathbf{u}}$. We prove this by showing that $0 <_\sigma 3$ and $1 <_\sigma 2$; then the claim follows from Lemma 12. If, contrary to what we want to show, we have $2 <_\sigma 1$, then $\mathbf{l}_{0,\sigma,\mathbf{u}}$ would begin with 02, contradicting the fact that $s_{0,\sigma,u}$ begins with 1. Consequently we have $1 <_\sigma 2$. Furthermore if $3 <_\sigma 0$, then $l_{0,\sigma,u}$ would begin with 013, contradicting the fact that $s_{0,\sigma,u}$ begins with 10. Therefore $s_{0,\sigma,u} = s_{0,\rho,u}$.

Now the identity $s_{0,\rho,u} = 1f(s_{0,\rho,u})$ implies $l_{0,\rho,u} = f(l_{0,\rho,u})$, so that $l_{0,\rho,u}$ is the unique iterative fixed point of f that starts with 0, that is $l_{0,\rho,u} = u$.

Finally, we are ready to prove the main result of this subsection.

Theorem 7. *Let* **w** *denote the Rudin–Shapiro word. Then* $l_{0,\rho,w} = 0w$.

Proof. Let $h = g \circ f$ be the composition of g and f. Then

$$h: \quad 0 \mapsto 00 \quad\quad 1 \mapsto 01 \quad\quad 2 \mapsto 10 \quad\quad 3 \mapsto 11.$$

According to Lemma 6, there exist a letter $a \in \Sigma_4$ and an infinite word $z \in \Sigma_4^{\mathbb{N}}$ such that $az \in \mathcal{S}_u$ and $l_{0,\rho,w} = uh(z)$, where u is a nonempty suffix of $h(a)$. Since $l_{0,\rho,w}$ clearly starts with 0000 and 00 does not occur in **u**, it follows that $u = 0$, $z = 01\cdots$, and $a = 2$.

On the other hand, it is easy to see that $2u \in \mathcal{S}_u$. Since $u = l_{0,\rho,u}$ by Lemma 13, we have $u \leq_\rho z$, and so $2u \leq_\rho 2z$. Furthermore, since h preserves ρ, that is to say if $x, y \in \{0,1,2,3\}^*$ with $x <_\rho y$, then $h(x) <_\rho h(y)$, we have $h(2u) \leq_\rho h(2z)$, which gives $0h(u) \leq_\rho 0h(z) = l_{0,\rho,w}$. Hence we must have $0h(u) = l_{0,\rho,w}$, and so

$$l_{0,\rho,w} = 0h(u) = 0g(u) = 0w.$$

8 Conclusion

The condition $f, g \in \mathcal{M}_x$ guaranteeing that the extremal words of a morphic word of the form $g(f^\omega(a))$ are morphic (Theorem 3) is quite powerful, as we have seen. Clearly, however, not all morphic sequences and the corresponding morphisms satisfy this condition. So does there exist a morphic sequence with an extremal sequence that is not morphic? All our failed attempts to produce such an example encourage us to conjecture that, in fact, all extremal words of all morphic sequences are morphic. Let us mention finally that Luca Q. Zamboni discovered an argument proving that the extremal words of all primitive morphic sequences are also primitive morphic. This proof will appear in a future joint work with him.

References

1. Allouche, J.-P.: Théorie des nombres et automates, Thèse d'État, Université Bordeaux I (1983)
2. Allouche, J.-P., Currie, J., Shallit, J.: Extremal infinite overlap-free binary words. The Electronic Journal of Combinatorics 5, #R27 (1998)
3. Allouche, J.-P., Cosnard, M.: Itérations de fonctions unimodales et suites engendrées par automates. C. R. Acad. Sci. Paris, Sér. A 296, 159–162 (1983)

4. Allouche, J.-P., Rampersad, N., Shallit, J.: Periodicity, repetitions, and orbits of an automatic sequence. Theoret. Comput. Sci. 410, 2795–2803 (2009)
5. Allouche, J.-P., Shallit, J.: Automatic Sequences: Theory, Applications, Generalizations. Cambridge University Press (2003)
6. Cassaigne, J., Nicolas, F.: Factor Complexity. In: Berthé, V., Rigo, M. (eds.) Combinatorics, Automata and Number Theory. Cambridge University Press, Cambridge (2010)
7. Currie, J.: Lexicographically least words in the orbit closure of the Rudin–Shapiro word. Theoret. Comput. Sci. 412, 4742–4746 (2011)
8. Damanik, D.: Local symmetries in the period-doubling sequence. Discrete Appl. Math. 100, 115–121 (2000)
9. Ferenczi, S.: Les transformations de Chacon: combinatoire, structure géométrique, lien avec les systèmes de complexité $2n + 1$. Bulletin de la S. M. F 123(2), 271–292 (1995)
10. Gan, S.: Sturmian sequences and the lexicographic world. Proc. Amer. Math. Soc. 129(5), 1445–1451 (2000) (electronic)
11. Head, T., Lando, B.: Fixed and stationary ω-words and ω-languages. In: Rozenberg, G., Salomaa, A. (eds.) The Book of L, pp. 147–156. Springer (1986)
12. Krieger, D.: On stabilizers of infinite words. Theoret. Comput. Sci. 400, 169–181 (2008)
13. Lothaire, M.: Algebraic Combinatorics on Words. In: Encyclopedia of Mathematics and its Applications, vol. 90, Cambride University Press, Cambridge (2002)
14. Makarov, M.: On the infinite permutation generated by the period doubling word. European J. Combin. 31, 368–378 (2010)
15. Pirillo, G.: Inequalities characterizing standard Sturmian and episturmian words. Theoret. Comput. Sci. 341, 276–292 (2005)
16. Pytheas Fogg, N., Berthé, V., Ferenczi, S., Mauduit, C., Siegel, A. (eds.): Substitutions in Dynamics, Arithmetics and Combinatorics. Lecture Notes in Mathematics, vol. 1794. Springer (2002)

Inner Palindromic Closure*

Jürgen Dassow[1], Florin Manea[2], Robert Mercaş[1], and Mike Müller[2]

[1] Otto-von-Guericke-Universität Magdeburg, Fakultät für Informatik,
PSF 4120, D-39016 Magdeburg, Germany
dassow@iws.cs.uni-magdeburg.de, robertmercas@gmail.com
[2] Christian-Albrechts-Universität zu Kiel, Institut für Informatik,
D-24098 Kiel, Germany
{flm,mimu}@informatik.uni-kiel.de

Abstract. We introduce the inner palindromic closure as a new operation ♠, which consists in expanding a factor u to the left or right by v such that vu or uv, respectively, is a palindrome of minimal length. We investigate several language theoretic properties of the iterated inner palindromic closure $♠^*(w) = \bigcup_{i \geq 0} ♠^i(w)$ of a word w.

1 Introduction

The investigation of repetitions of factors in a word is a very old topic in formal language theory. For instance, already in 1906, THUE proved that there exists an infinite word over an alphabet with three letters which has no factor of the form ww. Since the eighties a lot of papers on combinatorial properties concerning repetitions of factors were published (see [17] and the references therein).

The duplication got further interest in connection with its importance in natural languages [16] and in DNA sequences and chromosomes [18]. Motivated by these applications, grammars with derivations consisting in "duplications" (more precisely, a word $xuwvy$ is derived to $xwuwvy$ or $xuwvwy$ under certain conditions for w, u, and v) were introduced. We refer to [5,13].

Combining the combinatorial, linguistic and biological aspect, it is natural to introduce the duplication language $D(w)$ associated to a word $w \in \Sigma^+$, which is the language containing all words that double some factor of w, i.e., $D(w) = \{xuuy \mid w = xuy, x,y \in \Sigma^*, u \in \Sigma^+\}$ and its iterated version $D^*(w) = \bigcup_{i \geq 0} D^i(w)$. In the papers [1,4,6,19], the regularity of $D^*(w)$ was discussed; for instance, it was shown that, for any word w over a binary alphabet, $D^*(w)$ is regular and that $D^*(abc)$ is not regular. Further results on iterated duplication languages can be found in [10]. Also the case of bounded duplication, i.e., the length of the duplicated word is bounded by a constant, was studied, [11].

It was noted that words w containing hairpins, i.e., $w = xuyh(u^R)z$, and words w with $w = xuy$ and $u = h(u^R)$, where u^R is the mirror image of u and h is a letter-to-letter isomorphism, are of interest in DNA structures (see [8,9], where the Watson-Crick complementarity gives the isomorphism). Therefore, operations leading to words with hairpins as factors were studied (see [2,3]).

---

* The work of Florin Manea and Mike Müller is supported by the DFG grant 582014. The work of Robert Mercaş is supported by Alexander von Humboldt Foundation.

M.-P. Béal and O. Carton (Eds.): DLT 2013, LNCS 7907, pp. 155–166, 2013.
© Springer-Verlag Berlin Heidelberg 2013

In this paper, we consider the case where the operation leads to words which have palindromes (words with $w = w^R$) as factors (which is a restriction to the identity as the isomorphism). An easy step would be to obtain $xuu^R y$ from a word xuy in analogy to the duplication. But then all newly obtained palindromes are of even length. Thus it seems to be more interesting to consider the palindrome closure defined by DE LUCA [12]. Here a word is extended to a palindrome of minimal length. We allow this operation to be applied to factors and call it inner palindromic closure. We also study the case of iterated applications and a restriction bounding the increase of length.

The paper is organised as follows: After some preliminaries given in the following, we define the new operation, inner palindromic closure, and its versions in Section 2, where we also give some simple properties. In Sections 3 and 4, we discuss the regularity of the sets obtained by the inner palindromic closures. Finally, we present some language classes associated with the new operation.

Basic Definitions. For more details on the concepts we define here see [17].

A set $M \subseteq \mathbb{N}^m$ of vectors is called linear, if it can be represented as

$$M = \{B + \sum_{i=1}^{n} \alpha_i A_i \mid \alpha_i \in \mathbb{N}, 1 \le i \le n\}$$

for some vectors B and A_i, $1 \le i \le n$. It is called semi-linear if it can be represented as a finite union of linear sets.

An alphabet Σ is a non-empty finite set with the cardinality denoted by $\|\Sigma\|$, and the elements called letters. A sequence of letters constitute a word $w \in \Sigma^*$, and we denote the *empty word* by ε. The set of all finite words over Σ is denoted by Σ^*, and any subset of it is called a language. Moreover, for a language L, by $\mathrm{alph}(L)$ we denote the set of all symbols occurring in words of L.

If $w = u_1 v_1 u_2 v_2 \ldots u_n v_n$ and $u = u_i u_{i+1} \ldots u_j$ for $1 \le i \le j \le n$, we say that u is a *scattered factor* of w, denoted as $u \preccurlyeq w$. Consider now $v_k = \varepsilon$ for all $1 \le k \le n$. We say that u is a *factor* of w, and, if $i = 1$ we call u a *prefix*. If $j = n$ we call u a *suffix*. Whenever $i > 1$ or $j < |w|$, the factor u is called *proper*.

The length of a finite word w is the number of not necessarily distinct symbols it consists of, and is denoted by $|w|$. The number of occurrences of a certain letter a in w is designated by $|w|_a$. The *Parikh vector* of a word $w \in \Sigma^*$, denoted by $\Psi(w)$, is defined as $\Psi(w) = \langle |w|_{a_1}, |w|_{a_2}, \ldots, |w|_{a_{\|\Sigma\|}} \rangle$, where $\Sigma = \{a_1, a_2, \ldots, a_{\|\Sigma\|}\}$. A language L is called linear or semi-linear, if its set of Parikh vectors is linear or semi-linear, respectively.

For $i \ge 0$, the i-fold catenation of a word w with itself is denoted by w^i and is called the ith power of w. When $i = 2$, we call the word $w^2 = ww$ a square.

For a word $w \in \Sigma^*$, we denote its mirror image (or reversal) by w^R and say that w is a palindrome if $w = w^R$. For a language L, let $L^R = \{w^R \mid w \in L\}$.

We say that a language L is *dense*, if, for any word $w \in \Sigma^*$, $\Sigma^* w \Sigma^* \cap L$ is non-empty, i.e., each word occurs as a factor in L.

We recall Higman's Theorem.

Theorem 1 (Higman [7]). *If L is a language with the property that no word is a scattered factor of another one, then L is finite.*

2 Definitions and Preliminary Results

We now look at a word operation due to DE LUCA [12], which considers extensions to the left and right of words such that the newly obtained words are palindromes.

Definition 1. *For a word u, the left (right) palindromic closure of u is a word vu (uv) which is a palindrome for some non-empty word v such that any other palindromic word having u as proper suffix (prefix) has length greater than $|uv|$.*

Here the newly obtained words have length greater than the original one, but minimal among all palindromes that have the original word as prefix or suffix.

As for duplication and reversal, we can now define a further operation.

Definition 2. *For a word w, the left (right) inner palindromic closure of w is the set of all words $xvuy$ ($xuvy$) for any factorisation $w = xuy$ with possibly empty x, y and non-empty u, v, such that vu (uv) is the left (right) palindromic closure of u. We denote these operations by $\spadesuit_\ell(w)$ and $\spadesuit_r(w)$, respectively, and define the inner palindromic closure $\spadesuit(w)$ as the union of $\spadesuit_\ell(w)$ and $\spadesuit_r(w)$.*

The operation is extended to languages and an iterated version is introduced.

Definition 3. *For a language L, let $\spadesuit(L) = \bigcup_{w \in L} \spadesuit(w)$. We set $\spadesuit^0(L) = L$, $\spadesuit^n(L) = \spadesuit(\spadesuit^{n-1}(L))$ for $n \geq 1$, $\spadesuit^*(L) = \bigcup_{n \geq 0} \spadesuit^n(L)$. Any set $\spadesuit^n(L)$ is called a finite inner palindromic closure of L, and we say that $\spadesuit^*(L)$ is the iterated inner palindromic closure of L.*

We start with a simple observation.

Lemma 1. *For every word w, if $u \in \spadesuit^*(w)$, then $w \preccurlyeq u$.*

Remark 1. Obviously, for any language L, $\spadesuit_r^*(L) \subseteq \spadesuit^*(L)$ and $\spadesuit_\ell^*(L) \subseteq \spadesuit^*(L)$. In general, allowing both left and right operations is stronger than allowing them only in one direction. To see this we consider $L = \{ax \mid x \notin a^*\}$. The language $\spadesuit_r^*(L)$ contains only words of the form ay with $x \preccurlyeq y$, while the language $\spadesuit_\ell^*(L^R)$ contains only words of the form $y'a$ with $x^R \preccurlyeq y'$. This is not the case of the languages obtained by the application of $\spadesuit$, since we can insert either before or after the letter a a letter $b \neq a$. Thus, $\spadesuit^*(L)$ and $\spadesuit^*(L^R)$ also contain words starting and ending with $b \neq a$, respectively. Hence $\spadesuit_r^*(L) \subsetneq \spadesuit^*(L)$ and $\spadesuit_\ell^*(L^R) \subsetneq \spadesuit^*(L^R)$. ◁

The next results are in tone with the ones from [10, Proposition 3.1.1]:

Proposition 1. *For any semi-linear (linear) language, its iterated inner palindromic closure is semi-linear (respectively, linear).*

Since each word is described by a linear set, as consequence of the above we get the following assertion.

Corollary 1. *For any word, its iterated inner palindromic closure is linear.*

Furthermore, we have the following result. We omit its proof as it follows similarly to Proposition 3.

Proposition 2. *For any word w, the language $\spadesuit^*(w)$ is dense with respect to the alphabet $alph(w)$.*

We mention that Proposition 2 does not hold for languages. This can be seen from $L = \{ab, ac\}$. Obviously, any word in $\spadesuit^*(L)$ contains only a and b or only a and c. Therefore $abc \in \Sigma^*$ is not a factor of any word in $\spadesuit^*(L)$.

Lemma 2. *Let $\Sigma = \{a_1, a_2, \ldots, a_k\}$ and define the recursive sequences*

$$w_0' = \varepsilon \text{ and } w_0 = \varepsilon,$$
$$w_i' = w_{i-1} w_{i-1}' \text{ and } w_i = w_i' a_i \text{ for } 1 \leq i \leq k.$$

Then for $1 \leq i \leq k$, $alph(w_i)^* w_i \subseteq \spadesuit^*(w_i)$.

Proof. Note that, for $0 \leq j < i \leq k$, w_i' is a palindrome and w_j is a proper prefix of w_i. We want to generate $b_1 b_2 \ldots b_n w_i$ with $b_\ell \in alph(w_i)$ for $1 \leq \ell \leq n$. Let $b_1 = a_j$. Since w_j is a prefix of w_i, $w_i = w_j' a_j v$ for some v. Since w_j' is a palindrome, we obtain $a_j w_j' a_j v = b_1 w_i$ by an inner palindromic closure step. The conclusion follows after performing the procedure in succession for $b_2, \ldots, b_n$. $\square$

We now define a variant of the inner palindromic closure, where we restrict the length of the words which are involved in the palindromic closure. First we introduce a parametrised version of the palindromic closure operation from [12].

Definition 4. *For a word u and integers $m \geq 0$ and $n > 0$, we define the sets*

$$L_{m,n}(w) = \{u \mid u = u^R, u = xw \text{ for } x \neq \varepsilon, |x| \geq n, \ m \geq |w| - |x| \geq 0\},$$
$$R_{m,n}(w) = \{u \mid u = u^R, u = wx \text{ for } x \neq \varepsilon, |x| \geq n, \ m \geq |w| - |x| \geq 0\}.$$

The left (right) (m, n)-palindromic closure of w is the shortest word of $L_{m,n}(w)$ (resp., $R_{m,n}(w)$), or undefined if $L_{m,n}(w)$ (resp., $R_{m,n}(w)$) is empty.

The idea behind this new definition is that an element of $L_{m,n}(w)$ is a palindrome u obtained by extending the word w by adding a prefix x of length at least n such that the centre of the newly obtained palindrome u is inside the prefix of length $\lceil \frac{m}{2} \rceil$ of w. That is, $u = xvv^R x^R$ where $n \leq |x|$, $2|v| \leq m$, and $w = vv^R x^R$, or $u = xvav^R x^R$, where $n \leq |x|$, $2|v| + 1 \leq m$, and $w = vav^R x^R$. The left (m, n)-palindromic closure is the shortest such word u, obtained when the shortest v is chosen. The right (m, n)-palindromic closure is defined similarly.

We briefly describe the restrictions imposed by (m, n) on the left palindromic closure (similar explanations hold for the right variant). By (classical) left palindromic closure we added some letters to the left of a word that had a palindromic prefix to transform the entire initial word into a palindrome of minimal length. For the left (m, n)-palindromic closure we require that at least n letters should be added and that the palindromic prefix should not be longer than m.

We define now the parametrised version of the inner palindromic closure.

Definition 5. *For non-negative integers n, m with $n > 0$, we define the $\spadesuit_{(m,n)}$ one step inner palindromic closure of some word w as*

$$\spadesuit_{(m,n)}(w) = \{u \mid u = xy'z, \ w = xyz, \ and$$
$$y' \ is \ obtained \ by \ left \ or \ right \ (m,n)-palindromic \ closure \ from \ y\}.$$

This notion can be easily extended to languages, while its iterated version $\spadesuit^*_{(m,n)}$ is defined just as in the case of the inner palindromic closure.

Remark 2. Note that $L_{m,n}(w)$ and $R_{m,n}(w)$ are empty if and only if $|w| < n$; otherwise, $L_{m,n}(w)$ contains at least the word $w^R w$ and $R_{m,n}(w)$ contains the word ww^R. Therefore, $L_{m,n}(w)$ and $R_{m,n}(w)$ are either both empty or both non-empty; clearly, both sets are always finite.

Also, the length of the left (right) $(m, n+j)$-palindromic closure of w is greater or equal than both the length of the left (right) $(m+i, n+j)$-palindromic closure of w and the length of the left (right) (m, n)-palindromic closure of w for $i, j > 0$.

If $|w| < n$, then $\spadesuit_{(m,n)}(w) = \emptyset$. Further, if $|w| = n$ then $\spadesuit_{(m,n)}(w) = \{w^R w, ww^R\}$. Generally, for $|w| \geq n$, we have that $\spadesuit_{(m,n)}(w) \neq \emptyset$. Finally, it is not hard to see that $\spadesuit(w) = \spadesuit_{(|w|,1)}(w)$.   ◁

A statement similar to Proposition 2 also holds for the bounded operation.

Proposition 3. *For any word w with $|w| \geq n$ and positive integer m, the language $\spadesuit^*_{(m,n)}(w)$ is dense with respect to the alphabet $alph(w)$.*

Proof. We note that if u is a prefix of length at least n of w and u ends with a then there is a word w' starting with a in $\spadesuit_{(m,n)}(w)$. If the letter a appears only in the prefix of length $n-1$ of w, then we do as follows. Let $w_0 = w$ and let w_{i+1} the word obtained by left (m, n)-palindromic closure from w_i for $i \geq 0$. As w_i is a proper suffix of w_{i+1}, there exists i_a such that w_{i_a} has a prefix of length at least n that ends with a. Continuing this process, we derive a word w' that for each letter $s \in alph(w)$ has a prefix of length at least n ending with s.

Suppose we want to generate a word starting with $a_1 \cdots a_n$ by inner (m, n)-palindromic closure from w. First, we generate w' and let $v_0 = w'$. By the above, $v_0 = x_1 a_1 y_1$ for some $|x_1| \geq n - 1$. Then, applying a left (m, n)-palindromic closure to $x_1 a_1$ (which produces a word from the inner (m, n)-palindromic closure of v_0) we obtain from v_0 a palindrome $a_1 v_1$, where v_1 has v_0 as a proper suffix. Thus, v_1 also has prefixes of length greater than n that end with every letter in $alph(w)$. Next, to generate the word $a_1 a_2 v_2$ from $a_1 v_1 = a_1 x_2 a_2 y_2$ we apply a left (m, n)-palindromic closure operation to $x_2 a_2$. The process is repeated until we generate the word $a_1 \cdots a_n v_n$.   □

The next result is related to Proposition 3 and will be useful in the sequel.

Lemma 3. *Let Σ be an alphabet with $\|\Sigma\| \geq 2$, $a \notin \Sigma$, and m and n positive integers. Let $w = a^m y_1 a \cdots y_{p-1} a y_p$ be a word such that $alph(w) = \Sigma \cup \{a\}$, $m, p > 0$, $y_i \in \Sigma^*$ for $1 \leq i \leq p$, $|y_1| > 0$, and such that there exists $1 \leq j \leq p$ with $|y_j| \geq n$. Then, for each $v \in \Sigma^*$ with $|v| \geq n$, there exists $w' \in \spadesuit^*_{(m,n)}(w)$ such that v is a prefix of w' and $|w'|_a = |w|_a$.*

Proof. As a first step, for a word $z = z_1 a z_2 a \cdots a z_k$, where $a \notin \bigcup_{1 \le i \le k} \text{alph}(z_i)$ and $|z_1| \ge n$, we define $z_1' = z_1$ and $z_i' = (z_{i-1}')^R z_i$ for $1 < i \le k$. Let $z' = z_1' a \cdots a z_k'$. It is immediate that $z' \in \spadesuit^*_{(m,n)}(z)$, as it is obtained by applying iteratively right (m, n)-palindromic closure to the factors $z_i' a$ to get $z_i' a (z_i')^R$, for $i > 0$. Moreover, $\text{alph}(z_k') = \bigcup_{1 \le i \le k} \text{alph}(z_i)$ and $|z_i'| \ge n$ for all $1 \le i \le k$.

As a second step, for a word $v = v_\ell a v_{\ell-1} a \cdots a v_1$, where $a \notin \bigcup_{1 \le i \le \ell} \text{alph}(v_i)$ and $|v_1| \ge n$, we define $v_1' = v_1$ and $v_i' = v_i (v_{i-1}')^R$ for $1 < i \le \ell$. Let $v' = v_\ell' a \cdots a v_1'$. It is immediate that $v' \in \spadesuit^*_{(m,n)}(v)$, as it can be obtained by applying iteratively left (m, n)-palindromic closure to the factors $a v_i'$ to obtain $(v_i')^R a v_i'^R$, for $i > 0$. Moreover, $\text{alph}(y_\ell') = \bigcup_{1 \le i \le k} \text{alph}(y_i)$ and $|y_i'| \ge n$ for all $1 \le i \le \ell$.

Now we consider the word w from our hypothesis. We apply the first step described above to the factor $y_j a y_{j+1} \cdots a y_p$ to obtain $y_j' a \cdots a y_p'$, where $\text{alph}(y_p') = \bigcup_{j \le i \le p} \text{alph}(y_i')$ and $|y_p'| \ge n$. Afterwards, we apply the second step procedure to the factor $y_1 a y_2 a \cdots a y_{j-1} a y_j' a \cdots a y_p'$ to obtain $y_1'' a y_2'' a \cdots a y_{j-1}'' a y_j'' a \cdots a y_p''$, where $\text{alph}(y_1'') = \bigcup_{1 \le i \le p} \text{alph}(y_i) = \Sigma$ and $|y_1''| \ge n$. Accordingly, $w'' = a^m y_1'' a y_2'' a \cdots a y_p'' \in \spadesuit^*_{(m,n)}(w)$.

Now, for a word $v \in \Sigma^*$ we obtain the word $w_v'' = a^m v^R y_v a \cdots y_p'' a$ from w'', for some $y_v \in \Sigma^*$, just like in the proof of Proposition 3. If $|v| \ge n$, we can obtain from w_v'' the word $v w_v''$ by applying to $a^m v^R$ a left (m, n)-palindromic closure to get $v a^m v^R$. This concludes our proof. $\qquad\square$

3   On the Regularity of the Inner Palindromic Closure

We start with some facts on words over a binary alphabet.

Lemma 4. *[Propagation rule] For a word $w = a^m b^n$ with positive integers m and n, the set $\spadesuit(w)$ contains all words of length $m+n+1$ with a letter $x \in \{a, b\}$ inserted before or after any letter of w.*

Proof. To see this, assume we want to insert a letter a somewhere in w (the case of the insertion of a letter b is symmetric). To insert a between positions j and $j + 1$ with $j < m$ we just take the palindromic prefix a^{m-j} and perform a $\spadesuit_\ell$ step on it. This results in the word a^{m-j+1} which fulfils the conditions. When $m \le j \le m + n$, we perform a $\spadesuit_r$ step on the word ab^{j-m}, which produces the palindrome $ab^{j-m}a$. $\qquad\square$

As a consequence of the Propagation Rule, we can show that the necessary condition given in Lemma 1 is also sufficient in the case of binary alphabets.

Corollary 2. *For any binary words w and u, $w \preccurlyeq u$ if and only if $u \in \spadesuit^*(w)$.*

Proof. By Lemma 1, we have that $w \preccurlyeq u$ for all $u \in \spadesuit^*(w)$. Using Lemma 4, all words u with $w \preccurlyeq u$ are in fact in $\spadesuit^*(w)$ since in each of them we can insert a's and b's at arbitrary positions. $\qquad\square$

For the duplication operation, BOVET and VARRICCHIO [1] showed that for any binary language, its iterated duplication completion always gives a regular language. For the inner palindromic closure operation on such alphabets, the result is similar.

Theorem 2. *The iterated inner palindromic closure of a language over a binary alphabet is regular.*

Proof. According to Theorem 1, for a language L there exists a finite set L_0 with $L_0 \subseteq L$ such that for every word $w \in L$ there is a word $w_0 \in L_0$ with $w_0 \preccurlyeq w$. By Corollary 2, it follows that $\spadesuit^*(L)$ is the union of the sets $SW(w_0) = \{w' \in \text{alph}(w_0)^* \mid w_0 \preccurlyeq w'\}$, for all $w_0 \in L_0$. As all the sets $SW(w_0)$ are regular, it follows that $\spadesuit^*(L)$ is regular. $\qquad\square$

It is obvious that the finite inner palindromic closure of some finite language is always regular, since at each step we only obtain words which have at most twice the length of the longest word in the given language. However, when considering the entire class of regular languages the result is not necessarily regular.

Theorem 3. *The finite inner palindromic closure of a regular language is not necessarily regular.*

Proof. We take a positive integer k and a language $L = c_1 a_1^+ c_2 a_2^+ \ldots c_k a_k^+ b$. We intersect $\spadesuit^k(L)$ with the language given by the regular expression:

$$c_1 a_1^+ c_2 a_2^+ \ldots c_k a_k^+ b (a_k^+ c_k \ldots a_2^+ c_2 a_1^+ c_1)(a_k^+ c_k \ldots a_3^+ c_3 a_2^+ c_2) \ldots (a_k^+ c_k a_{k-1}^+ c_{k-1}) a_k^+ c_k$$

It is not hard to see that in any word of the intersection the number of a_i's in every maximal unary group adjacent to c_i is the same. Since this is a non-regular language and regular languages are closed under intersection, we conclude. $\qquad\square$

It remains an *open problem* whether or not the iterated inner palindromic closure of a regular language L, where $\|\text{alph}(L)\| \geq 3$, is also regular.

We mention that the non-regularity of $\spadesuit^*(L)$ with $\|\text{alph}(L)\| \geq 3$ cannot be obtained by a strategy similar to that by WANG [19], who showed the non-regularity of $D(L)$ with $\|\text{alph}(L)\| \geq 3$. There, the non-regularity of $D(L)$ comes as a consequence of a padding that needs to be added every time we want to construct a longer word as result of consecutive applications of our chosen rule. Consider now the word abc and the language $(abc)^*$ that contains no palindromes of length greater than one. However, $babc \in \spadesuit(abc)$, thus by Lemma 2 we can generate at the beginning as many abc's as we want, $(abc)^*babc$. Hence, we cannot use any more the argument that each palindromic step creates some extra padding at the end of the word whenever we investigate words that contain no palindromes.

4 Parametrised Inner Palindromic Closure

We now discuss the regularity of $\spadesuit^*_{(m,n)}(w)$. Before we state our results, we establish two facts on the avoidance of patterns.

Theorem 4. *There exist infinitely long binary words avoiding both palindromes of length 6 and longer, and squares of words with length 3 and longer.*

Proof. RAMPERSAD et al. [15] constructed an infinite word w, that is square-free and has no factors from the set $\{ac, ad, ae, bd, be, ca, ce, da, db, eb, ec, aba, ede\}$.

We can show that the morphism γ, defined by

$$\gamma(a) = abaabbab, \qquad \gamma(b) = aaabbbab, \qquad \gamma(c) = aabbabab,$$
$$\gamma(d) = aabbbaba, \qquad \gamma(e) = baaabbab,$$

maps this word w to a word with the desired properties.

As any palindrome of length $n > 2$ contains a shorter palindrome of length $n - 2$, a word avoiding palindromes of lengths 6 and 7 also avoids longer ones. Also, each palindrome of length 6 or 7 would occur in the image of some word of length 2. We see that no such palindromes occur in $\gamma(\{ab, ba, bc, cb, cd, dc, de, ea, ed\})$, therefore neither in $\gamma(w)$. We show that $\gamma(w)$ contains no squares other than $aa, bb, abab$ and $baba$ by applying methods used in [15]. $\qquad\square$

Theorem 5. *There exist infinitely long ternary words avoiding both palindromes of length 3 and longer, and squares of words with length 2 and longer.*

Proof. We claim that the morphism ψ, that is defined by

$$\psi(a) = abbccaabccab, \qquad \psi(b) = bccaabbcaabc, \qquad \psi(c) = caabbccabbca,$$

maps all infinite square-free ternary words h to words with the desired properties.

We see that $\psi(h)$ does not contain palindromes of length 3 or 4, since those would occur inside $\psi(u)$ for some square-free word u of length 2. We check that there are no squares other than aa, bb and cc in $\psi(h)$ using standard tools. $\qquad\square$

In the sequel, we exhibit a method to construct words whose iterated inner (m, n)-palindromic closure is not regular, for positive integers m, n. We first establish several notations. We associate to an integer $k \geq 2$ a pair of numbers (p_k, q_k) if there exists an infinite word over a k-letter alphabet avoiding both palindromes of length greater or equal to q_k and squares of words of length greater or equal to p_k. If more such pairs exist, we take (p_k, q_k) to be any of them.

Theorem 6. *Let $m > 0$ and $k \geq 2$ be two integers and define $n = \max\{\frac{q_k}{2}, p_k\}$. Let Σ be a k-letter alphabet with $a \notin \Sigma$ and $w = a^m y_1 a y_2 \cdots a y_{r-1} a y_r$ be a word such that $alph(w) = \Sigma \cup \{a\}$, $r > 0$, $y_i \in \Sigma^*$ for all $1 \leq i \leq r$, and there exists j with $1 \leq j \leq r$ and $|y_j| \geq n$. Then $\spadesuit^*_{(m,n)}(w)$ is not regular.*

Proof. Let α be an infinite word over Σ that avoids palindromes of length q_k and squares of words of length p_k. Note that due to Lemma 3, for each prefix u of α longer than n, there exists w_u with $|w_u|_a = r - 1$ such that $u a^m w_u \in \spadesuit^*_{(m,n)}(w)$.

We analyse how the words $u a^m v$ with u being a prefix of α and $|v|_a = r - 1$ are obtained by iterated (m, n)-palindromic closure steps from w. As u contains no a's, no squares of words of length p_k, as well as no palindromes with length greater than q_k, and the application of an (m, n)-palindromic closure step introduces a palindrome in the derived word, we get that the only possible cases of application of the operation in the derivation of $u a^m v$ are the following:

(1) $v = xyz$ and y is the (m, n)-palindromic closure of y' (implicitly, $|y'| < |y|$ and $|y|_a = |y'|_a$); in this case we have that $ua^m v$ is in $\spadesuit_{(m,n)}(ua^m xy'z)$.

(2) $u = u'x$, $v = yz$, and $xa^m y$ is the (m, n)-palindromic closure of $a^m y$ (implicitly, $x = y^R$ and neither x nor y contain any a's); in this case we have that $ua^m v$ is in $\spadesuit_{(m,n)}(u'a^m yz)$.

(3) $u = xyz$ and y is the (m, n)-palindromic closure of y' (implicitly, $|y'| < |y|$ and y' contains no a's); in this case we have that $ua^m v$ is in $\spadesuit_{(m,n)}(xy'za^m v)$.

Since we only apply (m, n)-palindromic closure operations, and the word we want to derive has the form $ua^m v$ with $|a^m v|_a = |w|_a$, it is impossible to apply any palindromic closure step that adds to the derived word more a symbols or splits the group a^m that occurs at the beginning of w. Intuitively, the palindromic closure operations that we apply are localised, due to the restricted form of the operation: they either occur inside u, or inside v, or are centred around a^m.

Moreover, by choosing $n \geq \frac{q_k}{2}$ if at any step we apply a palindromic closure operation of the type (3) above, then the final word u contains a palindrome of length greater than q_k. To see this, we assume, for the sake of a contradiction, that such an operation was applied. Then, we look at the last operation of this kind that was applied. Obviously, none of the operations of type (1) or (2) that were applied after that operation of type (3) could have modified the palindrome of length at least q_k introduced by it in the derived word before a^m. Therefore, that palindrome would also appear in u, a contradiction.

This means that all the intermediate words obtained during the derivation of $ua^m v$ from w have the form $u'a^m v'$ where u' is a prefix (maybe empty) of α and v' has exactly $|w|_a - m$ symbols a. We now look at the kind of operations that can be applied to such a word. In particular, we note that we cannot have more than $|v'| - n$ consecutive derivation steps in which the length of the word occurring after the first sequence of a's is preserved. In other words, we can apply at most $|v'| - n$ consecutive operations that fall in the situation (2).

Indeed, after ℓ such derivation steps one would obtain from $u'a^m v'$ a word $u'v_1 \cdots v_\ell a^m v'$ where v_i^R is a prefix of v' and $|v_i| \geq n$ for every $1 \leq i \leq \ell$. Assume, for the sake of a contradiction, that $\ell > |v'| - n$. Then, there exists j such that $1 \leq j < \ell$ and $|v_j| \geq |v_{j+1}|$. Therefore, $u'v_1 \cdots v_\ell$ contains a square of length at least $2n \geq 2p_k$. But such a square will remain in the derived word for the rest of the derivation, as neither an operation of type (1) nor one of type (2) could introduce letters inside it. Another contradiction with the form of u is reached.

We use this last remark to show by induction on the number of steps in the derivation, that if u is a finite prefix of α and $ua^m v \in \spadesuit^*_{(m,n)}(w)$, then $|u| \leq |v|^3$

If the derivation has one step, then the statement we want to show holds trivially, as the fact that the prefix u can be added to w implies that $|u| \leq |y_1|$.

Let us now assume that it holds for words obtained in at most k derivation steps, and show it for words obtained in $k+1$ derivation steps. If the last applied step to obtain $ua^m v$ is of type (1), then we obtained $ua^m v$ from $ua^m v'$ for some v' shorter than v. From the induction step we have that $|u| \leq |v'|^3$, and, consequently, $|u| \leq |v|^3$. According to the last made remark, we have that at most the last $|v| - n$ consecutive steps applied were of type (2). In these steps, the length of u increased by at most $\sum_{n \leq i \leq |v|} i \leq \frac{|v|(|v|+1)}{2}$. Therefore, we get $|u| - \frac{|v|(|v|+1)}{2} \leq (|v|-1)^3$; hence $|u| \leq |v|^3$. This concludes our induction proof.

We now show that the language

$$L = \{ua^m v \in \spadesuit^*_{(m,n)}(w) \mid |u| \geq n, |v|_a = r - 1\}$$

is not regular. Since this language is obtained from $\spadesuit^*_{(m,n)}(w)$ by intersection with a regular language, if L is not regular, then $\spadesuit^*_{(m,n)}(w)$ is not regular either.

We consider a word $u_0 a^m v_0 \in L$ such that u_0 is a prefix of α with $|u_0| \geq n$; clearly, L contains such a word. As we have shown above, $|u_0| \leq |v_0|^3$. We now take a prefix u_1 of α with $|u_1| > |v_0|^4$; it follows that $u_1 a^m v_0 \notin L$, thus u_0 and u_1 are in different equivalence classes with respect to the syntactic congruence defined by the language L. However, by the considerations made at the beginning of this proof, there exists v_1 such that $u_1 a^m v_1 \in L$. In the exact same manner we construct a word u_2, that is in a different equivalence class with respect to the syntactic congruence defined by the language L than both u_0 and u_1, and so on. This means we have an infinite sequence $(u_i)_{i \geq 0}$ where any two elements are in different equivalence classes with respect to the syntactic congruence defined by the language L. Thus, the syntactic congruence defined by L has an infinite number of equivalence classes, so L cannot be regular, and we conclude the proof.   □

The following theorem follows immediately from the previous results.

Theorem 7. *Let $w = a^p y_1 a \cdots y_{r-1} a y_r$, where $a \notin alph(y_i)$ for $1 \leq i \leq r$.*
(1) If $\|alph(w)\| \geq 3$ and $|y_j| \geq 3$ for some $1 \leq j \leq r$, then for every positive integer $m \leq p$ we have that $\spadesuit^*_{(m,3)}(w)$ is not regular.
(2) If $\|alph(w)\| \geq 4$ and $|y_j| \geq 2$ for some $1 \leq j \leq r$, then for every positive integer $m \leq p$ we have that $\spadesuit^*_{(m,2)}(w)$ is not regular.
(3) If $\|alph(w)\| \geq 5$, then for every positive integer $m \leq p$ we have that $\spadesuit^*_{(m,1)}(w)$ is not regular.
(4) For every positive integers m and n there exists u with $\spadesuit^*_{(m,n)}(u)$ not regular.

Proof. By Theorems 4 and 5 we can take $q_2 = 6$ and $p_2 = 3$, respectively, $q_3 = 3$ and $p_3 = 2$. Therefore, if we take $n = 3$, or $n = 2$, respectively, in the hypothesis of the theorem, then the results (1) and (2) follow for any positive $m \leq p$.

The third statement follows from [14, Theorem 4.15], where an infinite word avoiding both squares and palindromes is constructed. Thus, we can take $p_k = q_k = 1$, so n can be also taken to be 1. Finally, (4) is a consequence of (3).   □

In general, the regularity of the languages $\spadesuit^*_{(m,n)}(w)$ for positive integers m and n, and binary words w, $|w| \geq n$, is *left open*. We only show the following.

Theorem 8. *For any word $w \in \{a,b\}^+$ and integer $m \geq 0$, $\spadesuit^*_{(m,1)}(w)$ is regular.*

Proof. Due to the lack of space the technical details are skipped.

The general idea of the proof is to give a recursive definition of $\spadesuit^*_{(m,1)}(w)$. That is, $\spadesuit^*_{(m,1)}(w)$ is expressed as a finite union and concatenation of several languages $\spadesuit_{(m,1)}^*(w')$, with $|w'| < |w|$, and some other simple regular languages. To this end, we let $x \neq y \in \{a,b\}$ and identify a series of basic cases for which such a definition can be given easily: words that have no unary factor longer than m, words of the form xy^qx, and, finally, words of the form xy^q or y^qx. Building on these basic ingredients, we define $\spadesuit_{(m,1)}^*(w)$ for every word w by, basically, identifying a prefix of w that has one of these forms, separating it from the rest, and then computing, recursively, the iterated closure of the rest of the word.

In order to make this strategy work, one has to implement several steps. The first is to note that if a word w has no maximal unary factor longer than m, then $\spadesuit_{(m,1)}^*(w)$ contains all words that have w as scattered factor.

Further, if $uvxy^pxv^R \in \spadesuit_{(m,1)}(uvxy^q)$ for $q \leq p$, then we can find a sort of normal-form derivation of $uvxy^pxv^R$ by first deriving $uvxy^px$ in one step, and then appending any suffix (in particular v^R) by a process similar to propagation. Similar arguments hold when the factor is prefixed by palindromic closure. Intuitively, we can split the derivation of a word in separate parts and apply our operations only to maximal unary factors and the symbols that bound them (factors of the type xy^qx, y^qx, and xy^q, with the last two as suffixes or prefixes).

Next, the derivation of these basic factors on which the operation is applied can be further normalised. The basic idea is, intuitively, that whenever we start a derivation of a factor xy^qx, the first step that we should make is to split the group of y's in two smaller groups, and continue to derive each of them separately. More precisely, if $x^{\ell_1}y^{h_1}x^{\ell_2}y^{h_2}x^{\ell_3} \in \spadesuit_{(m,1)}^*(xy^qx)$ for some positive integers ℓ_1, ℓ_2, ℓ_3, h_1, and h_2, then there exist positive integers $p, r < q$ such that $x^{\ell_1}y^{h_1}x^{\ell_2}y^{h_2}x^{\ell_3} \in \spadesuit_{(m,1)}^*(xy^pxy^rx)$ and one of the following holds: $p \leq m$ or $r \leq m$ and $p = q - r$; or, $m < p, r$, and $p = m + 2k$ and $r = q - m - k$, or, vice-versa, $r = m + 2k$ and $p = q - m - k$, for some $k > 0$.

Similarly, when we start a derivation from a group xy^q, we first split the group of y's into xy^px and xy^r, with $r < q$, and then apply the above definition to these and repeat the process. Clearly, at every step we can lengthen the words by pumping x's in a group of x's, and by generating $\{a,b\}^*xy\{a,b\}^*$ from xy.

Using all the above, we can now find recursively the formula for $\spadesuit_{(m,1)}^*(w)$ by first separating a prefix having one of the basic forms, derive a word from it as we described, and then work, recursively, on the remaining suffix. $\square$

5   Final Remarks

Apart from solving the open problems stated in this article, the study of classes of languages obtained through these operations seems interesting to us. The following initial results show several possible directions for such investigations.

For a class $\mathcal{L}$ of languages, we set $\mathcal{L}^R = \{L^R \mid L \in \mathcal{L}\}$, and for a natural number $k \geq 1$, we define $\mathcal{L}_k = \{L \in \mathcal{L} \mid \|\mathrm{alph}(L)\| = k\}$. Consider the classes

$$\mathcal{P}_{\spadesuit_\ell} = \{L' \mid L' = \spadesuit_\ell^*(L) \text{ for some } L\}$$
$$\mathcal{P}_{\spadesuit_r} = \{L' \mid L' = \spadesuit_r^*(L) \text{ for some } L\}$$
$$\mathcal{P}_{\spadesuit} = \{L' \mid L' = \spadesuit^*(L) \text{ for some } L\}$$

Straightforward, for every language L, $\spadesuit_r(L) = (\spadesuit_\ell(L^R))^R$ and $\spadesuit_r^*(L) = (\spadesuit_\ell^*(L^R))^R$ hold (for both operations the propagation rule works in only one direction). Thus, we immediately get $\mathcal{P}_{\spadesuit_r} = (\mathcal{P}_{\spadesuit_\ell})^R$ and $\mathcal{P}_{\spadesuit_\ell} = (\mathcal{P}_{\spadesuit_r})^R$.

The following result is a consequence of Remark 1.

Lemma 5. *The classes* $\mathcal{P}_{\spadesuit_r} \setminus \mathcal{P}_\spadesuit$ *and* $\mathcal{P}_{\spadesuit_\ell} \setminus \mathcal{P}_\spadesuit$ *are both not empty.*

When we consider only binary alphabets, we have the following statement.

Proposition 4. $(\mathcal{P}_\spadesuit)_2 \subsetneq (\mathcal{P}_{\spadesuit_r})_2 = (\mathcal{P}_{\spadesuit_\ell})_2^R$ *and* $(\mathcal{P}_\spadesuit)_2 \subsetneq (\mathcal{P}_{\spadesuit_\ell})_2 = (\mathcal{P}_{\spadesuit_r})_2^R$.

References

1. Bovet, D.P., Varricchio, S.: On the regularity of languages on a binary alphabet generated by copying systems. Inf. Process. Lett. 44, 119–123 (1992)
2. Cheptea, D., Martín-Vide, C., Mitrana, V.: A new operation on words suggested by DNA biochemistry: Hairpin completion. Trans. Comput., 216–228 (2006)
3. Dassow, J., Holzer, M.: Language families defined by a ciliate bio-operation: hierarchies and decision problems. Int. J. Found. Comput. Sci. 16(4), 645–662 (2005)
4. Dassow, J., Mitrana, V., Păun, G.: On the regularity of duplication closure. Bulletin of the EATCS 69, 133–136 (1999)
5. Dassow, J., Mitrana, V., Salomaa, A.: Context-free evolutionary grammars and the structural language of nucleic acids. BioSystems 43, 169–177 (1997)
6. Ehrenfeucht, A., Rozenberg, G.: On regularity of languages generated by copying systems. Discrete Appl. Math. 8, 313–317 (1984)
7. Higman, G.: Ordering by divisibility in abstract algebras. Proc. London Math. Soc. 3(2), 326–336 (1952)
8. Kari, L., Konstantinidis, S., Losseva, E., Sosík, P., Thierrin, G.: Hairpin structures in DNA words. In: Carbone, A., Pierce, N.A. (eds.) DNA 2005. LNCS, vol. 3892, pp. 158–170. Springer, Heidelberg (2006)
9. Kari, L., Mahalingam, K.: Watson–Crick palindromes in DNA computing. Natural Computing 9(2), 297–316 (2010)
10. Leupold, P.: Languages Generated by Iterated Idempotencies. Ph.D. thesis, Univeritat Rovira y Virgili, Tarragona, Spain (2006)
11. Leupold, P., Mitrana, V.: Uniformly bounded duplication codes. RAIRO Theor. Inf. Appl. 41, 411–427 (2007)
12. de Luca, A.: Sturmian words: Structure, combinatorics, and their arithmetics. Theor. Comput. Sci. 183, 45–82 (1997)
13. Martín-Vide, C., Păun, G.: Duplication grammars. Acta Cybernet 14, 151–164 (1999)
14. Pansiot, J.J.: A propos d'une conjecture de F. Dejean sur les répétitions dans les mots. Discrete Appl. Math. 7, 297–311 (1984)
15. Rampersad, N., Shallit, J., Wang, M.-W.: Avoiding large squares in infinite binary words. Theor. Comput. Sci. 339(1), 19–34 (2005)
16. Rounds, W.C., Ramer, A.M., Friedman, J.: Finding natural languages a home in formal language theory. In: Mathematics of Languages, pp. 349–360. John Benjamins, Amsterdam (1987)
17. Rozenberg, G., Salomaa, A.: Handbook of Formal Languages. Springer-Verlag New York, Inc. (1997)
18. Searls, D.B.: The computational linguistics of biological sequences. In: Artificial Intelligence and Molecular Biology, pp. 47–120. AAAI Press, Cambridge (1993)
19. Wang, M.W.: On the irregularity of the duplication closure. Bulletin of the EATCS 70, 162–163 (2000)

On the Dual Post Correspondence Problem[*]

Joel D. Day[1], Daniel Reidenbach[1,**], and Johannes C. Schneider[2]

[1] Department of Computer Science, Loughborough University,
Loughborough, Leicestershire, LE11 3TU, UK
J.Day-10@student.lboro.ac.uk,
D.Reidenbach@lboro.ac.uk
[2] Fachbereich Informatik, Technische Universität Kaiserslautern,
Postfach 3049, 67653 Kaiserslautern, Germany
jschneider@informatik.uni-kl.de

Abstract. The Dual Post Correspondence Problem asks whether, for a given word α, there exists a pair of distinct morphisms σ, τ, one of which needs to be non-periodic, such that $\sigma(\alpha) = \tau(\alpha)$ is satisfied. This problem is important for the research on equality sets, which are a vital concept in the theory of computation, as it helps to identify words that are in trivial equality sets only.

Little is known about the Dual PCP for words α over larger than binary alphabets. In the present paper, we address this question in a way that simplifies the usual method, which means that we can reduce the intricacy of the word equations involved in dealing with the Dual PCP. Our approach yields large sets of words for which there exists a solution to the Dual PCP as well as examples of words over arbitrary alphabets for which such a solution does not exist.

Keywords: Morphisms, Equality sets, Dual Post Correspondence Problem, Periodicity forcing sets, Word equations, Ambiguity of morphisms.

1 Introduction

The *equality set* $E(\sigma, \tau)$ of two morphisms σ, τ is the set of all words α that satisfy $\sigma(\alpha) = \tau(\alpha)$. Equality sets were introduced by A. Salomaa [13] and Engelfriet and Rozenberg [4], and they can be used to characterise crucial concepts in the theory of computation, such as the recursively enumerable set (see Culik II [1]) and the complexity classes P and NP (see Mateescu et al. [10]). Furthermore, since the famous undecidable *Post Correspondence Problem* (PCP) by Post [11] asks whether, for given morphisms σ, τ, there *exists* a word α satisfying $\sigma(\alpha) = \tau(\alpha)$, it is simply the emptiness problem for equality sets.

Culik II and Karhumäki [2] study an alternative problem for equality sets, called the *Dual Post Correspondence Problem* (Dual PCP or DPCP for short): they ask whether, for any given word α, there exist a pair of distinct morphisms

---

[*] This work was supported by the London Mathematical Society, grant SC7-1112-02.
[**] Corresponding author.

M.-P. Béal and O. Carton (Eds.): DLT 2013, LNCS 7907, pp. 167–178, 2013.
© Springer-Verlag Berlin Heidelberg 2013

σ, τ (called a *solution* to the DPCP) such that $\sigma(\alpha) = \tau(\alpha)$. Note that, in order for this problem to lead to a rich theory, at least one of the morphisms needs to be non-periodic. If a word does not have such a pair of morphisms, then it is called *periodicity forcing*, since the only solutions to the corresponding instance of the DPCP are periodic.

The Dual Post Correspondence Problem is of particular interest for the research on equality sets as it helps to identify words that can only occur in *trivial* equality sets (i.e., equality sets $E(\sigma, \tau)$ where σ or τ are periodic). The existence of these words (namely the periodicity forcing ones) is a rather peculiar property of equality sets when compared to other types of formal languages, and it illustrates their combinatorial intricacy. In addition, the DPCP shows close connections to a special type of *word equations*, since a word α has a solution to the DPCP if and only there exists a non-periodic solution to the word equation $\alpha = \alpha'$, where α' is renaming of α. A further related concept is the *ambiguity of morphisms* (see, e.g., Freydenberger et al. [6,5], Schneider [14]). Research on this topic mainly asks whether, for a given word α, there exists a morphism σ that is *unambiguous* for it, i.e., there is no other morphism τ satisfying $\sigma(\alpha) = \tau(\alpha)$. Using this terminology, a word does not have a solution to the DPCP if *every* non-periodic morphism is unambiguous for it.

Previous research on the DPCP has established its decidability and numerous insights into words over *binary* alphabets that do or do not have a solution. In contrast to this, for larger alphabets, it is not even known whether the problem is nontrivial, i.e., whether there are periodicity forcing words, and if so, what they look like. It is the purpose of the present paper to study the DPCP for words over arbitrary alphabets. Our main results shall, firstly, establish an approach to the problem that reduces the complexity of the word equations involved, secondly, demonstrate that most words are not periodicity forcing and why that is the case and, thirdly, prove that the DPCP is nontrivial for all alphabet sizes.

Due to space constraints, all proofs have been omitted from this paper.

2   Definitions and Basic Observations

Let $\mathbb{N} := \{1, 2, \ldots\}$ be the set of natural numbers, and let $\mathbb{N}_0 := \mathbb{N} \cup \{0\}$. We often use $\mathbb{N}$ as an infinite alphabet of symbols. In order to distinguish between a word over $\mathbb{N}$ and a word over a (possibly finite) alphabet Σ, we call the former a *pattern*. Given a pattern $\alpha \in \mathbb{N}^*$, we call symbols occurring in α *variables* and denote the set of variables in α by $\mathrm{var}(\alpha)$. Hence, $\mathrm{var}(\alpha) \subseteq \mathbb{N}$. We use the symbol $\cdot$ to separate the variables in a pattern, so that, for instance, $1 \cdot 1 \cdot 2$ is not confused with $11 \cdot 2$. For a set X, the notation $|X|$ refers to the cardinality of X, and for a word X, $|X|$ stands for the length of X. By $|\alpha|_x$, we denote the number of occurrences of the variable x in the pattern α. Let $\alpha \in \{x_1, x_2, \ldots, x_n\}^*$ be a pattern. The *Parikh vector of* α, denoted by $\mathrm{P}(\alpha)$, is the vector $(|\alpha|_{x_1}, |\alpha|_{x_2}, \ldots, |\alpha|_{x_n})$.

Given arbitrary alphabets $\mathcal{A}, \mathcal{B}$, a *morphism* is a mapping $h : \mathcal{A}^* \to \mathcal{B}^*$ that is compatible with the concatenation, i.e., for all $v, w \in \mathcal{A}^*$, $h(vw) = h(v)h(w)$.

Hence, h is fully defined for all $v \in \mathcal{A}^*$ as soon as it is defined for all symbols in $\mathcal{A}$. Such a morphism h is called *periodic* if and only if there exists a $v \in \mathcal{B}^*$ such that $h(a) \in v^*$ for every $a \in \mathcal{A}$. For the *composition* of two morphisms $g, h : \mathcal{A}^* \to \mathcal{A}^*$, we write $g \circ h$, i.e., for every $w \in \mathcal{A}^*$, $g \circ h(w) = g(h(w))$. In this paper, we usually consider morphisms $\sigma : \mathbb{N}^* \to \{a, b\}^*$ and morphisms $\varphi : \mathbb{N}^* \to \mathbb{N}^*$. For a set $N \subseteq \mathbb{N}$, the morphism $\pi_N : \mathbb{N}^* \to \mathbb{N}^*$ is defined by $\pi_N(x) := x$ if $x \in N$ and $\pi_N(x) := \varepsilon$ if $x \notin N$. Thus, for a pattern $\alpha \in \mathbb{N}^+$, $\pi_N(\alpha)$ is the *projection* of α to its subpattern $\pi_N(\alpha)$ consisting of variables in N only. Let $\Delta \subset \mathbb{N}$ be a set of variables and Σ be an alphabet. Then two morphisms $\sigma, \tau : \Delta^* \to \Sigma^*$ are *distinct* if and only if there exists an $x \in \Delta$ such that $\sigma(x) \neq \tau(x)$.

Let $\alpha \in \mathbb{N}^+$. We call α *morphically imprimitive* if and only if there exist a pattern β with $|\beta| < |\alpha|$ and morphisms $\varphi, \psi : \mathbb{N}^* \to \mathbb{N}^*$ satisfying $\varphi(\alpha) = \beta$ and $\psi(\beta) = \alpha$. If α is not morphically imprimitive, we call α *morphically primitive*. As demonstrated by Reidenbach and Schneider [12], the partition of the set of all patterns into morphically primitive and morphically imprimitive ones is vital in several branches of combinatorics on words and formal language theory, and some of our results in the main part of the present paper shall again be based on this notion.

We now formally define the Dual PCP as a set:

Definition 1. *Let Σ be an alphabet. DPCP is the set of all $\alpha \in \mathbb{N}^+$ such that there exist a non-periodic morphism $\sigma : \mathbb{N}^* \to \Sigma^*$ and an (arbitrary) morphism $\tau : \mathbb{N}^* \to \Sigma^*$ satisfying $\sigma(\alpha) = \tau(\alpha)$ and $\sigma(x) \neq \tau(x)$ for an $x \in \mathrm{var}(\alpha)$.*

We wish to investigate what patterns $\alpha \in \mathbb{N}^+$ are contained in DPCP, and what patterns are not. Since all morphisms with unary target alphabets are periodic and since we can encode any Σ, $|\Sigma| \geq 2$, over $\{a, b\}$, we choose $\Sigma := \{a, b\}$ from now on.

The following proposition explains why in the definition of DPCP at least one morphism must be non-periodic.

Proposition 2. *For every $\alpha \in \mathbb{N}^+$ with $|\mathrm{var}(\alpha)| \geq 2$, there exist (periodic) morphisms $\sigma, \tau : \mathbb{N}^* \to \{a, b\}^*$ satisfying $\sigma(\alpha) = \tau(\alpha)$.*

Hence, allowing periodic morphisms would turn the Dual PCP into a trivial problem. Note that for patterns α with $|\mathrm{var}(\alpha)| = 1$, every morphism is unambiguous.

In the literature, patterns not in DPCP are called *periodicity forcing* since they force every pair of morphisms that agree on the pattern to be periodic. This notion can be extended to sets of patterns in a natural way: Let $\Delta \subset \mathbb{N}$ be a set of variables, and let $\beta_1, \beta_2, ..., \beta_n \in \Delta^+$ be patterns. The set $\{\beta_1, \beta_2, ..., \beta_n\}$ is *periodicity forcing* if, for every pair of distinct morphisms $\sigma, \tau : \Delta^* \to \{a, b\}^*$ which agree on every β_i for $1 \leq i \leq n$, σ and τ are periodic.

From Culik II and Karhumäki [2] it is known that DPCP is decidable. Furthermore, the following specific results on two-variable patterns that are or are not in DPCP can be derived from the literature on word equations and binary equality sets:

Proposition 3 ([2]). *Every two-variable pattern of length 4 or less is in* DPCP. *Every renaming or mirrored version of the patterns* $1 \cdot 2 \cdot 1 \cdot 1 \cdot 2$, $1 \cdot 2 \cdot 1 \cdot 2 \cdot 2$ *is not in* DPCP. *These are the only patterns of length 5 that are not in* DPCP. *In particular, the (morphically primitive) patterns* $1 \cdot 1 \cdot 2 \cdot 2 \cdot 2$, $1 \cdot 2 \cdot 1 \cdot 2 \cdot 1$, $1 \cdot 2 \cdot 2 \cdot 1 \cdot 1$ *and* $1 \cdot 2 \cdot 2 \cdot 2 \cdot 1$ *are in* DPCP.

Furthermore, we have the following examples of longer patterns.

Proposition 4 ([7]). *For any* $i \in \mathbb{N}$, $(1 \cdot 2)^i \cdot 1 \in$ DPCP.

Proposition 5 ([2]). *For any* $i, j \in \mathbb{N}$, $1^i \cdot 2^j \in$ DPCP.

Proposition 6 ([8]). *For any* $i \in \mathbb{N}$, $1 \cdot 2^i \cdot 1 \in$ DPCP.

Note that, for $i, j > 1$, the three propositions above give morphically primitive example patterns. Thus, the results are not trivially achievable by applying Corollary 18 in Section 4.

Proposition 7 ([3]). $1^2 \cdot 2^3 \cdot 1^2 \notin$ DPCP.

It is worth noting that the proof of the latter proposition takes about 9 pages. This illustrates how difficult it can be to show that certain example patterns do not belong to DPCP.

In [2], Culik II and Karhumäki state without proof that any ratio-primitive pattern $\alpha \in (1^3 \cdot 1^* \cdot 2^3 \cdot 2^*)^2$ is not in DPCP. A pattern $\alpha \in \{1, 2\}^+$ is called *ratio-primitive* if and only if, for every proper prefix β of α, it is $|\beta|_1 / |\beta|_2 \neq |\alpha|_1 / |\alpha|_2$. Otherwise, α is called *ratio-imprimitive*.

While the above examples are partly hard to find, some general statements on DPCP and its complement can be obtained effortlessly:

Proposition 8. *If* $\alpha \in$ DPCP, *then, for every* k, $\alpha^k \in$ DPCP. *If* $\alpha \notin$ DPCP, *then, for every* k, $\alpha^k \notin$ DPCP.

Proposition 9. *If* $\alpha, \beta \in$ DPCP *with* $\text{var}(\alpha) \cap \text{var}(\beta) = \emptyset$, *then* $\alpha\beta \in$ DPCP.

If we apply Proposition 8 to existing examples, then we can state the following insight:

Corollary 10. *There are patterns of arbitrary length in* DPCP. *There are patterns of arbitrary length not in* DPCP.

The existing literature on the Dual PCP mainly studies two-variable patterns. In contrast to this, as mentioned in Section 1, we wish to investigate the structure of DPCP for patterns over any numbers of variables. In this regard, we can state a number of immediate observations:

Proposition 11. *Let* $\alpha \in \mathbb{N}^+$, $|\text{var}(\alpha)| = 1$. *Then* $\alpha \notin$ DPCP.

It is easy to give example patterns with three or more variables that belong to DPCP. Proposition 19 in Section 4 gives a construction principle. Furthermore, as soon as a pattern α is projectable to a subpattern $\beta \in$ DPCP, also $\alpha \in$ DPCP.

Proposition 12. *Let* $\alpha \in \mathbb{N}^+$ *and* $V \subseteq \text{var}(\alpha)$ *with* $\pi_V(\alpha) \in \text{DPCP}$. *Then* $\alpha \in \text{DPCP}$.

On the other hand, this implies that every $\alpha \notin \text{DPCP}$ must not be projectable to a subpattern from DPCP.

Corollary 13. *Let* $\alpha \notin \text{DPCP}$. *Then for every* $V \subseteq \text{var}(\alpha)$, $\pi_V(\alpha) \notin \text{DPCP}$.

Consequently, on the one hand, the discovery of one pattern not in DPCP directly leads to a multitude of patterns not in DPCP (namely, all of its subpatterns). On the other hand, this situation makes it very difficult to find such example patterns since arbitrary patterns easily contain subpatterns from DPCP.

3   A Characteristic Condition

The most direct way to decide on whether a pattern α is in DPCP is to solve the word equation $\alpha = \alpha'$, where α' is a renaming of α such that $\text{var}(\alpha) \cap \text{var}(\alpha') = \emptyset$. Indeed, the set of solutions corresponds exactly to the set of all pairs of morphisms which agree on α. The pattern α is in DPCP if and only if there exists such a solution which is non-periodic. This explains why Culik II and Karhumäki [2] use Makanin's Algorithm for demonstrating the decidability of DPCP. Furthermore, it demonstrates why, in many respects, the more challenging questions often concern patterns *not* in DPCP. For such patterns, it is not enough to simply find a single non-periodic solution, but instead every single solution to the equation $\alpha = \alpha'$ must be accounted for. It is generally extremely difficult to determine the complete solution set to such an equation, and as a result, only a limited class of examples is known.

This section presents an alternative approach which attempts to reduce the difficulties associated with such equations. To this end, we apply a morphism $\varphi : \mathbb{N}^* \to \mathbb{N}^*$ to a pattern $\alpha \notin \text{DPCP}$, and we identify conditions that, if satisfied, yield $\varphi(\alpha) \notin \text{DPCP}$.

The main result of this section characterises such morphisms φ:

Theorem 14. *Let* $\alpha \in \mathbb{N}^+$ *be a pattern that is not in* DPCP, *and let* $\varphi : \text{var}(\alpha)^* \to \mathbb{N}^*$ *be a morphism. The pattern* $\varphi(\alpha)$ *is not in* DPCP *if and only if*

(i) *for every periodic morphism* $\rho : \text{var}(\alpha)^* \to \{\text{a}, \text{b}\}^*$ *and for all distinct morphisms* $\sigma, \tau : \text{var}(\varphi(\alpha))^* \to \{\text{a}, \text{b}\}^*$ *with* $\sigma \circ \varphi(\alpha) = \rho(\alpha) = \tau \circ \varphi(\alpha)$, σ *and* τ *are periodic and*

(ii) *for every non-periodic morphism* $\rho : \text{var}(\alpha)^* \to \{\text{a}, \text{b}\}^*$ *and for all morphisms* $\sigma, \tau : \text{var}(\varphi(\alpha))^* \to \{\text{a}, \text{b}\}^*$ *with* $\sigma \circ \varphi = \rho = \tau \circ \varphi$, $\sigma = \tau$.

As briefly mentioned above, Theorem 14 shows that insights into the structure of DPCP can be gained in a manner that partly circumvents the solution of word equations. Instead, we can make use of prior knowledge on patterns that are not in DPCP, which mainly exists for patterns over two variables, and expand this knowledge by studying the existence of morphisms φ that *preserve non-periodicity* (i. e., if certain morphisms σ are non-periodic, then $\sigma \circ \varphi$ needs to be non-periodic; see Condition (i)) and *preserve distinctness* (i. e., if certain morphisms σ, τ are distinct, then $\sigma \circ \varphi$ and $\tau \circ \varphi$ need to be distinct; see Condition (ii)).

Theorem 14 can be used to characterise the patterns in DPCP, but it is mainly suitable as a tool to find patterns that are *not* in DPCP. We shall study this option in Section 5, where we, due to our focus on the *if* direction of Theorem 14, can drop the additional conditions on non-periodicity and distinctness preserving morphisms φ that are postulated by the Theorem. In addition to reducing the need for studying word equations, the use of morphisms to generate examples not in DPCP shall prove to have another key benefit; since morphisms can be applied to infinitely many pre-image patterns, the construction of a single morphism automatically produces an infinite set of examples. This process can be applied iteratively – with morphisms providing new examples of patterns which can then potentially be used as the pre-images for the same, or other morphisms. Before we study this in more details, we wish to consider patterns that are in DPCP in the next section.

4   On Patterns in DPCP

In the present section, we wish to establish major sets of patterns over arbitrarily many variables that are in DPCP. Our first criterion is based on so-called *ambiguity factorisations*, which are a generalisation of imprimitivity factorisations used by Reidenbach and Schneider [12] to characterise the morphically primitive patterns.

Definition 15. *Let* $\alpha \in \mathbb{N}^+$. *An* ambiguity factorisation (of α) *is a mapping* $f : \mathbb{N}^+ \to \mathbb{N}^n \times (\mathbb{N}^+)^n$, $n \in \mathbb{N}$, *such that, for* $f(\alpha) = (x_1, x_2, \ldots, x_n; \gamma_1, \gamma_2, \ldots, \gamma_n)$, *there exist* $\beta_0, \beta_1, \ldots, \beta_n \in \mathbb{N}^*$ *satisfying* $\alpha = \beta_0\, \gamma_1\, \beta_1\, \gamma_2\, \beta_2 \ldots \gamma_n\, \beta_n$ *and*

(i) *for every* $i \in \{1, 2, \ldots, n\}$, $|\gamma_i| \geq 2$,
(ii) *for every* $i \in \{0, 1, \ldots, n\}$ *and for every* $j \in \{1, 2, \ldots, n\}$, $\mathrm{var}(\beta_i) \cap \mathrm{var}(\gamma_j) = \emptyset$,
(iii) *for every* $i \in \{1, 2, \ldots, n\}$, $|\gamma_i|_{x_i} = 1$ *and if* $x_i \in \mathrm{var}(\gamma_{i'})$ *for an* $i' \in \{1, 2, \ldots, n\}$, $\gamma_i = \delta_1\, x_i\, \delta_2$ *and* $\gamma_i' = \delta_1'\, x_i\, \delta_2'$, *then* $|\delta_1| = |\delta_1'|$ *and* $|\delta_2| = |\delta_2'|$.

Using this concept, we now can give a strong sufficient condition for patterns in DPCP:

Theorem 16. *Let* $\alpha \in \mathbb{N}^+$. *If there exists an ambiguity factorisation of* α, *then* $\alpha \in$ DPCP.

The following example illustrates Definition 15 and Theorem 16:

Example 17. Let the pattern α be given by

$$\alpha := \underbrace{1 \cdot 2 \cdot 2 \cdot 3}_{\gamma_1} \cdot \underbrace{2 \cdot 4 \cdot 5 \cdot 2}_{\gamma_2} \cdot \underbrace{5 \cdot 4 \cdot 2 \cdot 5}_{\gamma_3} \cdot 3 \cdot \underbrace{1 \cdot 2 \cdot 2}_{\gamma_4}$$

This pattern has an ambiguity partition, as is implied by the marked γ parts and the variables in bold face, which stand for the x_i.

We now consider two distinct non-periodic morphisms σ and τ, given by $\sigma(1) = \sigma(4) = \mathsf{a}$, $\sigma(2) = \sigma(5) = \mathsf{bb}$, $\sigma(3) = \varepsilon$ and $\tau(1) = \mathsf{abb}$, $\tau(4) = \mathsf{babb}$, $\tau(2) = \tau(5) = \mathsf{b}$, $\tau(3) = \varepsilon$. It can be verified with limited effort that σ and τ agree on α. ◊

Since ambiguity partitions are more general than imprimitivity partitions, we can immediately conclude that a natural set of patterns is included in DPCP:

Corollary 18. *Let $\alpha \in \mathbb{N}^+$. If α is morphically imprimitive, then $\alpha \in$ DPCP.*

Since most patterns are morphically imprimitive (see Reidenbach and Schneider [12]), this implies that most patterns are in DPCP, which confirms our intuitive considerations at the beginning of Section 3.

While ambiguity partitions are a powerful tool, they are technically rather involved. In this respect, our next sufficient condition on patterns in DPCP is much simpler, since it merely asks whether a pattern can be split in two factors that do not have any variables in common:

Proposition 19. *Let $\alpha \in \mathbb{N}^+$, $|\operatorname{var}(\alpha)| \geq 3$. If, for some $\alpha_1, \alpha_2 \in \mathbb{N}^+$ with $\operatorname{var}(\alpha_1) \cap \operatorname{var}(\alpha_2) = \emptyset$, $\alpha = \alpha_1 \alpha_2$, then $\alpha \in$ DPCP.*

Note that it is possible to extend Proposition 19 quite substantially since the same technique can be applied to, e.g., the pattern $\alpha := \alpha_1 \alpha_2 \alpha_1 \alpha_2$ and much more sophisticated types of patterns where certain factors have disjoint variable sets and can therefore be allocated to different periodic morphisms each. The following proposition is such an extension of Proposition 19.

Proposition 20. *Let $x, y, z \in \mathbb{N}$, and let $\alpha \in \{x, y, z\}^+$ be a pattern such that $\alpha = \alpha_0 z \alpha_1 z \ldots \alpha_{n-1} z \alpha_n$, $n \in \mathbb{N}$. If,*

- *for every $i \in \{0, 1, \ldots, n\}$, $\alpha_i = \varepsilon$ or $\operatorname{var}(\alpha_i) = \{x, y\}$, and*
- *for every $i, j \in \{0, 1, \ldots, n\}$ with $\alpha_i \neq \varepsilon \neq \alpha_j$, $\frac{|\alpha_i|_x}{|\alpha_i|_y} = \frac{|\alpha_j|_x}{|\alpha_j|_y}$,*

then $\alpha \in$ DPCP.

The following example pattern is covered by Proposition 20: $1 \cdot 1 \cdot 2 \cdot 2 \cdot 2 \cdot 3 \cdot 1 \cdot 2 \cdot 2 \cdot 1 \cdot 2 \cdot 3 \cdot 1 \cdot 1 \cdot 1 \cdot 1 \cdot 2 \cdot 2 \cdot 2 \cdot 2 \cdot 2 \cdot 2$. Although Proposition 20 is restricted to three-variable patterns, it is worth mentioning that we can apply it to arbitrary patterns that have a three-variable subpattern of this structure. This is a direct consequence of Proposition 12.

5   On Patterns Not in DPCP

As a result of the intensive research on binary equality sets, several examples of patterns over two variables are known not to be in DPCP (see Section 2). Hence, the most obvious question to ask is whether or not there exist such examples with more than two variables (and more generally, whether there exist examples for any given set of variables). The following results develop a structure for morphisms which map patterns not in DPCP to patterns with more variables which are also not in DPCP, ultimately allowing for the inductive proof of Theorem 33, which provides a strong positive answer.

As discussed in Section 3, this is accomplished by simplifying the conditions of Theorem 14, so that they ask the morphism φ to be (i) *non-periodicity preserving* and (ii) *distinctness-preserving*:

Lemma 21. *Let Δ_1, Δ_2 be sets of variables. Let $\varphi : \Delta_1{}^* \to \Delta_2{}^*$ be a morphism such that for every $x \in \Delta_2$, there exists a $y \in \Delta_1$ such that $x \in \mathrm{var}(\varphi(y))$, and*

(i) *for every non-periodic morphism $\sigma : \Delta_2{}^* \to \{\mathsf{a}, \mathsf{b}\}^*$, $\sigma \circ \varphi$ is non-periodic, and*

(ii) *for all distinct morphisms σ, $\tau : \Delta_2{}^* \to \{\mathsf{a}, \mathsf{b}\}^*$, where at least one is non-periodic, $\sigma \circ \varphi$ and $\tau \circ \varphi$ are distinct.*

Then for any $\alpha \notin \mathrm{DPCP}$ with $\mathrm{var}(\alpha) = \Delta_1$, $\varphi(\alpha) \notin \mathrm{DPCP}$.

Remark 22. Condition (i) of Lemma 21 is identical to asking that $\sigma \circ \varphi$ is periodic if and only if σ is periodic, since if σ is periodic, then $\sigma \circ \varphi$ will always be periodic as well.

While Lemma 21 provides a clear proof technique for demonstrating that a given pattern is not in DPCP, the conditions are abstract, and it does not directly lead to any new examples. The next step, therefore, is to investigate the existence and nature of morphisms φ which satisfy both conditions.

Since the main focus of the following results is concerned with properties of compositions of morphisms, the following two facts are included formally.

Fact 23. Let Δ_1, Δ_2 be sets of variables. let $\varphi : \Delta_1{}^* \to \Delta_2{}^*$ and $\sigma : \Delta_2{}^* \to \{\mathsf{a}, \mathsf{b}\}^*$ be morphisms. The morphism $\sigma \circ \varphi$ is periodic if and only if there exists a (primitive) word $w \in \Sigma^*$ such that for each $i \in \Delta_1$, there exists an $n \in \mathbb{N}_0$ with $\sigma(\varphi(i)) = w^n$.

Fact 24. Let Δ_1, Δ_2 be sets of variables. let $\varphi : \Delta_1{}^* \to \Delta_2{}^*$ and $\sigma : \Delta_2{}^* \to \{\mathsf{a}, \mathsf{b}\}^*$ be morphisms. The morphisms $\sigma \circ \varphi$ and $\tau \circ \varphi$ are distinct if and only if there exists a variable $i \in \Delta_1$ such that $\sigma(\varphi(i)) \neq \tau(\varphi(i))$.

Facts 23 and 24 highlight how properties such as periodicity and distinctness of a composition of two morphisms can be determined by observing certain properties of specific sets of patterns. Since the conditions in Lemma 21 rely only on these properties, it is apparent that, further than requiring that $\alpha \notin \mathrm{DPCP}$, the structure of α is not relevant. It is instead dependent on $\mathrm{var}(\alpha)$.

Each condition from Lemma 21 is relatively independent from the other, so it is appropriate to first establish classes of morphisms satisfying each one separately. Condition (i) is considered first. The satisfaction of Fact 23, and therefore Condition (i) of Lemma 21 relies on specific systems of word equations having only periodic solutions. The following proposition provides a tool for demonstrating exactly that.

Proposition 25. *(Lothaire [9]) All non-trivial, terminal-free word equations in two unknowns have only periodic solutions.*

In order to determine the satisfaction of Condition (i) of Lemma 21 for a particular morphism $\varphi : \Delta_1^* \to \Delta_2^*$, it is necessary to identify which morphisms $\sigma : \Delta_2^* \to \{a, b\}^*$ result in the composition $\sigma \circ \varphi$ being periodic. The next proposition gives the required characteristic condition on σ for $\sigma \circ \varphi$ to be periodic. Each term $\sigma(\gamma_i)$ in equality (1) below corresponds directly to a word $\sigma \circ \varphi(j)$, for some $j \in \Delta_1$. The satisfaction of the system of equalities is identical to each word $\sigma \circ \varphi(i)$ sharing a primitive root, allowing the relationship between σ and the periodicity of $\sigma \circ \varphi$ to be expressed formally.

Proposition 26. *Let Δ_1 and Δ_2 be sets of variables and let $\varphi : \Delta_1^* \to \Delta_2^*$, $\sigma : \Delta_2^* \to \{a, b\}^*$ be morphisms. For every $i \in \Delta_1$, let $\varphi(i) := \beta_i$, and let $\{\gamma_1, \gamma_2, \ldots \gamma_n\}$ be the set of all patterns β_j such that $\sigma(\beta_j) \neq \varepsilon$. If $n < 2$, the composition $\sigma \circ \varphi$ is trivially periodic. For $n \geq 2$, $\sigma \circ \varphi$ is periodic if and only if there exist $k_1, k_2, \ldots k_n \in \mathbb{N}$ such that*

$$\sigma(\gamma_1)^{k_1} = \sigma(\gamma_2)^{k_2} = \cdots = \sigma(\gamma_n)^{k_n}. \tag{1}$$

Corollary 27. *Let Δ_1 and Δ_2 be sets of variables, let $\varphi : \Delta_1^* \to \Delta_2^*$ be a morphism, and let $\varphi(i) := \beta_i$ for every $i \in \Delta_1$. The morphism φ satisfies Condition (i) of Lemma 21 if and only if, for every non-periodic morphism $\sigma : \Delta_2^* \to \{a, b\}^*$,*

(i) *There are at least two patterns β_i such that $\sigma(\beta_i) \neq \varepsilon$, and*
(ii) *there do not exist $k_1, k_2, \ldots, k_n \in \mathbb{N}$ such that*

$$\sigma(\gamma_1)^{k_1} = \sigma(\gamma_2)^{k_2} = \cdots = \sigma(\gamma_n)^{k_n} \tag{2}$$

where $\{\gamma_1, \gamma_2, \ldots, \gamma_n\}$ is the set of all patterns β_i such that $\sigma(\beta_i) \neq \varepsilon$.

Corollary 27 also provides a proof technique. Since there are finitely many combinations of $\beta_1, \beta_2, \ldots, \beta_m$, it is clear that the satisfaction of Condition (ii) of Corollary 27 will always rely on finitely many cases. By considering all possible sets $\{\gamma_1, \gamma_2, \ldots \gamma_n\}$, infinitely many morphisms can be accounted for in a finite and often very concise manner. Thus, it becomes much simpler to demonstrate that there cannot exist a non-periodic morphism σ such that $\sigma \circ \varphi$ is periodic, and therefore that Condition (i) of Lemma 21 is satisfied. We now give an example of such an approach.

Example 28. Let $\Delta_1 := \{1, 2, 3, 4\}$ and let $\Delta_2 := \{5, 6, 7, 8\}^*$. Let $\varphi : \Delta_1^* \to \Delta_2^*$ be the morphism given by $\varphi(1) := 5 \cdot 6$, $\varphi(2) := 6 \cdot 5$, $\varphi(3) := 5 \cdot 6 \cdot 7 \cdot 7$ and $\varphi(4) := 6 \cdot 8 \cdot 8 \cdot 5$. Consider all non-periodic morphisms $\sigma : \{5, 6, 7, 8\}^* \to \{a, b\}^*$. Note that if $\sigma(5 \cdot 6) \neq \varepsilon$ then $\sigma(6 \cdot 5) \neq \varepsilon$ and vice-versa. Also note that since σ is non-periodic, there must be at least two variables x such that $\sigma(x) \neq \varepsilon$. So if either $\sigma(5 \cdot 6 \cdot 7 \cdot 7) \neq \varepsilon$, or $\sigma(6 \cdot 8 \cdot 8 \cdot 5) \neq \varepsilon$, there must be at least one other pattern β_j with $\sigma(\beta_j) \neq \varepsilon$. Therefore, for any non-periodic morphism σ, there exists a minimum of two patterns β_i such that $\sigma(\beta_i) \neq \varepsilon$. Now consider all possible cases.

Assume first that $\sigma(5 \cdot 6) = \varepsilon$. Clearly $\sigma(5) = \sigma(6) = \varepsilon$, so $\sigma(6 \cdot 5) = \varepsilon$. Since σ is non-periodic, $\sigma(7) \neq \varepsilon$ and $\sigma(8) \neq \varepsilon$. By Proposition 26, $\sigma \circ \varphi$ is periodic if and only if there exist k_1, $k_2 \in \mathbb{N}$ such that $\sigma(7 \cdot 7)^{k_1} = \sigma(8 \cdot 8)^{k_2}$. By Proposition 25, this is the case only if σ is periodic and this is a contradiction, so $\sigma \circ \varphi$ is non-periodic.

Assume $\sigma(5 \cdot 6) \neq \varepsilon$ (so $\sigma(6 \cdot 5) \neq \varepsilon$, $\sigma(6 \cdot 8 \cdot 8 \cdot 5) \neq \varepsilon$, and $\sigma(5 \cdot 6 \cdot 7 \cdot 7) \neq \varepsilon$), then by Proposition 26, the composition $\sigma \circ \varphi$ is periodic if and only if there exist k_1, k_2, k_3, $k_4 \in \mathbb{N}$ such that

$$\sigma(5 \cdot 6)^{k_1} = \sigma(6 \cdot 5)^{k_2} = \sigma(6 \cdot 8 \cdot 8 \cdot 5)^{k_3} = \sigma(5 \cdot 6 \cdot 7 \cdot 7)^{k_4} \tag{3}$$

By Proposition 25, the first equality only holds if there exist a word $w \in \{a, b\}^*$ and numbers p, $q \in \mathbb{N}_0$ such that $\sigma(5) = w^p$ and $\sigma(6) = w^q$. Thus, equality (3) is satisfied if and only if $w^{k_1(p+q)} = (w^q \cdot \sigma(8 \cdot 8) \cdot w^p)^{k_3}$ and $w^{k_1(p+q)} = (w^{p+q} \cdot \sigma(7 \cdot 7))^{k_4}$. By Proposition 25, this is only the case if there exist r, $s \in \mathbb{N}$ such that $\sigma(7) = w^s$ and $\sigma(8) = w^r$. Thus, σ is periodic, which is a contradiction, so the composition $\sigma \circ \varphi$ is non-periodic.

All possibilities for non-periodic morphisms σ have been exhausted, so for any non-periodic morphism $\sigma : \{5, 6, 7, 8\}^* \to \{a, b\}^*$, the composition $\sigma \circ \varphi$ is also non-periodic and φ satisfies Condition (i) of Lemma 21.   ◇

Condition (ii) of Lemma 21 is now considered. Fact 24 shows that it relies on the (non-)existence of distinct, non-periodic morphisms which agree on a set of patterns (more precisely, the set of morphic images of single variables). The following proposition provides a characterisation for morphisms which satisfy the condition.

Proposition 29. *Let Δ_1, Δ_2 be sets of variables, and let $\varphi : \Delta_1^* \to \Delta_2^*$ be a morphism. For every $i \in \Delta_1$, let $\varphi(i) := \beta_i$. The morphism φ satisfies Condition (ii) of Lemma 21 if and only if $\{\beta_1, \beta_2, \ldots, \beta_n\}$ is a periodicity forcing set.*

Proposition 29 facilitates a formal comparison of the word equations involved in directly finding patterns not in DPCP and the word equations that need to be considered when using Lemma 21. Furthermore, it shows the impact of the choice of α on the complexity of applying the Lemma. However, it does not immediately provide a nontrivial morphism φ that satisfies Condition (ii) of Lemma 21. Therefore, we consider the following technical tool:

Proposition 30. *Let Δ_1, Δ_2 be sets of variables, and let $\varphi : \Delta_1^* \to \Delta_2^*$ be a morphism. For every $k \in \Delta_1$, let $\varphi(k) := \beta_k$ and let $\beta_i \notin \text{DPCP}$ for some $i \in \Delta_1$. For every $x \in \Delta_2 \setminus \text{var}(\beta_i)$, let there exist β_j and patterns γ_1, γ_2, such that $\beta_j = \gamma_1 \cdot \gamma_2$ and*

(i) $x \in \text{var}(\gamma_1)$, and for every $y \in \text{var}(\gamma_1)$ with $y \neq x$, $y \in \text{var}(\beta_i)$,
(ii) $\gamma_1 \notin \text{DPCP}$ with $|\text{var}(\gamma_1)| \geq |\text{var}(\beta_i)|$,
(iii) $\text{P}(\gamma_2)$ and $\text{P}(\beta_i)$ are linearly dependent.

Then φ satisfies Condition (ii) of Lemma 21.

The following example demonstrates the structure given in Proposition 30. It is chosen such that it also satisfies Corollary 27, allowing for the construction given in Proposition 32.

Example 31. Let $\Delta_1 := \{4, 5\}$, and let $\Delta_2 := \{1, 2, 3\}$. Let $\varphi : \Delta_1^* \to \Delta_2^*$ be the morphism given by $\varphi(4) = \beta_4 := 1 \cdot 2 \cdot 1 \cdot 1 \cdot 2$ and $\varphi(5) := \gamma_1 \cdot \gamma_2$ where $\gamma_1 := 1 \cdot 3 \cdot 1 \cdot 1 \cdot 3$ and $\gamma_2 := 2 \cdot 1 \cdot 1 \cdot 2 \cdot 1$. Notice that β_4 and γ_1 are not in DPCP. Let $\sigma, \tau : \{1, 2, 3\}^* \to \{\mathsf{a}, \mathsf{b}\}^*$ be distinct morphisms, at least one of which is non-periodic, that agree on β_4. By definition of DPCP, this is only possible if σ and τ agree on, or are periodic over $\{1, 2\}$.

If σ and τ agree on $\{1, 2\}$, then they agree on γ_2. This means that $\sigma(\gamma_1 \cdot \gamma_2) = \tau(\gamma_1 \cdot \gamma_2)$ if and only if $\sigma(1 \cdot 3 \cdot 1 \cdot 1 \cdot 3) = \tau(1 \cdot 3 \cdot 1 \cdot 1 \cdot 3)$. Furthermore σ and τ are distinct, so cannot agree on 3. However, since $\sigma(1) = \tau(1)$ but $\sigma(3) \neq \tau(3)$, this cannot be the case, therefore $\sigma \circ \varphi$ and $\tau \circ \varphi$ are distinct.

Note that if σ and τ agree on exactly one variable in $\{1, 2\}$, then they cannot agree on β_4. Consider the case that σ and τ do not agree on 1 or 2. Then they must be periodic over $\{1, 2\}$, so $\sigma(2 \cdot 1 \cdot 1 \cdot 2 \cdot 1) = \sigma(1 \cdot 2 \cdot 1 \cdot 1 \cdot 2)$ (and likewise for τ). It follows that $\sigma(2 \cdot 1 \cdot 1 \cdot 2 \cdot 1) = \tau(2 \cdot 1 \cdot 1 \cdot 2 \cdot 1)$ and, as a consequence, σ and τ agree on $\gamma_1 \cdot \gamma_2$ if and only if they agree on $\gamma_1 = 1 \cdot 3 \cdot 1 \cdot 1 \cdot 3$. However, due to the non-periodicity of σ or τ, $\sigma(3)$ or $\tau(3)$ must have a different primitive root to $\sigma(1)$ or $\tau(1)$, respectively. This means that σ and τ are distinct over $\{1, 3\}$, and at least one of them must be non-periodic over $\{1, 3\}$. This implies that σ and τ cannot agree on γ_1, and therefore $\sigma \circ \varphi$ and $\tau \circ \varphi$ are distinct.

Hence, there do not exist two distinct morphisms, at least one of which is non-periodic, that agree on $1 \cdot 2 \cdot 1 \cdot 1 \cdot 2$ and $1 \cdot 3 \cdot 1 \cdot 1 \cdot 3 \cdot 2 \cdot 1 \cdot 1 \cdot 2 \cdot 1$. These patterns, thus, form a periodicity forcing set, and, by Proposition 29, the morphism φ satisfies Condition (ii) of Lemma 21. $\Diamond$

The next proposition introduces a pattern over three variables which is not in DPCP. This not only demonstrates that this is possible for patterns over more than two variables, but provides the basis for the construction given in Theorem 33, which shows that there are patterns of arbitrarily many variables not in DPCP.

Proposition 32. *The pattern* $1 \cdot 2 \cdot 1 \cdot 1 \cdot 2 \cdot 1 \cdot 3 \cdot 1 \cdot 1 \cdot 3 \cdot 2 \cdot 1 \cdot 1 \cdot 2 \cdot 1 \cdot 1 \cdot 2 \cdot 1 \cdot 1 \cdot 2 \cdot 1 \cdot 1 \cdot 2 \cdot 1 \cdot 1 \cdot 2 \cdot 1 \cdot 3 \cdot 1 \cdot 1 \cdot 3 \cdot 2 \cdot 1 \cdot 1 \cdot 2 \cdot 1$ *is not in DPCP.*

It is now possible to state the following theorem, the proof of which provides a construction for a pattern not in DPCP over an arbitrary number of variables. This is achieved by considering, for any $n \geq 2$, the morphism $\varphi_n : \{1, 2, \ldots, n\}^* \to \{1, 2, \ldots, n+1\}^*$, given by $\varphi_n(1) := 1 \cdot 2 \cdot 1 \cdot 1 \cdot 2$, and for $2 \leq x \leq n$, $\varphi_n(x) := 1 \cdot (x+1) \cdot 1 \cdot 1 \cdot (x+1) \cdot 2 \cdot 1 \cdot 1 \cdot 2 \cdot 1$. This morphisms satisfies the conditions for Lemma 21, i. e., it maps any n-variable pattern that is not in DPCP to an $n + 1$-variable pattern that is also not in DPCP. It follows that if there exists a pattern with n variables not in DPCP, then there exists a pattern with n variables not in DPCP. Thus, by induction, there exist such patterns for any number of variables.

Theorem 33. *There are patterns of arbitrarily many variables not in DPCP.*

Hence, we may conclude that the Dual PCP is nontrivial for all alphabets with at least two variables, and we can show this in a constructive manner.

Acknowledgements. The authors wish to thank the anonymous referees for their helpful remarks and suggestions. Furthermore, numerous enlightening discussions with Paul C. Bell on an earlier version of this paper are gratefully acknowledged.

References

1. Culik II., K.: A purely homomorphic characterization of recursively enumerable sets. Journal of the ACM 26, 345–350 (1979)
2. Culik II, K., Karhumäki, J.: On the equality sets for homomorphisms on free monoids with two generators. Theoretical Informatics and Applications (RAIRO) 14, 349–369 (1980)
3. Czeizler, E., Holub, Š., Karhumäki, J., Laine, M.: Intricacies of simple word equations: An example. International Journal of Foundations of Computer Science 18, 1167–1175 (2007)
4. Engelfriet, J., Rozenberg, G.: Equality languages and fixed point languages. Information and Control 43, 20–49 (1979)
5. Freydenberger, D.D., Nevisi, H., Reidenbach, D.: Weakly unambiguous morphisms. Theoretical Computer Science 448, 21–40 (2012)
6. Freydenberger, D.D., Reidenbach, D., Schneider, J.C.: Unambiguous morphic images of strings. International Journal of Foundations of Computer Science 17, 601–628 (2006)
7. Hadravová, J., Holub, Š.: Large simple binary equality words. International Journal of Foundations of Computer Science 23, 1385–1403 (2012)
8. Karhumäki, J., Petre, E.: On some special equations on words. Technical Report 583, Turku Centre for Computer Science, TUCS (2003),
   http://tucs.fi:8080/publications/insight.php?id=tKaPe03a
9. Lothaire, M.: Combinatorics on Words. Addison-Wesley, Reading (1983)
10. Mateescu, A., Salomaa, A., Salomaa, K., Yu, S.: P, NP, and the Post Correspondence Problem. Information and Computation 121, 135–142 (1995)
11. Post, E.L.: A variant of a recursively unsolvable problem. Bulletin of the American Mathematical Society 52, 264–268 (1946)
12. Reidenbach, D., Schneider, J.C.: Morphically primitive words. Theoretical Computer Science 410, 2148–2161 (2009)
13. Salomaa, A.: Equality sets for homomorphisms of free monoids. Acta Cybernetica 4, 127–139 (1978)
14. Schneider, J.C.: Unambiguous erasing morphisms in free monoids. Theoretical Informatics and Applications (RAIRO) 44, 193–208 (2010)

Brzozowski Algorithm
Is Generically Super-Polynomial
for Deterministic Automata[*]

Sven De Felice and Cyril Nicaud

LIGM, Université Paris-Est & CNRS, 77454 Marne-la-Vallée Cedex 2, France
{sdefelic,nicaud}@univ-mlv.fr

Abstract. We study the number of states of the minimal automaton of the mirror of a rational language recognized by a random deterministic automaton with n states. We prove that, for any $d > 0$, the probability that this number of states is greater than n^d tends to 1 as n tends to infinity. As a consequence, the generic and average complexities of Brzozowski minimization algorithm are super-polynomial for the uniform distribution on deterministic automata.

1 Introduction

Brzozowski proved [5] that determinizing a trim co-deterministic automaton which recognizes a language $\mathcal{L}$ yields the minimal automaton of $\mathcal{L}$. This can be turned into a simple minimization algorithm: start with an automaton, compute its reversal, determinize it and reverse the result in order to obtain a co-deterministic automaton recognizing the same language. A last determinization gives the minimal automaton, by Brzozowski's property.

The determinization steps use the classical subset construction, which is well-known to be of exponential complexity in the worst-case. The co-deterministic automaton $\mathcal{A}_n$ of Fig. 1 is a classical example of such a combinatorial explosion: it has n states and its minimal automaton has 2^{n-1} states.

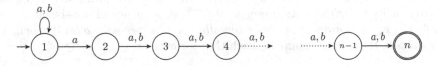

Fig. 1. Determinizing this co-deterministic automaton $\mathcal{A}_n$ with n states, which recognizes $A^* a A^{n-2}$, yields a minimal automaton with 2^{n-1} states

How good is Brzozowski minimization algorithm? If the input is a non-deterministic automaton, the combinatorial explosion can be unavoidable, as

---

[*] This work is supported by the French National Agency (ANR) through ANR-10-LABX-58 and through ANR-2010-BLAN-0204.

M.-P. Béal and O. Carton (Eds.): DLT 2013, LNCS 7907, pp. 179–190, 2013.
© Springer-Verlag Berlin Heidelberg 2013

for $\mathcal{A}_n$, and this algorithm can be a good solution (see [17] for an experimental comparison of Brzozowski algorithm versus determinization combined with Hopcroft algorithm). However, if the input is a deterministic automaton, Brzozowski algorithm still has exponential worst-case complexity, which is easily seen by taking the reverse of $\mathcal{A}_n$ as input. Since there exists polynomial solutions to minimize deterministic automata, such as Hopcroft algorithm [13] which runs in time $\mathcal{O}(n \log n)$, there is no use for Brzozowski algorithm in the deterministic case, unless the combinatorial explosion happens very rarely.

Let $\mathcal{L}$ be the language recognized by a n-state deterministic automaton taken uniformly at random. In this article we estimate the typical number of states of the minimal automaton of the mirror $\tilde{\mathcal{L}}$ of $\mathcal{L}$. More precisely, we prove that this quantity is generically super-polynomial, that is, for any $d > 0$, the probability that there are more than n^d states in the minimal automaton of $\tilde{\mathcal{L}}$ tends to 1 as n tends to infinity.

As a consequence, Brzozowski algorithm has super-polynomial generic and average complexity when used on deterministic automata, for the uniform distribution: the combinatorial explosion is almost always met during the process.

Some Related Works. The interest in statistical properties of random deterministic automata started with the work of Korshunov [14], who studied their combinatorics and exhibited some of their typical behavior. In recent years, an increased activity on the topic aimed at giving mathematical proofs for phenomena observed experimentally. For instance, it was proved in [1,8] that the average complexity of Moore algorithm, another minimization algorithm, is significantly better than its worst-case complexity, making this algorithm a reasonable solution in practice. The reader can find some results on the average state complexity of operations under different settings in [16,3]. Let us also mention the recent article [2], in the same area, which focus on quantifying the probability that a random deterministic automaton is minimal.

2 Preliminaries

For any $n \geq 1$, let $[n]$ denote the set $\{1, \ldots, n\}$. If E is a finite set, we denote its cardinality by $|E|$ and its power set by 2^E. A sequence of non-negative real numbers $(x_n)_{n \geq 1}$ *grows super-polynomially* (or is *super-polynomial*) when, for every $d > 0$, there exists some n_d such that for every $n \geq n_d$, $x_n \geq n^d$.

2.1 Automata

Basic Definitions. Let A be a finite alphabet, an automaton $\mathcal{A}$ is a tuple (Q, δ, I, F), where Q is its finite set of states, $I \subseteq Q$ is its set of initial states and $F \subseteq Q$ is its set of final states. Its transition function is a (partial) map from $Q \times A$ to 2^Q. A transition of $\mathcal{A}$ is a tuple $(p, a, q) \in Q \times A \times Q$, which we write $p \xrightarrow{a} q$, such that $q \in \delta(p, a)$. The map δ is classically extended by morphism to $Q \times A^*$. We denote by $\mathcal{L}(\mathcal{A})$ the set of words recognized by $\mathcal{A}$.

A deterministic and complete automaton is an automaton such that $|I| = 1$ and for every $p \in Q$ and $a \in A$, $|\delta(p,a)| = 1$; for such an automaton we consider that δ is a (total) map from $Q \times A^*$ to Q to simplify the notations.

A state p in an automaton is *accessible* (resp. *co-accessible*) when there is a path from an initial state to p (resp. from p to a final state). The *accessible part* (resp. *co-accessible* part) of an automaton is the set of its accessible states (resp. co-accessible states). A *trim* automaton is an automaton whose states are all accessible and co-accessible. If $\mathcal{A}$ is an automaton, we denote by $\mathrm{Trim}(\mathcal{A})$ the automaton obtained after removing states that are not accessible or not co-accessible.

For any automaton $\mathcal{A} = (Q, \delta, I, F)$, we denote by $\tilde{\mathcal{A}}$ the *reverse* of $\mathcal{A}$, which is the automaton $\tilde{\mathcal{A}} = (Q, \tilde{\delta}, F, I)$, where $p \xrightarrow{a} q$ is a transition of $\tilde{\mathcal{A}}$ if and only if $q \xrightarrow{a} p$ is a transition of $\mathcal{A}$. The automaton $\tilde{\mathcal{A}}$ recognizes the mirror[1] of $\mathcal{L}(\mathcal{A})$. An automaton is *co-deterministic* when its reverse is deterministic.

Recall that the *minimal automaton* of a rational language $\mathcal{L}$ is the smallest deterministic and complete automaton[2] that recognizes $\mathcal{L}$. To each rational language $\mathcal{L}$ corresponds a minimal automaton, which is unique up to isomorphism.

Subset Construction and Brzozowski Algorithm. If $\mathcal{A} = (Q, \delta, I, F)$ is a non-deterministic automaton, it is classical that the subset automaton of $\mathcal{A}$ defined by

$$\mathcal{B} = \left(2^Q, \gamma, \{I\}, \{X \in 2^Q \mid F \cap X \neq \emptyset\}\right)$$

is a deterministic automaton that recognizes the same language, where for every $X \in 2^Q$ and every $a \in A$, $\gamma(X, a) = \cup_{p \in X}\delta(p, a)$. This is of course still true if we only take the accessible part of $\mathcal{B}$, and this is not a difficulty when implementing it, since the accessible part of $\mathcal{B}$ can be built on the fly, using the rule for γ in a depth-first traversal of $\mathcal{B}$ starting from I. We denote by $\mathrm{Subset}(\mathcal{A})$ the accessible part of the subset automaton of $\mathcal{A}$.

In [5], Brzozowski established the following result:

Theorem 1 (Brzozowski). *If $\mathcal{A}$ is a trim co-deterministic automaton then $\mathrm{Subset}(\mathcal{A})$ is the minimal automaton of $\mathcal{L}(\mathcal{A})$.*

This theorem readily yields an algorithm to compute the minimal automaton of the language recognized by an automaton $\mathcal{A}$, based on the subset construction: since $\mathcal{B} = \mathrm{Subset}(\mathrm{Trim}(\tilde{\mathcal{A}}))$ is a deterministic automaton recognizing the mirror of $\mathcal{L}(\mathcal{A})$, then $\mathrm{Subset}(\mathrm{Trim}(\tilde{\mathcal{B}}))$ is the minimal automaton of $\mathcal{L}(\mathcal{A})$.

2.2 Combinatorial Structures

Permutations. A *permutation* of size n is a bijection from $[n]$ to $[n]$. A size-n permutation σ can be represented by a directed graph of set of vertices $[n]$, with an edge $i \to j$ whenever $\sigma(i) = j$. As σ is a bijection, such a graph is always a

---

[1] If $u = u_0 \ldots u_{n-1}$ is a word of length n, the *mirror* of u is the word $\tilde{u} = u_{n-1} \ldots u_0$.
[2] Minimal automata are not always required to be complete in the literature.

union of cycles. The *order* of a permutation is the smallest positive integer m such that σ^m the identity. It is equal to the least common multiple (lcm) of the lengths of its cycles.

Mappings. A *mapping* of size n is a total function from $[n]$ to $[n]$. As done for permutations, a mapping f can be seen as a directed graph with an edge $i \to j$ whenever $f(i) = j$. Such a graph is no longer a union of cycles, but a union of cycles of trees (trees whose roots are linked into directed cycles), as depicted in Fig. 2. Let f be a size-n mapping. An element $x \in [n]$ is a *cyclic point* of f when there exists an integer $i > 0$ such that $f^i(x) = x$. The *cyclic part* of a mapping f is the permutation obtained when restricting f on its set of cyclic points. The *normalized cyclic part* of f is obtained by relabelling the c cyclic points of f by elements of $[c]$ while keeping their relative order (see Fig 2).

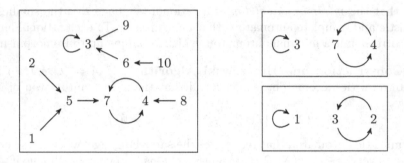

Fig. 2. A mapping of $\{1, \ldots, 10\}$ seen as a directed graph on the left. Its cyclic part is depicted on the upper right, and its normalized cyclic part on the lower right. The normalization is obtained by relabelling the 3 vertices with elements of $\{1, 2, 3\}$, while keeping the relative order; hence $3 \mapsto 1$, $4 \mapsto 2$ and $7 \mapsto 3$.

Automata as Combinatorial Structures. In the sequel, A is always a fixed alphabet with $k \geq 2$ letters. Let $\mathfrak{A}_n$ denote the set of all deterministic and complete automata on A whose set of states is $[n]$ and whose initial state is 1. Such an automaton $\mathcal{A}$ is characterized by the tuple (n, δ, F). A *transition structure* is an automaton without final states, and we denote by $\mathfrak{T}_n$ the set of n-state transition structures with the same label restrictions as for $\mathfrak{A}_n$. If $\mathcal{A} \in \mathfrak{A}_n$, an a-cycle of $\mathcal{A}$ is a cycle of the mapping induced by a, i.e. $p \mapsto \delta(p, a)$. If $\mathcal{C}$ is an a-cycle of length ℓ, the *word associated to* $\mathcal{C}$ is the word u of length ℓ on the alphabet $\{0, 1\}$ obtained as follows: if x is the smallest element of $\mathcal{C}$, $u_i = 1$ if and only if $\delta(x, a^i) \in F$, for $i \in \{0, \ldots, \ell-1\}$. In other words, one starts at x and follows the cycle, writing a 1 when the current state is final and a 0 otherwise. An a-cycle is *primitive* when its associated word u is primitive, that is, when u cannot be written $u = v^m$ for some word v and some integer $m \geq 2$.

2.3   Probabilities on Automata and Genericity

A *probabilistic model* is a sequence $(\mathbb{P}_n)_{n\geq 1}$ of probability measures on the same space. A property P is said to be *generic* for the probabilistic model $(\mathbb{P}_n)_{n\geq 1}$ when the probability that P is satisfied tends to 1 as n tends to infinity.

In our settings, we work on a set E of combinatorial objects with a notion of size, and we will only consider probabilistic models where the support of $\mathbb{P}_n$ is the finite set E_n of size-n objects. The *uniform model* (or *uniform distribution* which is a slight abuse of notation since there is one distribution for each n) on a set $E = \cup_{n\geq 1} E_n$ is defined for any $e \in E_n$ by $\mathbb{P}_n(\{e\}) = \frac{1}{|E_n|}$. The reader is referred to [12] for more information on combinatorial probabilistic models.

For any $0 < b < 1$, the *Bernoulli model of parameter b* (or just a *Bernoulli model* for short) on deterministic automata is the model where an automaton of size-n is obtained by first drawing an element of $\mathfrak{T}_n$ under the uniform distribution, then choosing whether each state is final or not with probability b, independently: the probability of an element $\mathcal{A} \in \mathfrak{A}_n$ with f final states is by definition $\frac{b^f(1-b)^{n-f}}{|\mathfrak{T}_n|}$. The uniform distribution on deterministic automata is obtained by choosing $b = \frac{1}{2}$.

3   Main Results

Our main result is Theorem 2 below, which gives a super-polynomial lower bound for the generic number of states of the minimal automaton of the mirror.

Theorem 2. *Consider a Bernoulli model for automata on an alphabet with at least two letters. For any $d > 0$, the minimal automaton of the mirror of $\mathcal{L}$, where $\mathcal{L}$ is the language recognized by a random deterministic n-state automaton, generically has a super-polynomial number of states.*

This directly yields the generic complexity of Brzozowski algorithm, and therefore its average case complexity. It also emphasizes that, in our case, the generic complexity analysis is more precise than the average case analysis: a negligible proportion of bad cases could also have lead to a bad average complexity.

Corollary 1 (Average complexity). *For any fixed alphabet with at least two letters, the generic and average complexity of Brzozowski algorithm is super-polynomial for Bernoulli models on deterministic automata.*

Proof. It is generically super polynomial by Theorem 2. Hence for any $d > 0$, the complexity is greater than n^{d+1} with probability more than $\frac{1}{2}$, for n large enough. Thus, the average complexity is bounded from below by $\frac{1}{2}n^{d+1} > n^d$ for n large enough.   □

Lemma 1 below is the main ingredient of the proof of Theorem 2, as it allows to focus on a-cycles only, which contains enough information to exhibit a lower bound. The other letters are necessary to prove that such a-cycles are accessible in a random automaton, as we shall see in Section 4.

Lemma 1. *Let $\mathcal{A} \in \mathfrak{A}_n$ be a deterministic automaton that contains m primitive a-cycles $\mathcal{C}_1, \ldots \mathcal{C}_m$ of length at least two that are all accessible. The minimal automaton of $\mathcal{L}(\tilde{\mathcal{A}})$ has at least $lcm(|\mathcal{C}_1|, \ldots, |\mathcal{C}_m|)$ states.*

Proof. By Theorem 1, the minimal automaton of the mirror of $\mathcal{L}(\mathcal{A})$ is obtained by determinizing the reverse of the accessible part of $\mathcal{A}$. Since the a-cycles are accessible, they are still there after removing the non-accessible part. Moreover, as they are primitive and of length at least two, they necessarily contain at least one final state. Hence, they are also co-accessible.

Let $\mathcal{C} = \cup_{j \in [m]} \mathcal{C}_j$ and let σ be the permutation of $\mathcal{C}$ defined by $\sigma(x) = y$ if and only if $\delta(y, a) = x$. This permutation is well defined, since every element of $\mathcal{C}$ has a unique preimage by a that lies in $\mathcal{C}$. We are interested in the natural action of σ on the subsets of $\mathcal{C}$: let F be the set of final states of $\mathcal{A}$, which is also the set of initial states of $\tilde{\mathcal{A}}$, and consider the set $X = \mathcal{C} \cap F$. Let ℓ be the size of the orbit of X under the action of $< \sigma >$. We have $\sigma^\ell(X) = X$. Let $\mathcal{C}_j$ be one of the cycles and let $X_j = \mathcal{C}_j \cap X$. The set $\mathcal{C}_j$ is stable under the action of σ, and $X_j \subseteq X$, thus $\sigma^\ell(X_j) = X_j$. Hence, the size of its orbit under the action of $< \sigma >$ divides ℓ. Moreover, since $\mathcal{C}_j$ is primitive, there are exactly $|\mathcal{C}_j|$ elements in the orbit of X_j, and thus $|\mathcal{C}_j|$ divides ℓ for every $j \in [m]$. Hence, ℓ is the lcm of the cycles' lengths. Therefore, by looking at the intersection of $\tilde{\delta}(F, a^i)$ with $\mathcal{C}$, for $i \geq 0$, there are at least $lcm(|\mathcal{C}_1|, \ldots, |\mathcal{C}_m|)$ accessible states in Subset$(\tilde{\mathcal{A}})$. $\square$

4   Accessibility in Random Transition Structures

The very first part of the algorithm is to remove useless states, and in particular states that are not accessible. The precise study of the number of accessible states in a random transition structure has been done in [6]: if X_n is the random variable associated with the number of accessible states, the expectation of X_n is equivalent to $v_k \cdot n$, for some explicit constant v_k, and the distribution is asymptotically Gaussian. In the sequel, we only need the following weaker result established in [6]:

Lemma 2. *There exists two real numbers α and β, with $0 < \alpha < \beta < 1$ such that the number of accessible states in a random transition structure of size n is generically in the interval $[\alpha n, \beta n]$.*

In order to use Lemma 1, we need to exhibit large enough primitive a-cycles in a random deterministic automaton in the proof of Theorem 2. This can only work if those cycles are in the accessible part of the automaton, which is established in Proposition 1 below. The proof directly follows a more general idea given by Andrea Sportiello in a private communication.

Proposition 1. *For the uniform distribution on transition structures of size n, all the a-cycles of lengths greater than $\log n$ are generically accessible.*

Proof. Let $i \in [n]$ and let $\mathcal{A}$ be an accessible transition structure with i states, whose states labels are in $[n]$ and such that 1 labels the initial state. By a direct counting argument [6], there are exactly $n^{k(n-i)}$ transition structures in $\mathfrak{T}_n$ whose accessible part is $\mathcal{A}$. Let us bound from above the number of such automata having a non-accessible a-cycle of size ℓ: to create such a cycle, one need to choose the ℓ state labels not in the accessible part and how these states are circularly linked using transitions labelled by a. Other transitions can end at any of the n states. There are therefore no more than $\binom{n-i}{\ell}(\ell-1)! \cdot n^{k(n-i)-\ell}$ possibilities. Hence, the probability that it happens, conditioned by having $\mathcal{A}$ as accessible part, is bounded from above by

$$\binom{n-i}{\ell}(\ell-1)!\, n^{-\ell} = \frac{n^{-\ell}}{\ell}(n-i)(n-i-1)\cdots(n-i-\ell+1) \le (n-i)^{\ell}n^{-\ell}.$$

This bound only depends on the size of the accessible part. Let X_n be the random variable associated with the number of states in the accessible part of a random transition structure. Using the formula above, the probability of having an a-cycle of length equal to ℓ that is not accessible and at least αn accessible states is bounded from above by[3]

$$\sum_{i=\alpha n}^{n}(n-i)^{\ell}n^{-\ell}\cdot \mathbb{P}(X_n=i) \le (1-\alpha)^{\ell}.$$

Hence the probability of having a non-accessible a-cycle of length at least ℓ and at least αn accessible states is bounded from above by $\sum_{j=\ell}^{(1-\alpha)n}(1-\alpha)^j$ which tends to 0 as ℓ tends to infinity, as the remainder of a converging series. This concludes the proof, using $\ell = \log n$, since by Lemma 2, the accessible part of a transition structure generically has more than αn states. $\qquad\square$

5   Proof of Theorem 2

Our proof of Theorem 2 relies on Lemma 1 and on a famous theorem of Erdős and Turán: let O_n be the random variable associated with the order of a random permutation of size n. Erdős and Turán theorem states that the mean value of $\log O_n$ is equivalent to $\frac{1}{2}\log^2 n$, and that when normalized[4], it converges in distribution to the normal law. In the sequel, we shall only need an intermediate result they use to establish their proof, which is the following [10, Eq. (14.3)]:

Proposition 2 (Erdős and Turán). *For the uniform distribution, the order of a random permutation of size n is generically greater than $\exp(\frac{1}{3}\log^2 n)$.*

The idea is to use Proposition 2 to quantify the lcm of the primitive accessible a-cycles in a random automaton, under the Bernoulli model. This requires some care, since not all a-cycles are necessarily accessible or primitive.

---

[3] For readability we have not use integer parts in the bounds, here and in the sequel; this does not change the results.

[4] Centered around its means and divided by its standard deviation.

5.1   Accessible a-Cycles

We first focus on the shape of random automata and therefore work on transition structures. By Proposition 1, all a-cycles of length greater than $\log n$ are generically accessible, and we need to exhibit enough such cycles.

The action of letter a in a uniform element of $\mathfrak{T}_n$ is a uniform size-n random mapping. These objects have been studied intensively, and their typical properties are well-known [11]. We shall need the two following results in the sequel.

Lemma 3. *For any $\epsilon > 0$, the number of cyclic points of a size-n random mapping is generically greater than $n^{\frac{1}{2}-\epsilon}$.*

Proof. Let α be a real number such that $\frac{1}{2} - \epsilon < \alpha < \frac{1}{2}$. Let f be a mapping of size n, and consider the sequence $1, f(1), f^2(1) = f(f(1)), \ldots$ At some point, $f^i(1)$ is for the first time equal to a $f^j(1)$ for $j < i$, and we have a cycle of length $i - j + 1$. This reduces the problem to the Birthday Paradox: we repeatedly draw a random number from $[n]$ (the image of the new iteration of f) until a number is seen twice. Let X_n be the random variable associated with the number of distinct numbers in the sequence, we classically have:

$$\mathbb{P}(X_n \geq m) = \left(1 - \frac{1}{n}\right)\left(1 - \frac{2}{n}\right)\cdots\left(1 - \frac{m-1}{n}\right).$$

Moreover, for $x \in (0, \frac{1}{2})$, $1 - x \geq \exp(-2x)$, and therefore, for $m \leq \frac{1}{2}n$ we have

$$\mathbb{P}(X_n \geq m) \geq \exp\left(-\frac{2}{n}\sum_{i=1}^{m-1} i\right) = \exp\left(-\frac{m(m-1)}{n}\right).$$

Hence $\mathbb{P}(X_n < n^\alpha) = \mathcal{O}(n^{2\alpha-1})$, and since $\alpha < \frac{1}{2}$, there are generically more than n^α distinct iterations. Moreover, by symmetry, if $f^i(1) = f^j(1)$ is the first collision, j is a uniform element of $\{0, \ldots, i-1\}$. Since $n^{\frac{1}{2}-\epsilon}$ is significantly smaller than n^α, the collision is generically not on one of the $n^{\frac{1}{2}-\epsilon}$ last iterations, and the cycle is of length greater than $n^{\frac{1}{2}-\epsilon}$. This concludes the proof, since the number of cyclic points is at least the length of this cycle.   □

Lemma 4. *Let $i \in [n]$ and let σ and τ be two permutations of $[i]$. The probability that the normalized cyclic permutation of a uniform size-n random mapping is σ is equal to the probability it is τ.*

Proof. (sketch) This is a folklore result. From its graph representation, one can see that a mapping is uniquely determined by its set $\mathcal{T} = \{T_1, \ldots, T_m\}$ of trees and the permutation of their roots. Conditioned to have $\mathcal{T}$ as set of trees, the normalized cyclic permutation of a random mapping is therefore a uniform permutation. The result follows directly, by the law of total probabilities.   □

Hence generically, the number of a-cyclic states is greater than, say, $n^{\frac{1}{3}}$, and conditioned by its size, the normalized cyclic permutation of a random mapping follows the uniform distribution. We can therefore use the statistical properties of random uniform permutations, which are very well-known as well. In particular, we shall need the following generic upper bound for the number of cycles.

Lemma 5. *For the uniform distribution, a size-n random permutation generically has less than $2\log n$ cycles.*

Proof. The expectation and standard deviation of the number of cycles in a random permutation are well-known (see for instance [12, Example IX.9 p. 644]) and are respectively equivalent to $\log n$ and $\sqrt{\log n}$. It implies by Chebyshev's inequality that a random permutation has generically less than $2\log n$ cycles.   □

The following proposition summarizes the results collected so far.

Proposition 3. *Generically in a random size-n transition structure, there are more than $n^{\frac{1}{3}}$ a-cyclic states, organized in less than $2\log n$ a-cycles, and all a-cycles of length greater than $\log n$ are accessible.*

Proof. Let ℓ be an integer such that $n^{\frac{1}{3}} < \ell \leq n$. Let C_n be the random variable associated with the number of a-cyclic points in a random size-n transition structure. Let also N_ℓ be the random variable associated with the number of cycles in a random permutation of size ℓ. By Lemma 5, there exists a non-increasing sequence $(\epsilon_n)_{n\geq 1}$ that tends to 0 such that

$$\mathbb{P}(N_\ell < 2\log \ell) \geq 1 - \epsilon_\ell.$$

Let $\mathfrak{G}_n \subseteq \mathfrak{T}_n$ denote the set of transition structures with more than $n^{\frac{1}{3}}$ a-cyclic states that are organized in less than $2\log n$ a-cycles. If T_n represents an element of $\mathfrak{T}_n$ taken uniformly at random, we have

$$\mathbb{P}(T_n \in \mathfrak{G}_n) = \sum_{\ell=n^{1/3}}^{n} \mathbb{P}(T_n \in \mathfrak{G}_n \mid C_n = \ell) \cdot \mathbb{P}(C_n = \ell).$$

By Lemma 4, $\mathbb{P}(T_n \in \mathfrak{G}_n \mid C_n = \ell) = \mathbb{P}(N_\ell < 2\log n) \geq \mathbb{P}(N_\ell < 2\log \ell)$, since under the condition $C_n = \ell$, the a-cyclic part of T_n is a uniform random permutation of length ℓ. Hence

$$\mathbb{P}(T_n \in \mathfrak{G}_n) \geq (1 - \epsilon_{n^{1/3}}) \sum_{\ell=n^{1/3}}^{n} \mathbb{P}(C_n = \ell) = (1 - \epsilon_{n^{1/3}}) \cdot \mathbb{P}(C_n \geq n^{1/3}).$$

Hence, by Lemma 3, a random transition structure is generically in $\mathfrak{G}_n$. This concludes the proof, since by Proposition 1, all a-cycles of length greater than $\log n$ are generically accessible.   □

5.2   Lcm of Truncated Random Permutations

Since we cannot guarantee that small cycles are accessible in a typical transition structure, we need to adapt Proposition 2 to obtain the needed lower bound for the lcm of the lengths of accessible a-cycle. In a size-n permutation, a *large cycle* (resp. *small cycle*) denote a cycle of length greater than (resp. at most) $3\log n$.

Lemma 6. *The lcm of the lengths of the large cycles in a uniform random permutation of size n is generically greater than $\exp(\frac{1}{4}\log^2 n)$.*

Proof. By Lemma 5 there are generically less than $2\log n$ cycles in a random permutation. The number of points in small cycles is therefore generically bounded from above by $6(\log n)^2$. For a given permutation, we split the lengths of its cycles into two sets L and S, whether they are greater than $3\log n$ or not. The order of the permutation is the lcm of the lengths of its cycles, and is therefore bounded from above by $\mathrm{lcm}(L) \cdot \mathrm{lcm}(S)$. Hence

$$\mathrm{lcm}(L) \geq \frac{\mathrm{lcm}(L \cup S)}{\mathrm{lcm}(S)}.$$

By Landau's theorem [15], the maximal order of a permutation of length ℓ is equivalent to $\exp(\sqrt{\ell \log \ell})$ and therefore bounded from above by $2\exp(\sqrt{\ell \log \ell})$ for large enough ℓ. Hence, the less than $6(\log n)^2$ points in small cycles form a permutation whose order, which is equal to $\mathrm{lcm}(S)$, is bounded from above by $2\exp(\sqrt{6\log^2 n}\log(6\log^2 n))$. Using this bound and Proposition 2 yields the result: for n large enough, we have a generic lower bound of

$$\frac{\exp(\frac{1}{3}\log^2 n)}{2\exp(\sqrt{6\log^2 n}\log(6\log^2 n))} = \frac{1}{2}\exp\left(\frac{1}{3}\log^2 n - \sqrt{6}\log n\log(6\log^2 n)\right)$$

$$\geq \exp\left(\frac{1}{4}\log^2 n\right).$$

5.3   Primitivity

One last effort is required, as we need to take final states into account and prove the generic primitivity of large cycles in a uniform random permutation under the Bernoulli model. Recall that an a-cycle of final and non-final states is encoded by a word with 1's and 0's (see Section 2.2). The following lemma establishes the needed result.

Lemma 7. *Generically, the a-cycles of length greater than $\log n$ in a random automaton with n states are all primitive.*

Proof. We first follow [7] for words on $\{0,1\}$ under the Bernoulli model of parameter $b \in (0,1)$: if a word u of length n is not primitive, there exist an integer $d \geq 2$ and a word v of length n/d such that $u = v^d$. For such a fixed v with z zeros, the probability that $u = v^d$ is $(1-b)^{dz}b^{n-dz}$. Since there are exactly $\binom{n/d}{z}$ such v, the probability that u is the d-power of a word, for any fixed $d \geq 2$ that divides n, is

$$\sum_{z=0}^{n/d}\binom{n/d}{z}(1-b)^{dz}b^{n-dz} = (b^d + (1-b)^d)^{\frac{n}{d}}.$$

Hence the probability that u is not primitive is bounded from above by the sum of $(b^d + (1-b)^d)^{\frac{n}{d}}$ for $2 \leq d \leq n$, which is smaller than $\alpha\lambda^n$, for $\lambda = \sqrt{b^2 + (1-b)^2}$

and for some constant $\alpha > 0$. Then, each a-cycle of length greater than $\log n$ is non-primitive with probability bounded from above by $\alpha \log n \cdot \lambda^{\log n}$. By Proposition 3, the probability ϵ_n that there are more than $2\log n$ a-cycles tends to 0. Hence, the probability of having a non-primitive a-cycle of length greater than $\log n$ is bounded from above by $2\alpha(\log n)^2 \lambda^{\log n} + \epsilon_n$, which tends to 0.   □

5.4   Conclusion of the Proof

We now have all the ingredients to establish the proof of Theorem 2. By Proposition 3, the a-cycles of a random automaton $\mathcal{A}$ generically form a random permutation of size greater that $n^{\frac{1}{3}}$. Therefore, the large a-cycles are generically of length greater than $3\log n^{\frac{1}{3}} = \log n$. Since a-cycles of size greater than $\log n$ are generically accessible and primitive by Proposition 1 and Lemma 7, the lcm of the large cycles' lengths is a lower bound for the number of states of the minimal automaton of $\mathcal{L}(\tilde{\mathcal{A}})$, by Lemma 1.

By lemma 4, conditioned by its size, the a-cyclic permutation is a uniform permutation. Using the law of total probability and Lemma 6 we therefore obtain that there are generically more than $\exp(\frac{1}{4}\log^2 n^{\frac{1}{3}})$ states in the minimal automaton of $\mathcal{L}(\tilde{\mathcal{A}})$, concluding the proof.

6   Conclusion and Perspectives

In this article we have found generic super-polynomial lower bounds for the mirror operator and for the complexity of Brzozowski algorithm. These results hold for deterministic automata under Bernoulli models, where the shape of the automaton is chosen uniformly at random, and where each state is final with a fixed probability $b \in (0, 1)$.

These probabilistic models are interesting since they contain the uniform distribution on deterministic automata. It is however natural to consider other distribution on automata, and we propose two directions.

The first idea is to change the distribution on final states, in order to have less final states in a typical automaton. The proof proposed in this article can be adapted to handle distributions where $b := b_n$ depends on n, provided there exists $0 < \alpha < \frac{1}{2}$ such that both b_n and $1 - b_n$ are in $\Omega(\frac{1}{n^\alpha})$. We cannot use exactly the same approach, since the large a-cycles are not all primitive generically, but we can still exhibit sufficiently many valid a-cycles to obtain the generic super-polynomial lower bound. However, it does not work anymore for smaller b_n (such as $b_n = \frac{1}{n^{2/3}}$) since generically all the a-cycles have no final states. Trying to handle such distributions is ongoing work. Note that the other works on random automata in the literature [1,8,2] also face the same limitations.

The other natural idea is to consider the uniform distribution on accessible deterministic automata and not on deterministic automata. The combinatorics of accessible deterministic automata is more involved [14,4], but it is sometimes possible to deduce generic properties for the distribution on accessible deterministic automata from the distribution on accessible automata [6]. In our case, this would require to prove that the error terms are all in $o(\frac{1}{\sqrt{n}})$.

Acknowledgment. We would like to thanks Andrea Sportiello for the fruitful technical discussion we had in Cluny, which, amongst many other things, lead to the proof of Proposition 1.

References

1. Bassino, F., David, J., Nicaud, C.: Average case analysis of Moore's state minimization algorithm. Algorithmica 63(1-2), 509–531 (2012)
2. Bassino, F., David, J., Sportiello, A.: Asymptotic enumeration of minimal automata. In: Dürr, Wilke (eds.) [9], pp. 88–99
3. Bassino, F., Giambruno, L., Nicaud, C.: The average state complexity of rational operations on finite languages. International Journal of Foundations of Computer Science 21(4), 495–516 (2010)
4. Bassino, F., Nicaud, C.: Enumeration and random generation of accessible automata. Theor. Comput. Sci. 381(1-3), 86–104 (2007)
5. Brzozowski, J.A.: Canonical regular expressions and minimal state graphs for definite events. In: Mathematical Theory of Automata. MRI Symposia Series, vol. 12, pp. 529–561. Polytechnic Press, Polytechnic Institute of Brooklyn, N.Y (1962)
6. Carayol, A., Nicaud, C.: Distribution of the number of accessible states in a random deterministic automaton. In: Dürr, Wilke (eds.) [9], pp. 194–205
7. Chassaing, P., Azad, E.Z.: Asymptotic behavior of some factorizations of random words (2010), arXiv:1004.4062v1
8. David, J.: Average complexity of Moore's and Hopcroft's algorithms. Theor. Comput. Sci. 417, 50–65 (2012)
9. Dürr, C., Wilke, T. (eds.): 29th International Symposium on Theoretical Aspects of Computer Science, STACS 2012, Paris, France, February 29 - March 3. LIPIcs, vol. 14. Schloss Dagstuhl - Leibniz-Zentrum fuer Informatik (2012)
10. Erdős, P., Turán, P.: On some problems of a statistical group-theory I. Z. Wahrscheinlichkeitstheorie und Verw. Gebiete 4, 175–186 (1965)
11. Flajolet, P., Odlyzko, A.M.: Random mapping statistics. In: Quisquater, J.-J., Vandewalle, J. (eds.) EUROCRYPT 1989. LNCS, vol. 434, pp. 329–354. Springer, Heidelberg (1990)
12. Flajolet, P., Sedgewick, R.: Analytic Combinatorics. Cambridge University Press (2009)
13. Hopcroft, J.E.: An $n \log n$ algorithm for minimizing the states in a finite automaton. In: Kohavi, Z. (ed.) The Theory of Machines and Computations, pp. 189–196. Academic Press (1971)
14. Korshunov, A.: Enumeration of finite automata. Problemy Kibernetiki 34, 5–82 (1978) (in Russian)
15. Landau, E.: Handbuch der lehre von der verteilung der primzahlen, vol. 2. B. G. Teubner (1909)
16. Nicaud, C.: Average state complexity of operations on unary automata. In: Kutylowski, M., Pacholski, L., Wierzbicki, T. (eds.) MFCS 1999. LNCS, vol. 1672, pp. 231–240. Springer, Heidelberg (1999)
17. Tabakov, D., Vardi, M.Y.: Experimental evaluation of classical automata constructions. In: Sutcliffe, G., Voronkov, A. (eds.) LPAR 2005. LNCS (LNAI), vol. 3835, pp. 396–411. Springer, Heidelberg (2005)

A Coloring Problem
for Sturmian and Episturmian Words

Aldo de Luca[1], Elena V. Pribavkina[2], and Luca Q. Zamboni[3]

[1] Dipartimento di Matematica
Università di Napoli Federico II, Italy
aldo.deluca@unina.it
[2] Ural Federal University, Ekaterinburg, Russia
elena.pribavkina@usu.ru
[3] Université Claude Bernard Lyon 1, France
and University of Turku, Finland
lupastis@gmail.com

Abstract. We consider the following open question in the spirit of Ramsey theory: Given an aperiodic infinite word w, does there exist a finite coloring of its factors such that no factorization of w is monochromatic? We show that such a coloring always exists whenever w is a Sturmian word or a standard episturmian word.

1 Introduction

Ramsey theory (including Van der Waerden's theorem) (see [6]) is a topic of great interest in combinatorics with connections to various fields of mathematics. A remarkable consequence of the Infinite Ramsey Theorem applied to combinatorics on words yields the following unavoidable regularity of infinite words[1]:

Theorem 1. *Let A be a non-empty alphabet, w be an infinite word over A, C a finite non-empty set (the set of colors), and $c : \mathrm{Fact}^+ w \to C$ any coloring of the set $\mathrm{Fact}^+ w$ of all non-empty factors of w. Then there exists a factorization of w of the form $w = V U_1 U_2 \cdots U_n \cdots$ such that for all positive integers i and j, $c(U_i) = c(U_j)$.*

One can ask whether given an infinite word there exists a suitable coloring map able to avoid the monochromaticity of *all* factors in all factorizations of the word. More precisely, the following variant of Theorem 1 was posed as a question by T.C. Brown [3] and, independently, by the third author [12]:

Question 1. *Let w be an aperiodic infinite word over a finite alphabet A. Does there exist a finite coloring $c : \mathrm{Fact}^+ w \to C$ with the property that for any factorization $w = U_1 U_2 \cdots U_n \cdots$, there exist positive integers i, j for which $c(U_i) \neq c(U_j)$?*

---

[1] Actually, the proof of Theorem 1 given by Schützenberger in [11] does not use Ramsey's theorem.

M.-P. Béal and O. Carton (Eds.): DLT 2013, LNCS 7907, pp. 191–202, 2013.
© Springer-Verlag Berlin Heidelberg 2013

Let us observe that for periodic words the answer to the preceding question is trivially negative. Indeed, let $w = U^\omega$, and $c : \mathrm{Fact}^+ w \to C$ be any finite coloring. By factoring w as $w = U_1 U_2 \cdots U_n \cdots$, where for all $i \geq 1$, $U_i = U$ one has $c(U_i) = c(U_j)$ for all positive integers i and j. It is easy to see that there exist non-recurrent infinite words w and finite colorings such that for any factorization $w = U_1 U_2 \cdots U_n \cdots$ there exist $i \neq j$ for which $c(U_i) \neq c(U_j)$. For instance, consider the infinite word $w = ab^\omega$ and define the coloring map as follows: for any non-empty factor U of w, $c(U) = 1$ if it contains a and $c(U) = 0$, otherwise. Then for any factoring $w = U_1 U_2 \cdots U_n \cdots$, $c(U_1) = 1$ and $c(U_i) = 0$ for all $i > 1$.

It is not very difficult to prove that there exist infinite recurrent words for which Question 1 has a positive answer, for instance square-free, overlap-free words, and standard Sturmian words [12].

In this paper we show that Question 1 has a positive answer for every Sturmian word where the number of colors is equal to 3. This solves a problem raised in both [3] and [12]. The proof requires some new combinatorial properties of Sturmian words. Moreover, we prove that the same result holds true for aperiodic standard episturmian words by using a number of colors equal to the number of distinct letters occurring in the word plus one.

For all definitions and notation not explicitly given in the paper, the reader is referred to the books [8,9]; for Sturmian words see [9, Chap. 2] and for episturmian words see [4,7] and the surveys of J. Berstel [1] and of A. Glen and J. Justin [5].

2   Sturmian Words

There exist several equivalent definitions of Sturmian words. In particular, we recall (see, for instance, Theorem 2.1.5 of [9]) that an infinite word $s \in \{a, b\}^\omega$ is Sturmian if and only if it is aperiodic and *balanced*, i.e., for all factors u and v of s such that $|u| = |v|$ one has:

$$||u|_x - |v|_x| \leq 1, \ x \in \{a, b\},$$

where $|u|_x$ denotes the number of occurrences of the letter x in u. Since a Sturmian word s is aperiodic, it must have at least one of the two factors aa and bb. However, from the balance property, it follows that a Sturmian word cannot have both the factors aa and bb.

Definition 1. *We say that a Sturmian word is of type a (resp. b) if it does not contain the factor bb (resp. aa).*

We recall that a factor u of a finite or infinite word w over the alphabet A is called *right special* (resp. *left special*) if there exist two different letters $x, y \in A$ such that ux, uy (resp. xu, yu) are factors of w.

A different equivalent definition of a Sturmian word is the following: A binary infinite word s is Sturmian if for every integer $n \geq 0$, s has a unique left (or equivalently right) special factor of length n. It follows from this that s is *closed under reversal*, i.e., if u is a factor of s so is its reversal $u^\sim$.

A Sturmian word s is called *standard* (or *characteristic*) if all its prefixes are left special factors of s. As is well known, for any Sturmian word s there exists a standard Sturmian word t such that $\mathrm{Fact}\, s = \mathrm{Fact}\, t$, where for any finite or infinite word w, $\mathrm{Fact}\, w$ denotes the set of all its factors including the empty word.

Definition 2. *Let $s \in \{a, b\}^\omega$ be a Sturmian word. A non-empty factor w of s is rich in the letter $z \in \{a, b\}$ if there exists a factor v of s such that $|v| = |w|$ and $|w|_z > |v|_z$.*

From the aperiodicity and the balance property of a Sturmian word one easily derives that any non-empty factor w of a Sturmian word s is rich either in the letter a or in the letter b but not in both letters. Thus one can introduce for any given Sturmian word s a map

$$r_s : \mathrm{Fact}^+ s \to \{a, b\}$$

defined as follows: for any non-empty factor w of s, $r_s(w) = z \in \{a, b\}$ if w is rich in the letter z. Clearly, $r_s(w) = r_s(w^\sim)$ for any $w \in \mathrm{Fact}^+ s$.

For any letter $z \in \{a, b\}$ we shall denote by $\bar{z}$ the complementary letter of z, i.e., $\bar{a} = b$ and $\bar{b} = a$.

Lemma 1. *Let w be a non-empty right special (resp. left special) factor of a Sturmian word s. Then $r_s(w)$ is equal to the first letter of w (resp. $r_s(w)$ is equal to the last letter of w).*

Proof. Write $w = zw'$ with $z \in \{a, b\}$ and $w' \in \{a, b\}^*$. Since w is a right special factor of s one has that $v = w'\bar{z}$ is a factor of s. Thus $|w| = |v|$ and $|w|_z > |v|_z$, whence $r_s(w) = z$. Similarly, if w is left special one deduces that $r_s(w)$ is equal to the last letter of w. $\square$

3   Preliminary Lemmas

Lemma 2. *Let s be a Sturmian word such that*

$$s = \prod_{i \geq 1} U_i,$$

where the U_i's are non-empty factors of s. If for every i and j, $r_s(U_i) = r_s(U_j)$, then for any $M > 0$ there exists an integer i such that $|U_i| > M$.

Proof. Suppose to the contrary that for some positive integer M we have that $|U_i| \leq M$ for each $i \geq 1$. This implies that the number of distinct U_i's in the sequence $(U_i)_{i \geq 1}$ is finite, say t. Let $r_s(U_i) = x \in \{a, b\}$ for all $i \geq 1$ and set for each $i \geq 1$:

$$f_i = \frac{|U_i|_x}{|U_i|}.$$

Thus $\{f_i \mid i \geq 1\}$ is a finite set of at most t rational numbers. We set $r = \min\{f_i \mid i \geq 1\}$.

Let $f_x(s)$ be the frequency of the letter x in s defined as

$$f_x(s) = \lim_{n \to \infty} \frac{|s_{[n]}|_x}{n},$$

where for every $n \geq 1$, $s_{[n]}$ denotes the prefix of s of length n. As is well known (see Prop. 2.1.11 of [9]), $f_x(s)$ exists and is an irrational number.

Let us now prove that $r > f_x(s)$. From Proposition 2.1.10 in [9] one derives that for all $V \in \text{Fact}\, s$

$$|V|f_x(s) - 1 < |V|_x < |V|f_x(s) + 1.$$

Now for any $i \geq 1$, U_i is rich in the letter x, so that there exists $V_i \in \text{Fact}\, s$ such that $|U_i| = |V_i|$ and $|U_i|_x > |V_i|_x$. From the preceding inequality one has:

$$|U_i|_x = |V_i|_x + 1 > |V_i|f_x(s) = |U_i|f_x(s),$$

so that for all $i \geq 1$, $f_i > f_x(s)$, hence $r > f_x(s)$.

For any $n > 0$, we can write the prefix $s_{[n]}$ of length n as:

$$s_{[n]} = U_1 \cdots U_k U'_{k+1},$$

for a suitable $k \geq 0$ and U'_{k+1} a prefix of U_{k+1}. Thus

$$|s_{[n]}|_x = \sum_{i=i}^{k} |U_i|_x + |U'_{k+1}|_x.$$

Since $|U_i|_x = f_i|U_i| \geq r|U_i|$ and $|U'_{k+1}| \leq M$, one has

$$|s_{[n]}|_x \geq r \sum_{i=1}^{k} |U_i| = r(n - |U'_{k+1}|) \geq rn - rM.$$

Thus

$$\frac{|s_{[n]}|_x}{n} \geq r - r\frac{M}{n},$$

and

$$f_x(s) = \lim_{n \to \infty} \frac{|s_{[n]}|_x}{n} \geq r,$$

a contradiction.   □

In the following we shall consider the Sturmian morphism R_a, that we simply denote R, defined as:

$$R(a) = a \text{ and } R(b) = ba. \tag{1}$$

For any finite or infinite word w, $\text{Pref}\, w$ will denote the set of all its prefixes. The following holds:

Lemma 3. *Let s be a Sturmian word and $t \in \{a, b\}^\omega$ such that $R(t) = s$. If either*

1) *the first letter of t (or, equivalently, of s) is b*

   or

2) *the Sturmian word s admits a factorization:*

$$s = U_1 \cdots U_n \cdots,$$

   where each U_i, $i \geq 1$, is a non-empty prefix of s terminating in the letter a and $r_s(U_i) = r_s(U_j)$ for all $i, j \geq 1$,

then t is also Sturmian.

Proof. Let us prove that in both cases t is balanced. Suppose to the contrary that t is unbalanced. Then (see Prop. 2.1.3 of [9]) there would exists v such that

$$ava, bvb \in \mathrm{Fact}\, t.$$

Thus

$$aR(v)a, baR(v)ba \in \mathrm{Fact}\, s.$$

If $ava \notin \mathrm{Pref}\, t$, then $t = \lambda ava\mu$, with $\lambda \in \{a, b\}^+$ and $\mu \in \{a, b\}^\omega$. Therefore $R(t) = R(\lambda)R(ava)R(\mu)$. Since the last letter of $R(\lambda)$ is a, it follows that $aaR(v)a \in \mathrm{Fact}\, s$. As $baR(v)b \in \mathrm{Fact}\, s$ we reach a contradiction with the balance property of s. In case 1), t begins in the letter b, so that $ava \notin \mathrm{Pref}\, t$ and then t is balanced. In case 2) suppose that $ava \in \mathrm{Pref}\, t$. This implies that $aR(v)a \in \mathrm{Pref}\, s$. From the preceding lemma in the factorization of s in prefixes there exists an integer $i > 1$ such that $|U_i| > |aR(v)a|$. Since U_{i-1} terminates in a and $U_{i-1}U_i \in \mathrm{Fact}\, s$, it follows that $aaR(v)a \in \mathrm{Fact}\, s$ and one contradicts again the balance property of s. Hence, t is balanced.

Trivially, in both cases t is aperiodic, so that t is Sturmian.   □

Let us remark that, in general, without any additional hypothesis, if $s = R(t)$, then t need not be Sturmian. For instance, if f is the Fibonacci word $f = abaababaaba \cdots$, then af is also a Sturmian word. However, it is readily verified that in this case the word t such that $R(t) = s$ is not balanced, so that t is not Sturmian.

For any finite or infinite word w over the alphabet A, alph w denotes the set of all distinct letters of A occurring in w. We will make use of the following lemma.

Lemma 4. *Let s be an aperiodic word having a factorization*

$$s = U_1 \cdots U_n \cdots,$$

where for $i \geq 1$, U_i are non-empty prefixes of s. Then for any $p \geq 1$, $U_1 \neq c^p$ where c is the first letter of s.

Proof. Suppose that $U_1 = c^p$. Since s is aperiodic there exists a minimal integer j such that $\mathrm{card}(\mathrm{alph}\, U_j) = 2$. Since U_j is a prefix of s, one has then $U_1 \cdots U_{j-1}U_j = U_j\xi$, with $\xi \in \{a, b\}^*$. As $U_1 \cdots U_{j-1} = c^q$ for a suitable $q \geq p$, it follows that $\xi = c^q$ and $U_j \in cc^*$, a contradiction.   □

4   Main Results

Proposition 1. *Let s be a Sturmian word of type a having a factorization*

$$s = U_1 \cdots U_n \cdots ,$$

where for $i \geq 1$, U_i are non-empty prefixes of s such that $r_s(U_i) = r_s(U_j)$ for all $i, j \geq 1$. Then one of the following two properties holds:

i) *All U_i, $i \geq 1$, terminate in the letter a.*
ii) *For all $i \geq 1$, $U_i a \in \text{Pref } s$.*

Proof. Let us first suppose that s begins in the letter b. All prefixes U_i, $i \geq 1$, of s begin in the letter b and, as s is of type a, have to terminate in the letter a. Thus in this case Property i) is satisfied.

Let us then suppose that s begins in the letter a. Now either all prefixes U_i, $i \geq 1$, terminate in the letter a or all prefixes U_i, $i \geq 1$, terminate in the letter b or some of the prefixes terminate in the letter a and some in the letter b. We have then to consider the following cases:

Case 1. All prefixes U_i, $i \geq 1$, terminate in the letter b.

Since s is of type a, and all prefixes U_i, $i \geq 1$ terminate in the letter b, one has $U_i a \in \text{Pref } s$ and Property ii) is satisfied.

Case 2. Some of the prefixes U_i, $i \geq 1$, terminate in the letter a and some in the letter b.

We have to consider two subcases:

a) $r_s(U_i) = b$, for all $i \geq 1$.

As all U_i, $i \geq 1$, begin in a, if any U_i were right special, then by Lemma 1, $r_s(U_i) = a$, a contradiction. It follows that for all $i \geq 1$, $U_i a \in \text{Pref } s$.

b) $r_s(U_i) = a$, for all $i \geq 1$.

Some of the prefixes U_j, $j \geq 1$, terminate in a (since otherwise we are in Case 1). Let U_k be a prefix terminating in a for a suitable $k \geq 1$. If a prefix U_i terminates in b, then aU_i is not a factor of s. Indeed, otherwise, the word $aU_i b^{-1}$ is such that $|aU_i b^{-1}| = |U_i|$ and $|aU_i b^{-1}|_b < |U_i|_b$, so that $r_s(U_i) = b$ a contradiction. Thus one derives that all U_l with $l \geq k$ terminate in a. Moreover, if some U_i terminate in b, by Lemma 2 there exists $j > k$ such that U_j has the prefix U_i, so that $U_{j-1} U_i \in \text{Fact } s$. Since U_{j-1} terminates in a, one has that aU_i is a factor of s, a contradiction. Thus all U_i, $i \geq 1$, terminate in a.   □

Proposition 2. *Let s be a Sturmian word having a factorization*

$$s = U_1 \cdots U_n \cdots ,$$

where for $i \geq 1$, U_i are non-empty prefixes of s such that $r_s(U_i) = r_s(U_j)$ for all $i, j \geq 1$. Then there exists a Sturmian word t such that

$$t = V_1 \cdots V_n \cdots ,$$

where for all $i \geq 1$, V_i are non-empty prefixes of t, $r_t(V_i) = r_t(V_j)$ for all $i, j \geq 1$, and $|V_1| < |U_1|$.

Proof. We can suppose without loss of generality that s is a Sturmian word of type a. From Proposition 1 either all U_i, $i \geq 1$, terminate in the letter a or for all $i \geq 1$, $U_i a \in$ Pref s. We consider two cases:

Case 1. For all $i \geq 1$, $U_i a \in$ Pref s.

We can suppose that s begins in the letter a. Indeed, otherwise, if the first letter of s is b, then all U_i, $i \geq 1$, begin in the letter b and, as s is of type a, they have to terminate in the letter a. Thus the case that the first letter of s is b will be considered when we will analyze Case 2.

We consider the injective endomorphism of $\{a, b\}^*$, L_a, or simply L, defined by

$$L(a) = a \text{ and } L(b) = ab.$$

Since s is of type a, the first letter of s is a, and $X = \{a, ab\}$ is a code having a finite deciphering delay (cf. [2]), the word s can be uniquely factorized by the elements of X. Thus there exists a unique word $t \in \{a, b\}^\omega$ such that $s = L(t)$. The following holds:

1. The word t is a Sturmian word.
2. For any $i \geq 1$ there exists a non-empty prefix V_i of t such that $L(V_i) = U_i$.
3. The word t can be factorized as $t = V_1 \cdots V_n \cdots$.
4. $|V_1| < |U_1|$.
5. For all $i, j \geq 1$, $r_t(V_i) = r_t(V_j)$.

Point 1. This is a consequence of the fact that L is a standard Sturmian morphism (see Corollary 2.3.3 in Chap. 2 of [9]).

Point 2. For any $i \geq 1$, since $U_i a \in$ Pref s and any pair (c, a) with $c \in \{a, b\}$ is synchronizing for $X^\infty = X^* \cup X^\omega$ (cf. [2]), one has that $U_i \in X^*$, so that there exists $V_i \in$ Pref t such that $L(V_i) = U_i$.

Point 3. One has $L(V_1 \cdots V_n \cdots) = U_1 \cdots U_n \cdots = s = L(t)$. Thus $t = V_1 \cdots V_n \cdots$.

Point 4. By Lemma 4, U_1 is not a power of a so that in U_1 there must be at least one occurrence of the letter b. This implies that $|V_1| < |U_1|$.

Point 5. We shall prove that for all $i \geq 1$, $r_t(V_i) = r_s(U_i)$. From this one has that for all $i, j \geq 1$, $r_t(V_i) = r_t(V_j)$.

Since t is a Sturmian word, there exists $V_i' \in$ Fact t such that

$$|V_i| = |V_i'| \text{ and either } |V_i|_a > |V_i'|_a \text{ or } |V_i|_a < |V_i'|_a.$$

In the first case $r_t(V_i) = a$ and in the second case $r_t(V_i) = b$. Let us set

$$F_i = L(V_i').$$

Since $U_i = L(V_i)$, from the definition of the morphism L one has:

$$|F_i|_a = |V_i'|_a + |V_i'|_b = |V_i'|, \ |F_i|_b = |V_i'|_b. \tag{2}$$

$$|U_i|_a = |V_i|_a + |V_i|_b = |V_i|, \quad |U_i|_b = |V_i|_b. \tag{3}$$

Let us first consider the case $r_t(V_i) = a$, i.e., $|V_i|_a = |V_i'|_a + 1$ and $|V_i|_b = |V_i'|_b - 1$. From the preceding equations one has:

$$|F_i| = |U_i| + 1.$$

Moreover, from the definition of L one has that F_i begins in the letter a. Hence, $|a^{-1}F_i| = |U_i|$ and $|a^{-1}F_i|_a = |F_i|_a - 1 = |U_i|_a - 1$. Thus $|U_i|_a > |a^{-1}F_i|_a$. Since $a^{-1}F_i \in$ Fact s, one has

$$r_s(U_i) = r_t(V_i) = a.$$

Let us now consider the case $r_t(V_i) = b$, i.e., $|V_i|_a = |V_i'|_a - 1$ and $|V_i|_b = |V_i'|_b + 1$. From (2) and (3) one derives:

$$|U_i| = |F_i| + 1,$$

and $|U_i|_b > |F_i|_b$. Now $F_i a$ is a factor of s. Indeed, $F_i = L(V_i')$ and for any $c \in \{a, b\}$ such that $V_i'c \in$ Fact t one has $L(V_i'c) = F_iL(c)$. Since for any letter c, $L(c)$ begins in the letter a it follows that $F_i a \in$ Fact s. Since $|F_i a| = |U_i|$ and $|U_i|_b > |F_i|_b = |F_i a|_b$, one has that U_i is rich in b. Hence, $r_s(U_i) = r_t(V_i) = b$.

Case 2. All U_i, $i \geq 1$, terminate in the letter a.

We consider the injective endomorphism of $\{a, b\}^*$, R_a, or simply R, defined in (1). Since s is of type a and $X = \{a, ba\}$ is a prefix code, the word s can be uniquely factorized by the elements of X. Thus there exists a unique word $t \in \{a, b\}^\omega$ such that $s = R(t)$. The following holds:

1. The word t is a Sturmian word.
2. For any $i \geq 1$ there exists a non-empty prefix V_i of t such that $R(V_i) = U_i$.
3. The word t can be factorized as $t = V_1 \cdots V_n \cdots$.
4. $|V_1| < |U_1|$.
5. For all $i, j \geq 1$, $r_t(V_i) = r_t(V_j)$.

Point 1. From Lemma 3, since $R(t) = s$ it follows that t is Sturmian.

Point 2. For any $i \geq 1$, since U_i terminates in the letter a and any pair (a, c) with $c \in \{a, b\}$ is synchronizing for X^∞, one has that $U_i \in X^*$, so that there exists $V_i \in$ Pref t such that $R(V_i) = U_i$.

Point 3. One has $R(V_1 \cdots V_n \cdots) = U_1 \cdots U_n \cdots = s = R(t)$. Thus $t = V_1 \cdots V_n \cdots$.

Point 4. By Lemma 4, U_1 is not a power of the first letter c of s, so that in U_1 there must be at least one occurrence of the letter $\bar{c}$. This implies that $|V_1| < |U_1|$.

Point 5. We shall prove that for all $i \geq 1$, $r_t(V_i) = r_s(U_i)$. From this one has that for all $i, j \geq 1$, $r_t(V_i) = r_t(V_j)$.

Since t is a Sturmian word, there exists $V_i' \in$ Fact t such that

$$|V_i| = |V_i'| \text{ and either } |V_i|_a > |V_i'|_a \text{ or } |V_i|_a < |V_i'|_a.$$

In the first case $r_t(V_i) = a$ and in the second case $r_t(V_i) = b$. Let us set

$$F_i = R(V_i').$$

Since $U_i = R(V_i)$, from the definition of the morphism R one has that equations (2) and (3) are satisfied.

Let us first consider the case $r_t(V_i) = a$, i.e., $|V_i|_a = |V_i'|_a + 1$ and $|V_i|_b = |V_i'|_b - 1$. From the preceding equations one has:

$$|F_i| = |U_i| + 1.$$

From the definition of the morphism R one has that $F_i = R(V_i')$ terminates in the letter a. Hence, $|F_i a^{-1}| = |U_i|$ and $|F_i a^{-1}|_a = |F_i|_a - 1 = |U_i|_a - 1$. Thus $|U_i|_a = |F_i a^{-1}|_a + 1$, so that U_i is rich in a and $r_s(U_i) = r_t(V_i) = a$.

Let us now suppose that $r_t(V_i) = b$, i.e., $|V_i|_a = |V_i'|_a - 1$ and $|V_i|_b = |V_i'|_b + 1$. From (2) and (3) one derives:

$$|U_i| = |F_i| + 1,$$

and $|U_i|_b > |F_i|_b$. We prove that $aF_i \in \text{Fact } s$. Indeed, $F_i = R(V_i')$ and for any $c \in \{a, b\}$ such that $cV_i' \in \text{Fact } t$ one has $R(c)R(V_i') = R(c)F_i$. Note that such a letter c exists always as t is recurrent. Since for any letter c, $R(c)$ terminates in the letter a it follows that $aF_i \in \text{Fact } s$. Since $|aF_i| = |U_i|$ and $|U_i|_b > |aF_i|_b = |F_i|_b$, one has that U_i is rich in b. Hence, $r_s(U_i) = r_t(V_i) = b$.   □

Theorem 2. *Let s be a Sturmian word having a factorization*

$$s = U_1 \cdots U_n \cdots,$$

where each U_i, $i \geq 1$, is a non-empty prefix of s. Then there exist integers $i, j \geq 1$ such that $r_s(U_i) \neq r_s(U_j)$.

Proof. Let s be a Sturmian word and suppose that s admits a factorization

$$s = U_1 \cdots U_n \cdots,$$

where for $i \geq 1$, U_i are non-empty prefixes such that for all $i, j \geq 1$, $r_s(U_i) = r_s(U_j)$. Among all Sturmian words having this property we can always consider a Sturmian word s such that $|U_1|$ is minimal. Without loss of generality we can suppose that s is of type a. By Proposition 2 there exists a Sturmian word t such that

$$t = V_1 \cdots V_n \cdots,$$

where for all $i \geq 1$, V_i are non-empty prefixes, $r_t(V_i) = r_t(V_j)$ for all $i, j \geq 1$, and $|V_1| < |U_1|$, that contradicts the minimality of the length of U_1.   □

Theorem 3. *Let s be a Sturmian word. There exists a coloring c of the non-empty factors of s, $c : \text{Fact}^+ s \to \{0, 1, 2\}$ such that for any factorization*

$$s = V_1 \cdots V_n \cdots$$

in non-empty factors V_i, $i \geq 1$, there exist integers i, j such that $c(V_i) \neq c(V_j)$.

Proof. Let us define the coloring c as follows: for any $V \in \text{Fact}^+ s$

$$c(V) = \begin{cases} 0 \text{ if } V \text{ is not a prefix of } s \\ 1 \text{ if } V \text{ is a prefix of } s \text{ and } r_s(V) = a \\ 2 \text{ if } V \text{ is a prefix of } s \text{ and } r_s(V) = b \end{cases}$$

Let us suppose to contrary that for all i, j, $c(V_i) = c(V_j) = x \in \{0, 1, 2\}$. If $x = 0$ we reach a contradiction as V_1 is a prefix of s so that $c(V_1) \in \{1, 2\}$. If $x = 1$ or $x = 2$, then all V_i have to be prefixes of s having the same richness, but this contradicts Theorem 2.   □

5   The Case of Standard Episturmian Words

An infinite word s over the alphabet A is called *standard episturmian* if it is closed under reversal and every left special factor of s is a prefix of s. A word $s \in A^\omega$ is called *episturmian* if there exists a standard episturmian $t \in A^\omega$ such that $\text{Fact } s = \text{Fact } t$. We recall the following facts about episturmian words [4,7]:

Fact 1. Every prefix of an aperiodic standard episturmian word s is a left special factor of s. In particular an aperiodic standard episturmian word on a two-letter alphabet is a standard Sturmian word.

Fact 2. If s is a standard episturmian word with first letter a, then a is *separating*, i.e., for any $x, y \in A$ if $xy \in \text{Fact } s$, then $a \in \{x, y\}$.

For each $x \in A$, let L_x denote the standard episturmian morphism [7] defined for any $y \in A$ by $L_x(y) = x$ if $y = x$ and $L_x(y) = xy$ for $x \neq y$.

Fact 3. The infinite word $s \in A^\omega$ is standard episturmian if and only if there exist a standard episturmian word t and $a \in A$ such that $s = L_a(t)$. Moreover, t is unique and the first letter of s is a.

The following was proved in [10]:

Fact 4. A recurrent word w over the alphabet A is episturmian if and only if for each factor u of w, a letter b exists (depending on u) such that $AuA \cap \text{Fact } w \subseteq buA \cup Aub$.

Definition 3. *We say that a standard episturmian word s is of type a, $a \in A$, if the first letter of s is a.*

Theorem 4. *Let s be an aperiodic standard episturmian word over the alphabet A and let $s = U_1 U_2 \cdots$ be any factorization of s with each U_i, $i \geq 1$, a non-empty prefix of s. Then there exist indices $i \neq j$ for which U_i and U_j terminate in a different letter.*

Proof. Suppose to the contrary that there exists an aperiodic standard episturmian word s over the alphabet A admitting a factorization $s = U_1 U_2 \cdots$ in which all U_i are non-empty prefixes of s ending in the same letter. Amongst all aperiodic standard episturmian words over the alphabet A having the preceding factorization, we may choose one such s for which $|U_1|$ is minimal. Let $a \in A$ be the first letter of s, so that s is of type a.

Let us now prove that for every $i \geq 0$, one has that $U_i a$ is a prefix of s. Let us first suppose that for all $i \geq 1$, U_i ends in a letter $x \neq a$. Since a is separating (s is of type a), x can be followed only by a, so that the prefix U_i can be followed only by a. This implies that $U_i a$ is a prefix of s.

Let us then suppose that for all $i \geq 1$, U_i ends in a. Since U_1 is a prefix of s, and all U_i, $i \geq 1$, begin in a one has that $U_1 a$ is a prefix of s. Now let $i > 1$. Since U_{i-1} ends in a it follows that $a U_i a$ is a factor of s.

Let $U_i x$ be a prefix of s; we want to show that $x = a$. Since $U_i x$ is left special (as it is a prefix of s), there exists a letter $y \neq a$ such that $y U_i x$ is a factor of s. Now from this and by Fact 4, there exists a letter b (depending only on U_i) such that either $x = b$ or $y = b$.

So now, by Fact 4, since $a U_i a$ and $y U_i x$ are both factors of s, we deduce $b = a$ and either $x = a$ or $y = a$. Since $y \neq a$, it follows that $x = a$. Therefore, we have proved that for every $i \geq 1$, $U_i a$ is a prefix of s.

Let us now observe that U_1 must contain the occurrence of a letter $x \neq a$. Indeed, otherwise, suppose that $U_1 = a^k$ and consider the least $i > 1$ such that x occurs in U_i. This implies, by using an argument similar to that of the proof of Lemma 4, that U_i cannot be a prefix of s.

By Fact 3, one has that there exists a unique standard episturmian word s' such that $s = L_a(s')$ and alph $s' \subseteq$ alph $s \subseteq A$. Moreover, since s is aperiodic, trivially one has that also s' is aperiodic.

Let us observe that the set $X = \{a\} \cup \{ax \mid x \in A\}$ is a code having deciphering delay equal to 1 and that any pair (x, a) with $x \in A$ is synchronizing for X^∞. This implies that s can be uniquely factored by the words of X. Moreover, since $U_i a$ is a prefix of s, from the synchronization property of X^∞, it follows that for each $i \geq 1$,

$$U_i = L_a(U_i'),$$

where U_i' is a prefix of s'. From the definition of L_a and the preceding formula, one has that the last letter of U_i is equal to the last letter of U_i'.

Moreover,

$$L_a(U_1' \cdots U_n' \cdots) = U_1 \cdots U_n \cdots = s = L_a(s').$$

Thus $s' = U_1' \cdots U_n' \cdots$, where each U_i', $i \geq 1$, is a non-empty prefix of s' and for all $i, j \geq 1$, U_i' and U_j' terminate in the same letter. Since in $U_1 = L_a(U_1')$ there is the occurrence of a letter different from a one obtains that $|U_1'| < |U_1|$ which is a contradiction. □

Let us observe that in the case of a standard Sturmian word, Theorem 4 is an immediate consequence of Theorem 2 and Lemma 1.

Theorem 5. *Let s be an aperiodic standard episturmian word and let $k = \mathrm{card}(\mathrm{alph}\, s)$. There exists a coloring c of the non-empty factors of s, $c : \mathrm{Fact}^+ s \to \{0, 1, \ldots, k\}$ such that for any factorization $s = V_1 \cdots V_n \cdots$ in non-empty factors V_i, $i \geq 1$, there exist integers i, j such that $c(V_i) \neq c(V_j)$.*

Proof. Let alph $s = \{a_1, \ldots, a_k\}$. We define the coloring c as follows: for any $V \in \mathrm{Fact}^+ s$

$$
c(V) = \begin{cases}
0 \text{ if } V \text{ is not a prefix of } s \\
1 \text{ if } V \text{ is a prefix of } s \text{ terminating in } a_1 \\
\vdots \quad \vdots \\
k \text{ if } V \text{ is a prefix of } s \text{ terminating in } a_k
\end{cases}
$$

Let us suppose by contradiction that for all i, j, $c(V_i) = c(V_j) = x \in \{0, 1, \ldots, k\}$. If $x = 0$ we reach a contradiction as V_1 is a prefix of s so that $c(V_1) \in \{1, \ldots, k\}$. If $x \in \{1, \ldots, k\}$, then all V_i have to be prefixes of s terminating in the same letter, but this contradicts Theorem 4.   □

Acknowledgments. The authors are indebted to Tom Brown for his suggestions and comments. The second author acknowledges support from the Presidential Program for young researchers, grant MK-266.2012.1. The third author is partially supported by a FiDiPro grant from the Academy of Finland.

References

1. Berstel, J.: Sturmian and Episturmian words (A survey of some recent results). In: Bozapalidis, S., Rahonis, G. (eds.) CAI 2007. LNCS, vol. 4728, pp. 23–47. Springer, Heidelberg (2007)
2. Berstel, J., Perrin, D., Reutenauer, C.: Codes and Automata. Cambridge University Press (2010)
3. Brown, T.C.: Colorings of the factors of a word, preprint Department of Mathematics, Simon Fraser University, Canada (2006)
4. Droubay, X., Justin, J., Pirillo, G.: Episturmian words and some constructions of de Luca and Rauzy. Theoret. Comput. Sci. 255, 539–553 (2001)
5. Glen, A., Justin, J.: Episturmian words: a survey. RAIRO-Theoret. Inform. Appl. 43, 403–442 (2009)
6. Graham, R., Rothshild, B.L., Spencer, J.H.: Ramsey Theory, 2nd edn. J. Wiley, New York (1990)
7. Justin, J., Pirillo, G.: Episturmian words and episturmian morphisms. Theoret. Comput. Sci. 276, 281–313 (2002)
8. Lothaire, M.: Combinatorics on Words. Addison-Wesley, Reading (1983); (reprinted by Cambridge University Press, Cambridge, 1997)
9. Lothaire, M.: Algebraic Combinatorics on Words. Cambridge University Press, Cambridge (2002)
10. Richomme, G.: A Local Balance Property of Episturmian Words. In: Harju, T., Karhumäki, J., Lepistö, A. (eds.) DLT 2007. LNCS, vol. 4588, pp. 371–381. Springer, Heidelberg (2007)
11. Schützenberger, M.P.: Quelques problèmes combinatoires de la théorie des automates, Cours professé à l'Institut de Programmation en 1966/67, notes by J.-F. Perrot
12. Zamboni, L.Q.: A Note on Coloring Factors of Words, in Oberwolfach Report 37/2010, Mini-workshop: Combinatorics on Words, pp. 42–44, August 22-27 (2010)

The Chomsky-Schützenberger Theorem for Quantitative Context-Free Languages

Manfred Droste[1] and Heiko Vogler[2]

[1] Institute of Computer Science, Leipzig University, D-04109 Leipzig, Germany
droste@informatik.uni-leipzig.de
[2] Department of Computer Science, Technische Universität Dresden,
D-01062 Dresden, Germany
Heiko.Vogler@tu-dresden.de

Abstract. Weighted automata model quantitative aspects of systems like the consumption of resources during executions. Traditionally, the weights are assumed to form the algebraic structure of a semiring, but recently also other weight computations like average have been considered. Here, we investigate quantitative context-free languages over very general weight structures incorporating all semirings, average computations, lattices. In our main result, we derive the Chomsky-Schützenberger Theorem for such quantitative context-free languages, showing that each arises as the image of the intersection of a Dyck language and a recognizable language under a suitable morphism. Moreover, we show that quantitative context-free languages are expressively equivalent to a model of weighted pushdown automata. This generalizes results previously known only for semirings.

1   Introduction

The Chomsky-Schützenberger Theorem forms a famous cornerstone in the theory of context-free languages [9] relating arbitrary context-free languages to Dyck languages and recognizable languages. A weighted version of this result was presented in [25] where the weights are taken from a commutative semiring. For surveys on this we refer the reader to [1, 23].

Recently, in [5-7] new models of quantitative automata for technical systems have been investigated describing, e.g., the average consumption of resources. In [8] pushdown automata with mean-payoff cost functions were considered which comprize a quantitative modelling of sequential programs with recursion. These cost functions cannot be computed in semirings. Automata over the general algebraic structure of valuation monoids were investigated in [11, 12]. In valuation monoids, each sequence of weights gets assigned a single weight; examples include products in semirings as well as average computations on the reals. Hence automata over valuation monoids include both semiring-weighted automata and quantitative automata.

It is the goal of this paper to investigate weighted context-free grammars over valuation monoids. Hence we may associate to each derivation, for instance, the

M.-P. Béal and O. Carton (Eds.): DLT 2013, LNCS 7907, pp. 203–214, 2013.
© Springer-Verlag Berlin Heidelberg 2013

average of the costs of the involved productions. We could also associate with each production its degree of sharpness or truth, as in multi-valued logics, using bounded lattices as valuation monoids. Thereby, we can associate to each word over the underlying alphabet Σ such a value (real number, element of a lattice, etc.) indicating its total cost or degree of truth, and any function from Σ^* into the value set is called a quantitative language or series. Note that by the usual identification of sets with $\{0, 1\}$-valued functions, classical languages arise as particular quantitative languages.

Now we give a summary of our results. We prove the equivalence of weighted context-free grammars and weighted pushdown automata over arbitrary valuation monoids (cf. Theorem 1). In our main result we derive a weighted version of the Chomsky-Schützenberger Theorem over arbitrary valuation monoids (cf. Theorem 2). In particular, we show that any quantitative context-free language arises as the image of the intersection of a Dyck language and a recognizable language under a suitable weighted morphism, and also as the image of a Dyck language and a recognizable series under a free monoid morphism. Conversely, each quantitative language arising as such an image is a quantitative context-free language. This shows that the weighted Chomsky-Schützenberger Theorem holds for much more general weighted structures than commutative semirings, in particular, neither associativity, nor commutativity, nor distributivity of the multiplication are needed. In our proofs, due to the lack of the above properties, we cannot use the theory of semiring-weighted automata (cf. [20, 25]); instead we employ explicit constructions of weighted automata taking care of precise calculations of the weights to deduce our results from the classical, unweighted Chomsky-Schützenberger Theorem. The classical Chomsky-Schützenberger Theorem is contained in the weighted result by considering the Boolean semiring $\{0, 1\}$.

In the rest of this paper we will abbreviate "context-free" by "CF".

2   Valuation Monoids and Series

We define a *unital valuation monoid* to be a tuple $(K, +, \text{val}, 0, 1)$ such that (i) $(K, +, 0)$ is a commutative monoid, (ii) $\text{val} : K^* \to K$ is a mapping such that $\text{val}(a) = a$ for each $a \in K$, (iii) $\text{val}(a_1, \ldots, a_n) = 0$ whenever $a_i = 0$ for some $1 \leq i \leq n$, and (iv) $\text{val}(a_1, \ldots, a_{i-1}, 1, a_{i+1}, \ldots, a_n) = \text{val}(a_1, \ldots, a_{i-1}, a_{i+1}, \ldots, a_n)$ for any $1 \leq i \leq n$, and (v) $\text{val}(\varepsilon) = 1$.

Note that, similarly to products where the element 1 is neutral and can be left out, val can be considered as a very general product operation in which the unit 1 is neutral as reflected by requirements (iv) and (v). The concept of *valuation monoid* was introduced in [11, 12] as a structure $(K, +, \text{val}, 0)$ with a mapping val $: K^+ \to K$ satisfying requirements (i)-(iii) correspondingly. In [11, 12, 21], also many examples of valuation monoids were given. For this paper, it will be important that the valuation monoids contain a unit 1. We will see below in Example 1 that this means no restriction of generality.

Example 1. 1. Let $(K, +, \text{val}, 0)$ be a valuation monoid and let 1 be an element not in K. We put $K' = K \cup \{1\}$ and define $(K', +', \text{val}', 0, 1)$ such that $+'$ extends $+$ by $x +' 1 = 1 +' x = 1$ for each $x \in K'$, $\text{val}'(\varepsilon) = 1$, and $\text{val}'(a_1, \ldots, a_n) = \text{val}(b_1, \ldots, b_m)$ where $b_1 \ldots b_m$ is the subsequence of $a_1, \ldots, a_n$ excluding 1's. Then $(K', +', \text{val}', 0, 1)$ is a unital valuation monoid.

2. The structure $(\mathbb{R} \cup \{-\infty\}, \sup, \text{avg}, -\infty)$ with $\text{avg}(a_1, \ldots, a_n) = \frac{1}{n} \cdot \sum_{i=1}^{n} a_i$ is a valuation monoid (with the usual laws for $-\infty$). Applying the procedure of Example 1 to it, we could add ∞ as the unit 1, disregarding ∞ when calculating averages. This leads to a unital valuation monoid $(\mathbb{R} \cup \{-\infty, \infty\}, \sup, \text{avg}, -\infty, \infty)$.

3. A *strong bimonoid* is a tuple $(K, +, \cdot, 0, 1)$ such that $(K, +, 0)$ is a commutative monoid, $(K, \cdot, 1)$ is a monoid, and $a \cdot 0 = 0 \cdot a = 0$. Then we can consider K as the particular unital valuation monoid $(K, +, \text{val}, 0, 1)$ where $\text{val}(a_1, \ldots, a_n) = a_1 \cdot a_2 \cdot a_3 \cdot \ldots \cdot a_n$.

Now we list three examples of (classes of) strong bimonoids.

(a) A strong bimonoid is a *semiring*, if the multiplication is distributive (from both sides) over addition. For a range of examples of strong bimonoids which are not semirings we refer the reader to [13].

(b) The Boolean semiring $\mathbb{B} = (\{0, 1\}, \vee, \wedge, 0, 1)$ allows us to give exact translations between unweighted and $\mathbb{B}$-weighted settings. The semiring $(\mathbb{N}, +, \cdot, 0, 1)$ of natural numbers permits counting.

(c) Each bounded lattice $(L, \vee, \wedge, 0, 1)$ (i.e. $0 \leq x \leq 1$ for each $x \in L$) is a strong bimonoid. There is a wealth of lattices [4, 17] which are not distributive, hence strong bimonoids but not semirings.

The importance of infinitary sum operations was observed early on in weighted automata theory, cf. [15]. In our context, they will arise for ambiguous CF grammars if a given word has infinitely many derivations.

A monoid $(K, +, 0)$ is *complete* [15] if it has an infinitary sum operation $\sum_I : K^I \to K$ for any index set I such that $\sum_{i \in \emptyset} a_i = 0$, $\sum_{i \in \{k\}} a_i = a_k$, $\sum_{i \in \{j,k\}} a_i = a_j + a_k$ for $j \neq k$, and $\sum_{j \in J} \left(\sum_{i \in I_j} a_i \right) = \sum_{i \in I} a_i$ if $\bigcup_{j \in J} I_j = I$ and $I_j \cap I_k = \emptyset$ for $j \neq k$. A monoid $(K, +, 0)$ is *idempotent* if $a + a = a$ for each $a \in K$, and a complete monoid is *completely idempotent* if $\sum_I a = a$ for each $a \in K$ and any index set I. We call a unital valuation monoid $(K, +, \text{val}, 0, 1)$ complete, idempotent, or completely idempotent if $(K, +, 0)$ has the respective property.

Example 2. 1. The Boolean semiring $\mathbb{B}$ and the tropical semiring $(\mathbb{N} \cup \{\infty\}, \min, +, \infty, 0)$ are complete and completely idempotent. For a wealth of further examples of complete semirings see [16, Ch.22].

2. The unital valuation monoid $(\mathbb{R} \cup \{-\infty, \infty\}, \sup, \text{avg}, -\infty, \infty)$ (cf. Example 1(2)) is complete and completely idempotent.

3. Consider the commutative monoid $(\{0, 1, \infty\}, +, 0)$ with $1 + 1 = 1$, $1 + \infty = \infty + \infty = \infty$, and $\sum_I 1 = \infty$ for any infinite index set I and corresponding natural laws for infinite sums involving the other elements. This monoid is complete and idempotent, but not completely idempotent.

Let Σ be an alphabet and K a unital valuation monoid. A *series* or *quantitative language* over Σ and K is a mapping $s : \Sigma^* \to K$. As usual, we denote $s(w)$ by (s, w). The *support* of s is the set $supp(s) = \{w \in \Sigma^* \mid (s, w) \neq 0\}$. The *image* of s is the set $im(s) = \{(s, w) \mid w \in \Sigma^*\}$. The class of all series over Σ and K is denoted by $K\langle\langle \Sigma^* \rangle\rangle$.

Let $s, s' \in K\langle\langle \Sigma^* \rangle\rangle$ be series. We define the sum $s + s'$ by letting $(s + s', w) = (s, w) + (s', w)$ for each $w \in \Sigma^*$. A family of series $(s_i \mid i \in I)$ is *locally finite* if for each $w \in \Sigma^*$ the set $I_w = \{i \in I \mid (s_i, w) \neq 0\}$ is finite. In this case or if K is complete, we define $\sum_{i \in I} s_i \in K\langle\langle \Sigma^* \rangle\rangle$ by letting $(\sum_{i \in I} s_i, w) = \sum_{i \in I_w} (s_i, w)$ for every $w \in \Sigma^*$. For $L \subseteq \Sigma^*$, we define the *characteristic series* $\mathbb{1}_L \in K\langle\langle \Sigma^* \rangle\rangle$ by $(\mathbb{1}_L, w) = 1$ if $w \in L$, and $(\mathbb{1}_L, w) = 0$ otherwise for $w \in \Sigma^*$.

In the rest of this paper, let $(K, +, \mathrm{val}, 0, 1)$ *denote an arbitrary unital valuation monoid, unless specified otherwise.*

3   Weighted Context-Free Grammars

In this section, we introduce our notion of weighted CF grammars and we present basic properties. A *CF grammar* (CFG) is a tuple $\mathcal{G} = (N, \Sigma, Z, P)$ where N is a finite set (*nonterminals*), Σ is an alphabet (*terminals*), $Z \in N$ (*initial nonterminal*), and $P \subseteq N \times (N \cup \Sigma)^*$ is a finite set (*productions*).

The leftmost derivation relation of $\mathcal{G}$ is the binary relation on the set $(N \cup \Sigma)^*$ of *sentential forms* defined as follows. For every production $\rho = (A \to \xi) \in P$ we define the binary relation $\overset{\rho}{\Rightarrow} (N \cup \Sigma)^* \times (N \cup \Sigma)^*$ such that for every $w \in \Sigma^*$ and $\zeta \in (N \cup \Sigma)^*$, we have $w A \zeta \overset{\rho}{\Rightarrow} w \xi \zeta$. The *(leftmost) derivation relation* of $\mathcal{G}$ is the binary relation $\Rightarrow = \bigcup_{\rho \in P} \overset{\rho}{\Rightarrow}$. A *derivation* of $\mathcal{G}$ is a sequence $d = \rho_1 \ldots \rho_n$ of productions $\rho_i \in P$ such that there are sentential forms $\xi_0, \ldots, \xi_n$ with $\xi_{i-1} \overset{\rho_i}{\Rightarrow} \xi_i$ for every $1 \leq i \leq n$. We abbreviate this derivation by $\xi_0 \overset{d}{\Rightarrow} \xi_n$. Let $A \in N$ and $w \in \Sigma^*$. An *A-derivation of w* is a derivation d such that $A \overset{d}{\Rightarrow} w$. We let $D(A, w)$ denote the set of all A-derivations of w. And we let $D(w)$ denote the set $D(Z, w)$ of all *derivations of w*. The *language generated by* $\mathcal{G}$ is the set $L(\mathcal{G}) = \{w \in \Sigma^* \mid D(w) \neq \emptyset\}$.

We say that $\mathcal{G}$ is *ambiguous* if there is a $w \in L(\mathcal{G})$ such that $|D(w)| \geq 2$; otherwise $\mathcal{G}$ is *unambiguous*. A CF language L is *inherently ambiguous* if every CFG $\mathcal{G}$ with $L = L(\mathcal{G})$ is ambiguous.

Next let K be a unital valuation monoid. A *CF grammar with weights in K* is a tuple $\mathcal{G} = (N, \Sigma, Z, P, \mathrm{wt})$ where (N, Σ, Z, P) is a CFG and $\mathrm{wt} : P \to K$ is a mapping (*weight assignment*). We say that $\mathcal{G}$ is unambiguous if the underlying CFG is unambiguous.

The *weight* of a derivation $d = \rho_1 \ldots \rho_n$ is the element in K defined by

$$\mathrm{wt}(d) = \mathrm{val}(\mathrm{wt}(\rho_1), \ldots, \mathrm{wt}(\rho_n)) \ .$$

We say that $\mathcal{G}$ is a *weighted CF grammar* (WCFG) if $D(w)$ is finite for every $w \in \Sigma^*$ or if K is complete. In this case we define the *quantitative language* of $\mathcal{G}$ to be the series $\|\mathcal{G}\| \in K\langle\langle \Sigma^* \rangle\rangle$ given for every $w \in \Sigma^*$ by

$$(\|\mathcal{G}\|, w) = \sum_{d \in D(w)} \mathrm{wt}(d) \; .$$

Note that this sum exists by our assumptions on a WCFG. A series $s \in K\langle\!\langle \Sigma^* \rangle\!\rangle$ is a *quantitative CF language* if there is a WCFG $\mathcal{G}$ such that $s = \|\mathcal{G}\|$. The class of all quantitative CF languages over Σ and K is denoted by $\mathrm{CF}(\Sigma, K)$. Moreover, we let $\mathrm{uCF}(\Sigma, K)$ comprise all series $\|\mathcal{G}\|$ where $\mathcal{G}$ is an unambiguous WCFG. We say that two WCFG are *equivalent*, if they have the same quantitative language.

Clearly, any CFG $\mathcal{G}$ can be transformed into a WCFG over the Boolean semiring $\mathbb{B}$ by adding the weight assignment $\mathrm{wt}\colon P \to \mathbb{B}$ such that $\mathrm{wt}(\rho) = 1$ for each $\rho \in P$. Then for each $w \in \Sigma^*$ we have $w \in L(\mathcal{G})$ if and only if $(\|\mathcal{G}\|, w) = 1$, i.e., $\|\mathcal{G}\| = \mathbb{1}_{L(\mathcal{G})}$. Consequently, a language L is CF if and only if $\mathbb{1}_L \in \mathrm{CF}(\Sigma, \mathbb{B})$. This shows that WCFG form a generalization of CFG.

A WCFG $\mathcal{G}$ is in *head normal form* if every production has the form $A \to xB_1 \ldots B_k$ where $x \in \Sigma \cup \{\varepsilon\}$, $k \geq 0$, and $A, B_1, \ldots, B_k \in N$. By standard construction we obtain that, for every (unambiguous) WCFG there is an equivalent (unambiguous) WCFG in head normal form.

Example 3. We consider the set of all arithmetic expressions over addition, multiplication, and the variable x. Assuming that the calculation of the addition (and multiplication) of two values needs $n \in \mathbb{N}$ (resp., $m \in \mathbb{N}$) machine clock cycles, we might wish to know the average number of clock cycles the machine needs to calculate any of the operations occurring in an expression.

For this we consider the unital valuation monoid $(\mathbb{R} \cup \{-\infty, \infty\}, \sup, \mathrm{avg}, -\infty, \infty)$ as in Example 1(2) and the WCFG $\mathcal{G} = (N, \Sigma, E, P, \mathrm{wt})$ with the following set of productions

$$\rho_1 : E \to (E + E), \quad \rho_2 : E \to (E * E), \quad \rho_3 : E \to x$$

and $\mathrm{wt}(\rho_1) = n$, $\mathrm{wt}(\rho_2) = m$, $\mathrm{wt}(\rho_3) = 1$. For the expression $w = ((x*x)+(x*x))$, we have that $D(w) = \{d\}$ with $d = \rho_1 d'd'$ and $d' = \rho_2\rho_3\rho_3$. In fact, $\mathcal{G}$ is unambiguous. Then

$$(\|\mathcal{G}\|, w) = \mathrm{wt}(d) = \mathrm{val}(n, \underbrace{m, 1, 1}_{d'}, \underbrace{m, 1, 1}_{d'}) = \mathrm{avg}(n, m, m) = \frac{n + 2 \cdot m}{3} \; .$$

4 Weighted Pushdown Automata

In this section, we introduce our notion of weighted pushdown automata, and we derive a few basic properties. First let us fix our notation for pushdown automata. A *pushdown automaton* (PDA) over Σ is a tuple $\mathcal{M} = (Q, \Sigma, \Gamma, q_0, \gamma_0, F, T)$ where Q is a finite set (*states*), Σ is an alphabet (*input symbols*), Γ is an alphabet (*pushdown symbols*), $q_0 \in Q$ (*initial state*), $\gamma_0 \in \Gamma$ (*initial pushdown symbol*), $F \subseteq Q$ (*final states*), and $T \subseteq Q \times (\Sigma \cup \{\varepsilon\}) \times \Gamma \times Q \times \Gamma^*$ is a finite set (*transitions*). For a transition (q, x, γ, p, π), we call q, x, and p its *source state*, *label*, and *target state*, respectively.

The computation relation of $\mathcal{M}$ is the binary relation on the set $Q \times \Sigma^* \times \Gamma^*$ of *configurations* defined as follows. For every transition $\tau = (q, x, \gamma, p, \pi) \in T$ we define the binary relation $\overset{\tau}{\vdash} \subseteq (Q \times \Sigma^* \times \Gamma^*) \times (Q \times \Sigma^* \times \Gamma^*)$ such that for every $w \in \Sigma^*$ and $\mu \in \Gamma^*$, we have $(q, xw, \gamma\mu) \overset{\tau}{\vdash} (p, w, \pi\mu)$. The *computation relation of* $\mathcal{M}$ is the binary relation $\vdash = \bigcup_{\tau \in T} \overset{\tau}{\vdash}$. A *computation* is a sequence $\theta = \tau_1 \ldots \tau_n$ of transitions τ_i such that there are configurations $c_0, \ldots, c_n$ with $c_{i-1} \overset{\tau_i}{\vdash} c_i$ for every $1 \leq i \leq n$. We abbreviate this computation by $c_0 \overset{\theta}{\vdash} c_n$. The *label* of a computation $\tau_1 \ldots \tau_n$ is the sequence of labels of the involved transitions. Let $w \in \Sigma^*$ and $q \in Q$. A *q-computation on* w is a computation θ such that $(q, w, \gamma_0) \overset{\theta}{\vdash} (p, \varepsilon, \varepsilon)$ for some $p \in F$. We let $\Theta(q, w)$ denote the set of all q-computations on w, and we let $\Theta(w) = \Theta(q_0, w)$. Moreover, we let $\Theta = \bigcup_{w \in \Sigma^*} \Theta(w)$. The *language recognized by* $\mathcal{M}$ is the set $L(\mathcal{M}) = \{w \in \Sigma^* \mid \Theta(w) \neq \emptyset\}$. That means, we consider acceptance of words by final state and empty pushdown.

Let $\mathcal{M}$ be any PDA. We say that $\mathcal{M}$ is *ambiguous* if there is a $w \in L(\mathcal{M})$ such that $|\Theta(w)| \geq 2$; otherwise $\mathcal{M}$ is *unambiguous*.

Next let K be a unital valuation monoid. A *pushdown automaton with weights in* K is a tuple $\mathcal{M} = (Q, \Sigma, \Gamma, q_0, \gamma_0, F, T, \mathrm{wt})$ where $(Q, \Sigma, \Gamma, q_0, \gamma_0, F, T)$ is a PDA and $\mathrm{wt} \colon T \to K$ is a mapping (*weight assignment*). We say that $\mathcal{M}$ is unambiguous if the underlying PDA is unambiguous.

The *weight* of a computation $\theta = \tau_1 \ldots \tau_n$ is the element in K defined by

$$\mathrm{wt}(\theta) = \mathrm{val}(\mathrm{wt}(\tau_1), \ldots, \mathrm{wt}(\tau_n)) \ .$$

We say that $\mathcal{M}$ is a *weighted pushdown automaton* (WPDA) if $\Theta(w)$ is finite for every $w \in \Sigma^*$ or if K is complete. In this case we define the *quantitative behavior* of $\mathcal{M}$ to be the series $\|\mathcal{M}\| \in K\langle\!\langle \Sigma^* \rangle\!\rangle$ given for every $w \in \Sigma^*$ by

$$(\|\mathcal{M}\|, w) = \sum_{\theta \in \Theta(w)} \mathrm{wt}(\theta) \ .$$

The class of quantitative behaviors of all WPDA over Σ and K is denoted by $\mathrm{PDA}(\Sigma, K)$. Moreover, we let $\mathrm{uPDA}(\Sigma, K)$ comprise all series $\|\mathcal{M}\|$ where $\mathcal{M}$ is an unambiguous WPDA. We say that two WPDA are *equivalent* if they have the same quantitative behavior.

Clearly, any PDA $\mathcal{M}$ can be transformed into a WPDA over the Boolean semiring $\mathbb{B}$ by adding the weight assignment $\mathrm{wt} \colon T \to \mathbb{B}$ such that $\mathrm{wt}(\tau) = 1$ for each $\tau \in T$. Then for each $w \in \Sigma^*$ we have $w \in L(\mathcal{M})$ if and only if $(\|\mathcal{M}\|, w) = 1$, i.e., $\|\mathcal{M}\| = \mathbb{1}_{L(\mathcal{M})}$. Consequently, a language L is recognized by a PDA if and only if $\mathbb{1}_L \in \mathrm{PDA}(\Sigma, \mathbb{B})$. This shows that WPDA form a generalization of PDA.

A WPDA $\mathcal{M} = (Q, \Sigma, \Gamma, q_0, \gamma_0, F, T, \mathrm{wt})$ is *state normalized* if there is no transition in T with q_0 as target state, F is a singleton, say, $F = \{q_f\}$, and there is no transition in T with q_f as source state. By a standard construction we obtain that, for every (unambiguous) WPDA there is an equivalent state normalized (unambiguous) WPDA.

Again by a standard construction [19, Lecture 25] we obtain that for every (unambiguous) WPDA there is an equivalent (unambiguous) WPDA with just one state.

Lemma 1. *Let* $s_1, s_2 \in \mathrm{PDA}(\Sigma, K)$. *Then* $s_1 + s_2 \in \mathrm{PDA}(\Sigma, K)$.

We mention that in [8] pushdown games with quantitative objectives were investigated. Such games are formalized on the basis of paths through pushdown systems where the latter are particular pushdown automata with weights: the input alphabet Σ is a singleton and no ε-transition occurs. Moreover, as weight structure, pushdown systems employ the set of integers with mean-payoff. Roughly, the mean-payoff of a computation is the average of its transition weights (taking the limit superior of the averages of finite prefixes on infinite computations). Then in [8] game-theoretic problems on the set of all paths for which the mean-payoff is above a given threshold are investigated.

Finally, we note that weighted pushdown systems over bounded idempotent semirings were used in interprocedural dataflow analysis [24].

5   Equivalence of WCFG and WPDA

A classical result says that a language L is CF iff L is accepted by a pushdown automaton. This was extended to algebraic series and weighted pushdown automata with weights taken in semirings in [20, Cor. 14.16]. The goal of this small section is to prove the generalization to arbitrary unital valuation monoids.

For this we use the following concept. Let $\mathcal{M} = (\{*\}, \Sigma, \Gamma, *, \gamma_0, \{*\}, T, \mathrm{wt}_{\mathcal{M}})$ be a WPDA over K with one state and $\mathcal{G} = (N, \Sigma, Z, P, \mathrm{wt}_{\mathcal{G}})$ be a WCFG over K in head normal form. We say that $\mathcal{M}$ and $\mathcal{G}$ are *related* if $\Gamma = N$, $\gamma_0 = Z$, $\tau = (*, x, A, *, B_1 B_2 \ldots B_n) \in T$ iff $\rho = (A \to x B_1 B_2 \ldots B_n)$ is in P; $\mathrm{wt}_{\mathcal{M}}(\tau) = \mathrm{wt}_{\mathcal{G}}(\rho)$ if τ and ρ correspond to each other as above. Then the following lemma is easy to see (cf. e.g. [19, Lecture 25]).

Lemma 2. *Let* $\mathcal{M}$ *be a WPDA with one state and* $\mathcal{G}$ *be a WCFG in head normal form. If* $\mathcal{M}$ *and* $\mathcal{G}$ *are related, then* $\|\mathcal{M}\| = \|\mathcal{G}\|$. *Moreover,* $\mathcal{M}$ *is unambiguous iff* $\mathcal{G}$ *is unambiguous.*

The previous lemma and the normal forms of WPDA and WCFG imply the following theorem.

Theorem 1. *For every alphabet* Σ *and unital valuation monoid* K *we have* $\mathrm{PDA}(\Sigma, K) - \mathrm{CF}(\Sigma, K)$ *and* $\mathrm{uPDA}(\Sigma, K) = \mathrm{uCF}(\Sigma, K)$.

6   Theorem of Chomsky-Schützenberger

In this section let K again be a unital valuation monoid. The goal of this section will be to prove a quantitative version of the Chomsky-Schützenberger Theorem. Recently, in [18] the Chomsky-Schützenberger Theorem has been used as a pattern for a parsing algorithm of probabilistic context-free languages.

Let Y be an alphabet. Then we let $\overline{Y} = \{\overline{y} \mid y \in Y\}$. The *Dyck language over* Y, denoted by D_Y, is the language which is generated by the CFG $\mathcal{G}_Y = (N, Y \cup \overline{Y}, Z, P)$ with $N = \{Z\}$ and the rules $Z \to yZ\overline{y}$ for any $y \in Y$, $Z \to ZZ$, and $Z \to \varepsilon$.

Next we introduce monomes and alphabetic morphisms. A series $s \in K\langle\!\langle \Sigma^* \rangle\!\rangle$ is called a *monome* if $supp(s)$ is empty or a singleton. If $supp(s) = \{w\}$, then we also write $s = (s, w).w$. We let $K[\Sigma \cup \{\varepsilon\}]$ denote the set of all monomes with support in $\Sigma \cup \{\varepsilon\}$.

Let Δ be an alphabet and $h : \Delta \to K[\Sigma \cup \{\varepsilon\}]$ be a mapping. The *alphabetic morphism induced by* h is the mapping $h' : \Delta^* \to K\langle\!\langle \Sigma^* \rangle\!\rangle$ such that for every $n \geq 0$, $\delta_1, \ldots, \delta_n \in \Delta$ with $h(\delta_i) = a_i.y_i$ we have

$$h'(\delta_1 \ldots \delta_n) = val(a_1, \ldots, a_n).y_1 \ldots y_n .$$

Note that $h'(v)$ is a monome for every $v \in \Delta^*$, and $h'(\varepsilon) = 1.\varepsilon$. If $L \subseteq \Delta^*$ such that the family $(h'(v) \mid v \in L)$ is locally finite or if K is complete, we let $h'(L) = \sum_{v \in L} h'(v)$. In the sequel we identify h' and h.

We also call a mapping $h : \Delta \to \Sigma \cup \{\varepsilon\}$ and its unique extension to a morphism from Δ^* to Σ^* an *alphabetic morphism*. In this case, if $r \in K\langle\!\langle \Delta^* \rangle\!\rangle$ is such that $\{v \in h^{-1}(w) \mid (r, v) \neq 0\}$ is finite for each $w \in \Sigma^*$, or if K is complete, we define $h(r) \in K\langle\!\langle \Sigma^* \rangle\!\rangle$ by letting $(h(r), w) = \sum_{v \in \Delta^*, h(v) = w} (r, v)$.

Next we introduce the intersection of a series with a language as follows. Let $s \in K\langle\!\langle \Sigma^* \rangle\!\rangle$ and $L \subseteq \Sigma^*$. We define the series $s \cap L \in K\langle\!\langle \Sigma^* \rangle\!\rangle$ by letting $(s \cap L, w) = (s, w)$ if $w \in L$, and $(s \cap L, w) = 0$ otherwise.

Finally, a *weighted finite automaton* over K and Σ (for short: WFA) is a tuple $\mathcal{A} = (Q, q_0, F, T, wt)$ where Q is a finite set (states), $q_0 \in Q$ (initial state), $F \subseteq Q$ (final states), $T \subseteq Q \times \Sigma \times Q$ (transitions), and $wt : T \to K$ (transition weight function). We call $\mathcal{A}$ *deterministic* if for every $q \in Q$ and $\sigma \in \Sigma$, there is at most one $p \in Q$ with $(q, \sigma, p) \in T$.

If $w = \sigma_1 \ldots \sigma_n \in \Sigma^*$ where $n \geq 0$ and $\sigma_i \in \Sigma$, a *path* P over w is a sequence $P = (q_0, \sigma_1, q_1) \ldots (q_{n-1}, \sigma_n, q_n) \in T^*$. The path P is *successful* if $q_n \in F$. The *weight of* P is the value

$$wt(P) = val(wt((q_0, \sigma_1, q_1)), \ldots, wt((q_{n-1}, \sigma_n, q_n))) .$$

The *behavior* of $\mathcal{A}$ is the series $\|\mathcal{A}\| \in K\langle\!\langle \Sigma^* \rangle\!\rangle$ such that for every $w \in \Sigma^*$, $(\|\mathcal{A}\|, w) = \sum_{P \text{ succ. path over } w} wt(P)$. A series $s \in K\langle\!\langle \Sigma^* \rangle\!\rangle$ is called *deterministically recognizable* if $s = \|\mathcal{A}\|$ for some deterministic WFA $\mathcal{A}$.

Our main result will be:

Theorem 2. *Let* K *be a unital valuation monoid and* $s \in K\langle\!\langle \Sigma^* \rangle\!\rangle$ *be a series. Then the following four statements are equivalent.*

1. $s \in CF(\Sigma, K)$.
2. *There are an alphabet* Y, *a recognizable language* R *over* $Y \cup \overline{Y}$, *and an alphabetic morphism* $h : Y \cup \overline{Y} \to K[\Sigma \cup \{\varepsilon\}]$ *such that* $s = h(D_Y \cap R)$.

3. There are an alphabet Δ, an unambiguous CFG $\mathcal{G}$ over Δ, and an alphabetic morphism $h : \Delta \to K[\Sigma \cup \{\varepsilon\}]$ such that $s = h(L(\mathcal{G}))$.

4. There are an alphabet Y, a deterministically recognizable series $r \in K\langle\!\langle (Y \cup \overline{Y})^* \rangle\!\rangle$, and an alphabetic morphism $h : Y \cup \overline{Y} \to \Sigma \cup \{\varepsilon\}$ such that $s = h(r \cap D_Y)$.

Moreover, if K is complete and completely idempotent, then 1-4 are also equivalent to:

5. There are an alphabet Δ, a context-free language L over Δ, and an alphabetic morphism $h : \Delta \to K[\Sigma \cup \{\varepsilon\}]$ such that $s = h(L)$.

We split the proof of Theorem 2 into several parts. The following lemma proves the implication $1 \Rightarrow 3$ of Theorem 2.

Lemma 3. *Let $s \in \mathrm{CF}(\Sigma, K)$. Then there are an alphabet Δ, an unambiguous CFG $\mathcal{G}$ over Δ, and an alphabetic morphism $h : \Delta \to K[\Sigma \cup \{\varepsilon\}]$ such that $s = h(L(\mathcal{G}))$.*

Proof. We can assume that $s = \|\mathcal{H}\|$ for some WCFG $\mathcal{H} = (N, \Sigma, Z, P, \mathrm{wt})$ in head normal form. We let $\Delta = P$, and we construct the CFG $\mathcal{G} = (N, P, Z, P')$ and the mapping $h : P \to K[\Sigma \cup \{\varepsilon\}]$ such that, if $\rho = (A \to xB_1 \ldots B_k)$ is in P, then $A \to \rho B_1 \ldots B_k$ is in P' and we define $h(\rho) = \mathrm{wt}(\rho).x$. Obviously, $\mathcal{G}$ is unambiguous. By definition of h, we have that $h(d) = \mathrm{val}(\mathrm{wt}(\rho_1), \ldots, \mathrm{wt}(\rho_n)).w$ for every $w \in \Sigma^*$ and $d = \rho_1 \ldots \rho_n \in D_{\mathcal{H}}(w)$. Hence $\mathrm{wt}(d) = (h(d), w)$. Also $(h(d) \mid d \in L(\mathcal{G}))$ is locally finite if K is not complete. Then for every $w \in \Sigma^*$ we have $(\|\mathcal{H}\|, w) = \sum_{d \in D_{\mathcal{H}}(w)} \mathrm{wt}(d) = \sum_{d \in D_{\mathcal{H}}(w)} (h(d), w) = \sum_{d \in L(\mathcal{G})} (h(d), w) = \left(\sum_{d \in L(\mathcal{G})} h(d), \, w \right) = (h(L(\mathcal{G})), w)$. Thus $s = h(L(\mathcal{G}))$.

Lemma 4. *Let L be a CF language over Δ and $h : \Delta \to K[\Sigma\cup\{\varepsilon\}]$ an alphabetic morphism such that $(h(v) \mid v \in L)$ is locally finite in case K is not complete. If L can be generated by some unambiguous CFG or if K is complete and completely idempotent, then $h(L) \in \mathrm{CF}(\Sigma, K)$.*

Proof. Let $\mathcal{M} = (Q, \Delta, \Gamma, q_0, \gamma_0, F, T)$ be a PDA with $L(\mathcal{M}) = L$. Moreover, by Theorem 1, if $L = L(\mathcal{G})$ for some unambiguous CFG $\mathcal{G}$, then we can assume that $\mathcal{M}$ is unambiguous. Let $\overline{\delta} \in \Delta$ be an arbitrary, but fixed element.

The following construction employs the same technique as in [14, Lemma 5.7] of coding the preimage of h into the state set; thereby non-injectivity of h is handled appropriately. We construct the PDA with weights $\mathcal{M}' = (Q', \Sigma, \Gamma, q_0', \gamma_0, F', T', \mathrm{wt})$ where $Q' = \{q_0'\} \cup Q \times (\Delta \cup \{\varepsilon\})$ for some element q_0' with $q_0' \notin Q \times (\Delta \cup \{\varepsilon\})$, $F' = F \times \{\overline{\delta}\}$, and T' and wt are defined as follows.

– For every $x \in \Delta \cup \{\varepsilon\}$, the rule $\tau = (q_0', \varepsilon, \gamma_0, (q_0, x), \gamma_0)$ is in T' and $\mathrm{wt}(\tau) = 1$.

– Let $\tau = (q, x, \gamma, p, \pi) \in T$ and $x' \in \Delta \cup \{\varepsilon\}$.

- If $x \in \Delta$ and $h(x) = a.y$, then $\tau' = ((q, x), y, \gamma, (p, x'), \pi) \in T'$ and $\mathrm{wt}(\tau') = a$.
- If $x = \varepsilon$, then $\tau' = ((q, \varepsilon), \varepsilon, \gamma, (p, x'), \pi) \in T'$ and $\mathrm{wt}(\tau') = 1$.

Let $w \in \Sigma^*$. First, let $v \in \Delta^*$ with $h(v) = z.w$ for some $z \in K$. We write $v = \delta_1 \ldots \delta_n \in \Delta^*$ with $n \geq 0$ and $\delta_i \in \Delta$. Let $h(\delta_i) = a_i.y_i$ for every $1 \leq i \leq n$. Thus $h(v) = \mathrm{val}(a_1, \ldots, a_n).y_1 \ldots y_n$ and $w = y_1 \ldots y_n$.

Let $\theta = \tau_1 \ldots \tau_m$ be a q_0-computation in $\Theta_{\mathcal{M}}(v)$; note that $m \geq \max\{n, 1\}$ because at least γ_0 has to be popped. Let x_i be the second component of τ_i, so, $x_i \in \Delta \cup \{\varepsilon\}$, and $v = x_1 \ldots x_m$. Then we construct the q_0'-computation $\theta' = \tau_0' \tau_1' \ldots \tau_m'$ in $\Theta_{\mathcal{M}'}(y_1 \ldots y_n)$ as follows:

- $\tau_0' = (q_0', \varepsilon, \gamma_0, (q_0, x_1), \gamma_0)$.
- If $1 \leq i \leq m$ and $\tau_i = (q, x_i, \gamma, p, \pi)$, then $\tau_i' = ((q, x_i), y', \gamma, (p, x_{i+1}), \pi)$ where $y' = y$ if $x_i \in \Delta$ and $h(x_i) = a.y$, and $y' = \varepsilon$ if $x_i = \varepsilon$, and $x_{m+1} = \overline{\delta}$.

Note that if $x_i \in \Delta$ and $h(x_i) = a.y$, then $\mathrm{wt}(\tau_i') = a$, and if $x_i = \varepsilon$, then $\mathrm{wt}(\tau_i') = 1$ for each $1 \leq i \leq m$, by definition of wt. Consequently

$$(h(v), w) = \mathrm{val}(a_1, \ldots, a_n) = \mathrm{val}(\mathrm{wt}(\tau_0'), \mathrm{wt}(\tau_1'), \ldots, \mathrm{wt}(\tau_m')) = \mathrm{wt}(\theta').$$

In particular, $\mathrm{wt}(\theta_1') = (h(v), w) = \mathrm{wt}(\theta_2')$ for every $\theta_1, \theta_2 \in \Theta_{\mathcal{M}}(v)$.

Conversely, for every q_0'-computation $\theta' = \tau_0' \tau_1' \ldots \tau_m'$ in $\Theta_{\mathcal{M}'}(w)$ by definition of T' there are a uniquely determined $v \in \Delta^*$ and a uniquely determined q_0-computation $\theta = \tau_1 \ldots \tau_m$ in $\Theta_{\mathcal{M}}(v)$ such that θ' is the computation constructed above. Since $\mathcal{M}$ is unambiguous or K is complete, it follows that $\mathcal{M}'$ is a WPDA.

So, for every $w \in \Sigma^*$ we obtain

$$(h(L(\mathcal{M})), w) = \left(\sum_{v \in L(\mathcal{M})} h(v), w\right) = \sum_{\substack{v \in L(\mathcal{M}): \\ (h(v), w) \neq 0}} (h(v), w)$$

$$=^* \sum_{\substack{v \in L(\mathcal{M}), \theta \in \Theta_{\mathcal{M}}(v): \\ (h(v), w) \neq 0}} \mathrm{wt}(\theta') = \sum_{\theta' \in \Theta_{\mathcal{M}'}(w)} \mathrm{wt}(\theta') = (\|\mathcal{M}'\|, w)$$

where the $*$-marked equality holds because (1) K is complete and completely idempotent or (2) $\mathcal{M}$ is unambiguous. Thus $\|\mathcal{M}'\| = h(L(\mathcal{M}))$ and the result follows from Theorem 1.

Proof of Theorem 2: $1 \Leftrightarrow 3$: By Lemmas 3 and 4. $2 \Rightarrow 3$: There is an unambiguous CFG $\mathcal{G}$ with $L(\mathcal{G}) = D_Y \cap R$, cf. [9, Prop.1, p.145] and [2, Lm.4.1].

$3 \Rightarrow 2$: By the classical result of Chomsky-Schützenberger (cf. e.g. [19, Thm.G1]) there are an alphabet Y, a recognizable language R over $Y \cup \overline{Y}$, and an alphabetic morphism $g : Y \cup \overline{Y} \to \Delta \cup \{\varepsilon\}$ such that $L(\mathcal{G}) = g(D_Y \cap R)$. By analysis of the construction, we have that the set $g^{-1}(v) \cap D_Y \cap R$ is in a one-to-one correspondence with $D_{\mathcal{G}}(v)$, for every $v \in L(\mathcal{G})$. Since $\mathcal{G}$ is unambiguous, we have that $|g^{-1}(v) \cap D_Y \cap R| = 1$. It follows that $(h \circ g(v') \mid v' \in D_Y \cap R)$ is locally finite. Thus $h \circ g : Y \cup \overline{Y} \to K[\Sigma \cup \{\varepsilon\}]$ is an alphabetic morphism, $(h \circ g)(D_Y \cap R)$ is well defined, and $s = (h \circ g)(D_Y \cap R)$.

$2 \Rightarrow 4$: Let $\widetilde{Y} = Y \cup \overline{Y}$. Recall that $h(v) \in K\langle\langle \Sigma^* \rangle\rangle$ is a monome for every $v \in \widetilde{Y}^*$. We define $h' : \widetilde{Y}^* \to \Sigma^*$ by letting $h'(v) = w$ if $h(v) = a.w$. Clearly, h' is a morphism. Choose a deterministic finite automaton $\mathcal{A}' = (Q, q_0, F, T)$ over $\widetilde{Y}$ recognizing R. We define a deterministic WFA $\mathcal{A} = (Q, q_0, F, T, \mathrm{wt})$ over $\widetilde{Y}$

by putting $\mathrm{wt}(t) = a$ if $t = (q, z, p)$ and $h(z) = a.x$. Let $r = \|\mathcal{A}\| \in K\langle\langle \widetilde{Y}^* \rangle\rangle$. Note that $(r, v) = (h(v), w)$ if $v \in R$ and $h'(v) = w$, and $(r, v) = 0$ otherwise.

By assumption $s = h(D_Y \cap R) = \sum_{v \in D_Y \cap R} h(v)$. Hence, for $w \in \Sigma^*$: $(s, w) = \sum_{v \in D_Y \cap R}(h(v), w) = \sum_{v \in D_Y, h'(v) = w}(r, v) = (h'(r \cap D_Y), w)$ where the sums exist because they have only finitely many nonzero entries if K is not complete. Thus $s = h'(r \cap D_Y)$.

$4 \Rightarrow 3$: We put $\widetilde{Y} = Y \cup \overline{Y}$. Also, let $\widetilde{Y}_0 = \widetilde{Y} \cup \{\gamma_0\}$ with an element $\gamma_0 \notin \widetilde{Y}$. By assumption, there is a deterministic WFA $\mathcal{A} = (Q, q_0, F, T, \mathrm{wt})$ over $\widetilde{Y}$ with $\|\mathcal{A}\| = r$. We let $\mathcal{A}' = (Q, q_0, F, T)$, a deterministic finite automaton over $\widetilde{Y}$.

Next, we wish to define a PDA $\mathcal{M}$ over $\widetilde{Y}$ recognizing $L(\mathcal{A}') \cap D_Y$. Let $\mathcal{M} = (Q, \widetilde{Y}, \widetilde{Y}_0, q_0, \gamma_0, F, T')$ such that $(q, \varepsilon, \gamma_0, q, \varepsilon) \in T'$ for each $q \in F$, and for every $x \in \widetilde{Y}$ and $\gamma \in \widetilde{Y}_0$, $(q, x, \gamma, p, \pi) \in T'$ iff $(q, x, p) \in T$ and ($\pi = x\gamma$ if $x \in Y$, and $\pi = \varepsilon$ if $\gamma \in Y$ and $x = \overline{\gamma}$). Since $\mathcal{A}'$ is deterministic, $\mathcal{M}$ is an unambiguous PDA and $L(\mathcal{M}) = L(\mathcal{A}') \cap D_Y$.

Next, we extend $\mathcal{M}$ to a PDA $\mathcal{M}_T = (Q, T, \widetilde{Y}_0, q_0, \gamma_0, F, \overline{T})$ by letting $(q, t, \gamma, p, \pi) \in \overline{T}$ iff $(q, x, \gamma, p, \pi) \in T'$ and either $t = (q, x, p) \in T$ or $t = x = \varepsilon$.

Clearly, $\mathcal{M}_T$ is unambiguous. Moreover, since $\mathcal{A}'$ is deterministic, for each $v \in L(\mathcal{M}) \subseteq L(\mathcal{A}')$ there is a unique successful path $p_v \in T^*$ on v in $\mathcal{A}'$. Then $p_v \in L(\mathcal{M}_T)$. Conversely, each $v' \in L(\mathcal{M}_T)$ arises as $v' = p_v$ for a uniquely determined word $v \in L(\mathcal{M})$ in this way.

We let $\mathrm{lab}: T^* \to \widetilde{Y}^*$ be the alphabetic morphism mapping each transition to its label, i.e., $\mathrm{lab}(q, x, p) = x$. Finally we define an alphabetic morphism $h_K: T \to K[\Sigma \cup \{\varepsilon\}]$ by letting $h_K(t) = \mathrm{wt}(t).h(\mathrm{lab}(t))$.

We claim that $h_K(L(\mathcal{M}_T)) = h(r \cap D_Y)$. Let $w \in \Sigma^*$. Note that if $v \in L(\mathcal{M})$ and $v' = p_v$ as above, then $\mathrm{lab}(v') = v$ and $h_K(v') = \mathrm{wt}(v').h(v)$. Since $v' = p_v$ is the unique successful path in $\mathcal{A}$ on v, we obtain $\mathrm{wt}(v') = (\|\mathcal{A}\|, v)$. Moreover, $(h_K(v'), w) \neq 0$ implies $w = h(v)$. Also, $(\|\mathcal{A}\|, v) = 0$ if $v \notin L(\mathcal{A}')$. Hence:

$$(h_K(L(\mathcal{M}_T)), w) = \sum_{v' \in L(\mathcal{M}_T)}(h_K(v'), w) = \sum_{\substack{v \in L(\mathcal{M}) \\ h(v) = w}} \mathrm{wt}(v')$$

$$= \sum_{\substack{v \in L(\mathcal{A}') \cap D_Y \\ h(v) = w}}(\|\mathcal{A}\|, v) = \sum_{\substack{v \in D_Y \\ h(v) = w}}(\|\mathcal{A}\|, v) = \sum_{\substack{v \in \widetilde{Y}^* \\ h(v) = w}}(r \cap D_Y, v)$$

$$= (h(r \cap D_Y), w).$$

$3 \Rightarrow 5$: trivial. $5 \Rightarrow 1$: by Lemma 4. $\square$

References

1. Autebert, J., Berstel, J., Boasson, L.: Context-free languages and pushdown automata. In: Rozenberg, G., Salomaa, A. (eds.) Handbook of Formal Languages. Word, Language, Grammar, vol. 1, pp. 111–174. Springer (1997)
2. Bar–Hillel, Y., Perles, M., Shamir, E.: On formal properties of simple phrase structure grammars. Z. Phonetik. Sprach. Komm. 14, 143–172 (1961)
3. Berstel, J., Reutenauer, C.: Rational Series and Their Languages. EATCS Monographs on Theoretical Computer Science, vol. 12. Springer (1988)
4. Birkhoff, G.: Lattice Theory. AMS (1967)

214   M. Droste and H. Vogler

5. Chatterjee, K., Doyen, L., Henzinger, T.: Quantitative languages. ACM Transactions on Computational Logic 11(4), Article 23 (2010)
6. Chatterjee, K., Doyen, L., Henzinger, T.: Expressiveness and closure properties for quantitative languages. In: LICS 2009, pp. 199–208. IEEE Comp. Soc. (2009)
7. Chatterjee, K., Doyen, L., Henzinger, T.A.: Probabilistic weighted automata. In: Bravetti, M., Zavattaro, G. (eds.) CONCUR 2009. LNCS, vol. 5710, pp. 244–258. Springer, Heidelberg (2009)
8. Chatterjee, K., Velner, Y.: Mean-payoff pushdown games. In: 27th Annual ACM/IEEE Symposium on Logic in Computer Science, pp. 195–204 (2012)
9. Chomsky, N., Schützenberger, M.P.: The algebraic theory of context-free languages. In: Computer Programming and Formal Systems, pp. 118–161. North-Holland (1963)
10. Droste, M., Gastin, P.: Weighted automata and weighted logics. Theor. Comput. Sci. 380(1-2), 69–86 (2007)
11. Droste, M., Meinecke, I.: Describing average- and longtime-behavior by weighted MSO logics. In: Hliněný, P., Kučera, A. (eds.) MFCS 2010. LNCS, vol. 6281, pp. 537–548. Springer, Heidelberg (2010)
12. Droste, M., Meinecke, I.: Weighted automata and regular expressions over valuation monoid. Intern. J. of Foundations of Comp. Science 22, 1829–1844 (2011)
13. Droste, M., Stüber, T., Vogler, H.: Weighted finite automata over strong bimonoids. Information Sciences 180, 156–166 (2010)
14. Droste, M., Vogler, H.: Weighted automata and multi-valued logics over arbitrary bounded lattices. Theoretical Computer Science 418, 14–36 (2012)
15. Eilenberg, S.: Automata, Languages, and Machines – Volume A. Pure and Applied Mathematics, vol. 59. Academic Press (1974)
16. Golan, J.S.: Semirings and their Applications. Kluwer Acad. Publ. (1999)
17. Grätzer, G.: General Lattice Theory. Birkhäuser, Basel (2003)
18. Hulden, M.: Parsing CFGs and PCFGs with a Chomsky-Schützenberger representation. In: Vetulani, Z. (ed.) LTC 2009. LNCS, vol. 6562, pp. 151–160. Springer, Heidelberg (2011)
19. Kozen, D.: Automata and Computability. Springer (1997)
20. Kuich, W., Salomaa, A.: Semirings, Automata, Languages. Monogr. Theoret. Comput. Sci. EATCS Ser., vol. 5. Springer (1986)
21. Meinecke, I.: Valuations of weighted automata: Doing it in a rational way. In: Kuich, W., Rahonis, G. (eds.) Algebraic Foundations in Computer Science. LNCS, vol. 7020, pp. 309–346. Springer, Heidelberg (2011)
22. Okhotin, A.: Non-erasing variants of the Chomsky-Schützenberger theorem. In: Yen, H.-C., Ibarra, O.H. (eds.) DLT 2012. LNCS, vol. 7410, pp. 121–129. Springer, Heidelberg (2012)
23. Petre, I., Salomaa, A.: Algebraic systems and pushdown automata. In: Droste, M., Kuich, W., Vogler, H. (eds.) Handbook of Weighted Automata, ch. 7, pp. 257–311. Springer (2009)
24. Reps, T., Schwoon, S., Jha, S., Melski, D.: Weighted pushdown systems and their application to interprocedural dataflow analysis. Science of Programming 58, 206–263 (2005)
25. Salomaa, A., Soittola, M.: Automata-Theoretic Aspects of Formal Power Series. Texts and Monographs in Computer Science. Springer (1978)
26. Schützenberger, M.P.: On the definition of a family of automata. Inf. and Control 4, 245–270 (1961)

Operational Characterization
of Scattered MCFLs*

Zoltán Ésik and Szabolcs Iván

University of Szeged, Hungary

Abstract. We give a Kleene-type operational characterization of Muller context-free languages (MCFLs) of well-ordered and scattered words.

1 Introduction

A word, called 'arrangement' in [12], is an isomorphism type of a countable labeled linear order. They form a generalization of the classic notions of finite and ω-words.

Finite automata on ω-words have by now a vast literature, see [21] for a comprehensive treatment. Finite automata acting on well-ordered words longer than ω have been investigated in [2,9,10,23,24], to mention a few references. In the last decade, the theory of automata on well-ordered words has been extended to automata on all countable words, including scattered and dense words. In [3,5,8], both operational and logical characterizations of the class of languages of countable words recognized by finite automata were obtained.

Context-free grammars generating ω-words were introduced in [11] and subsequently studied in [7,20]. Context-free grammars generating arbitrary countable words were defined in [14,15]. Actually, two types of grammars were defined, context-free grammars with Büchi acceptance condition (BCFG), and context-free grammars with Muller acceptance condition (MCFG). These grammars generate the Büchi and the Muller context-free languages of countable words, abbreviated as BCFLs and MCFLs. Every BCFL is clearly an MCFL, but there exists an MCFL of well-ordered words that is not a BCFL, for example the set of all countable well-ordered words over some alphabet. In fact, it was shown in [14] that for every BCFL L of well-ordered words there is an integer n such that the order type of the underlying linear order of every word in L is bounded by ω^n.

A Kleene-type characterization of BCFLs of well-ordered and scattered words was given in [17]. Here we provide a Kleene-type characterization of MCFLs of well-ordered and scattered words. Before presenting the necessary preliminaries in detail, we give a formulation of our main result, at least in the well-ordered case.

* The publication is supported by the European Union and co-funded by the European Social Fund. Project title: "Telemedicine-focused research activities on the field of Matematics, Informatics and Medical sciences", Project number: TAMOP-4.2.2.A-11/1/KONV-2012-0073.

M.-P. Béal and O. Carton (Eds.): DLT 2013, LNCS 7907, pp. 215–226, 2013.
© Springer-Verlag Berlin Heidelberg 2013

Suppose that Σ is an alphabet, and let $\Sigma^\sharp$ denote the set of all (countable) words over Σ. Let $P(\Sigma^\sharp)$ be the set of all subsets of $\Sigma^\sharp$. The set of $\mu\omega T_w$-expressions over Σ is defined by the following grammar:

$$T ::= a \mid \varepsilon \mid x \mid T+T \mid T \cdot T \mid \mu x.T \mid T^\omega$$

Here, each letter $a \in \Sigma$ denotes the language containing a as its unique word, while ε denotes the language containing only the empty word. The symbols $+$ and $\cdot$ are interpreted as set union and concatenation over $P(\Sigma^\sharp)$, and the variables x range over languages in $\Sigma^\sharp$. The μ-operator corresponds to taking least fixed points. Finally, $^\omega$ is interpreted as the ω-power operation over $P(\Sigma^\sharp)$: $L \mapsto L \cdot L \cdots$. An expression is closed if each variable occurs in the scope of a least fixed-point operator. Each closed expression denotes a language in $P(\Sigma^\sharp)$. Our main result in the well-ordered case, which is a corollary of Theorem 2 is:

Theorem 1. *A language $L \subseteq \Sigma^\sharp$ is an MCFL of well-ordered words iff it is denoted by some closed $\mu\omega T_w$-expression.*

Example 1. The expression $\mu x.(x^\omega + a + b + \varepsilon)$ denotes the set of all well-ordered words over the alphabet $\{a, b\}$.

It was shown in [17] that the syntactic fragment of the above expressions, with the ω-power operation restricted to closed expressions, characterizes the BCFLs of well-ordered words. A similar, but more involved result holds for MCFLs of scattered words, cf. Theorem 2. Both theorems were conjectured by the authors of [17].

2   Notation

2.1   Linear Orderings

A *linear ordering* is a pair $(I, <)$, where I is a set and $<$ is an irreflexive transitive trichotomous relation (i.e. a strict total ordering) on I. If I is finite or countable, we say that the ordering is finite or countable as well. *In this paper, all orderings are assumed to be countable.* A good reference for linear orderings is [22].

An *embedding* of the linear ordering $(I, <)$ into $(J, \prec)$ is an order preserving function $f : I \to J$, i.e. $x < y$ implies $f(x) \prec f(y)$ for each $x, y \in I$. If f is surjective, we call it an *isomorphism*. Two linear orderings are said to be *isomorphic* if there exists an isomorphism between them. Isomorphism between linear orderings is an equivalence relation; classes of this equivalence relation are called *order types*. If $I \subseteq J$ and $<$ is the restriction of $\prec$ onto I, then we say that $(I, <)$ is a *sub-ordering* of $(J, \prec)$.

Examples of linear orderings are the ordering $(\mathbb{N}, <)$ of the positive integers, the ordering $(\mathbb{N}_-, <)$ of the negative integers, the ordering $(\mathbb{Z}, <)$ of the integers and the ordering $(\mathbb{Q}, <)$ of the rationals. The respective order types are denoted ω, $-\omega$, ζ and η. In order to ease notation, we write simply I for $(I, <)$ if the ordering $<$ is standard or known from the context.

An ordering is *scattered* if it does not have a sub-ordering of order type η, otherwise it is *quasi-dense*. An ordering is a *well-ordering* if it does not have a sub-ordering of order type $-\omega$. Order types of well-orderings are called *ordinals*.

When $(I, <)$ is an ordering and for each $i \in I$, $(J_i, <_i)$ is an ordering, then the *generalized sum* $\sum_{i \in I}(J_i, <_i)$ is the disjoint union $\{(i, j) : i \in I, j \in J_i\}$ equipped with the lexicographic ordering $(i, j) < (i', j')$ iff $i < i'$, or $i = i'$ and $j <_i j'$. It is known that if $(I, <)$ and the $(J_i, <_i)$ are scattered or well-ordered, then so is the generalized sum. The operation of generalized sum can be extended to order types since it preserves isomorphisms. For example, $\zeta = -\omega + \omega$. Ordinals are also equipped with an exponentiation operator.

Hausdorff classified linear orderings into an infinite hierarchy. Following [18], we present a variant of this hierarchy. Let VD_0 be the collection of all finite linear orderings, and when α is some ordinal, let VD_α be the collection of all finite sums of linear orderings of the form $\sum_{i \in \mathbb{Z}}(I_i, <_i)$, where for each integer $i \in \mathbb{Z}$, $(I_i, <_i)$ is a member of VD_{α_i} for some ordinal $\alpha_i < \alpha$. According to a theorem of Hausdorff (see e.g. [22], Thm. 5.24), a (countable) linear ordering $(I, <)$ is scattered if and only if it belongs to VD_α for some (countable) ordinal α; the least such α is called the *rank* of $(I, <)$, denoted $\mathrm{rank}(I, <)$.

2.2   Words, Tree Domains, Trees

An *alphabet* is a finite nonempty set Σ of symbols, usually called *letters*. A *word* over Σ is a linear ordering $(I, <)$ equipped with a *labeling function* $\lambda : I \to \Sigma$. An *embedding of words* is a mapping preserving the order and the labeling; a surjective embedding is an *isomorphism*. Order theoretic properties of the underlying linear ordering of a word are transferred to the word. A word is finite if its underlying linear order is finite, and an ω-word, if its underlying linear order is a well-order of order type ω. We usually identify isomorphic words and denote by $\Sigma^\sharp$ the set of all words over Σ. As usual, we denote the collection of finite and ω-words over Σ by Σ^* and Σ^ω, respectively. The length of a word $u \in \Sigma^*$ is denoted $|u|$. A language over Σ is a subset of $\Sigma^\sharp$. As in the introduction, we let $P(\Sigma^\sharp)$ denote the collection of all languages over Σ.

When $(I, <)$ is a linear ordering and $w_i = (J_i, <_i, \lambda_i)$ for $i \in I$ are words, then we define their *concatenation* $\prod_{i \in I} w_i$ as the word with underlying linear order $\sum_{i \in I}(J_i, <_i)$ and labeling $\lambda(i, j) = \lambda_i(j)$. When I has two elements, we obtain the usual notion of concatenation, denoted $u \cdot v$, or just uv. The operation of concatenation is extended to languages in $P(\Sigma^\sharp)$: $\prod_{i \in I} L_i = \{\prod_{i \in I} w_i : w_i \in L_i\}$. When $L, L_1, L_2 \subseteq \Sigma^\sharp$, then we define $L_1 + L_2$ to be the set union and $L_1 L_2 = \{uv : u \in L_1, v \in L_2\}$. Moreover, we define $L^\omega = \prod_{i \in \mathbb{N}} L$.

The set $P(\Sigma^\sharp)$ of languages over Σ, equipped with the inclusion order, is a complete lattice. When A is a set, a function $f : P(A)^n \to P(A)$ is *monotone* if $A_i \subseteq A'_i$ for each $i \in [n]$ implies $f(A_1, \ldots, A_n) \subseteq f(A'_1, \ldots, A'_n)$. The following fact is clear.

Lemma 1. *The functions* $+, \cdot : P(\Sigma^\sharp)^2 \to P(\Sigma^\sharp)$ *and* $^\omega : P(\Sigma^\sharp) \to P(\Sigma^\sharp)$ *are monotone.*

We will also consider *pairs* of words over an alphabet Σ, equipped with a finite concatenation and an ω-product operation. For pairs (u,v), (u',v') in $\Sigma^\sharp \times \Sigma^\sharp$, we define the product $(u,v) \cdot (u',v')$ to be the pair $(uu', v'v)$, and when for each $i \in \mathbb{N}$, (u_i, v_i) is in $\Sigma^\sharp \times \Sigma^\sharp$, then we let $\prod_{i \in \mathbb{N}} (u_i, v_i)$ be the word $\left(\prod_{i \in \mathbb{N}} u_i \right) \left(\prod_{i \in \mathbb{N}_-} v_i \right)$.

Let $P(\Sigma^\sharp \times \Sigma^\sharp)$ denote the set of all subsets of $\Sigma^\sharp \times \Sigma^\sharp$. Then $P(\Sigma^\sharp \times \Sigma^\sharp)$ is naturally equipped with the operations of set union $L + L'$, concatenation $L \cdot L' = \{(u,v) \cdot (u',v') : (u,v) \in L, \ (u',v') \in L'\}$ and Kleene star $L^* = \{\varepsilon\} \cup L \cup L^2 \cup \cdots$. We also define an ω-power operation $P(\Sigma^\sharp \times \Sigma^\sharp) \to P(\Sigma^\sharp)$ by $L^\omega = \{ \prod_{i \in \mathbb{N}} (u_i, v_i) : (u_i, v_i) \in L \}$. When $L_1, L_2 \subseteq \Sigma^\sharp$, let $L_1 \times L_2 = \{(u,v) : u \in L_1, \ v \in L_2\} \subseteq \Sigma^\sharp \times \Sigma^\sharp$.

Lemma 2. *The functions*

$$\times : P(\Sigma^\sharp)^2 \to P(\Sigma^\sharp \times \Sigma^\sharp)$$
$$+, \cdot : P(\Sigma^\sharp \times \Sigma^\sharp)^2 \to P(\Sigma^\sharp \times \Sigma^\sharp)$$
$$* : P(\Sigma^\sharp \times \Sigma^\sharp) \to P(\Sigma^\sharp \times \Sigma^\sharp)$$
$$\omega : P(\Sigma^\sharp \times \Sigma^\sharp) \to P(\Sigma^\sharp)$$

are monotone.

We will use Lemma 1 and Lemma 2 in the following context. Suppose that for each $i \in [n] = \{1, \ldots, n\}$, $f_i : P(\Sigma^\sharp)^{n+p} \to P(\Sigma^\sharp)$ is a function that can be constructed by function composition from the above functions, the projection functions and constant functions. Let $f = \langle f_1, \ldots, f_n \rangle : P(\Sigma^\sharp)^{n+p} \to P(\Sigma^\sharp)^n$ be the target tupling of the f_i. Then f is a monotone function, and by Tarski's fixed point theorem, for each $y \in P(\Sigma^\sharp)^p$ there is a least solution of the fixed point equation $x = f(x, y)$ in the variable x ranging over $P(\Sigma^\sharp)^n$. This least fixed point, denoted $\mu x. f(x, y)$, gives rise to a function $P(\Sigma^\sharp)^p \to P(\Sigma^\sharp)^n$ in the parameter y. It is known that this function is also monotone, see e.g. [6].

A *tree domain* is a prefix closed nonempty (but possibly infinite) subset of $\mathbb{N}^*$. Elements of a tree domain T are also called nodes of T. When x and $x \cdot i$ are nodes of T for $x \in \mathbb{N}^*$ and $i \in \mathbb{N}$, then $x \cdot i$ is a *child* of x. A *descendant* of a node x is a node of the form $x \cdot y$, where $y \in \mathbb{N}^*$. Nodes of T having no child are the *leaves* of T. The leaves, equipped with order inherited from the lexicographic ordering of $\mathbb{N}^*$ form the *frontier* of T, denoted $\mathrm{fr}(T)$. An *inner node* of T is a non-leaf node. Subsets of a tree domain T which themselves are tree domains are called *prefixes* of T. A *path* of a tree domain T is a prefix of T such that each node has at most one child. A path can be identified with the unique sequence w in $\mathbb{N}^{\leq \omega}$ of all sequences over $\mathbb{N}$ of length at most ω such that the set of nodes of the path consists of the finite prefixes of w. A path π of T is *maximal* if no path of T contains π properly. When T is a tree domain and $x \in T$ is a node of T, then the *sub-tree domain* $T|_x$ of T is the set $\{y : xy \in T\}$. A tree domain T is *locally finite* if each node has a descendant which is a leaf.

A *tree* over an alphabet Δ is a mapping $t : \mathrm{dom}(t) \to \Delta \cup \{\varepsilon\}$, where $\mathrm{dom}(t)$ is a tree domain, such that inner vertices are mapped to letters in Δ. Notions such as nodes, paths etc. of tree domains are lifted to trees. When π is a path of the tree t, then $\text{labels}(\pi) = \{t(u) : u \in \pi\}$ is the set of labels of the nodes of π, and $\text{infLabels}(\pi)$ is the set of labels occurring infinitely often. For a path π, $\text{head}(\pi)$ denotes the minimal node x of π (with respect to the prefix order) with $\text{infLabels}(\pi) = \text{labels}(\pi|_x)$, if π is infinite; otherwise $\text{head}(\pi)$ is the last node of π. The labeled *frontier word* $\text{lfr}(t)$ of a tree t is determined by the leaves *not* labeled by ε, which is equipped with the ordering inherited from the lexicographic ordering of $\mathbb{N}^*$ and labeling function inherited from the labeling function of t. It is worth observing that when $\pi = x_0, x_1, \ldots$ is an infinite path of a tree t and for each i, α_i (β_i, resp.) is the word determined by the leaf labels of the descendants of x_i to the left (right, resp.) of x_{i+1} (i.e. if x_{i+1} is the jth child of x_i, then $\alpha_i = \text{lfr}(t|_{x1}) \cdot \text{lfr}(t|_{x2}) \cdot \ldots \cdot \text{lfr}(t|_{x(j-1)})$ and similarly for β_i), then $\text{lfr}(t) = \prod_{i \in \mathbb{N}} (\alpha_i, \beta_i)$.

2.3 Muller Context-Free Languages of Scattered Words

A *Muller context-free grammar*, or MCFG, is a system $G = (V, \Sigma, R, S, \mathcal{F})$, where V is the alphabet of nonterminals, Σ is the alphabet of terminals, $\Sigma \cap V = \emptyset$, R is the finite set of productions of the form $A \to \alpha$ with $A \in V$ and $\alpha \in (\Sigma \cup V)^*$, $S \in V$ is the start symbol and $\mathcal{F} \subseteq P(V)$ is the set of *nonempty* accepting sets.

A *derivation tree* of the above grammar G is a tree $t : \text{dom}(t) \to V \cup \Sigma \cup \{\varepsilon\}$ satisfying the following conditions:

1. For each inner node x of t there exists a rule $X \to X_1 \ldots X_n$ in R with $X_i \in \Sigma \cup V$ such that $t(x) = X$, the children of x are exactly $x \cdot 1, \ldots, x \cdot n$, and for each $i \in [n]$, $t(x \cdot i) = X_i$, so that when when $n = 0$, x has a single child $x \cdot 1$ labeled ε;
2. For each infinite path π of t, $\text{infLabels}(\pi)$ is an accepting set of G.

A derivation tree is *complete* if its leaves are all labeled in $\Sigma \cup \{\varepsilon\}$. If t is a derivation tree having *root symbol* $t(\varepsilon) = A$, then we say that t is an A-tree. The language $L(G, A) \subseteq \Sigma^\sharp$ *generated* from $A \in V$ is the set of frontier words of complete A-trees. The language $L(G)$ generated by G is $L(G, S)$. An MCFL is a language generated by some MCFG.

Example 2. If $G = (\{S, I\}, \{a, b\}, R, S, \{\{I\}\})$, with

$$R = \{S \to a, S \to b, S \to c, S \to I, I \to SI\},$$

then $L(G)$ consists of all the well-ordered words over $\{a, b\}$.

Example 3. If $G = (\{S, I\}, \{a, b\}, R, S, \{\{I\}\})$, with

$$R = \{S \to a, S \to b, S \to \varepsilon, S \to I, I \to SIS\},$$

then $L(G)$ consists of all the scattered words over $\{a, b\}$.

Let $L \subseteq \Sigma^{\sharp}$ be an MCFL consisting of scattered words and $G = (V, \Sigma, R, S, \mathcal{F})$ an MCFG with $L(G) = L$. We may assume that G is in *normal form* [15] – among the properties of this normal form we will use the following ones (see [15], Prop. 14) frequently:

- For every derivation tree there is a locally finite derivation tree with the same root symbol and same labeled frontier.
- The frontier of each derivation tree is scattered.

In the rest of the paper, we fix an MCFG $G = (V, \Sigma, R, S, \mathcal{F})$ in normal form generating only scattered words.

When t is a derivation tree, then we define $\mathrm{rank}(t) = \mathrm{rank}(\mathrm{fr}(t))$. For a derivation tree t, let $\mathrm{maxNodes}(t)$ be the prefix of $\mathrm{dom}(t)$ consisting of the nodes having maximal rank, i.e. $\mathrm{maxNodes}(t) = \{x \in \mathrm{dom}(t) : \mathrm{rank}(t|_x) = \mathrm{rank}(t)\}$. Suppose that t is locally finite. It is known, (see e.g. [16], proof of Proposition 1, paragraph 4) that in this case $\mathrm{maxNodes}(t)$ is the union of finitely many maximal paths. Clearly, the set $\{\pi_1, \ldots, \pi_n\}$ of these paths is unique. Let $\mathrm{level}(t)$ stand for the above n, the number of maximal paths covering $\mathrm{maxNodes}(t)$. Also, let $\mathrm{branch}(t)$ stand for the longest common prefix of the paths $\pi_1, \ldots, \pi_n$ (which is a finite word if $\mathrm{level}(t) > 1$ and is π_1 if $\mathrm{level}(t) = 1$).

We call a (not necessarily locally finite) derivation tree t *simple* if $\mathrm{maxNodes}(t)$ contains a single infinite path π and if $\mathrm{infLabels}(\pi) = \mathrm{labels}(\pi)$, i.e. $\mathrm{head}(\pi) = \varepsilon$. (When t is additionally locally finite, then this path π contains all nodes of $\mathrm{maxNodes}(t)$.) Such a path is called the *central path* of t. If t is a simple A-tree and F is the set of labels of its central path, then we call t an *F-simple A-tree*.

3   The Main Result

For locally finite complete derivation trees t' and t, let $t' \prec t$ if one of the following conditions holds:

1. $\mathrm{rank}(t') < \mathrm{rank}(t)$;
2. $\mathrm{rank}(t') = \mathrm{rank}(t)$ and $\mathrm{level}(t') < \mathrm{level}(t)$;
3. $\mathrm{rank}(t') = \mathrm{rank}(t)$, $\mathrm{level}(t') = \mathrm{level}(t) > 1$ and $|\mathrm{branch}(t')| < |\mathrm{branch}(t)|$;
4. $\mathrm{rank}(t') = \mathrm{rank}(t)$, $\mathrm{level}(t') = \mathrm{level}(t) = 1$, that is, the set of nodes of maximal rank is a path π in t and a path π' in t'. Then let $t' \prec t$ iff $|\mathrm{head}(\pi')| < |\mathrm{head}(\pi)|$.

Lemma 3. *The relation $\prec$ is a well-partial order (wpo) of locally finite complete derivation trees. The minimal elements of this wpo are the one-node trees corresponding to the elements of $\Sigma \cup \{\varepsilon\}$. Suppose that t is a locally finite complete derivation tree and $t' = t|_x$ is a proper subtree of t, so that $x \neq \varepsilon$. If t is not simple, or if t is simple but x does not belong to the central path of t, then $t' \prec t$.*

Proof. It is clear that $\prec$ is irreflexive. To prove that it is transitive, suppose that $t'' \prec t'$ and $t' \prec t$. If $\mathrm{rank}(t'') < \mathrm{rank}(t)$, then clearly $t'' \prec t$. Suppose that $\mathrm{rank}(t'') = \mathrm{rank}(t)$. Then also $\mathrm{rank}(t'') = \mathrm{rank}(t') = \mathrm{rank}(t)$. If $\mathrm{level}(t'') < \mathrm{level}(t)$ then $t'' \prec t$ again. Thus, we may suppose that $\mathrm{level}(t'') = \mathrm{level}(t)$, so that $\mathrm{level}(t'') = \mathrm{level}(t') = \mathrm{level}(t) = n$. Now there are two cases. If $n > 1$, then, since $t'' \prec t'$ and $t' \prec t$, we know that $|\mathrm{branch}(t'')| < |\mathrm{branch}(t')| < |\mathrm{branch}(t)|$ and thus $t'' \prec t$. If $n = 1$, then the maximal nodes form a single maximal path in each of the trees t'', t' and t. Let us denote these paths by π'', π' and π, respectively. As $t'' \prec t'$ and $t' \prec t$, we have that $|\mathrm{head}(\pi'')| < |\mathrm{head}(\pi')| < |\mathrm{head}(\pi)|$, so that $t'' \prec t$ again.

The fact that there is no infinite decreasing sequence of locally finite complete derivation trees with respect to the relation $\prec$ is clear, since every set of ordinals is well-ordered.

Suppose now that t is a locally finite complete derivation tree which has at least two nodes. By assumption, t has a leaf node x. Let $t' = t|_x$. If $\mathrm{rank}(t') < \mathrm{rank}(t)$ then $t' \prec t$. Otherwise, $\mathrm{rank}(t') = \mathrm{rank}(t) = 0$ and t is necessarily finite (since the frontier of an infinite locally finite derivation tree is infinite). Clearly, $\mathrm{maxNodes}(t)$ is the set of all nodes of t, and either $\mathrm{level}(t') = 1 < \mathrm{level}(t)$, or $\mathrm{level}(t') = \mathrm{level}(t) = 1$. In the latter case, t has a single maximal path π, and $|\mathrm{head}(\pi')| = 0 < |\mathrm{head}(\pi)|$ for the single maximal path π' of t'. In either case, $t' \prec t$. Thus, no locally finite complete derivation tree having more than one node is minimal. On the other hand, all one-node complete derivation trees corresponding to the elements of $\Sigma \cup \{\varepsilon\}$ are clearly minimal (and locally finite).

To prove the last claim, suppose that t is a locally finite complete derivation tree and $t' = t|_x$. If $\mathrm{rank}(t') < \mathrm{rank}(t)$, we are done. Otherwise, $\mathrm{rank}(t') = \mathrm{rank}(t)$ and x is a member of $\mathrm{maxNodes}(t)$. Thus, if π is a maximal path of $\mathrm{maxNodes}(t')$, then $x\pi$ is a maximal path of $\mathrm{maxNodes}(t)$. Hence $\mathrm{level}(t') \leq \mathrm{level}(t)$. If $\mathrm{level}(t') < \mathrm{level}(t)$, we are done. Otherwise, $\mathrm{level}(t') = \mathrm{level}(t)$ and $\mathrm{maxNodes}(t) = x\mathrm{maxNodes}(t')$.

Now there are two cases.

1. If $\mathrm{level}(t) > 1$, then $\mathrm{branch}(t) = x\mathrm{branch}(t')$, thus $|\mathrm{branch}(t')| < |\mathrm{branch}(t)|$ and $t' \prec t$.

2. Suppose that $\mathrm{level}(t) = 1$, and let π denote the unique maximal path of t whose nodes form the set $\mathrm{maxNodes}(t)$. Since $\mathrm{rank}(t') = \mathrm{rank}(t)$, we have that x belongs to π and, by assumption, t is not simple. Since t is not simple and has at least two nodes, $\mathrm{head}(\pi) \neq \varepsilon$ and $|\mathrm{head}(\pi')| < |\mathrm{head}(\pi)|$, where π' is the unique maximal path of t' whose nodes form the set $\mathrm{maxNodes}(t')$. (Actually π' is determined by the proper suffix $\pi|_x$ of π.) $\qquad\square$

Now we define certain ordinary ω-regular languages [19,21] corresponding to central paths of simple derivation trees. Let Γ stand for the (finite) set consisting of those triplets

$$(\alpha, B, \beta) \in (V \cup \Sigma)^* \times V \times (V \cup \Sigma)^*$$

for which $\alpha B\beta$ occurs as the right-hand side of a production of G. For any nonterminal $A \in V$ and accepting set $F \in \mathcal{F}$, let $R_{A,F} \subseteq \Gamma^\omega$ stand for the set of

ω-words over Γ accepted by the deterministic (partial) Muller (word) automaton $(F, \Gamma, \delta, A, \{F\})$, with $B = \delta(C, (\alpha, D, \beta))$ if and only if $D = B$ and $C \to \alpha B \beta$ is a production of G. By definition, each $R_{A,F}$ is an ω-regular set which can be built from singleton sets corresponding to the elements of Γ by the usual regular operations and the ω-power operation (actually, since every state has to be visited infinitely many times, $R_{A,F}$ can be written as the ω-power of a regular language of finite words over Γ).

Members of $R_{A,F}$ correspond to central paths of F-simple A-trees in the following sense. Given $w = (\alpha_1, A_1, \beta_1)(\alpha_2, A_2, \beta_2) \ldots \in R_{A,F}$, we define an F-simple A-tree t_w of G as follows. The nodes $x_0, x_1, \ldots$ of the central path of t_w are $x_0 = \varepsilon$, and $x_i = x_{i-1} \cdot (|\alpha_i| + 1)$, for $i > 0$. Each x_i has $|\alpha_{i+1} A_{i+1} \beta_{i+1}|$ children, respectively labeled by the letters of the word $\alpha_{i+1} A_{i+1} \beta_{i+1}$. Nodes not on the central path of t_w are leaf nodes.

It is straightforward to see the following claims:

1. For each $w \in R_{A,F}$, t_w is an F-simple A-tree.
2. Every F-simple A-tree has a prefix of the form t_w, for some $w \in R_{A,F}$. Thus, every such tree can be constructed by choosing an appropriate $w \in R_{A,F}$, and substituting a derivation tree t_x with root symbol $t_w(x)$ for each leaf x of t_w.

Moreover, it is clear that when $w = (\alpha_1, A_1, \beta_1)(\alpha_2, A_2, \beta_2) \ldots$, then $\mathrm{lfr}(t_w)$ is $(\prod_{i \in \mathbb{N}} \alpha_i) \cdot (\prod_{i \in \mathbb{N}_-} \beta_i)$.

Let us assign a variable X_A to each $A \in V$, and let $\mathcal{X}$ be the set of all variables. For each ordinary regular expression r over Γ, we define an expression (term) $\bar{r}$ over $\Sigma \cup \mathcal{X}$ involving the function symbols $\times, +, \cdot$. To this end, when α is a word in $(\Sigma \cup V)^*$, let $\bar{\alpha}$ be the word in $(\mathcal{X} \cup \Sigma)^*$ obtained by replacing each occurrence of a nonterminal A by the variable X_A. Then, for a letter $\gamma = (\alpha, A, \beta) \in \Gamma$, define $\bar{\gamma} = \bar{\alpha} \times \bar{\beta}$. To obtain $\bar{r}$, we replace each occurrence of a letter γ in r by $\bar{\gamma}$.

When A is a nonterminal and $A \in F$ for some $F \in \mathcal{F}$, consider an ordinary regular expression $r_{A,F}$ over Γ such that $r_{A,F}^\omega$ denotes the set $R_{A,F}$ (defined above) of all ω-words corresponding to central paths of F-simple A-trees. Then consider the following system of equations E_G associated with G in the variables $\mathcal{X}$:

$$X_A = \sum_{A \to u \in R} \bar{u} + \sum_{A \in F \in \mathcal{F}} (\overline{r_{A,F}})^\omega.$$

Example 4. The system of equations E_G associated with the grammar in Example 3 is:

$$X_S = a + b + \varepsilon + X_I$$
$$X_I = (X_S \times X_S)^\omega$$

As usual, we can associate a function $f_G : P(\Sigma^\sharp)^{\mathcal{X}} \to P(\Sigma^\sharp)^{\mathcal{X}}$ with E_G. By Lemmas 1 and 2 and using the facts that the projections are monotone and that monotone functions are closed under function composition, we have that f_G is monotone. Thus, f_G has a least fixed point.

Proposition 1. *For each $A \in V$, the corresponding component of the least fixed point solution of the system E_G is the language $L(G, A)$ of all words derivable from A.*

Proof. The fact that the languages $L(G, A)$, $A \in V$, form a solution is clear from the definition of E_G. Let us also define $L(G, a) = \{a\}$, for each $a \in \Sigma \cup \{\varepsilon\}$. Suppose that the family of languages L_A, $A \in V$ is another solution, and let $L_a = \{a\}$ for $a \in \Sigma \cup \{\varepsilon\}$. We want to show that if t is a locally finite complete A-tree with $\mathrm{lfr}(t) = u$, then $u \in L_A$, for each $A \in \Sigma \cup \{\varepsilon\} \cup V$. We apply well-founded induction with respect to the wpo $\prec$.

For the base case, if t consists of a single node, then $A = a \in \Sigma \cup \{\varepsilon\}$, $u = a$, and our claim is clear. Otherwise, there are two cases: either t is a simple tree, or not.

If $t = A(t_1, \ldots, t_n)$ is not simple, then we have $t_i \prec t$ for each $i \in [n]$ by Lemma 3. Let A_i be the root symbol of t_i and u_i the labeled frontier word of t_i for each i. By the induction hypothesis, each u_i is a member of L_{A_i}. Since t is a derivation tree, $A \to A_1 \ldots A_n$ is a production of G. Thus, by the construction of E_G, $u = u_1 \ldots u_n \in L_A$.

Otherwise, if t is an F-simple A-tree for some $F \in \mathcal{F}$ and $A \in V$, then t can be constructed from a tree t_w with $w \in R_{A,F}$ by replacing each leaf node x of t_w by some locally finite complete derivation tree t_x with root symbol $t_w(x)$. Since such leaves are not on the central path of t, we have $t_x \prec t$ for each x, again by Lemma 3. Applying the induction hypothesis, we get that the labeled frontier word u_x of each t_x is a member of $L_{t_w(x)}$. Thus, by the construction of E_G, u is a member of L_A. $\qquad\square$

It is well-known, cf. [4,1] or [6], Chapter 8, Theorem 2.15 and Chapter 6, Section 8.1, Equation (3.2), that when $\mathcal{L}, \mathcal{L}', \mathcal{L}''$ are complete lattices and $f : \mathcal{L} \times \mathcal{L}' \times \mathcal{L}'' \to \mathcal{L}$ and $g : \mathcal{L} \times \mathcal{L}' \times \mathcal{L}'' \to \mathcal{L}'$ are monotone functions, then the least solution (in the parameter z) of the system of equations

$$x = f(x, y, z)$$
$$y = g(x, y, z)$$

can be obtained by Gaussian elimination as

$$x = \mu x.f(x, \mu y.g(x, y, z), z)$$
$$y = \mu y.g(\mu x.f(x, \mu y.g(x, y, z), z), y, z)$$

Using this fact and Proposition 1, we obtain our final result.

Let the set of $\mu \omega T_s$-expressions over the alphabet Σ be defined by the following grammar (with T being the initial nonterminal):

$$T ::= a \mid \varepsilon \mid x \mid T + T \mid T \cdot T \mid \mu x.T \mid P^\omega$$
$$P ::= T \times T \mid P + P \mid P \cdot P \mid P^*$$

Here, $a \in \Sigma$ and $x \in \mathcal{X}$ for an infinite countable set $\mathcal{X}$ of variables. An occurrence of a variable is *free* if it is not in the scope of a μ-operation, and bound, if it is not free. A *closed expression* does not have free variable occurrences. The semantics of these expressions are defined as expected using the monotone functions over $P(\Sigma^\sharp)$ and $P(\Sigma^\sharp \times \Sigma^\sharp)$ introduced earlier. When the free variables of an expression form the set $\mathcal{Y}$, then an expression denotes a language in $P((\Sigma \cup \mathcal{Y})^\sharp)$.

Remark 1. Actually, ε is redundant, as it is expressible by $((\mu x.x \times \mu x.x)^*)^\omega$. We do not need a constant 0 denoting the empty set of pairs since it is expressible by $(\mu x.x) \times (\mu x.x)$.

Theorem 2. *A language $L \subseteq \Sigma^\sharp$ is an MCFL of scattered words if and only if it can be denoted by a closed $\mu\omega T_s$-expression.*

Proof. It is easy to show that each expression denotes an MCFL of scattered words. One uses the following facts, where Δ denotes an alphabet and $x, \# \notin \Delta$.

- The set of MCFLs (of scattered words) over Δ is closed under $+$ and $\cdot$.
- If $L, L' \subseteq \Delta^\sharp$ are MCFLs (of scattered words), then $L\#L' \subseteq (\Delta \cup \{\#\})^\sharp$ is an MCFL (of scattered words).
- Suppose that $L, L' \subseteq \Delta^\sharp \# \Delta^\sharp$ are MCFLs (of scattered words). Then

$$\{uv\#v'u' : u\#u' \in L, \ v\#v' \in L'\} \subseteq \Delta^\sharp \# \Delta^\sharp$$

  is an MCFL (of scattered words).
- Suppose that $L \subseteq \Delta^\sharp \# \Delta^\sharp$ is an MCFL (of scattered words). Then

$$\{u_1 \ldots u_n \# v_n \ldots v_1 : n \geq 0, \ u_i \# v_i \in L\} \subseteq \Delta^\sharp \# \Delta^\sharp$$

  is an MCFL (of scattered words).
- Suppose that $L \subseteq \Delta^\sharp \# \Delta^\sharp$ is an MCFL (of scattered words). Then

$$\{(u_1 u_2 \ldots)(\ldots v_2 v_1) : u_i \# v_i \in L\} \subseteq \Delta^\sharp$$

  is an MCFL (of scattered words).
- Suppose that $L \subseteq (\Delta \cup \{x\})^\sharp$ is an MCFL (of scattered words). Then, with respect to set inclusion, there is a least language $L' \subseteq \Delta^\sharp$ such that $L[x \mapsto L'] = L'$, and this language L' is an MCFL (of scattered words). (Here, $L[x \mapsto L']$ is the language obtained from L by 'substituting' L' for x.)

For detailed proofs of the above facts see [13]. The other direction follows from Proposition 1 using Gaussian elimination. □

It is easy to show that Theorem 1 follows from Theorem 2.

Example 5. The expression $\mu x.((x \times x)^\omega + a + b + \varepsilon)$ denotes the set of all scattered words over the alphabet $\{a, b\}$.

Example 6. Let $L \subseteq \{a, b\}^\sharp$ be the language of all words w such that the word obtained from w by removing all occurrences of letter b is well-ordered, as is the 'mirror image' of the word obtained by removing all occurrences of letter *a.* It is not difficult to show that each word in L contains only a finite number of 'alternations' between a and b. Using this fact, an MCFG generating L is: $G = (\{S, A, B, I, J\}, \Sigma, R, S, \{\{I\}, \{J\}\})$ with R consisting of the productions

$$S \to AS \mid BS \mid \varepsilon$$
$$A \to a \mid \varepsilon \mid I$$
$$I \to AI$$
$$B \to b \mid \varepsilon \mid J$$
$$J \to JB$$

Using the algorithm described above (with some simplification), an expression for L is:

$$t_S = \mu x_S.((t_A + t_B)x_S + \varepsilon)$$

with

$$t_A = \mu x_A.(a + \varepsilon + (x_A \times \varepsilon)^\omega)$$
$$t_B = \mu x_B.(b + \varepsilon + (\varepsilon \times x_B)^\omega).$$

References

1. de Bakker, J.W., Scott, D.: A theory of programs. IBM Seminar Vienna (August 1969)
2. Bedon, N.: Finite automata and ordinals. Theoretical Computer Science 156, 119–144 (1996)
3. Bedon, N., Bès, A., Carton, O., Rispal, C.: Logic and rational languages of words indexed by linear orderings. In: Hirsch, E.A., Razborov, A.A., Semenov, A., Slissenko, A. (eds.) Computer Science – Theory and Applications. LNCS, vol. 5010, pp. 76–85. Springer, Heidelberg (2008)
4. Bekić, H.: Definable operations in general algebras, and the theory of automata and flowcharts. IBM Seminar Vienna (December 1969)
5. Bès, A., Carton, O.: A Kleene theorem for languages of words indexed by linear orderings. In: De Felice, C., Restivo, A. (eds.) DLT 2005. LNCS, vol. 3572, pp. 158–167. Springer, Heidelberg (2005)
6. Bloom, S.L., Ésik, Z.: Iteration Theories. EATCS Monograph Series in Theoretical Computer Science. Springer (1993)
7. Boasson, L.: Context-free sets of infinite words. In: Weihrauch, K. (ed.) GI-TCS 1979. LNCS, vol. 67, pp. 1–9. Springer, Heidelberg (1979)
8. Bruyère, V., Carton, O.: Automata on linear orderings. J. Computer and System Sciences 73, 1–24 (2007)
9. Büchi, J.R.: The monadic second order theory of ω_1. In: Decidable theories, II. Lecture Notes in Math, vol. 328, pp. 1–127. Springer, Heidelberg (1973)
10. Choueka, Y.: Finite automata, definable sets, and regular expressions over ω^n-tapes. J. Computer and System Sciences 17(1), 81–97 (1978)
11. Cohen, R.S., Gold, A.Y.: Theory of ω-languages, parts one and two. J. Computer and System Sciences 15, 169–208 (1977)

12. Courcelle, B.: Frontiers of infinite trees. Theoretical Informatics and Applications 12, 319–337 (1978)
13. Ésik, Z., Iván, S.: Operational characterization of scattered MCFLs. arXiv:1304.6388 [cs.FL]
14. Ésik, Z., Iván, S.: Büchi context-free languages. Theoretical Computer Science 412, 805–821 (2011)
15. Ésik, Z., Iván, S.: On Muller context-free grammars. Theoretical Computer Science 416, 17–32 (2012)
16. Ésik, Z., Iván, S.: Hausdorff rank of scattered context-free linear orders. In: Fernández-Baca, D. (ed.) LATIN 2012. LNCS, vol. 7256, pp. 291–302. Springer, Heidelberg (2012)
17. Ésik, Z., Okawa, S.: On context-free languages of scattered words. In: Yen, H.-C., Ibarra, O.H. (eds.) DLT 2012. LNCS, vol. 7410, pp. 142–153. Springer, Heidelberg (2012)
18. Khoussainov, B., Rubin, S., Stephan, F.: Automatic linear orders and trees. ACM Transactions on Computational Logic (TOCL) 6, 675–700 (2005)
19. Muller, R.: Infinite sequences and finite machines. In: 4th Annual Symposium on Switching Circuit Theory and Logical Design, pp. 3–16. IEEE Computer Society (1963)
20. Nivat, M.: Sur les ensembles de mots infinis engendrés par une grammaire algébrique (French). Theoretical Informatics and Applications 12, 259–278 (1978)
21. Perrin, D., Pin, J.-E.: Infinite Words. Elsevier (2004)
22. Rosenstein, J.G.: Linear Orderings. Academic Press (1982)
23. Wojciechowski, J.: Classes of transfinite sequences accepted by finite automata. Fundamenta Informaticae 7, 191–223 (1984)
24. Wojciechowski, J.: Finite automata on transfinite sequences and regular expressions. Fundamenta Informaticae 8, 379–396 (1985)

Abelian Repetitions in Sturmian Words

Gabriele Fici[1], Alessio Langiu[2], Thierry Lecroq[3], Arnaud Lefebvre[3],
Filippo Mignosi[4], and Élise Prieur-Gaston[3]

[1] Dipartimento di Matematica e Informatica, Università di Palermo, Italy
Gabriele.Fici@unipa.it
[2] Department of Informatics, King's College London, London, UK
Alessio.Langiu@kcl.ac.uk
[3] Normandie Université, LITIS EA4108, Université de Rouen, 76821
Mont-Saint-Aignan Cedex, France
{Thierry.Lecroq,Arnaud.Lefebvre,Elise.Prieur}@univ-rouen.fr
[4] Dipartimento di Informatica, Università dell'Aquila, L'Aquila, Italy
Filippo.Mignosi@di.univaq.it

Abstract. We investigate abelian repetitions in Sturmian words. We exploit a bijection between factors of Sturmian words and subintervals of the unitary segment that allows us to study the periods of abelian repetitions by using classical results of elementary Number Theory. If k_m denotes the maximal exponent of an abelian repetition of period m, we prove that $\limsup k_m/m \geq \sqrt{5}$ for any Sturmian word, and the equality holds for the Fibonacci infinite word. We further prove that the longest prefix of the Fibonacci infinite word that is an abelian repetition of period F_j, $j > 1$, has length $F_j(F_{j+1} + F_{j-1} + 1) - 2$ if j is even or $F_j(F_{j+1} + F_{j-1}) - 2$ if j is odd. This allows us to give an exact formula for the smallest abelian periods of the Fibonacci finite words. More precisely, we prove that for $j \geq 3$, the Fibonacci word f_j has abelian period equal to F_n, where $n = \lfloor j/2 \rfloor$ if $j = 0, 1, 2 \mod 4$, or $n = 1 + \lfloor j/2 \rfloor$ if $j = 3 \mod 4$.

1 Introduction

The study of repetitions in words is a classical subject in Theoretical Computer Science both from the combinatorial and the algorithmic point of view. Repetitions are strictly related to the notion of periodicity. Recall that a word w of length $|w|$ has a *period* $p > 0$ if $w[i] = w[i+p]$ for any $1 \leqslant i \leqslant |w| - p$, where $w[i]$ is the symbol in position i of w. Every word w has a minimal period $p \leq |w|$. If $|w|/p \geq 1$, then w is called a *repetition* of period p and *exponent* $|w|/p$. When $|w|/p = k$ is an integer, the word w is called an *integer power*, since it can be written as $w = u^k$, i.e., w is the concatenation of k copies of a word u of length p. If instead $|w|/p$ is not an integer, the word w is called a *fractional power*. So one can write $w = u^k v$, where v is the prefix of u such that $|w|/p = k + |v|/|u|$. For example, the word $w = aabaaba$ is a 7/3-power since it has minimal period 3 and length 7. A classical reference on periodicity is [22, Chap. 7].

M.-P. Béal and O. Carton (Eds.): DLT 2013, LNCS 7907, pp. 227–238, 2013.
© Springer-Verlag Berlin Heidelberg 2013

Abelian properties concerning words have been studied since the very beginning of Formal Languages and Combinatorics on Words. The notion of Parikh vector has become a standard and is often used without an explicit reference to the original 1966 Parikh's paper [27]. Abelian powers were first considered in 1961 by Erdös [13] as a natural generalization of usual powers. Research concerning abelian properties of words and languages developed afterwards in different directions. In particular, there is a recent increasing of interest on abelian properties of words linked to periodicity (see, for example, [2,6,12,29,31,32]), and on the algorithmic search of abelian periodicities in strings [7,10,14,15,20].

Recall that the Parikh vector $\mathcal{P}_w$ of a finite word w enumerates the cardinality of each letter of the alphabet in w. Therefore, two words have the same Parikh vector if one can be obtained from the other by permuting letters. We say that the word w is an *abelian repetition* of (abelian) period m and exponent $|w|/m$ if w can be written as $w = u_0u_1 \cdots u_{j-1}u_j$ for words u_i and an integer $j > 2$, where for $0 < i < j$ all the u_i's have the same Parikh vector $\mathcal{P}$ whose sum of components is m and the Parikh vectors of u_0 and u_j are contained in $\mathcal{P}$ (see [8]). When u_0 and u_j are empty, w is called an *abelian power* or *weak repetition* [11]. For example, the word $w = abaab$ is an abelian repetition of period 2, since one can set $u_0 = a$, $u_1 = ba$, $u_2 = ab$ and $u_3 = \varepsilon$, where ε denotes the empty word.

It is well known that Sturmian words and Fibonacci words, in particular, are extremal cases for several problems related to repetitions (see for example [9,18,26]) and are worst-case examples for classical pattern matching algorithms, e.g. Knuth-Morris-Pratt [1,21]. There exists a huge bibliography concerning Sturmian words (see for instance the survey papers [3,4], [22, Chap. 2], [30, Chap. 6] and references therein). In particular, there is an analogous result to the one presented in this paper concerning classical repetitions in the Fibonacci infinite word [25]. In [23], a bijection between factors of Sturmian words and subintervals of the unitary segment is described. We show in this paper that this bijection preserves abelian properties of factors (see Proposition 4). Therefore, we are able to apply techniques of Number Theory coupled with Combinatorics on Words to obtain our main results. More precisely, if k_m denotes the maximal exponent of an abelian repetition of period m, we prove that $\limsup k_m/m \geq \sqrt{5}$ for any Sturmian word, and the equality holds for the Fibonacci infinite word.

We further prove that for any Fibonacci number F_j, $j > 1$, the longest prefix of the Fibonacci infinite word that is an abelian repetition of period F_j has length $F_j(F_{j+1} + F_{j-1} + 1) - 2$ if j is even or $F_j(F_{j+1} + F_{j-1}) - 2$ if j is odd (Theorem 7). This allows us to give an exact formula for the smallest abelian periods of the Fibonacci finite words. More precisely, we prove, in Theorem 8, that for $j \geq 3$, the Fibonacci word f_j has abelian period equal to F_n, where $n = \lfloor j/2 \rfloor$ if $j = 0, 1, 2 \mod 4$, or $n = 1 + \lfloor j/2 \rfloor$ if $j = 3 \mod 4$.

Due to space constraints the proofs are omitted, but they will be included in an upcoming full version of the paper.

2   Preliminaries

Let $\Sigma = \{a_1, a_2, \ldots, a_\sigma\}$ be a finite ordered alphabet of cardinality σ and Σ^* the set of words over Σ. We denote by $|w|$ the length of the word w. We write $w[i]$ the i-th symbol of w and $w[i..j]$ the factor of w from the i-th symbol to the j-th symbol, with $1 \leqslant i \leqslant j \leqslant |w|$. We denote by $|w|_a$ the number of occurrences of the symbol $a \in \Sigma$ in the word w.

The *Parikh vector* of a word w, denoted by $\mathcal{P}_w$, counts the occurrences of each letter of Σ in w, i.e., $\mathcal{P}_w = (|w|_{a_1}, \ldots, |w|_{a_\sigma})$. Given the Parikh vector $\mathcal{P}_w$ of a word w, we denote by $\mathcal{P}_w[i]$ its i-th component and by $|\mathcal{P}_w|$ the sum of its components. Thus, for a word w and $1 \leqslant i \leqslant \sigma$, we have $\mathcal{P}_w[i] = |w|_{a_i}$ and $|\mathcal{P}_w| = \sum_{i=1}^{\sigma} \mathcal{P}_w[i] = |w|$. Finally, given two Parikh vectors $\mathcal{P}, \mathcal{Q}$, we write $\mathcal{P} \subset \mathcal{Q}$ if $\mathcal{P}[i] \leqslant \mathcal{Q}[i]$ for every $1 \leqslant i \leqslant \sigma$ and $|\mathcal{P}| < |\mathcal{Q}|$.

Following [8], we give the definition below.

Definition 1. *A word w is an abelian repetition of period $m > 0$ and exponent $|w|/m = k$ if one can write $w = u_0 u_1 \cdots u_{j-1} u_j$ for some $j > 2$ such that $\mathcal{P}_{u_0} \subset \mathcal{P}_{u_1} = \ldots = \mathcal{P}_{u_{j-1}} \supset \mathcal{P}_{u_j}$, and $|\mathcal{P}_{u_1}| = \ldots = |\mathcal{P}_{u_{j-1}}| = m$.*

An abelian power is an abelian repetition in which $u_0 = u_j = \varepsilon$.

We call u_0 and u_j the *head* and the *tail* of the abelian repetition, respectively. Notice that the length $t = |u_j|$ of the tail is uniquely determined by $h = |u_0|$, m and $|w|$, namely $t = (|w| - h) \bmod m$.

Example 1. The word $w = abaababa$ is an abelian repetition of period 2 and exponent 4, since one can write $w = a \cdot ba \cdot ab \cdot ab \cdot a$. Notice that w is also an abelian repetition of period 3 and exponent 8/3, since $w = \varepsilon \cdot aba \cdot aba \cdot ba$.

In the rest of the paper, when we refer to an abelian repetition of period m, we always suppose that m is the minimal abelian period of w.

Remark 1. We adopt the convention that an abelian repetition of exponent $k \geq 2$ has also exponent k' for any real number k' such that $2 \leq k' \leq k$. This is a standard convention widely adopted in the classical case.

2.1   Sturmian Words

From now on, we fix the alphabet $\Sigma = \{a, b\}$. We start by recalling a bijection between factors of Sturmian words and subintervals of the unitary segment introduced in [23].

Let α and ρ be two real numbers with $\alpha \in (0, 1)$. Following the notations of [17], the fractional part of a number r is defined by $\{r\} = r - \lfloor r \rfloor$, where $\lfloor r \rfloor$ is the greatest integer smaller than or equal to r. Therefore, for $\alpha \in (0, 1)$, one has that $\{-\alpha\} = 1 - \alpha$.

The sequence $\{n\alpha + \rho\}, n > 0$, defines an infinite word $s_{\alpha, \rho} = a_1(\alpha, \rho) a_2(\alpha, \rho) \cdots$ by the rule

$$a_n(\alpha, \rho) = \begin{cases} \mathbf{b} & \text{if } \{n\alpha + \rho\} \in [0, \{-\alpha\}), \\ \mathbf{a} & \text{if } \{n\alpha + \rho\} \in [\{-\alpha\}, 1). \end{cases}$$

See Fig. 1 for a graphical illustration.

We will write a_n instead of $a_n(\alpha, \rho)$ whenever there is no possibility of mistake. If α is rational, i.e. $\alpha = n/m$, with n and m coprime integers, then it is easy to prove that the word $s_{\alpha,\rho}$ is periodic and m is its minimal period. In this case, $s_{\alpha,\rho}$ is also periodic in the abelian sense, since it trivially has abelian period m.

If instead α is irrational, then $s_{\alpha,\rho}$ is not periodic and is called a *Sturmian word*. Therefore, in the rest of the paper, we always suppose α irrational.

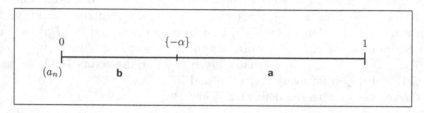

Fig. 1. An application of Proposition 1 when $\alpha = \phi - 1 \approx 0.618$ (thus $\{-\alpha\} \approx 0.382$) for $i = 0$. If $\{n\alpha + \rho\} \in [\{-\alpha\}, 1)$, then $a_n = \mathbf{a}$; otherwise $a_n = \mathbf{b}$.

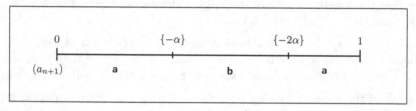

Fig. 2. An application of Proposition 1 when $\alpha = \phi - 1 \approx 0.618$ (thus $\{-\alpha\} \approx 0.382$) for $i = 1$. If $\{n\alpha + \rho\} \in [0, \{-\alpha\}) \cup [\{-2\alpha\}, 1)$, then $a_{n+1} = \mathbf{a}$; otherwise $a_{n+1} = \mathbf{b}$.

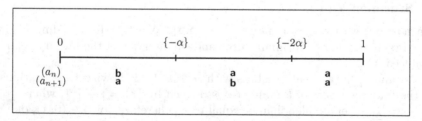

Fig. 3. A single graphic representation of the information given in Fig. 1 and 2. If $\{n\alpha + \rho\} \in [0, \{-\alpha\}) = L_0(\alpha, 2)$, then $a_n = \mathbf{b}$, $a_{n+1} = \mathbf{a}$. If $\{n\alpha + \rho\} \in [\{-\alpha\}, \{-2\alpha\}) = L_1(\alpha, 2)$, then $a_n = \mathbf{a}$, $a_{n+1} = \mathbf{b}$. If $\{n\alpha + \rho\} \in [\{-2\alpha\}, 1) = L_2(\alpha, 2)$, then $a_n = \mathbf{a}$, $a_{n+1} = \mathbf{a}$.

Example 2. For $\alpha = \phi - 1$ and $\rho = 0$, where $\phi = (1 + \sqrt{5})/2$ is the golden ratio, one obtains the Fibonacci infinite word

$$f = \mathbf{abaababaabaababaababa} \cdots$$

Remark 2. Since $\alpha \in (0,1)$, we have $\{-i\alpha\} \neq \{-(i+1)\alpha\}$ for any natural number i. We shall use this fact freely and with no explicit mention.

It is possible to prove (see [23, Corollary 2.3]) that the following result holds.

Proposition 1. *Let* α *and* ρ *be real numbers, with* $\alpha \in (0,1)$ *irrational. For any natural numbers* n, i, *with* $n > 0$, *if* $\{-(i+1)\alpha\} < \{-i\alpha\}$ *then*

$$a_{n+i} = \mathbf{a} \iff \{n\alpha + \rho\} \in [\{-(i+1)\alpha\}, \{-i\alpha\}),$$

whereas if $\{-i\alpha\} < \{-(i+1)\alpha\}$ *then*

$$a_{n+i} = \mathbf{a} \iff \{n\alpha + \rho\} \in [0, \{-i\alpha\}) \cup [\{-(i+1)\alpha\}, 1).$$

In Fig. 1 and 2 we display a graphical representation of the formula given in Proposition 1 for $\alpha = \phi - 1$ when $i = 0$ and $i = 1$, respectively. In Fig. 3 we present within a single graphic the situations illustrated in Fig. 1 and 2.

Let m be a positive integer. Consider the $m + 2$ points $0, 1, \{-i\alpha\}$, for $1 \leq i \leq m$. Rearranging these points in increasing order one has:

$$0 = c_0(\alpha, m) < c_1(\alpha, m) < \ldots < c_k(\alpha, m) < \ldots < c_m(\alpha, m) < c_{m+1}(\alpha, m) = 1.$$

One can therefore define the $m + 1$ non-empty subintervals

$$L_k(\alpha, m) = [c_k(\alpha, m), c_{k+1}(\alpha, m)), \ 0 < k \leq m.$$

By using Proposition 1, it is possible to associate with each interval $L_k(\alpha, m)$ a factor of length m of the word $s_{\alpha, \rho}$, and this correspondence is bijective (see [24]). We call this correspondence the *Sturmian bijection*.

Proposition 2. *Each factor of* $s_{\alpha, \rho}$ *of length* m, $a_n a_{n+1} \cdots a_{n+m-1}$, *depends only on the interval* $L_k(\alpha, m)$ *containing the point* $\{n\alpha + \rho\}$; *more precisely, it depends only on the set* $I_k(\alpha, m)$ *of integers* $i \in \{0, 1, \ldots, m-1\}$ *such that either* $\{-(i+1)\alpha\} < \{-i\alpha\}$ *and* $c_k(\alpha, m) \in [\{-(i+1)\alpha\}, \{-i\alpha\})$ *or* $\{-(i+1)\alpha\} > \{-i\alpha\}$ *and* $c_k(\alpha, m) \notin [\{-i\alpha\}, \{-(i+1)\alpha\})$. *The set* $I_k(\alpha, m)$ *is the set of the integers* i, *with* $0 \leq i \leq m - 1$, *such that* $a_{n+i} = \mathbf{a}$.

Corollary 1. *Since the set of factors of* $s_{\alpha, \rho}$ *depends only on the sequence* $\{-i\alpha\}$, $i > 0$, *it does not depend on* ρ. *In particular, then, for any* ρ *the word* $s_{\alpha, \rho}$ *has the same set of factors of the word* $s_{\alpha, 0}$.

Example 3. Let $\alpha = \phi - 1$. In Fig. 3 we show an example of the Sturmian bijection when $m = 2$. The ordered sequence of points defining the subintervals $L_k(\alpha, 2)$ is

$$c_0(\alpha, 2) = 0, \ c_1(\alpha, 2) = \{-\alpha\} \approx 0.382, \ c_2(\alpha, 2) = \{-2\alpha\} \approx 0.764, \ c_3(\alpha, 2) = 1.$$

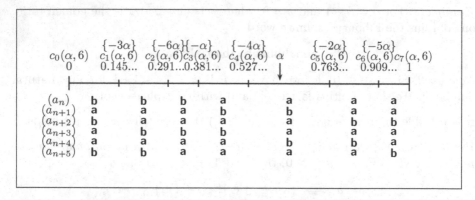

Fig. 4. The subintervals $L_k(\alpha, m)$ of the Sturmian bijection obtained for $\alpha = \phi - 1$ and $m = 6$. Below each interval there is the factor of s_α of length 6 associated with that interval. For $\rho = 0$ and $n = 1$, the prefix of length 6 of the Fibonacci word is associated with $L_4(\alpha, 6) = [c_4(\alpha, 6), c_5(\alpha, 6))$, which is the interval containing α.

In Fig. 4 we show an example of the Sturmian bijection when $\alpha = \phi - 1$ and $m = 6$. Below each interval there is the factor of s_α of length $m = 6$ associated with that interval. The prefix of length 6 of the Fibonacci word corresponds to the factor below the interval containing α (so, for $n = 1$ and $\rho = 0$). Notice that all the factors of length 6 of the Fibonacci word appear, and moreover they are lexicographically ordered from right to left. This property concerning lexicographic order holds for any Sturmian word and any length m of factors, and is stated in next proposition, which is of independent interest and is related to some recent research on Sturmian words and the lexicographic order (see [5,16,19,28]).

Proposition 3. *Let $m \geq 1$ and k, k' such that $0 \leq k, k' \leq m$. Then $k < k'$ if and only if the factor $t_{\alpha,\rho,m}$ associated to $L_k(\alpha, m)$ in the Sturmian bijection is lexicographically greater than the factor $t'_{\alpha,\rho,m}$ associated to $L_{k'}(\alpha, m)$.*

In the next section we present a new property of the Sturmian bijection, that will allow us to use some standard Number Theory techniques to deal with abelian repetitions in Sturmian words and, in particular, in the Fibonacci infinite word. Similar techniques are used in [31] to derive some other interesting results on abelian powers in Sturmian words.

3   Sturmian Bijection and Parikh Vectors

Let $s_{\alpha,\rho}$ be a Sturmian word. Since we are mainly interested in the set of factors of $s_{\alpha,\rho}$, we do not lose generality, by Corollary 1, supposing $\rho = 0$. The Sturmian words with $\rho = 0$ are called *characteristic*, and have been the object of deep studies within the field of Sturmian words. For simplicity of notation, we will write s_α instead of $s_{\alpha,0}$.

We now describe some properties of the Sturmian bijection between the factors of length m of s_α and the subintervals $L_k(\alpha, m)$, that we will use to prove the main results of the paper.

Proposition 4. *Under the Sturmian bijection, all the factors corresponding to an interval $c_k(\alpha, m) = [x, y)$ with $x \geq \{-m\alpha\}$ have the same Parikh vector $v_1(\alpha, m)$ and all the factors corresponding to an interval $[x, y)$ with $y \leq \{-m\alpha\}$ have the same Parikh vector $v_2(\alpha, m)$. Moreover, one has $v_1(\alpha, m)[1] = v_2(\alpha, m)[1] + 1$.*

The reader can see in Fig. 4 that the factors of length 6 corresponding to an interval to the left of $\{-6(\phi - 1)\}$ have Parikh vector $(3, 3)$, while the other ones have Parikh vector $(4, 2)$.

We now address the following questions:

1. Given m, how large can be the exponent of an abelian repetition of period m in s_α?
2. What can we say in the particular case of the Fibonacci word, i.e., when $\alpha = \phi - 1$?

The next result follows straightforwardly from Proposition 4.

Corollary 2. *Let w be an abelian power of period m and exponent $k + 1$ appearing in s_α in position n. Then all the points in the sequence $\{n\alpha\}, \{(n + m)\alpha\}, \{(n + 2m)\alpha\}, \ldots, \{(n + km)\alpha\}$ are in the same subinterval in which $[0, 1)$ is subdivided by the point $\{-m\alpha\}$, i.e., either $[0, \{-m\alpha\})$ or $[\{-m\alpha\}, 1)$.*

The next proposition is a technical step to prove the following theorem.

Proposition 5. *If $k \geq 1$, the $k + 1$ points of Corollary 2 are naturally ordered. That is to say, if $\{m\alpha\} < 0.5$, then they are all in the subinterval $[0, \{-m\alpha\})$ and one has $\{n\alpha\} < \{(n + m)\alpha\} < \ldots < \{(n + km)\alpha\}$; if instead $\{m\alpha\} > 0.5$ then they are all in the interval $[\{-m\alpha\}, 1)$ and one has $\{(n + km)\alpha\} < \{(n + (k - 1)m)\alpha\} < \ldots < \{n\alpha\}$.*

Theorem 1. *Let m be a positive integer such that $\{m\alpha\} < 0.5$ (resp. $\{m\alpha\} > 0.5$). Then:*

1. *In s_α there is an abelian power of period m and exponent $k \geq 2$ if and only if $\{m\alpha\} < \frac{1}{k}$ (resp. $\{-m\alpha\} < \frac{1}{k}$).*
2. *If in s_α there is an abelian power of period m and exponent $k \geq 2$ starting in position i with $\{i\alpha\} \geq \{m\alpha\}$ (resp. $\{i\alpha\} \leq \{m\alpha\}$), then $\{m\alpha\} < \frac{1}{k+1}$ (resp. $\{-m\alpha\} < \frac{1}{k+1}$). Conversely, if $\{m\alpha\} < \frac{1}{k+1}$ (resp. $\{-m\alpha\} < \frac{1}{k+1}$), then there is an abelian power of period m and exponent $k \geq 2$ starting in position m.*

The previous theorem allows us to deal with abelian repetitions in a Sturmian word s_α by using classical results on the approximation of the irrational α by rationals. This is a classical topic in Number Theory. Since the number $\phi - 1$ has special properties within this topic, we have in turn specific results for the Fibonacci infinite word.

4 Approximating Irrationals by Rationals and Abelian Repetitions

We recall some classical results of Number Theory. For any notation not explicitly defined in this section we refer to [17, Chap. X, XI].

The sequence $F_0 = 1, F_1 = 1, F_{j+1} = F_j + F_{j-1}$ for $j \geq 1$ is the well known sequence of Fibonacci numbers. The sequence of fractions $\frac{F_{j+1}}{F_j}$ converges to $\phi = \frac{\sqrt{5}+1}{2}$, while the sequence $\frac{F_j}{F_{j+1}}$ converges to $\phi - 1 = \frac{\sqrt{5}-1}{2}$. Moreover, the sequences $\frac{F_{j+1}}{F_j}$ and $0 = \frac{0}{1}, \frac{F_j}{F_{j+1}}, j = 0, 1, \ldots$, are the sequences of convergents, in the development in continued fractions, of ϕ and $\phi - 1$ respectively.

Concerning the approximation given by the above convergents, the following result holds (see [17, Chap. X, Theorem 171] and [17, Chap. XI, Section 11.8]).

Theorem 2. *For any $j > 0$,*

$$\phi - \frac{F_{j+1}}{F_j} = (\phi - 1) - \frac{F_{j-1}}{F_j} = \frac{(-1)^j}{F_j(\phi F_j + F_{j-1})}.$$

We also report the following theorems (see [17, Chap. XI, Theorem 193 and the proof of Theorem 194]).

Theorem 3. *Any irrational α has an infinity of approximations which satisfy*

$$\left| \frac{n}{m} - \alpha \right| < \frac{1}{\sqrt{5}m^2}.$$

Theorem 4. *Let $\alpha = \phi - 1$. If $A > \sqrt{5}$, then the inequality*

$$\left| \frac{n}{m} - \alpha \right| < \frac{1}{Am^2}$$

has only a finite number of solutions.

The last two theorems, coupled with the first part of Theorem 1, allow us to derive the next result.

Theorem 5. *Let s_α be a Sturmian word. For any integer $m > 1$, let k_m be the maximal exponent of an abelian repetition of period m in s_α. Then*

$$\limsup_{m \to \infty} \frac{k_m}{m} \geq \sqrt{5},$$

and the equality holds if $\alpha = \phi - 1$.

5 Prefixes of the Fibonacci Infinite Word

We now study the abelian repetitions that are prefixes of the Fibonacci infinite word. For this, we will make use of the second part of Theorem 1. Notice that an abelian repetition of period m appearing as a prefix of the Fibonacci word can have a head of length equal to $m - 1$ at most. Therefore, we have to check all the abelian powers that start in position i for every $i = 1, \ldots, m$. In order to do this, we report here another result (see [17, Chap. X, Theorem 182]).

Theorem 6. *Let n_i/m_i be the i-th convergent to α. If $i > 1$, $0 < m \leq m_i$ and $n/m \neq n_i/m_i$, then $|n_i - m_i\alpha| < |n - m\alpha|$.*

The previous theorem implies the following result.

Corollary 3. *Suppose that $m > 1$ is the denominator of a convergent to α and that $\{m\alpha\} < 0.5$ (resp. $\{m\alpha\} > 0.5$). Then for any i such that $1 \leq i < m$, one has $\{i\alpha\} \geq \{m\alpha\}$ (resp. $\{i\alpha\} \leq \{m\alpha\}$).*

From the previous corollary, we have that if $m > 1$ is a Fibonacci number and $\alpha = \phi - 1$, then the hypotheses of the second part of Theorem 1 are satisfied. The next proposition is a direct consequence of Corollary 3, Theorem 1 and Theorem 2.

Proposition 6. *Let $j > 1$. In the Fibonacci infinite word, the longest abelian power having period F_j and starting in a position $i \leq F_j$ has an occurrence starting in position F_j, and has exponent equal to*

$$\lfloor \phi F_j + F_{j-1} \rfloor - 1 = \begin{cases} F_{j+1} + F_{j-1} - 1 & \text{if } j \text{ is even;} \\ F_{j+1} + F_{j-1} - 2 & \text{if } j \text{ is odd.} \end{cases}$$

The following theorem provides a formula for computing the length of the longest abelian repetition occurring as a prefix in the Fibonacci infinite word.

Theorem 7. *Let $j > 1$. The longest prefix of the Fibonacci infinite word that is an abelian repetition of period F_j has length $F_j(F_{j+1} + F_{j-1} + 1) - 2$ if j is even or $F_j(F_{j+1} + F_{j-1}) - 2$ if j is odd.*

Corollary 4. *Let $j > 1$ and k_j be the maximal exponent of a prefix of the Fibonacci word that is an abelian repetition of period F_j. Then*

$$\lim_{j \to \infty} \frac{k_j}{F_j} = \sqrt{5}.$$

In Fig. 5 we give a graphical representation of the longest prefix of the Fibonacci infinite word that is an abelian repetition of period m for $m = 2, 3$ and 5. In Table 1 we give the length $lp(F_j)$ of the longest prefix of the Fibonacci infinite word that is an abelian repetition of period F_j, for $j = 2, \ldots, 11$, computed using the formula of Theorem 7. We also show the values of the distance between $\sqrt{5}$

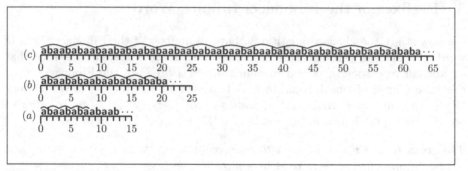

Fig. 5. Longest abelian repetition of period m that is a prefix of the Fibonacci word for $m = 2, 3, 5$. (a) For $m = 2$, the longest abelian repetition has length $8 = 1 + 3p + 1$. (b) For $m = 3$, the longest abelian repetition has length $19 = 2 + 5p + 2$. (c) For $m = 5$, the longest abelian repetition has length $58 = 4 + 10p + 4$.

Table 1. The length of the longest prefix ($lp(F_j)$) of the Fibonacci word having abelian period F_j for $j = 2, \ldots, 11$. The table also reports rounded distances (multiplied by 10^2) between $\sqrt{5}$ and the ratio between the exponent $k_j = lp(F_j)/F_j$ of the longest prefix of the Fibonacci word having abelian period F_j and F_j (see Corollary 4).

j	2	3	4	5	6	7	8	9	10	11
F_j	2	3	5	8	13	21	34	55	89	144
$lp(F_j)$	8	19	58	142	388	985	2616	6763	17798	46366
$\lvert\sqrt{5} - k_j/F_j\rvert \times 10^2$	23.6	12.5	8.393	1.732	5.98	0.25	2.69	0.037	1.087	0.005

and the ratio between the maximal exponent $k_j = lp(F_j)/F_j$ of a prefix of the Fibonacci infinite word having abelian period F_j and F_j.

Recall that the Fibonacci (finite) words are defined by $f_0 = \mathbf{b}$, $f_1 = \mathbf{a}$, and for every $j > 1$, $f_{j+1} = f_j f_{j-1}$. So, for every j, one has $|f_j| = F_j$. As a consequence of the formula given in Theorem 7, we have the following result on the smallest abelian periods of the Fibonacci words.

Theorem 8. *For $j \geq 3$, the (smallest) abelian period of the word f_j is the n-th Fibonacci number F_n, where $n = \lfloor j/2 \rfloor$ if $j = 0, 1, 2 \mod 4$, or $n = 1 + \lfloor j/2 \rfloor$ if $j = 3 \mod 4$.*

For example, the abelian period of the word $f_4 = \mathbf{abaab}$ is $2 = F_2 = \lfloor 4/2 \rfloor$, since one can write $f_4 = \mathbf{a} \cdot \mathbf{ba} \cdot \mathbf{ab}$; the abelian period of $f_5 = \mathbf{abaababa}$ is $2 = F_2$; the abelian period of $f_6 = \mathbf{abaababaabaab}$ is $3 = F_3$; the abelian period of $f_7 = \mathbf{abaababaabaababaababa}$ is $5 = F_4$. In Table 2 we report the abelian periods of the first Fibonacci words.

Table 2. The (smallest) abelian periods of the Fibonacci words f_j for $j = 3, \ldots, 16$

j	3	4	5	6	7	8	9	10	11	12	13	14	15	16
a. p. of f_j	F_2	F_2	F_2	F_3	F_4	F_4	F_4	F_5	F_6	F_6	F_6	F_7	F_8	F_8

We conclude the paper with the following open problems:

1. Is it possible to find the exact value of $\limsup \frac{k_m}{m}$ for other Sturmian words s_α with slope α different from $\phi - 1$?
2. Is it possible to give the exact value of this superior limit when α is an algebraic number of degree 2?

References

1. Aho, A.: Algorithms for Finding Patterns in Strings. In: van Leeuwen, J. (ed.) Handbook of Theoret. Comput. Sci, pp. 257–300. Elsevier Science Publishers B. V, Amsterdam (1990)
2. Avgustinovich, S., Karhumäki, J., Puzynina, S.: On abelian versions of Critical Factorization Theorem. RAIRO Theor. Inform. Appl. 46, 3–15 (2012)
3. Berstel, J.: Sturmian and episturmian words (a survey of some recent results). In: Bozapalidis, S., Rahonis, G. (eds.) CAI 2007. LNCS, vol. 4728, pp. 23–47. Springer, Heidelberg (2007)
4. Berstel, J., Lauve, A., Reutenauer, C., Saliola, F.: Combinatorics on Words: Christoffel Words and Repetition in Words. CRM monograph series, vol. 27. American Mathematical Society (2008)
5. Bucci, M., De Luca, A., Zamboni, L.: Some characterizations of Sturmian words in terms of the lexicographic order. Fundamenta Informaticae 116(1-4), 25–33 (2012)
6. Cassaigne, J., Richomme, G., Saari, K., Zamboni, L.: Avoiding Abelian powers in binary words with bounded Abelian complexity. Int. J. Found. Comput. Sci. 22(4), 905–920 (2011)
7. Christou, M., Crochemore, M., Iliopoulos, C.S.: Identifying all abelian periods of a string in quadratic time and relevant problems. Int. J. Found. Comput. Sci. 23(6), 1371–1384 (2012)
8. Constantinescu, S., Ilie, L.: Fine and Wilf's theorem for abelian periods. Bull. Eur. Assoc. Theoret. Comput. Sci. EATCS 89, 167–170 (2006)
9. Crochemore, M., Ilie, L., Rytter, W.: Repetitions in strings: Algorithms and combinatorics. Theoret. Comput. Sci. 410(50), 5227–5235 (2009)
10. Crochemore, M., Iliopoulos, C.S., Kociumaka, T., Kubica, M., Pachocki, J., Radoszewski, J., Rytter, W., Tyczynski, W., Walen, T.: A note on efficient computation of all abelian periods in a string. Inf. Process. Lett. 113(3), 74–77 (2013)
11. Cummings, L.J., Smyth, W.F.: Weak repetitions in strings. J. Combin. Math. Combin. Comput. 24, 33–48 (1997)
12. Domaratzki, M., Rampersad, N.: Abelian primitive words. Int. J. Found. Comput. Sci. 23(5), 1021–1034 (2012)
13. Erdős, P.: Some unsolved problems. Magyar Tud. Akad. Mat. Kutato. Int. Kozl. 6, 221–254 (1961)

14. Fici, G., Lecroq, T., Lefebvre, A., Prieur-Gaston, E.: Computing Abelian Periods in Words. In: Proceedings of the Prague Stringology Conference, PSC 2011, pp. 184–196. Czech Technical University in Prague (2011)
15. Fici, G., Lecroq, T., Lefebvre, A., Prieur-Gaston, E., Smyth, W.F.: Quasi-Linear Time Computation of the Abelian Periods of a Word. In: Proceedings of the Prague Stringology Conference, PSC 2012, pp. 103–110. Czech Technical University in Prague (2012)
16. Glen, A., Justin, J., Pirillo, G.: Characterizations of finite and infinite episturmian words via lexicographic orderings. European Journal of Combinatorics 29(1), 45–58 (2008)
17. Hardy, G.H., Wright, E.M.: An Introduction to the Theory of Numbers, 5th edn. Clarendon Press, Oxford (1979)
18. Iliopoulos, C.S., Moore, D., Smyth, W.F.: A Characterization of the Squares in a Fibonacci String. Theoret. Comput. Sci. 172(1-2), 281–291 (1997)
19. Jenkinson, O., Zamboni, L.Q.: Characterisations of balanced words via orderings. Theoret. Comput. Sci. 310(1-3), 247–271 (2004)
20. Kociumaka, T., Radoszewski, J., Rytter, W.: Fast algorithms for abelian periods in words and greatest common divisor queries. In: STACS 2013. LIPIcs, vol. 20, pp. 245–256. Schloss Dagstuhl - Leibniz-Zentrum fuer Informatik (2013)
21. Kolpakov, R., Kucherov, G.: Finding Maximal Repetitions in a Word in Linear Time. In: Proceedings of the 40th Annual Symposium on Foundations of Computer Science, FOCS 1999, pp. 596–604. IEEE Computer Society (1999)
22. Lothaire, M.: Algebraic Combinatorics on Words. Cambridge University Press, Cambridge (2002)
23. Mignosi, F.: Infinite Words with Linear Subword Complexity. Theoret. Comput. Sci. 65(2), 221–242 (1989)
24. Mignosi, F.: On the number of factors of Sturmian words. Theoret. Comput. Sci. 82, 71–84 (1991)
25. Mignosi, F., Pirillo, G.: Repetitions in the Fibonacci infinite word. RAIRO Theor. Inform. Appl. 26, 199–204 (1992)
26. Mignosi, F., Restivo, A.: Characteristic Sturmian words are extremal for the critical factorization theorem. Theoret. Comput. Sci. 454(0), 199–205 (2012)
27. Parikh, R.J.: On context-free languages. J. Assoc. Comput. Mach. 13(4), 570–581 (1966)
28. Perrin, D., Restivo, A.: A note on Sturmian words. Theoret. Comput. Sci. 429, 265–272 (2012)
29. Puzynina, S., Zamboni, L.Q.: Abelian returns in Sturmian words. J. Comb. Theory, Ser. A 120(2), 390–408 (2013)
30. Pytheas Fogg, N.: Substitutions in Dynamics, Arithmetics and Combinatorics. Lecture Notes in Math, vol. 1794. Springer, Heidelberg (2002)
31. Richomme, G., Saari, K., Zamboni, L.: Abelian complexity of minimal subshifts. Journal of the London Mathematical Society 83(1), 79–95 (2011)
32. Samsonov, A., Shur, A.: On Abelian repetition threshold. RAIRO Theor. Inform. Appl. 46, 147–163 (2012)

Composition Closure
of ε-Free Linear Extended
Top-Down Tree Transducers

Zoltán Fülöp[1],* and Andreas Maletti[2],**

[1] Department of Foundations of Computer Science, University of Szeged
Árpád tér 2, H-6720 Szeged, Hungary
fulop@inf.u-szeged.hu
[2] Institute for Natural Language Processing, University of Stuttgart
Pfaffenwaldring 5b, 70569 Stuttgart, Germany
maletti@ims.uni-stuttgart.de

Abstract. The expressive power of compositions of linear extended top-down tree transducers with and without regular look-ahead is investigated. In particular, the restrictions of ε-freeness, strictness, and nondeletion are considered. The composition hierarchy is finite for all ε-free variants of these transducers except for ε-free nondeleting linear extended top-down tree transducers. The least number of transducers needed for the full expressive power of arbitrary compositions is presented.

1 Introduction

The top-down tree transducer is a simple formal model that encodes a tree transformation. It was introduced in [21,22] and intensively studied thereafter (see [13,14,12] for an overview). Roughly speaking, a top-down tree transducer processes the input tree symbol-by-symbol and specifies in its rules, how to translate an input symbol into an output tree fragment together with instructions on how to process the subtrees of the input symbol. This asymmetry between input (single symbol) and output (tree fragment) was removed in extended top-down tree transducers (xt), which were introduced and studied in [1,2]. In an xt the left-hand side of a rule now contains an input tree fragment, in which each variable can occur at most once as a placeholder for a subtree. In particular, the input tree fragment can even be just a variable, which matches every tree, and such rules are called ε-rules. We consider linear xt (l-xt), in which the right-hand side of each rule contains each variable at most once as well. Restricted variants of l-xt are used in most approaches to syntax-based machine translation [16,17].

We also add regular look-ahead [6] (i.e., the ability to check a regular property for the subtrees in an input tree fragment) to l-xt, so our most expressive model

---

* Supported by the program TÁMOP-4.2.1/B-09/1/KONV-2010-0005 of the Hungarian National Development Agency.
** Supported by the German Research Foundation (DFG) grant MA/4959/1-1.

M.-P. Béal and O. Carton (Eds.): DLT 2013, LNCS 7907, pp. 239–251, 2013.
© Springer-Verlag Berlin Heidelberg 2013

is the linear extended top-down tree transducer with regular look-ahead (l-xt^R). Instead of variables in the left-hand side and a state-variable combination in the right-hand side of a rule, we only use states with the restriction that each state can occur at most once in the left-hand side and at most once in the right-hand side. Moreover, all states that occur in the right-hand side must also occur in the left-hand side. In this way, for each rule the states establish implicit links (a state links its occurrence in the right-hand side with its occurrence in the left-hand side), which form a bijection between a subset of the state occurrences in the left-hand side and all state occurrences in the right-hand side. The state occurrences (in the left-hand side) that do not participate in the bijection (i.e., those states that exclusively occur in the left-hand side) can restrict the acceptable subtrees at their position with the help of regular look-ahead. The implicit links in a rule are made explicit in a derivation, and a rule application expands (explicitly) linked state occurrences at the same time. Example 2 shows an l-xt^R, for which we illustrate a few derivation steps in Fig. 1. We use l-XT^R and l-XT to denote the class of all tree transformations computed by l-xt^R and l-xt, respectively.

The expressive power of the various subclasses of l-XT^R is already well understood [15,11]. However, in practice complex systems are often specified with the help of compositions of tree transformations [20] because it is much easier to develop (or train) small components that manage a part of the overall transformation. Consequently, [17] and others declare that closure under composition is a very desirable property for classes of tree transformations (especially in the area of natural language processing). If C represents the class of all tree transformations computable by a device, then the fact that C is closed under composition means that we can replace any composition chain specified by several devices by just a single device, which enables an efficient modular development. Unfortunately, neither l-XT^R nor l-XT is closed under composition [2,3,15].

For a class C of tree transformations we obtain a composition hierarchy $C \subseteq C^2 \subseteq C^3 \subseteq \cdots$, where C^n denotes the n-fold composition of C. The class C might be closed under composition at a power n (i.e., $C^n = C^{n+1}$) or its hierarchy might be infinite (i.e., $C^n \subsetneq C^{n+1}$ for all n). The classes that are closed at a low power are also important in practice. We investigate the composition hierarchy of the classes l-XT^R and l-XT together with various subclasses determined by the properties: ε-freeness, strictness, and nondeletion, abbreviated by '$\not\varepsilon$', 's', and 'n', respectively. We use these symbols in front of l-XT^R and l-XT to obtain the class of all tree transformations computable by the corresponding restricted l-xt^R and l-xt, respectively. In this paper we consider in detail the closure of the classes $\not\varepsilon l$-XT^R, $\not\varepsilon l$-XT, $\not\varepsilon sl$-XT^R, and $\not\varepsilon sl$-XT under composition.

It is known that none of our considered classes is closed under composition [3]. In addition, it is known [3] that $\not\varepsilon snl$-XT $= \not\varepsilon snl$-XT^R is closed at power 2. We complete the picture by providing the least power at which the above classes are closed under composition in the following table.

Class	Least power of closedness	Proved in
$\not{s}$l-XT, $\not{s}$l-XT^R	2	Theorem 14
$\not{l}$-XT	3 or 4 (4)	Theorem 17 (Conjecture)
$\not{l}$-XT^R	3	Theorem 17
otherwise	∞	[9, Theorem 34]

2   Notation

We denote the set of all nonnegative integers by $\mathbb{N}$. Every subset of $S \times T$ is a *relation* from S to T. Given relations $R_1 \subseteq S \times T$ and $R_2 \subseteq T \times U$, the *inverse* of R_1 and the *composition* of R_1 and R_2 are denoted by R_1^{-1} and $R_1 ; R_2$, respectively. These notions and notations are lifted to classes of relations in the usual manner. Moreover, the *powers* of a class $\mathcal{C}$ are defined by $\mathcal{C}^1 = \mathcal{C}$ and $\mathcal{C}^{n+1} = \mathcal{C}^n ; \mathcal{C}$ for $n \geq 1$. The *composition hierarchy* [resp. *composition closure*] of $\mathcal{C}$ is the family $(\mathcal{C}^n \mid n \geq 1)$ [resp. the class $\bigcup_{n \geq 1} \mathcal{C}^n$]. If $\mathcal{C}^{n+1} = \mathcal{C}^n$, then $\mathcal{C}$ *is closed under composition at power* n. A *ranked alphabet* is a finite set Σ, which is partitioned by $\Sigma = \bigcup_{k \in \mathbb{N}} \Sigma_k$ into subsets Σ_k containing the elements of rank k. We also write $\sigma^{(k)}$ to indicate that $\sigma \in \Sigma_k$. For the rest of this paper, Σ, Δ, and Γ will denote arbitrary ranked alphabets. For every set T, let $\Sigma(T) = \{\sigma(t_1, \ldots, t_k) \mid \sigma \in \Sigma_k, t_1, \ldots, t_k \in T\}$. Let S be a set with $S \cap \Sigma = \emptyset$. The set $T_\Sigma(S)$ of Σ-*trees with leaf labels* S is the smallest set U such that $S \subseteq U$ and $\Sigma(U) \subseteq U$. We write T_Σ for $T_\Sigma(\emptyset)$, and we use $\mathrm{pos}(t) \subseteq \mathbb{N}^*$ to denote the *positions* of $t \in T_\Sigma(S)$. For words $v, w \in \mathbb{N}^*$, we denote the longest common prefix of v and w by $\mathrm{lcp}(v, w)$. The positions $\mathrm{pos}(t)$ are partially ordered by the prefix order $\preceq$ on $\mathbb{N}^*$ [i.e., $v \preceq w$ if and only if $v = \mathrm{lcp}(v, w)$]. The *size* $|t|$ of the tree $t \in T_\Sigma(S)$ is $|\mathrm{pos}(t)|$. Let $t \in T_\Sigma(S)$ and $w \in \mathrm{pos}(t)$. We denote the *label* of t at w by $t(w)$, and the w-*rooted subtree* of t by $t|_w$. For every $U \subseteq S$, we let $\mathrm{pos}_U(t) = \{w \in \mathrm{pos}(t) \mid t(w) \in U\}$ and $\mathrm{pos}_s(t) = \mathrm{pos}_{\{s\}}(t)$ for every $s \in S$. The tree t is *linear* (resp. *nondeleting*) in U if $|\mathrm{pos}_u(t)| \leq 1$ (resp. $|\mathrm{pos}_u(t)| \geq 1$) for every $u \in U$. Moreover, $\mathrm{var}(t) = \{s \in S \mid \mathrm{pos}_s(t) \neq \emptyset\}$. We write $t[u]_w$ for the tree obtained from $t \in T_\Sigma(S)$ by replacing the subtree $t|_w$ at w by $u \in T_\Sigma(S)$.

For every $n \in \mathbb{N}$ we fix the set $X_n = \{x_1, \ldots, x_n\}$ of variables. Given $t \in T_\Sigma(X_n)$ and $t_1, \ldots, t_n \in T_\Sigma(S)$, we write $t[t_1, \ldots, t_n]$ for the tree obtained from t by replacing each occurrence of x_i by t_i for all $1 \leq i \leq n$. A *tree homomorphism from* Σ *to* Δ is a family of mappings $(h_k \mid k \in \mathbb{N})$ such that $h_k \colon \Sigma_k \to T_\Delta(X_k)$ for every $k \in \mathbb{N}$. Such a tree homomorphism is *linear* (resp. *nondeleting*) if for every $\sigma \in \Sigma_k$ the tree $h_k(\sigma)$ is linear (resp. nondeleting) in X_k. Moreover, it is *strict* [resp. *delabeling*] if $h_k \colon \Sigma_k \to \Delta(T_\Delta(X_k))$ [resp. $h_k \colon \Sigma_k \to X_k \cup \Delta(X_k)$] for every $k \in \mathbb{N}$. We abbreviate the above restrictions by 'l', 'n', 's', and 'd'. The tree homomorphism $(h_k \mid k \in \mathbb{N})$ induces a mapping $h \colon T_\Sigma(S) \to T_\Delta(S)$ defined in the usual way. We denote by H the class of all tree homomorphisms, and for any combination w of '1', 'n', 's', and 'd' we denote by w-H the class of all w-tree homomorphisms. The set $\text{Reg}(\Gamma)$ contains all *regular tree languages* $L \subseteq T_\Gamma$ [13,14] over the ranked alphabet Γ. Finally, let $\text{FTA}(\Gamma) = \{\text{id}_L \mid L \in \text{Reg}(\Gamma)\}$, where $\text{id}_L = \{(t,t) \mid t \in L\}$, and let $\text{FTA} = \bigcup_\Gamma \text{FTA}(\Gamma)$ be the class of all partial identities induced by $\bigcup_\Gamma \text{Reg}(\Gamma)$.

3   Linear Extended Top-Down Tree Transducers

Our main model is the linear extended top-down tree transducer [1,2,17,16] with regular look-ahead (l-xt^R), which is based on the non-extended variant [21,22,6]. We will present it in a form that is closer to synchronous grammars [4].

Definition 1 (see [15, Section 2.2]). A *linear extended top-down tree transducer* with regular look-ahead (l-xt^R) is a tuple $M = (Q, \Sigma, \Delta, I, R, c)$, where

- Q is a finite set of *states*, of which those in $I \subseteq Q$ are *initial*,
- Σ and Δ are ranked alphabets of *input* and *output* symbols,
- $R \subseteq T_\Sigma(Q) \times Q \times T_\Delta(Q)$ is a finite set of *rules* such that ℓ and r are linear in Q and $\text{var}(r) \subseteq \text{var}(\ell)$ for every $(\ell, q, r) \in R$, and
- $c \colon Q^{\text{la}} \to \text{Reg}(\Sigma)$ assigns regular look-ahead to each (potentially) deleted state, where $Q^{\text{la}} = \{q' \in Q \mid \exists (\ell, q, r) \in R \colon q' \in \text{var}(\ell), q' \notin \text{var}(r)\}$.

Next, we recall some important syntactic properties of our model. To this end, let $M = (Q, \Sigma, \Delta, I, R, c)$ be an l-xt^R for the rest of the paper. It is

- a *linear extended tree transducer* [l-xt], if $c(q) = T_\Sigma$ for every $q \in Q^{\text{la}}$,
- a *linear top-down tree transducer with regular look ahead* [l-t^R] if $\ell \in \Sigma(Q)$ for every $(\ell, q, r) \in R$,
- a *linear top-down tree transducer* [l-t] if it is both an l-xt and an l-t^R,
- ε-*free* [$\sharp$] (resp. *strict* [s]) if $\ell \notin Q$ (resp. $r \notin Q$) for every $(\ell, q, r) \in R$,
- a *delabeling* [d] if $\ell \in \Sigma(Q)$ and $r \in Q \cup \Delta(Q)$ for every $(\ell, q, r) \in R$,
- *nondeleting* [n] if $\text{var}(r) = \text{var}(\ell)$ for every $(\ell, q, r) \in R$ (i.e., $Q^{\text{la}} = \emptyset$), and
- a *finite-state relabeling* [qr] if it is a nondeleting, strict delabeling l-t such that $\text{pos}_p(\ell) = \text{pos}_p(r)$ for every $(\ell, q, r) \in R$ and $p \in \text{var}(r)$.

For example, dl-t stands for "delabeling linear top-down tree transducer". We write $\ell \xrightarrow{q_1, \ldots, q_k} r$ for the rules $(\ell, q_1, r), \ldots, (\ell, q_k, r)$. For every $p \in Q$ and $(\ell, q, r) \in R$ we identify $\text{pos}_p(\ell)$ and $\text{pos}_p(r)$ with their unique element if the sets are non-empty. Finally, for every $q \in Q$, we let $R_q = \{\rho \in R \mid \rho = (\ell, q, r)\}$.

Example 2. Let us consider the dl-t^R $M_1 = (Q, \Sigma, \Sigma, \{\star\}, R, c)$ with the states $Q = \{\star, p, q, q^{\text{la}}, \text{id}, \text{id}'\}$, the symbols $\Sigma = \{\sigma^{(2)}, \sigma_1^{(2)}, \sigma_2^{(2)}, \gamma^{(1)}, \alpha^{(0)}\}$, and the following rules in R:

$$\sigma_1(p,q) \xrightarrow{\star,p} \sigma_1(p,q) \qquad \sigma(q,\text{id}) \xrightarrow{q} q \qquad \gamma(\text{id}) \xrightarrow{\text{id},\text{id}'} \gamma(\text{id})$$

$$\sigma_2(\text{id},\text{id}') \xrightarrow{p,q} \sigma_2(\text{id},\text{id}') \qquad \sigma(q^{\text{la}},q) \xrightarrow{q} q \qquad \alpha \xrightarrow{\text{id},\text{id}'} \alpha \ .$$

Since $Q^{\text{la}} = \{q^{\text{la}}, \text{id}\}$, we set $c(q^{\text{la}}) = \{t \in T_\Sigma \mid \text{pos}_{\sigma_2}(t) = \emptyset\}$ and $c(\text{id}) = T_\Sigma$.

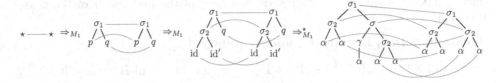

Fig. 1. Derivation using the dl-t^R M_1 of Example 2

Next, we recall the semantics of the l-xt^R M, which is given by synchronous substitution. Let $\mathcal{L} = \{D \mid D \subseteq \mathbb{N}^* \times \mathbb{N}^*\}$ be the set of all *link structures*.

Definition 3 (see [10, Section 3]). A triple $\langle \xi, D, \zeta \rangle \in T_\Sigma(Q) \times \mathcal{L} \times T_\Delta(Q)$ is a *sentential form* (for M) if $v \in \mathrm{pos}(\xi)$ and $w \in \mathrm{pos}(\zeta)$ for every $(v, w) \in D$. For a set $\mathcal{S}$ of sentential forms we define $\mathrm{links}(\mathcal{S}) = \{D \mid \langle \xi, D, \zeta \rangle \in \mathcal{S}\}$. Let $\rho \in R$ be the rule (ℓ, q, r), and let $v, w \in \mathbb{N}^*$. The *explicit link structure* of ρ for the positions v and w is $\mathrm{links}_{v,w}(\rho) = \{(v.\,\mathrm{pos}_p(\ell), w.\,\mathrm{pos}_p(r)) \mid p \in \mathrm{var}(r)\}$.

Definition 4 (see [10, Section 3]). Given two sentential forms $\langle \xi, D, \zeta \rangle$ and $\langle \xi', D', \zeta' \rangle$, we write $\langle \xi, D, \zeta \rangle \Rightarrow_M \langle \xi', D', \zeta' \rangle$ if

- there are $\rho = (\ell, q, r) \in R$ and $(v, w) \in D$ with $v \in \mathrm{pos}_q(\xi)$ and $w \in \mathrm{pos}_q(\zeta)$ such that $\xi' = \xi[\ell]_v$, $\zeta' = \zeta[r]_w$, and $D' = D \cup \mathrm{links}_{v,w}(\rho)$, or
- there are $v \in \mathrm{pos}_Q(\xi)$ and $t \in c(\xi(v))$ with $w \notin \mathrm{pos}_Q(\zeta)$ for all $(v, w) \in D$ such that $\xi' = \xi[t]_v$, $\zeta' = \zeta$, and $D' = D$.

The l-xt^R M computes the dependencies

$$\mathrm{dep}(M) = \{(t, D, u) \in T_\Sigma \times \mathcal{L} \times T_\Delta \mid \exists q \in I : \langle q, \{(\varepsilon, \varepsilon)\}, q \rangle \Rightarrow^*_M (t, D, u)\} ,$$

where $\varepsilon \in \mathbb{N}^*$ is the empty word and $\Rightarrow^*_M$ is the reflexive, transitive closure of $\Rightarrow_M$. It also computes the link structures $\mathrm{links}(M) = \mathrm{links}(\mathrm{dep}(M))$ and the tree transformation $M = \{(t, u) \mid (t, D, u) \in \mathrm{dep}(M)\}$.

A few derivation steps using M_1 of Example 2 are illustrated in Fig. 1. Since every translation $(t, u) \in M$ is ultimately created by (at least) one successful derivation, we can inspect the links in the derivation process to obtain dependencies, which were called *contributions* in [7]. We use stem-capitalized versions of the abbreviations for the corresponding classes of computed tree transformations. For instance, dnl-XT is the class of all tree transformations computable by dnl-xt. The regular look-ahead is useless for nondeleting l-xt^R (because $Q^{\mathrm{la}} = \emptyset$), and thus nl-XT^R = nl-XT and similarly for the non-extended case and for all defined subclasses. Finally, we use the brackets '[' and ']' for optional use of the restrictions $\notin$, 's', 'd', and 'n' that have to be consistently applied.

Next, we relate the class l-XT^R to l-T^R, which tells us how to emulate linear extended top-down tree transducers with regular look-ahead by linear top-down tree transducers with regular look-ahead. To illustrate the consistent application of optional restrictions, we observe that $\notin$l-XT^R = snl-H^{-1} ; l-T^R

and $\not{c}$sdl-XT^R = snl-H^{-1} ; $\not{c}$sdl-T^R are instances of the first result of the next theorem.

Theorem 5 ([11, Lemma 4.1 and Corollary 4.1]).

$$\not{c}[s][d][n]l\text{-}XT^R = snl\text{-}H^{-1} \, ; [s][d][n]l\text{-}T^R \quad [s][d][n]l\text{-}XT^R = nl\text{-}H^{-1} \, ; [s][d][n]l\text{-}T^R$$

4   Our Classes Are Closed at a Finite Power

In this section, we show that the classes $\not{c}$l-XT^R, $\not{c}$l-XT, $\not{c}$sl-XT^R, and $\not{c}$sl-XT are closed under composition at a finite power. We first recall a central result of [3]. Note that [3] expresses this result in terms of a class B of bimorphisms, but $\not{c}$snl-XT = B by [2] and [18, Theorem 4].

Theorem 6 ([3, Theorem 6.2]). $\not{c}$snl-XT $\subsetneq$ $\not{c}$snl-XT^2 = $\not{c}$snl-XT^n for $n \geq 3$.

Now we establish our first composition result, which is analogous to the classical composition result for linear top-down tree transducers with regular look-ahead [6, Theorem 2.11]. The only difference is that our first transducer has extended left-hand sides (i.e., it is an l-xt^R instead of just an l-t^R).

Lemma 7. $[\not{c}][s][d][n]l\text{-}XT^R \, ; [s][d][n]l\text{-}T^R = [\not{c}][s][d][n]l\text{-}XT^R$

Proof. Immediate, from Theorem 5 and the classical composition result for l-T^R in [6, Theorem 2.11][1], which states $[s][d][n]l\text{-}T^R \, ; [s][d][n]l\text{-}T^R = [s][d][n]l\text{-}T^R$. □

Next, we present a decomposition that corresponds to property P of [5, Section II-2-2-3-2]. It demonstrates how to simulate an $\not{c}$l-xt^R by a delabeling l-t^R and an $\not{c}$snl-xt, for which we have the composition closure result in Theorem 6. We immediately combine the result with Lemma 7 to demonstrate, how we can shift an [s]dl-t^R from the back to the front.

Lemma 8. $\not{c}[s]l\text{-}XT^R \, ; [s]dl\text{-}T^R \subseteq \not{c}[s]l\text{-}XT^R \subseteq [s]dl\text{-}T^R \, ; \not{c}$snl-XT

Proof. The first inclusion is due to Lemma 7. For the second inclusion, assume that M is ε-free. Moreover, let $m \in \mathbb{N}$ be such that $m \geq |\text{var}(r)|$ for every $(\ell, q, r) \in R$. For every rule $\rho = (\ell, q, r) \in R$ and non-state position $w \in \text{pos}_\Sigma(\ell)$, let $\text{used}_\rho(w) = \{i \in \mathbb{N} \mid wi \in \text{pos}(\ell), \text{var}(\ell|_{wi}) \cap \text{var}(r) \neq \emptyset\}$. We construct a dl-xt^R $M' = (Q', \Sigma, \Gamma, I', R', c')$ such that

- $Q' = \{\langle \rho, w \rangle \mid \rho = (\ell, q, r) \in R, w \in \text{pos}(\ell)\}$ and $I' = \{\langle \rho, \varepsilon \rangle \mid q \in I, \rho \in R_q\}$,
- $\Gamma = \{\rho^{(|\text{used}_\rho(\varepsilon)|)} \mid \rho \in R\} \cup \{@_i^{(i)} \mid 0 \leq i \leq m\}$,
- for every rule $\rho = (\ell, q, r) \in R$ and non-state position $w \in \text{pos}_\Sigma(\ell)$, the rule

$$\ell(w)(\langle \rho, w1 \rangle, \ldots, \langle \rho, wk \rangle) \xrightarrow{\langle \rho, w \rangle} \begin{cases} \langle \rho, wi_1 \rangle & \text{if } r \in Q \\ \rho(\langle \rho, wi_1 \rangle, \ldots, \langle \rho, wi_n \rangle) & \text{if } r \notin Q, w = \varepsilon \\ @_n(\langle \rho, wi_1 \rangle, \ldots, \langle \rho, wi_n \rangle) & \text{otherwise,} \end{cases}$$

is in R', where $\ell(w) \in \Sigma_k$ and $\{i_1, \ldots, i_n\} = \text{used}_\rho(w)$ with $i_1 < \cdots < i_n$,

---

[1] The abbreviation 'd' has a completely different meaning in [6].

- for every rule $\rho = (\ell, q, r) \in R$, (non-deleted) state position $w \in \mathrm{pos}_{\mathrm{var}(r)}(\ell)$, and rule $\rho' \in R_{\ell(w)}$, the rule $\langle \rho', \varepsilon \rangle \xrightarrow{\langle \rho, w \rangle} \langle \rho', \varepsilon \rangle$ is in R', and
- $c'(\langle \rho, w \rangle) = \ell|_w[q \leftarrow c(q) \mid q \in \mathrm{var}(\ell|_w)]$ for every potentially deleted state $\langle \rho, w \rangle \in \{\langle \rho, w \rangle \in Q \mid \mathrm{used}_\rho(w) = \emptyset\}$, where $\leftarrow$ denotes the standard OI-substitution [8].

To obtain the desired dl-t^R we simply eliminate the ε-rules using standard methods.[2] Intuitively speaking, the transducer M' processes the input and deletes subtrees that are not necessary for further processing. Moreover, it executes nonstrict rules of M and marks the positions in the input where a strict rule application would be possible. It remains to construct the l-xt M''. Let $m'' \geq |\ell|$ for all $(\ell, q, r) \in R$, and let $M'' = (\{\star\}, \Gamma, \Delta, \{\star\}, R'')$ such that R'' contains all valid rules $\rho(t_1, \ldots, t_k) \xrightarrow{*} r[q \leftarrow \star \mid q \in Q]$ of a strict nondeleting l-xt with $\rho = (\ell, q, r) \in R$, $\mathrm{pos}_R(t_i) = \emptyset$, and $|t_i| \leq m''$ for every $1 \leq i \leq k$, where k is the rank of ρ.   □

Example 9. Let $\rho = \sigma(p, \sigma(\alpha, q)) \xrightarrow{q} \sigma(\alpha, \sigma(q, \alpha))$ be a rule with non-trivial look-ahead $c(p) = L$. We illustrate the construction of M' (in Lemma 8):

$$\sigma(\langle \rho, 1 \rangle, \langle \rho, 2 \rangle) \xrightarrow{\langle \rho, \varepsilon \rangle} \rho(\langle \rho, 2 \rangle) \qquad\qquad \alpha \xrightarrow{\langle \rho, 21 \rangle} @_0$$

$$\sigma(\langle \rho, 21 \rangle, \langle \rho, 22 \rangle) \xrightarrow{\langle \rho, 2 \rangle} @_1(\langle \rho, 22 \rangle) \qquad\qquad \langle \rho', \varepsilon \rangle \xrightarrow{\langle \rho, 22 \rangle} \langle \rho', \varepsilon \rangle$$

for all rules $\rho' \in R_q$. Moreover, the look-ahead c' of M' is such that $c'(\langle \rho, 1 \rangle) = L$ and $c'(\langle \rho, 21 \rangle) = \{\alpha\}$.

Theorem 10. $(\not\varepsilon[\mathrm{s}]\mathrm{l}\text{-}\mathrm{XT}^R)^n \subseteq [\mathrm{s}]\mathrm{dl}\text{-}\mathrm{T}^R \,;\, \not\varepsilon\mathrm{snl}\text{-}\mathrm{XT}^2 \subseteq (\not\varepsilon[\mathrm{s}]\mathrm{l}\text{-}\mathrm{XT}^R)^3$ for $n \geq 1$.

Proof. The second inclusion is trivial. We prove the first inclusion by induction over n. For $n = 1$, it follows from Lemma 8, and in the induction step, we obtain

$$(\not\varepsilon[\mathrm{s}]\mathrm{l}\text{-}\mathrm{XT}^R)^{n+1} \quad \subseteq \not\varepsilon[\mathrm{s}]\mathrm{l}\text{-}\mathrm{XT}^R \,;\, [\mathrm{s}]\mathrm{dl}\text{-}\mathrm{T}^R \,;\, \not\varepsilon\mathrm{snl}\text{-}\mathrm{XT}^2$$
$$\subseteq [\mathrm{s}]\mathrm{dl}\text{-}\mathrm{T}^R \,;\, \not\varepsilon\mathrm{snl}\text{-}\mathrm{XT}^3 = [\mathrm{s}]\mathrm{dl}\text{-}\mathrm{T}^R \,;\, \not\varepsilon\mathrm{snl}\text{-}\mathrm{XT}^2$$

by the induction hypothesis, then Lemma 8, and lastly Theorem 6.   □

It is known [6, Theorem 2.6] that we can simulate every l-t^R (with look-ahead) by a composition of two l-t (without look-ahead). This allows us to conclude that the class $\not\varepsilon$l-XT is closed under composition at the fourth power.

Corollary 11. $\not\varepsilon[\mathrm{s}]\mathrm{l}\text{-}\mathrm{XT}^n \subseteq \mathrm{QR} \,;\, [\mathrm{s}]\mathrm{dl}\text{-}\mathrm{T} \,;\, \not\varepsilon\mathrm{snl}\text{-}\mathrm{XT}^2 \subseteq \not\varepsilon[\mathrm{s}]\mathrm{l}\text{-}\mathrm{XT}^4$ for every $n \geq 1$.

Proof. The second inclusion is trivial, and for the first inclusion we use Theorem 10 and $[\mathrm{s}]\mathrm{dl}\text{-}\mathrm{T}^R \subseteq \mathrm{QR} \,;\, [\mathrm{s}]\mathrm{dl}\text{-}\mathrm{T}$.   □

---

[2] Note that due to the ε-freeness of M, we have $w \neq \varepsilon$ in the ε-rules of the fourth item. Since these rules are the only constructed ε-rules, we cannot chain two ε-rules.

In the rest of the section, we will show that the (strict) classes $\not\!\!s$l-XT^R and $\not\!\!s$l-XT are closed under composition already at the second power. This time, the main lemma demonstrates how to shift a strict delabeling linear homomorphism from the front to the back again creating a nondeleting transducer (cf. Lemma 8).

Lemma 12. sdl-H ; $\not\!\!s$l-XT $\subseteq$ $\not\!\!s$l-XT $\subseteq$ $\not\!\!s$nl-XT ; sdl-H

Proof. For the first inclusion, let $d\colon T_\Gamma \to T_\Sigma$ be a strict delabeling linear tree homomorphism. Moreover, assume that M is a strict and ε-free l-xt, and let $m \in \mathbb{N}$ be such that $m \geq |\ell|$ for every $(\ell, q, r) \in R$. We construct the l-xt $M' = (Q', \Gamma, \Delta, I, R', c')$ with $Q' = Q \cup \{1, \ldots, m\}$ such that for every rule $(\ell, q, r) \in R$ we have each valid rule (ℓ', q, r) in R' where $\ell' \in d^{-1}(\ell)$ and $|\mathrm{pos}_\Gamma(\ell')| = |\mathrm{pos}_\Sigma(\ell)|$. Recall that d also defines a tree transformation $d\colon T_\Gamma(Q') \to T_\Sigma(Q')$, which acts as an identity on states; i.e., $d(q') = q'$ for every $q' \in Q'$. Moreover, $c'(q') = T_\Gamma$ for all $q' \in (Q')^{\mathrm{la}}$. Finally, we observe that M' is strict because it has the same right-hand sides as M, and it is ε-free because h is strict. For the second inclusion,

$$\not\!\!s\text{l-XT} \subseteq \text{snl-H}^{-1} \,;\, \text{FTA} \,;\, \text{sl-H} \subseteq \text{snl-H}^{-1} \,;\, \text{FTA} \,;\, \text{snl-H} \,;\, \text{sdl-H} \subseteq \not\!\!s\text{nl-XT} \,;\, \text{sdl-H} \ ,$$

where the first and the last inclusion are by [18, Theorem 4] and the second inclusion is due to [5, Section I-2-1-3-5]. $\qquad\square$

In contrast to Theorem 10 and Corollary 11, look-ahead does not increase the power of closedness in the strict case. In fact, the next theorem shows that $(\not\!\!s\text{l-XT}^R)^n = \not\!\!s\text{l-XT}^n$ for all $n \geq 2$.

Theorem 13. $(\not\!\!s\text{l-XT}^R)^n \subseteq \not\!\!s\text{nl-XT} \,;\, \not\!\!s\text{l-XT} \subseteq \not\!\!s\text{l-XT}^2$ for every $n \geq 1$.

Proof. Again, the second inclusion is trivial. For the first inclusion, we first prove that $\not\!\!s[n]$l-XT^R ; $\not\!\!s$l-XT^R = $\not\!\!s[n]$l-XT^R ; $\not\!\!s$l-XT, which we call (†), as follows:

$$\not\!\!s[n]\text{l-XT}^R \,;\, \not\!\!s\text{l-XT}^R \subseteq \not\!\!s[n]\text{l-XT}^R \,;\, \text{QR} \,;\, \not\!\!s\text{l-XT} \subseteq \not\!\!s[n]\text{l-XT}^R \,;\, \not\!\!s\text{l-XT} \ ,$$

where we used [6, Theorem 2.6] in the first step and Lemma 7 in the second step.[3] Now we prove the first inclusion of our main statement by induction on n. The induction basis ($n = 1$) follows from $\not\!\!s$l-XT$^R \subseteq$ QR ; $\not\!\!s$l-XT [6, Theorem 2.6], and the induction step is proved as follows

$$(\not\!\!s\text{l-XT}^R)^{n+1} \subseteq (\not\!\!s\text{l-XT}^R)^n \,;\, \not\!\!s\text{l-XT} \subseteq \not\!\!s\text{nl-XT} \,;\, \not\!\!s\text{l-XT}^2 \subseteq \not\!\!s\text{nl-XT}^3 \,;\, \text{sdl-H}$$

$$\subseteq \not\!\!s\text{nl-XT}^2 \,;\, \text{sdl-H} \qquad \subseteq \not\!\!s\text{nl-XT} \,;\, \not\!\!s\text{l-XT}^R \subseteq \not\!\!s\text{nl-XT} \,;\, \not\!\!s\text{l-XT}$$

using, in sequence, statement (†), the induction hypothesis, Lemma 12 twice, Theorem 6, Lemma 7, and statement (†) again. $\qquad\square$

---

[3] The converse inclusion is trivial.

5   Least Power of Closedness

In this section, we will determine the least power at which the class is closed under composition for the classes $l\text{-XT}^R$, $sl\text{-XT}^R$, and $sl\text{-XT}$. In addition, we conjecture the least power for the class $l\text{-XT}$.

Theorem 14. For every $n \geq 3$

$$sl\text{-XT} \subsetneq sl\text{-XT}^R \subsetneq sl\text{-XT}^2 = (sl\text{-XT}^R)^2 = sl\text{-XT}^n = (sl\text{-XT}^R)^n .$$

Proof. Theorem 13 proves the final three equalities. The first inclusion is trivial and strictness follows from the proof of [15, Lemma 4.3]. The second inclusion is also trivial (given the previous equalities) and the strictness follows from [18, Theorem 4] and [3, Section 3.4], which show that class $sl\text{-XT}^R$ is not closed under composition at power 1.[4]   □

Definition 15 ([19, Definitions 8 and 10]). A set $\mathcal{D} \subseteq \mathcal{L}$ of link structures

- is *input hierarchical*[5] if for every $D \in \mathcal{D}$ and $(v_1, w_1), (v_2, w_2) \in D$ we have
  (i) if $v_1 \prec v_2$, then $w_1 \preceq w_2$, and (ii) if $v_1 = v_2$, then $w_1 \preceq w_2$ or $w_2 \preceq w_1$.
- has *bounded distance in the input* if there exists an integer $k \in \mathbb{N}$ such that for every $D \in \mathcal{D}$ and all $(v, w), (vv'', w'') \in D$ there exists $(vv', w') \in D$ with $v' \prec v''$ and $|v'| \leq k$.

Moreover, $\mathcal{D}$ is *output hierarchical* (resp. has *bounded distance in the output*) if $\mathcal{D}^{-1}$ is input hierarchical (resp. has bounded distance in the input). If $\mathcal{D}$ fulfills both versions of the property, then we just call it *hierarchical* or *bounded distance*.

Corollary 16 (of Def. 4). links(M) is hierarchical with bounded distance.

We will consider the problem whether a tree transformation can be computed by two l-xt^R. For this we specify certain links that are intuitively clear and necessary between nodes of input-output tree pairs. Then we consider whether this specification can be implemented by two l-xt^R. Often we cannot identify the nodes of a link exactly. In such cases, we use splines with inverted arrow heads, which indicate that there is a link to some position of the subtree pointed to.

Theorem 17. For every $n \geq 4$,

$$l\text{-XT} \subsetneq l\text{-XT}^R \subsetneq l\text{-XT}^2 \subseteq (l\text{-XT}^R)^2 \subsetneq l\text{-XT}^3 \subseteq (l\text{-XT}^R)^3$$
$$= l\text{-XT}^4 = (l\text{-XT}^R)^n = l\text{-XT}^{n+1} .$$

Proof. We have $(l\text{-XT}^R)^n \subseteq l\text{-XT}^{n+1}$ for all $n \geq 1$ by repeated application of Lemma 8. The equalities follow from Theorem 10, so we only have to prove strictness. The first inclusion is strict by [15, Lemma 4.3] and the strictness of the second inclusion follows from that of the fourth. Finally, we prove the

---

[4] In fact, Theorem 17 reproves this statement.
[5] This notion is called *strictly input hierarchical* in [19].

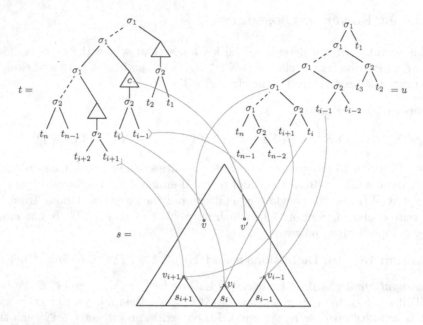

Fig. 2. The relevant part of the specification used in the proof of Theorem 17

strictness of the fourth inclusion. For this, recall the l-t^R M_1 of Example 2. In addition, we use the two bimorphisms B_2 and B_3 of [5, Section II-2-2-3-1], which are in the class B mentioned before Theorem 6, and hence can also be defined by some $\notin$snl-xt M_2 and M_3, respectively. For convenience, we present M_2 and M_3 explicitly before we show that $\tau = M_1 ; M_2 ; M_3$ cannot be computed by a composition of two $\notin$l-xt^R.

Let $M_2 = (\{\star, \mathrm{id}, \mathrm{id}'\}, \Sigma, \Sigma, \{\star\}, R_2)$ be the $\notin$snl-xt with the rules

$$\sigma_1(\star, \sigma_2(\mathrm{id}, \mathrm{id}')) \xrightarrow{\star} \sigma(\sigma(\star, \mathrm{id}), \mathrm{id}') \qquad \sigma_2(\mathrm{id}, \mathrm{id}') \xrightarrow{\star} \sigma(\mathrm{id}, \mathrm{id}')$$

$$\gamma(\mathrm{id}) \xrightarrow{\mathrm{id}, \mathrm{id}'} \gamma(\mathrm{id}) \qquad\qquad\qquad \alpha \xrightarrow{\mathrm{id}, \mathrm{id}'} \alpha .$$

Moreover, let $M_3 = (\{\star, p, \mathrm{id}, \mathrm{id}'\}, \Sigma, \Sigma, \{\star\}, R_3)$ be the $\notin$snl-xt with the rules

$$\sigma(p, \mathrm{id}) \xrightarrow{\star} \sigma_1(p, \mathrm{id}) \qquad\qquad \sigma(\sigma(p, \mathrm{id}), \mathrm{id}') \xrightarrow{p} \sigma_1(p, \sigma_2(\mathrm{id}, \mathrm{id}'))$$

$$\gamma(\mathrm{id}) \xrightarrow{p} \gamma(\mathrm{id}) \qquad\qquad \gamma(\mathrm{id}) \xrightarrow{\mathrm{id}, \mathrm{id}'} \gamma(\mathrm{id}) \qquad \alpha \xrightarrow{p, \mathrm{id}, \mathrm{id}'} \alpha .$$

We present a proof by contradiction, hence we assume that $\tau = N_1 ; N_2$ for some $\notin$l-xt^R $N_1 = (P_1, \Sigma, \Delta, I_1, R_1')$ and $N_2 = (P_2, \Delta, \Sigma, I_2, R_2')$. Using a standard construction, we can construct an ε-cycle free l-xt^R $N_2' = (P_2', \Delta, \Sigma, I_2', R_2'')$ such that $N_2' = N_2$ and each rule $\ell \xrightarrow{p} r \in R_2''$ that contains γ in its right-hand side r obeys $r = \gamma(p)$ with $p \in P_2'$. With the help of Corollary 16, we can further conclude that $\mathrm{links}(N_1)$ and $\mathrm{links}(N_2')$ are hierarchical with bounded distance k_1 and k_2, respectively. Moreover, let

$$m \geq \max \{k_1, k_2, |\ell|, |r| \mid \ell \xrightarrow{p} r \in R_1' \cup R_2''\} .$$

Clearly, all $(t, u) \in \tau$ have the shape shown in Fig. 2. Next, we will make an assumption, derive the contradiction, and then prove the assumption. Suppose that there exists $(t, u) \in \tau$ such that (see Fig. 2 for the named subtrees)

- the left σ_1-spine of t is longer than m,
- for all trees $c' \in T_\Sigma(\{x_1\})$ indicated by small triangles in t (like c in Fig. 2) the only element of $\mathrm{pos}_{x_1}(c')$ is longer than m, and
- for all $(t, D_1, s) \in \mathrm{dep}(N_1)$, $(s, D_2, u) \in \mathrm{dep}(N_2')$, and $1 \leq j \leq n$ we have
  - there exists $(v_j, w_j) \in D_2$ such that w_j is the root of t_j in u, and
  - there exists $(y_j, v_j') \in D_1$ such that y_j is a position inside t_j and $v_j \preceq v_j'$.

Since the left σ_1-spine in u is longer than k_2 and there are links at the root (i.e., $(\varepsilon, \varepsilon) \in D_2$) and at w_n, there must be a linking point at position $w \in \mathrm{pos}_{\sigma_1}(u)$ along the left σ_1-spine with $w \neq \varepsilon$, which links to position v in the intermediate tree s (i.e., $(v, w) \in D_2$). Let $u|_w = \sigma_1(u', \sigma_2(t_{i+1}, t_i))$ for some $2 \leq i \leq n-2$. By our assumption, there exist links $(v_{i+1}, w_{i+1}), (v_i, w_i), (v_{i-1}, w_{i-1}) \in D_2$. Since D_2 is hierarchical and w_{i+1} and w_i are below w in u, we know that v_{i+1} and v_i are below v in s (i.e., $v \preceq v_{i+1}, v_i$), whereas $v \npreceq v_{i-1}$. Next, we locate t_i in the input tree t. By the general shape of t, the subtree t_i occurs in a subtree $\sigma_1(t', c[\sigma_2(t_i, t_{i-1})])$ for some tree $c \in T_\Sigma(\{x_1\})$ with exactly one occurrence of x_1. We know that c is suitably large, which forces a linking point y inside c in addition to those in t_{i+1}, t_i, and t_{i-1}, which exist by the assumption. Note that y is a proper prefix of the root position of the subtree $\sigma_2(t_i, t_{i-1})$. Let $(y, v') \in D_1$ be the link linking c to s, which dominates the links $(y_i, v_i'), (y_{i-1}, v_{i-1}') \in D_1$ linking t_i and t_{i-1} to s, respectively. Thus, $v' \preceq v_i', v_{i-1}'$ and $v' \npreceq v_{i+1}'$ because $y \npreceq y_{i+1}$. Obviously, $v' \npreceq v_{i+1}$, and moreover, $v' \preceq v_i, v_{i-1}$ because otherwise the positions v_{i+1}, v_i, v_{i-1} would not be incomparable, which is required because $\mathrm{links}(N_2')$ is hierarchical. We have either $\mathrm{lcp}(v_{i+1}, v_i) \preceq \mathrm{lcp}(v_i, v_{i-1})$ or $\mathrm{lcp}(v_i, v_{i-1}) \preceq \mathrm{lcp}(v_{i+1}, v_i)$. We either get $v \preceq \mathrm{lcp}(v_{i+1}, v_i) \preceq \mathrm{lcp}(v_i, v_{i-1}) \preceq v_{i-1}$ or $v' \preceq \mathrm{lcp}(v_i, v_{i-1}) \preceq \mathrm{lcp}(v_{i+1}, v_i) \preceq v_{i+1}$, which are both contradictions.

It remains to show the assumption. Obviously, the first two items can be satisfied simply by a proper selection of $(t, u) \in \tau$. For every $1 \leq j \leq n$, we know that there exists a link $(v_j, w_j) \in D_2$ to the root of t_j in u due to the special shape of the right-hand sides of N_2'. We note that all $v_1, \ldots, v_n$ are pairwise incomparable. Moreover, we observe that there is a linear height (and size) relation between input and output trees related by a link, which is true for all ε-cycle free l-xt. Consequently, there is a linear height relation between $s_j = s|_{v_j}$ and $t_j = u|_{w_j}$. Thus by selecting each t_j suitably tall, we can enforce that each s_j is taller than m, which yields that there is a link $(y_j, v_j') \in D_1$ such that $v_j \preceq v_j'$. Exploiting the linear height (and size) relation between linked subtrees again, we can additionally show that (i) y_j is a position inside t_j in t, in which case we are done, or (ii) y_j is a prefix of the root position of t_j in t. In the latter case, the size of t_j can be chosen such that there is also a link (y_j', v_j'') with $y_j \prec y_j'$ and $v_j' \preceq v_j''$. Moreover, this can be iterated until y_j' points to a position inside t_j. A detailed proof of these statements can be found in [9]. $\square$

We conjecture that $\not\negthinspace l\text{-XT}^3 \subsetneq \not\negthinspace l\text{-XT}^4 = \not\negthinspace l\text{-XT}^n$ for every $n \geq 4$. The inclusion is trivial and the equality follows from Corollary 11. For the strictness, the proof of Theorem 17 essentially shows that in the first step we must delete the contexts indicated by triangles (such as c) in Fig. 2 because otherwise we can apply the method used in the proof to derive a contradiction (it relies on the existence of a linking point inside such a context c). Thus, in essence we must first implement a variant of the $\not\negthinspace l\text{-xt}^R$ M_1 of Example 2. It is a simple exercise to show that the deletion of the excess material cannot be done by a single l-xt as it cannot reliably determine the left-most occurrence of σ_2 without the look-ahead. Thus, if we only have l-xt to achieve the transformation, then we already need a composition of two l-xt to perform the required deletion.

For the sake of completeness we mention the following. In the full version [9] of this paper we prove that the composition hierarchy is infinite for all other combinations of '$\not\negthinspace$', 's', and 'n'.

Theorem 18 ([9, Theorem 34]). The composition hierarchy of the classes $\not\negthinspace nl\text{-XT}$, $[s][n]l\text{-XT}^R$, and $[s][n]l\text{-XT}$ is infinite.

Acknowledgment. The authors are indebted to an anonymous referee for his valuable report.

References

1. Arnold, A., Dauchet, M.: Transductions inversibles de forêts. Thèse 3ème cycle M. Dauchet, Université de Lille (1975)
2. Arnold, A., Dauchet, M.: Bi-transductions de forêts. In: ICALP, pp. 74–86. Edinburgh University Press (1976)
3. Arnold, A., Dauchet, M.: Morphismes et bimorphismes d'arbres. Theoret. Comput. Sci. 20(1), 33–93 (1982)
4. Chiang, D.: An introduction to synchronous grammars. In: ACL, Association for Computational Linguistics (2006); part of a tutorial given with K. Knight
5. Dauchet, M.: Transductions de forêts — Bimorphismes de magmoïdes. Première thèse, Université de Lille (1977)
6. Engelfriet, J.: Top-down tree transducers with regular look-ahead. Math. Systems Theory 10(1), 289–303 (1977)
7. Engelfriet, J., Maneth, S.: Macro tree translations of linear size increase are MSO definable. SIAM J. Comput. 32(4), 950–1006 (2003)
8. Engelfriet, J., Schmidt, E.M.: IO and OI I. J. Comput. System Sci. 15(3), 328–353 (1977)
9. Fülöp, Z., Maletti, A.: Composition closure of linear extended top-down tree transducers (2013), manuscript available at http://arxiv.org/abs/1301.1514
10. Fülöp, Z., Maletti, A., Vogler, H.: Preservation of recognizability for synchronous tree substitution grammars. In: ATANLP, pp. 1–9. Association for Computational Linguistics (2010)
11. Fülöp, Z., Maletti, A., Vogler, H.: Weighted extended tree transducers. Fundam. Inform. 111(2), 163–202 (2011)

12. Fülöp, Z., Vogler, H.: Syntax-Directed Semantics — Formal Models Based on Tree Transducers. Springer (1998)
13. Gécseg, F., Steinby, M.: Tree Automata. Akadémiai Kiadó, Budapest (1984)
14. Gécseg, F., Steinby, M.: Tree languages. In: Rozenberg, G., Salomaa, A. (eds.) Handbook of Formal Languages, vol. 3, ch. 1, pp. 1–68. Springer (1997)
15. Graehl, J., Hopkins, M., Knight, K., Maletti, A.: The power of extended top-down tree transducers. SIAM J. Comput. 39(2), 410–430 (2009)
16. Graehl, J., Knight, K., May, J.: Training tree transducers. Comput. Linguist. 34(3), 391–427 (2008)
17. Knight, K., Graehl, J.: An overview of probabilistic tree transducers for natural language processing. In: Gelbukh, A. (ed.) CICLing 2005. LNCS, vol. 3406, pp. 1–24. Springer, Heidelberg (2005)
18. Maletti, A.: Compositions of extended top-down tree transducers. Inform. and Comput. 206(9–10), 1187–1196 (2008)
19. Maletti, A.: Tree transformations and dependencies. In: Kanazawa, M., Kornai, A., Kracht, M., Seki, H. (eds.) MOL 12. LNCS, vol. 6878, pp. 1–20. Springer, Heidelberg (2011)
20. May, J., Knight, K., Vogler, H.: Efficient inference through cascades of weighted tree transducers. In: ACL, pp. 1058–1066. Association for Computational Linguistics (2010)
21. Rounds, W.C.: Mappings and grammars on trees. Math. Systems Theory 4(3), 257–287 (1970)
22. Thatcher, J.W.: Generalized^2 sequential machine maps. J. Comput. System Sci. 4(4), 339–367 (1970)

Subword Complexity and k-Synchronization

Daniel Goč, Luke Schaeffer, and Jeffrey Shallit

School of Computer Science, University of Waterloo, Waterloo, ON N2L 3G1 Canada
{dgoc,l3schaef,shallit}@uwaterloo.ca

Abstract. We show that the subword complexity function $\rho_{\mathbf{x}}(n)$, which counts the number of distinct factors of length n of a sequence $\mathbf{x}$, is k-synchronized in the sense of Carpi if $\mathbf{x}$ is k-automatic. As an application, we generalize recent results of Goldstein. We give analogous results for the number of distinct factors of length n that are primitive words or powers. In contrast, we show that the function that counts the number of unbordered factors of length n is *not* necessarily k-synchronized for k-automatic sequences.

1   Introduction

Let $k \geq 2$ be an integer, let $\Sigma_k = \{0, 1, 2, \ldots, k-1\}$, and let $(n)_k$ denote the canonical representation of n in base k, starting with the most significant digit, *without* leading zeroes. If $x \in \Sigma_k^*$, we let $[x]_k$ denote the integer represented by x (where x is allowed to have leading zeroes). To represent a pair of integers (m, n), we use words over the alphabet $\Sigma_k \times \Sigma_k$. For such a word x, we let $\pi_i(x)$ to be the projection onto the i'th coordinate. The canonical representation $(m, n)_k$ is defined to be the word x such that $[\pi_1(x)]_k = m$ and $[\pi_2(x)]_k = n$, and having no leading $[0, 0]$'s. For example $(43, 17)_2 = [1, 0][0, 1][1, 0][0, 0][1, 0][1, 1]$.

Recently, Arturo Carpi and his co-authors [6,4,5] introduced a very interesting class of sequences that are computable by automata in a novel fashion: the class of k-synchronized sequences. Let $(f(n))_{n \geq 0}$ be a sequence taking values in $\mathbb{N}$. They call such a sequence k-*synchronized* if there is a deterministic finite automaton M accepting the base-k representation of the graph of f, namely $\{(n, f(n))_k : n \geq 0\}$.

Sequences that are k-synchronized are "halfway between" the class of k-automatic sequences, introduced by Cobham [9] and studied in many papers; and the class of k-regular sequences, introduced by Allouche and Shallit [1,2]. They are particularly interesting for two reasons. If a sequence $(f(n))$ is k-synchronized, then

(a) we immediately get a bound on its growth rate: $f(n) = O(n)$;
(b) we immediately get a linear-time algorithm for efficiently calculating $f(n)$.

Result (a) can be found in [6, Prop. 2.5]. We now state and prove result (b).

Theorem 1. *Suppose $(f(n))_{n \geq 0}$ is k-synchronized. Then there is an algorithm that, given the base-k representation of n, will compute the base-k representation of $f(n)$ in $O(\log n)$ time.*

M.-P. Béal and O. Carton (Eds.): DLT 2013, LNCS 7907, pp. 252–263, 2013.
© Springer-Verlag Berlin Heidelberg 2013

Proof. We know there is a DFA $M = (Q, \Sigma_k \times \Sigma_k, \delta, q_0, F)$ accepting $L = \{(n, f(n))_k : n \geq 0\}$. Let $w = (n)_k$. It is easy to construct a $(|w| + 2)$-state DFA that accepts words with 0^*w in the first coordinate and no leading $[0, 0]$'s. Such a DFA accepts the language $K = \{(n, m)_k : m \geq 0\}$. We now construct a DFA H for the language $K \cap L$, where H has $|Q|(|w| + 2) = \Theta(|w|)$ states, using the familiar direct product construction from automata theory.

The only word in $K \cap L$ is $(n, f(n))_k$, so we apply a linear-time directed graph reachability algorithm (such as breadth-first or depth-first search) to the underlying transition graph of H. This finds the unique path $x \in (\Sigma_k \times \Sigma_k)^*$ from the initial state in W to an accepting state. Then x is labeled $(n, f(n))_k$, so reading the second coordinate yields the base-k representation of $f(n)$. $\square$

In this paper, we are concerned with infinite words over a *finite* alphabet. Let $\mathbf{x} = a_0a_1a_2\cdots$ be an infinite word. By $\mathbf{x}[n]$ we mean a_n and by $\mathbf{x}[m..n]$ we mean the factor $a_m a_{m+1} \cdots a_n$ of $\mathbf{x}$ of length $n - m + 1$. The subword complexity function $\rho_{\mathbf{x}}(n)$ counts the number of distinct factors of length n.

An infinite word or sequence $\mathbf{x}$ is said to be k-automatic if there is an automaton with outputs associated with the states that, on input $(n)_k$, reaches a state with output $\mathbf{x}[n]$. In this paper we show that if $\mathbf{x}$ is a k-automatic sequence, then the subword complexity $\rho_{\mathbf{x}}(n)$ is k-synchronized. As an application, we generalize and simplify recent results of Goldstein [14,15]. Furthermore, we obtain analogous results for the number of length-n primitive words and the number of length-n powers.

We remark that there are a number of quantities about k-automatic sequences already known to be k-synchronized. These include the separator sequence [6], the repetitivity index [4], the recurrence function [8], and the "appearance" function [8]. The latter two examples were not explicitly stated to be k-synchronized in [8], but the result follows immediately from the proofs in that paper.

2   Subword Complexity

Cobham [9] proved that if $\mathbf{x}$ is a k-automatic sequence, then $\rho_{\mathbf{x}}(n) = O(n)$. Cassaigne [7] proved that any infinite word $\mathbf{x}$ satisfying $\rho_{\mathbf{x}}(n) = O(n)$ also satisfies $\rho_{\mathbf{x}}(n + 1) - \rho_{\mathbf{x}}(n) = O(1)$. Carpi and D'Alonzo [5] showed that the subword complexity function $\rho_{\mathbf{x}}(n)$ is a k-regular sequence.

Charlier, Rampersad, and Shallit [8] found this result independently, using a somewhat different approach. They used the following idea. Call an occurrence of the factor $t = \mathbf{x}[i..i+n-1]$ "novel" if t does not appear as a factor of $\mathbf{x}[0..i+n-2]$. In other words, the leftmost occurrence of t in $\mathbf{x}$ begins at position i. Then the number of factors of length n in $\mathbf{x}$ is equal to the number of novel occurrences of factors of length n. The property that $\mathbf{x}[i..i+n-1]$ is novel can be expressed as a predicate, as follows:

$$\{(n, i)_k \ : \ \forall j, 0 \leq j < i \quad \mathbf{x}[i..i + n - 1] \neq \mathbf{x}[j..j + n - 1]\} =$$
$$\{(n, i)_k \ : \ \forall j, 0 \leq j < i \ \exists m, 0 \leq m < n \quad \mathbf{x}[i + m] \neq \mathbf{x}[j + m]\}. \quad (1)$$

As shown in [8], the base-k representation of the integers satisfying any predicate of this form (expressible using quantifiers, integer addition and subtraction, indexing into a k-automatic sequence $\mathbf{x}$, logical operations, and comparisons) can be accepted by an explicitly-constructable deterministic finite automaton. From this, it follows that the sequence $\rho_{\mathbf{x}}(n)$ is k-regular, and hence can be computed explicitly, in polynomial time, in terms of the product of certain matrices and vectors depending on the base-k expansion of n [8].

We show that, in fact, the subword complexity function $\rho_{\mathbf{x}}(n)$ is k-synchronized. The main observation needed is the following (Theorem 3): in any sequence of linear complexity, the starting positions of novel occurrences of factors are "clumped together" in a bounded number of contiguous blocks. This makes it easy to count them.

More precisely, let $\mathbf{x}$ be an infinite word and for any n consider the set of novel occurrences $E_{\mathbf{x}}(n) := \{i \ : \ \text{the occurrence } \mathbf{x}[i..i + n - 1] \text{ is novel }\}$. We consider how $E_{\mathbf{x}}(n)$ evolves with increasing n.

As an example, consider the Thue-Morse sequence

$$\mathbf{t} = t_0 t_1 t_2 \cdots = 0110100110010110 \cdots ,$$

defined by letting t_n be the number of 1's in the binary expansion of n, taken modulo 2. The gray squares in the rows of Figure 1 depict the members of $E_{\mathbf{t}}(n)$ for the Thue-Morse sequence for $1 \leq n \leq 9$.

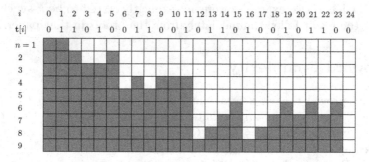

Fig. 1. Evolution of novel occurrences of factors in the Thue-Morse sequence

Lemma 2. *Let $\mathbf{x}$ be an infinite word. If the factor of length n beginning at position i is a novel occurrence, so is*

(a) *the factor of length $n + 1$ beginning at position i;*
(b) *the factor of length $n + 1$ beginning at position $i - 1$ (for $i \geq 1$).*

Proof. (a) Suppose the factor of length $n+1$ also occurs at some position $j < i$. Then the factor of length n also occurs at position j, contradicting the fact that it was a novel occurrence at i.

(b) Suppose the factor of length $n+1$ beginning at position $i-1$ occurs at some earlier position $j < i-1$. We can write the factor as ax, where a is a single letter and x is a word, so the factor of length n beginning at position i must also occur at position $j+1 < i$. But then it is not a novel occurrence. □

Theorem 3. *Let* $\mathbf{x}$ *be an infinite word. For* $n \geq 1$, *the number of contiguous blocks in* $E_{\mathbf{x}}(n)$ *is at most* $\rho_{\mathbf{x}}(n) - \rho_{\mathbf{x}}(n-1) + 1$.

Proof. We prove the claim by induction on n. For $n = 1$ the claim says there are at most $\rho_{\mathbf{x}}(1)$ contiguous blocks, which is evidently true, since there are at most $\rho_{\mathbf{x}}(1)$ novel factors of length 1.

Now assume the claim is true for all $n' < n$; we prove it for n. Consider the evolution of the novel occurrences of factors in going from length $n-1$ to n. Every occurrence that was previously novel is still novel, and furthermore in every contiguous block except the first, we get novel occurrences at one position to the left of the beginning of the block. So if row $n-1$ has t contiguous blocks, then we get $t-1$ novel occurrences at the beginning of each block, except the first. (Of course, the first block begins at position 0, since any factor beginning at position 0 is novel, no matter what the length is.) The remaining $\rho_{\mathbf{x}}(n) - \rho_{\mathbf{x}}(n-1) - (t-1)$ novel occurrences could be, in the worst case, in their own individual contiguous blocks. Thus row n has at most $t + \rho_{\mathbf{x}}(n) - \rho_{\mathbf{x}}(n-1) - (t-1) = \rho_{\mathbf{x}}(n) - \rho_{\mathbf{x}}(n-1) + 1$ contiguous blocks. □

In our Thue-Morse example, it is well-known that $\rho_{\mathbf{t}}(n) - \rho_{\mathbf{t}}(n-1) \leq 4$, so the number of contiguous blocks in any row is at most 5. This is achieved, for example, for $n = 6$.

Example 4. We give an example of a recurrent infinite word over a finite alphabet where the number of contiguous blocks in $E_{\mathbf{x}}(n)$ is unbounded. Consider the word

$$\mathbf{w} = \prod_{n \geq 1} (n)_2 = 110111001011101111000 \cdots.$$

Then for each $n \geq 5$ the first occurrence of each of the words $0^{n-1}1, 0^{n-2}11, \ldots,$ $0^2 1^{n-2}$ have a non-novel occurrence immediately following them, which shows there at at least $n-2$ blocks in $E_{\mathbf{w}}(n)$.

Corollary 5. *If* $\rho_{\mathbf{x}}(n) = O(n)$, *then there is a constant* C *such that every row* $E_{\mathbf{x}}(n)$ *in the evolution of novel occurrences consists of at most* C *contiguous blocks.*

Proof. By the result of Cassaigne [7], we know that there exists a constant C such that $\rho_{\mathbf{x}}(n) - \rho_{\mathbf{x}}(n-1) \leq C - 1$. By Theorem 3, we know there are at most C contiguous blocks in any $E_{\mathbf{x}}(n)$. □

Theorem 6. *Let* **x** *be a k-automatic sequence. Then its subword complexity function $\rho_{\mathbf{x}}(n)$ is k-synchronized.*

Proof. Following [8], it suffices to show how to accept the language

$$\{(n, m)_k \; : \; n \geq 0 \text{ and } m = \rho_{\mathbf{x}}(n)\}$$

with a finite automaton. Here is a sketch of the argument. From our results above, we know that there is a finite constant $C \geq 1$ such that the number of contiguous blocks in any row of the factor evolution diagram is bounded by C. So we simply "guess" the endpoints of every block and then verify that each factor of length n starting at the positions inside blocks is a novel occurrence, while all other factors are not. Finally, we verify that m is the sum of the sizes of the blocks.

To fill in the details, we observe above in (1) that the predicate "the factor of length n beginning at position i of **x** is a novel occurrence" is solvable by a finite automaton. Similarly, given endpoints a, b and n, the predicates "every factor of length n beginning at positions a through b is a novel occurrence", "no factor of length n beginning at positions a through b is a novel occurrence" and "no factor of length n after position a is novel" are also solvable by a finite automaton. The length of each block is just $b - a + 1$, and it is easy to create an automaton that will check if the sums of the lengths of the blocks equals m, which is supposed to be $\rho_{\mathbf{x}}(n)$. □

Applying Theorem 1 we get

Corollary 7. *Given a k-automatic sequence* **x***, there is an algorithm that, on input n in base k, will produce $\rho_{\mathbf{x}}(n)$ in base k in time $O(\log n)$.*

As another application, we can recover and improve some recent results of Goldstein [14,15]. He showed how to compute the quantities $\limsup_{n \geq 1} \rho_{\mathbf{x}}(n)/n$ and $\liminf_{n \geq 1} \rho_{\mathbf{x}}(n)/n$ for the special case of k-automatic sequences that are the fixed points of k-uniform morphisms related to certain groups. Corollary 8 below generalizes these results to all k-automatic sequences.

Corollary 8. *There is an algorithm, that, given a k-automatic sequence* **x***, will compute $\sup_{n \geq 1} \rho_{\mathbf{x}}(n)/n$, $\limsup_{n \geq 1} \rho_{\mathbf{x}}(n)/n$, and $\inf_{n \geq 1} \rho_{\mathbf{x}}(n)/n$, $\liminf_{n \geq 1} \rho_{\mathbf{x}}(n)/n$.*

Proof. We already showed how to construct an automaton accepting $\{(n, \rho_{\mathbf{x}}(n))_k \; : \; n \geq 1\}$. Now we just use the results from [18,17]. Notice that the lim sup corresponds to what is called the largest "special point" in [17]. □

Example 9. Continuing our example of the Thue-Morse sequence, Figure 2 displays a DFA accepting $\{(n, \rho_{\mathbf{t}}(n))_k \; : \; n \geq 0\}$. Inputs are given with the most significant digit first; the "dead" state and transitions leading to it are omitted.

Given an infinite word **x**, we can also count the number of contiguous blocks in each $E_{\mathbf{x}}(n)$ for $n \geq 0$. (For the Thue-Morse sequence this gives the sequence $1, 1, 2, 1, 3, 1, 5, 3, 3, 1, \ldots$) If **x** is k-automatic, then this sequence is also, as the following theorem shows:

Theorem 10. *If* **x** *is k-automatic then the sequence $(e(n))_{n\geq 0}$ counting the number of contiguous blocks in the n'th step $E_\mathbf{x}(n)$ of the evolution of novel occurrences of factors in* **x** *is also k-automatic.*

Proof. Since we have already shown that the number of contiguous blocks is bounded by some constant C if **x** is k-automatic, it suffices to show for each $i \leq C$ we can create an automaton to accept the language

$$\{(n)_k \; : \; E_\mathbf{x}(n) \text{ has exactly } i \text{ contiguous blocks }\}.$$

To do so, on input n in base k we guess the endpoints of the i contiguous nonempty blocks, verify that the length-n occurrences at those positions are novel, and that all other occurrences are not novel. □

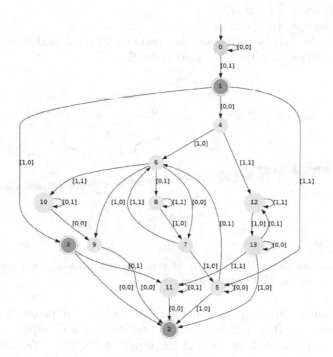

Fig. 2. Automaton computing the subword complexity of the Thue-Morse sequence

Example 11. Figure 3 below gives the automaton computing the number $e(n)$ of contiguous blocks of novel occurrences of length-n factors for the Thue-Morse sequence. Here is a brief table:

n	0	1	2	3	4	5	6	7	8	9	10	11	12	13	14	15	16	17	18	19	20	21	22	23	24	25	26
$e(n)$	1	1	2	1	3	1	5	3	3	1	5	5	5	3	3	3	3	1	5	5	5	5	5	5	5	3	3

3   Implementation

We wrote a program that, given an automaton generating a k-automatic sequence $\mathbf{x}$, will produce a deterministic finite automaton accepting the language $\{(n, \rho_{\mathbf{x}}(n))_k \; : \; n \geq 0\}$. We used the following variant which does not require advance knowledge of the bound on the first difference of $\rho_{\mathbf{x}}(n)$:

1. Construct an automaton R that accepts (n, s, e, ℓ) if, for factors of length n, the next contiguous block of novel occurrences after position s ends at position e and has length ℓ. If there are no blocks past s, accept $(n, s, s, 0)$.
2. Construct an automaton M_0 that accepts $(n, 0, 0)$.
3. Construct an automaton M_{j+1} that accepts (n, S, e) if there exist s and S' such that
    (i) M_j accepts (n, S', s)
    (ii) R accepts $(n, s, e, S - S')$.
4. If $M_{j+1} = M_j$ then we are done. We create an automaton that accepts (n, S) if there exists e such that M_j accepts (n, S, e).

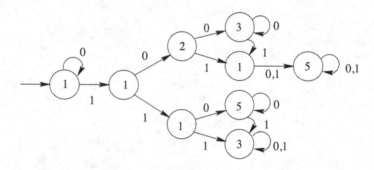

Fig. 3. Automaton computing number of contiguous blocks of novel occurrences of length-n factors in the Thue-Morse sequence

Besides the automaton depicted in Figure 1, we ran our program on the paperfolding sequence [11] and the so-called "period-doubling sequence" [10]. The former gives a DFA of 20 states and the latter of 7 states. We omit them for space considerations.

4   Powers and Primitive Words

We say a nonempty word w is a *power* if there exists a word x and an integer $k \geq 2$ such that $w = x^k$; otherwise we say w is *primitive*. Given a nonempty word z, there is a unique way to write it as y^i, where y is primitive and i is an integer ≥ 1; this y is called the *primitive root* of z. Thus, for example, the primitive root of murmur is mur.

We say $w = a_1 \cdots a_n$ has a *period* p if $a_i = a_{i+p}$ for $1 \le i \le n - p$. Thus, for example, `alfalfa` has period 3. It is easy to see that a word w is a power if and only if it has a period p such that $p < |w|$ and $p \mid |w|$.

Two finite words x, y are conjugates if one is a cyclic shift of the other; in other words, if there exist words u, v such that $x = uv$ and $y = vu$. For example, `enlist` is a conjugate of `listen`. As is well-known, every conjugate of a power of a word x is a power of a conjugate of x. The lexicographically least conjugate of a primitive word is called a *Lyndon word*. We call the lexicographically least conjugate of the primitive root of x the *Lyndon root* of x.

The following lemma says that if we consider the starting positions of length-n powers in a word x, then there must be large gaps between contiguous blocks of such starting positions.

Lemma 12. *Let z be a finite or infinite word, and let $n \ge 2$ be an integer. Suppose there exist integers i, j such that*

(a) $w_1 := z[i..i + n - 1]$ is a power;
(b) $w_2 := z[j..j + n - 1]$ is a power;
(c) $i < j \le i + n/3$.

Then $z[t..t - n - 1]$ is a power for $i \le t \le j$. Furthermore, if u_1 is the Lyndon root of w_1, then u_1 is also the Lyndon root of each word $z[t..t - n - 1]$.

Proof. Let x_1 be the primitive root of w_1 and x_2 be the primitive root of w_2. Since w_1 and w_2 are powers, there exist integers $e_1, e_2 \ge 2$ such that $w_1 = x_1^{e_1}$ and $w_2 = x_2^{e_2}$.

Since w_1 and w_2 are both of length n, and since their starting positions are related by $i < j \le i + n/3$, it follows that the word $v := z[j..i + n - 1]$ is common to both w_1 and w_2, and $|v| = i + n - j \ge i + 2n/3 + n/3 - j \ge 2n/3$.

Now there are three cases to consider:

(a) $|x_1| > |x_2|$;   (b) $|x_1| < |x_2|$;   (c) $|x_1| = |x_2|$.

Case (a): We must have $e_2 > e_1 \ge 2$, so $e_2 \ge 3$. Since v is a suffix of w_1, it has period $|x_1| \le n/2$. Since v is a prefix of w_2, it has period $|x_2| \le n/3$. We note that $\gcd(|x_1|, |x_2|) = n/\mathrm{lcm}(e_1, e_2) \le n/6$, so

$$|x_1| + |x_2| - \gcd(|x_1|, |x_2|) \le \frac{2n}{3} \le |v|.$$

By a theorem of Fine and Wilf [12], it now follows that v, and hence x_1, has period $p := \gcd(|x_1|, |x_2|) \le |x_2| < |x_1|$. Now p is less than $|x_1|$ and also divides it, so this means x_1 is a power, a contradiction, since we assumed x_1 is primitive. So this case cannot occur.

Case (b) gives a similar contradiction.

Case (c): We have $e_1 = e_2 \ge 2$. Then the last occurrence of x_1 in w_1 lies inside x_2^2, and so x_1 is a conjugate of x_2. Hence w_1 is a conjugate of w_2. It now follows that $z[t..t + n - 1]$ is a conjugate of w_1 for every t, $i \le t \le j$. But the conjugate of a power is itself a power, and we are done. $\square$

Remark 13. The bound of $n/3$ in the statement of Lemma 12 is best possible, as shown by the following class of examples. Let h be the morphism that maps $1 \to 21$ and $2 \to 22$, and consider the word $h^i(122122121212)$. This word is of length $12 \cdot 2^i$, and contains squares of length $3 \cdot 2^{i+1}$ starting in the first $3 \cdot 2^i$ positions, and cubes of length $3 \cdot 2^{i+1}$ ending in the last $2^i + 1$ positions. This achieves a gap of $n/3 + 1$ infinitely often.

Now, given an infinite word $\mathbf{x}$, we define a function $\alpha_{\mathbf{x}}(n)$, the *appearance function*, to be the least index i such that every length-n factor of $\mathbf{x}$ appears in the prefix $\mathbf{x}[0..i + n - 1]$; see [3, §10.10].

Theorem 14. *If $\mathbf{x}$ is a k-automatic sequence, then $\alpha_{\mathbf{x}}(n) = O(n)$.*

Proof. First, we show that the appearance function is k-synchronized. It suffices to show that there is an automaton accepting $\{(n, m)_k : m = \alpha_{\mathbf{x}}(n)\}$. To see this, note that on input $(n, m)_k$ we can check that

- for all $i \geq 0$ there exists j, $0 \leq j \leq m$ such that $\mathbf{x}[i..i+n-1] = \mathbf{x}[j..j+n-1]$; and
- for all $l < m$ we have $\mathbf{x}[m..m + n - 1] \neq \mathbf{x}[l..l + n - 1]$.

From [6, Prop. 2.5] we know k-synchronized functions are $O(n)$.   □

As before, we now consider the occurrences of length-n powers in $\mathbf{x}$:

Lemma 15. *If $\mathbf{x}$ is k-automatic, then there are only a constant number of maximal blocks of novel occurrences of length-n powers in $\mathbf{x}$.*

Proof. To begin with, we consider maximal blocks of length-n powers in $\mathbf{x}$ (not considering whether they are novel occurrences). From Theorem 14 we know that every length-n factor must occur at a position $< Cn$, for some constant C (depending on $\mathbf{x}$). We first argue that the number of maximal blocks of length-n powers, up to the position of the last length-n power to occur for the first time, is at most $3C$.

Suppose there are $\geq 3C + 1$ such blocks. Then Lemma 12 says that any two such blocks must be separated by at least $n/3$ positions. So the first occurrence of the last factor to occur occurs at a position $\geq (3C)(n/3) = Cn$, a contradiction.

So using a constant number of blocks, in which each position of each block starts a length-n factor that is a power, we cover the starting positions of all such factors. It now remains to process these blocks to remove occurrences of length-n powers that are not novel.

The first thing we do is remove from each block the positions starting length-n factors that have already occurred *in that block*. This has the effect of truncating long blocks. The new blocks have the property that each factor occurring at the starting positions in the blocks never appeared before in that block.

Above we already proved that inside each block, the powers that begin at each position are all powers of some conjugate of a fixed Lyndon word. Now we process the blocks associated with the same Lyndon root together, from the first (leftmost) to the last. At each step, we remove from the current block all the positions where length-n factors begin that have appeared in any previous block. When all blocks have been processed, we need to see that there are still at most a constant number of contiguous blocks remaining.

Suppose the associated Lyndon root is y, with $|y| = d$. Each position in a block is the starting position of a power of a conjugate of y, and hence corresponds to a right rotation of y by some integer i, $0 \le i < d$. Thus each block B_j actually corresponds to some I_j that is a contiguous subblock of $0, 1, \ldots, d-1$ (thought of as arranged in a circle).

As we process the blocks associated with y from left to right we replace I_j with $I'_j := I_j - (I_1 \cup \cdots \cup I_{j-1})$. Now if $I \subseteq \{0, 1, \ldots, d-1\}$ is a union of contiguous subblocks, let $\#I$ be the number of contiguous subblocks making up I. We claim that

$$\#I'_1 + \#I'_2 + \cdots + \#I'_n + \#(\bigcup_{1 \le i \le n} I'_i) \le 2n. \tag{2}$$

To see this, suppose that when we set $I'_n := I_n - (I_1 \cup \cdots \cup I_{n-1})$, the subblock I_n has an intersection with t of the lower-numbered subblocks. Forming the union $(\bigcup_{1 \le i \le n} I'_i)$ then obliterates t subblocks and replaces them with 1. But I'_n has $t-1$ new subblocks, plus at most 2 at either edge (see Figure 4). This means that the left side of (2) increases by at most $(1-t) + (t-1) + 2 = 2$. Doing this n times gives the result.

$$\bigcup_{1 \le i \le j-1} I'_i$$

$$I_j$$

$$I'_j$$

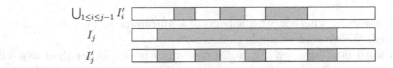

Fig. 4. How the number of blocks changes

Now at the end of the procedure there will be at least one interval in the union of all the I_i, so $\#I'_1 + \#I'_2 + \cdots + \#I'_n \le 2n - 1$.

Earlier we showed that there are at most $3C$ maximal blocks of length-n powers, up to the position of the last length-n power to occur for the first time. Then, after processing these blocks to remove positions corresponding to factors that occurred earlier, we will have at most $2(3C) = 6C$ blocks remaining. $\square$

Corollary 16. *If* $\mathbf{x}$ *is* k-*automatic, then the following are* k-*synchronized:*

- *the function counting the number of distinct length-n factors that are powers;*
- *the function counting the number of distinct length-n factors that are primitive words.*

Proof. Suppose $\mathbf{x}$ is k-automatic, and generated by the automaton M. From the Lyndon-Schützenberger theorem [16], we know that a word x is a power if and only if there exist nonempty words y, z such that $x = yz = zy$. Thus, we can express the predicate $P(i, j) :=$ "$\mathbf{x}[i..j]$ is a power" as follows: "there exists d, $0 < d < j{-}i{+}1$, such that $\mathbf{x}[i..j{-}d] = \mathbf{x}[i{+}d..j]$ and $\mathbf{x}[j{-}d{+}1..j] = \mathbf{x}[i..i{+}d{-}1]$". Furthermore, we can express the predicate $P'(i, n) :=$ "$\mathbf{x}[i..i{+}n{-}1]$ is a length-n power and the first occurrence of that power in $\mathbf{x}$", as

$$P(i, i + n - 1) \wedge (\forall i', \ 0 \leq i' < i, \ x[i'..i' + n - 1] \neq x[i..i + n - 1]).$$

From Lemma 15 we know that the novel occurrences of length-n powers are clustered into a finite number of blocks. Then, as in the proof of Theorem 6, we can guess the endpoints of these blocks, and verify that the length-n factors beginning at the positions inside the blocks are novel occurrences of powers, while those outside are not, and sum the lengths of the blocks, using a finite automaton built from M. Thus, the function counting the number of length-n powers in $\mathbf{x}$ is k-synchronized.

The number of length-n primitive words in $\mathbf{x}$ is then also k-synchronized, since it is expressible as the total number of words of length n, minus the number of length-n powers.   □

Remark 17. Using the technique above, we can prove analogous results for the functions counting the number of length-n words that are α-powers, for any fixed rational number $\alpha > 1$.

5   Unsynchronized Sequences

It is natural to wonder whether other aspects of k-automatic sequences are always k-synchronized. We give an example that is not.

We say a word w is *bordered* if it has a nonempty prefix, other than w itself, that is also a suffix. Alternatively, w is bordered if it can be written in the form $w = tvt$, where t is nonempty. Otherwise a word is *unbordered*.

Charlier et al. [8] showed that $u_\mathbf{x}(n)$, the number of unbordered factors of length n of a sequence $\mathbf{x}$, is k-regular if $\mathbf{x}$ is k-automatic. They also gave a conjecture for recursion relations defining $u_\mathbf{t}(n)$ where $\mathbf{t}$ is the Thue-Morse sequence; this conjecture has recently been verified [13].

We give here an example of a k-automatic sequence where the number of unbordered factors of length n is not k-synchronized.

Consider the characteristic sequence of the powers of 2: $\mathbf{c} := 0110100010 \cdots$.

Theorem 18. *The sequence $\mathbf{c}$ is 2-automatic, but the function $u_\mathbf{c}(n)$ counting the number of unbordered factors is not 2-synchronized.*

Proof. It is not hard to verify that $\mathbf{c}$ is 2-automatic and that $\mathbf{c}$ has exactly $r + 2$ unbordered factors of length $2^r + 1$, for $r \geq 2$ — namely, the factors beginning at positions 2^i for $0 \leq i \leq r - 1$, and the factors beginning at positions 2^{r+1} and $3 \cdot 2^r$. However, if $u_\mathbf{c}(n)$ were 2-synchronized, then reading an input where the first component looks like $0^i10^{r-1}1$ (and hence a representation of $2^r + 1$)

for large r would force the transitions to enter a cycle. If the transitions in or before the cycle contained a nonzero entry in the second component, this would force $u_c(n)$ to grow linearly with n when n is of the form $2^r + 1$. Otherwise, the corresponding transitions for the second component are just 0's, in which case $u_c(n)$ is bounded above by a constant, for n of the form $2^r + 1$. Both cases lead to a contradiction. $\qquad\square$

Acknowledgments. We thank the referees for a careful reading of the paper.

References

1. Allouche, J.-P., Shallit, J.: The ring of k-regular sequences. Theoret. Comput. Sci. 98, 163–197 (1992)
2. Allouche, J.-P., Shallit, J.: The ring of k-regular sequences, II. Theoret. Comput. Sci. 307, 3–29 (2003)
3. Allouche, J.-P., Shallit, J.: Automatic Sequences: Theory, Applications, Generalization. Cambridge University Press (2003)
4. Carpi, A., D'Alonzo, V.: On the repetitivity index of infinite words. Internat. J. Algebra Comput. 19, 145–158 (2009)
5. Carpi, A., D'Alonzo, V.: On factors of synchronized sequences. Theoret. Comput. Sci. 411, 3932–3937 (2010)
6. Carpi, A., Maggi, C.: On synchronized sequences and their separators. RAIRO Inform. Théor. App. 35, 513–524 (2001)
7. Cassaigne, J.: Special factors of sequences with linear subword complexity. In: Dassow, J., Rozenberg, G., Salomaa, A. (eds.) Developments in Language Theory II, pp. 25–34. World Scientific (1996)
8. Charlier, É., Rampersad, N., Shallit, J.: Enumeration and decidable properties of automatic sequences. In: Mauri, G., Leporati, A. (eds.) DLT 2011. LNCS, vol. 6795, pp. 165–179. Springer, Heidelberg (2011)
9. Cobham, A.: Uniform tag sequences. Math. Systems Theory 6, 164–192 (1972)
10. Damanik, D.: Local symmetries in the period-doubling sequence. Disc. Appl. Math. 100, 115–121 (2000)
11. Dekking, F.M., Mendès France, M., van der Poorten, A.J.: Folds! Math. Intelligencer 4, 130–138, 173–181, 190–195 (1982); Erratum 5, 5 (1983)
12. Fine, N.J., Wilf, H.S.: Uniqueness theorems for periodic functions. Proc. Amer. Math. Soc. 16, 109–114 (1965)
13. Goč, D., Mousavi, H., Shallit, J.: On the number of unbordered factors. In: Dediu, A.-H., Martín-Vide, C., Truthe, B. (eds.) LATA 2013. LNCS, vol. 7810, pp. 299–310. Springer, Heidelberg (2013)
14. Goldstein, I.: Asymptotic subword complexity of fixed points of group substitutions. Theoret. Comput. Sci. 410, 2084–2098 (2009)
15. Goldstein, I.: Subword complexity of uniform D0L words over finite groups. Theoret. Comput. Sci. 412, 5728–5743 (2011)
16. Lyndon, R.C., Schützenberger, M.P.: The equation $a^M = b^N c^P$ in a free group. Michigan Math. J. 9, 289–298 (1962)
17. Schaeffer, L., Shallit, J.: The critical exponent is computable for automatic sequences. Int. J. Found. Comput. Sci. 23, 1611–1626 (2012)
18. Shallit, J.: The critical exponent is computable for automatic sequences. In: Ambroz, P., Holub, S., Másaková, Z. (eds.) Proceedings 8th International Conference Words 2011. Elect. Proc. Theor. Comput. Sci., vol. 63, pp. 231–239 (2011)

Some Decision Questions Concerning the Time Complexity of Language Acceptors

Oscar H. Ibarra[1,*] and Bala Ravikumar[2]

[1] Department of Computer Science
University of California, Santa Barbara, CA 93106, USA
ibarra@cs.ucsb.edu
[2] Department of Computer & Engineering Science
Sonoma State University, Rohnert Park, CA 94928 USA
ravi@cs.sonoma.edu

Abstract. Almost all the decision questions concerning the resource requirements of a computational device are undecidable. Here we want to understand the exact boundary that separates the undecidable from the decidable cases of such problems by considering the time complexity of very simple devices that include NFAs (1-way and 2-way), NPDAs and NPDAs augmented with counters - and their unambiguous restrictions. We consider several variations - based on whether the bound holds exactly or as an upper-bound and show decidability as well as undecidability results. We also introduce a stronger version of machine equivalence (known as run-time equivalence) and identify classes of machines for which run-time equivalence is decidable (undecidable). In the case of decidable problems, we also attempt to determine more precisely the complexity class to which the problem belongs.

Keywords: NFA, NPDA, counter machine, GSM, reversal-bounded counters, time complexity, decidable, undecidable, run-time equivalence, unambiguous, k-ambiguous.

1   Introduction

Decision questions concerning the time complexity of a computational device are very basic to understanding its performance (correctness, efficiency, optimality) and hence such questions have been addressed from the beginning of computation theory. The earliest such questions can be traced back to Turing's original paper in which halting problem on blank tape was addressed [20]. Most such questions are undecidable, however. It is therefore natural to explore the simplest of such questions for which decision algorithms exist. For example, consider an NFA M with ε-moves. Such an NFA may require, on an input w, an accepting path that is much longer than $|w|$ because of ε-moves. One basic question is: Is it true that every string w in $L(M)$ can be accepted within $2|w|$ steps? We address such questions in this paper. Most of the problems we consider involve determining if

---

* Supported in part by NSF Grants CCF-1143892 and CCF-1117708.

M.-P. Béal and O. Carton (Eds.): DLT 2013, LNCS 7907, pp. 264–276, 2013.
© Springer-Verlag Berlin Heidelberg 2013

the time complexity of all the accepted input strings can be upper-bounded in a certain way. Note that many of the devices we consider are nondeterministic and thus the time complexity we consider is the *shortest* among all possible accepting computations. The precise nature of the problem considered leads to questions such as: (a) Is the time bound at most $|w| + k$ for some k independent of the length of $|w|$)? This means the number of ε-moves is bounded above by a constant. (b) Is the time complexity at most $k|w|$, where k could be fixed, or be part of the input. (c) Is the time complexity exactly $k|w|$ for a given k?

The primary motivation for this work is to understand the boundary between decidable and undecidable cases of such analysis questions since we feel that these are fundamental questions. The problems we consider have the same flavor as those studied in [4,3]. The research in [4] investigated the mean-payoff in NFA models in which the edges are weighted by real numbers and the goal is to compute for a given string w, the minimum average weight of computation which is defined as the total weight divided by the length of string, minimized over all paths leading to acceptance of the string. The paper [3] studied the "amplitude" of an automaton, which intuitively characterizes how much a typical execution of the automaton fluctuates. We note that similar constructions in some of our results in this paper have also been used in [3]. Our work is also related to studies on measures of nondeterminism in which the number of nondeterministic moves on an accepting computation is used as a complexity measure [6].

We use a wide range of undecidable (e.g. [14], [8]) and decidable problems (e.g. [11], [19], [15]) to establish the new results. **Because of page limitation, many proofs are omitted. They will appear in a full version of the paper.**

2   Preliminaries

We use the standard machine models such as DFA, NFA, DPDA, NPDA etc. (with ε-moves) as for example, in [12]. We will also look at machines with a two-way read-only input. These devices have left and right end markers, and acceptance is when the input head falls off the right end marker in an accepting state. At each move, the input head can move left or right or remain stationary. A two-way device M is k-crossing (where $k \geq 1$) if every accepted input can be accepted in an accepting computation in which the input head crosses the boundary between any pair of adjacent cells at most k times. Note that the head can stay on a cell an unbounded (but a finite) number of steps, and these "stays" are not counted in the crossing bound. Staying is like ε-move. M is finite-crossing if it is k-crossing for some given k.

A fundamental model that we will use is a one-way or two-way NFA or NPDA augmented with a finite number of reversal-bounded counters. At each step, each counter (which is initially set to zero and can never become negative) can be incremented by 1, decremented by 1, or left unchanged and can be tested for zero. The counters are reversal-bounded in the sense that there is a specified r such that during any computation, no counter can change mode from increasing to decreasing and vice-versa more than r times. A counter is 1-reversal if once it decrements, it can no longer increment. Clearly, an r-reversal counter can be simulated by $\lceil (r+1)/2 \rceil$ 1-reversal counters. Machines augmented with reversal-bounded counters have been extensively studied (see, e.g, [1,13,8]).

We will need the following results throughout the paper.

Theorem 1

1. *The emptiness (Is $L(M)$ empty?) and infiniteness (Is $L(M)$ infinite?) problems for NPDAs augmented with reversal-bounded counters are decidable [13].*
2. *The emptiness and infiniteness problems for finite-crossing two-way NFAs augmented with reversal-bounded counters are decidable [13,8].*

3 Unambiguous Machines

Let $\mathcal{M}$ be a class of nondeterministic (one-way or two-way) machines, possibly augmented with infinite storage units, and $\mathcal{M}_c$ be the class of machines obtained by augmenting the machines in $\mathcal{M}$ with reversal-bounded counters.

Theorem 2. *If $\mathcal{M}_c$ has a decidable emptiness problem, then for any k, it is decidable, given an arbitrary unambiguous machine M in $\mathcal{M}$, whether every string w in $L(M)$ can be accepted by a computation of length (i.e., number of moves) exactly $k|w|$ (resp., at most $k|w|$).*

Proof. First consider the case when the machines are one-way. Given $k \geq 1$ and an unambiguous machine M in $\mathcal{M}$, we construct a machine M' in $\mathcal{M}_c$ which, given input w, simulates the computation of M on w. M' uses two new 1-reversal counters, X_1 and X_2. M' uses counter X_1 to store the number of moves of M and counter X_2 to store $|w|$. When M accepts w, M' checks and accepts if value$(X_1) \neq k \cdot$value(X_2) (resp., if value$(X_1) > k \cdot$value(X_2)). This can be done by decrementing counter X_1 by k for every decrement of 1 in counter X_2. It follows that M does not make exactly $k|w|$ (resp., makes more than $k|w|$) moves if and only if $L(M')$ is not empty, which is decidable.

If M is a two-way machine, M' first makes a left-to-right sweep of the input w and stores $|w|$ in counter X_2. Then M' returns the input head to the left end marker and simulates M while storing the number of moves of M in counter X_1. When M accepts, M' carries out the same procedure as above.   □

Corollary 1. *If $\mathcal{M}_c$ has a decidable emptiness problem, then for any k, it is decidable, given an arbitrary unambiguous machine M in $\mathcal{M}$ and an integer $d \geq 0$, whether every string w in $L(M)$ can be accepted by a computation of length exactly $k|w| + d$ (resp., at most $k|w| + d$).*

A variation of Theorem 2 is the following:

Theorem 3. *If $\mathcal{M}_c$ is effectively closed under homomorphism and has a decidable emptiness and infiniteness problems, then for any k, it is decidable, given an arbitrary unambiguous machine M in $\mathcal{M}$, whether there is an integer $d \geq 0$, such that every string w in $L(M)$ can be accepted by a computation of length exactly $k|w| + d$ (resp., at most $k|w| + d$). Moreover, if such a d exists, the smallest such d can be effectively computed.*

Proof. Again, consider first the case when M is one-way. Let M be an unambiguous machine in $\mathcal{M}$. Let $\#$ be a new symbol, and let $L = \{w\#^d \mid w \in L(M), d \geq 0, w$ is accepted in no less than $k|w| + d$ moves $\}$.

We construct a machine M' in $\mathcal{M}_c$ which, when given input $w\#^d$, simulates the computation of M on w. M' uses two 1-reversal counters, X_1 and X_2. M' uses counter X_1 to store the number of moves of M and counter X_2 to store $|w|$ during the simulation. When M accepts w, M' reads $\#^d$ and decrements counter X_1 by d. M' then checks and accepts if value$(X_1) \neq k\cdot$value(X_2) (resp., if value$(X_1) > k\cdot$value(X_2)). Next, we construct a machine M'' in $\mathcal{M}_c$ that accepts $L'' = \{\#^d \mid w\#^d \in L(M')\}$. (We can do this, since $\mathcal{M}_c$ is effectively closed under homomorphism.) It follows that there is no d such that every string w in $L(M)$ can be accepted by a computation of length exactly $k|w| + d$ (resp., at most $k|w| + d$) if and only if $L(M'')$ is not empty (resp., infinite), which is decidable. If d exists, the smallest such d can be found exhaustively (starting with $d = 0$) by Corollary 1.

The case when M is two-way is similar, except that M' first stores $|w|$ in counter X_2. The simulation of M and and comparison of the counters X_1 and X_2 proceed as above. $\qquad\square$

Theorem 2, Corollary 1, and Theorem 3 apply to each class of $\mathcal{M}$ below, since $\mathcal{M}_c$ has decidable emptiness and infiniteness problems by Theorem 1. Clearly, the first two classes are effectively closed under homomorphism. That the third class is also effectively closed under homomorphism, follows from the following result in [8]: Given a finite-crossing two-way NFA augmented with reversal-bounded counters, we can effectively construct an equivalent (one-way) NFA augmented with reversal-bounded counters.

1. Unambiguous NFAs
2. Unambiguous NPDAs augmented with reversal-bounded counters
3. Unambiguous finite-crossing two-way NFAs augmented with reversal-bounded counters.

When $k = 1$, the requirement in Theorem 2 can be weakened:

Proposition 1. *If M has a decidable emptiness problem, then it is decidable, given an arbitrary unambiguous machine M in $\mathcal{M}$ and $d \geq 0$, whether every string w in $L(M)$ can be accepted by a computation of length exactly $|w| + d$ (resp., at most $|w| + d$).*

The following classes of machines have decidable emptiness problem, so the decidability in Proposition 1 applies. Note that these classes were not listed in the examples following Theorem 2, since the emptiness problem for these machines when augmented with reversal-bounded counters is undecidable.

1. Two-way DFA augmented with *one* reversal-bounded counter. The emptiness problem for these machines is decidable [15]. (However, when augmented with one more reversal-bounded counter, emptiness becomes undecidable, since two-way DFAs with two reversal-bounded counters have undecidable emptiness problem [13].)
2. Unambiguous one-way stack machines [5]. Informally, a one-way stack machine is a one-way pushdown automaton with the additional power that the stack head can enter the stack in a "read-only" mode; however, the stack head can return to the top of the stack to push and pop. This class of machines has decidable emptiness problem. (However, when augmented with two reversal-bounded counters, their emptiness becomes undecidable. This is because such a machine can copy the one-way input on the stack and then use the stack in a read-only mode and the two reversal-bounded counters to simulate a two-way DFA augmented with two reversal-bounded counters, and emptiness of the latter is undecidable [13].)

4 2-Ambiguous Machines

The results of the previous section are not valid for ambiguous machines, even when the degree of ambiguity is 2 and the machines are quite simple.

Theorem 4. *For any $k \geq 1$, it is undecidable to determine, given a 2-ambiguous nondeterministic one-counter machine M, whether every w in $L(M)$ is accepted by some computation of length exactly $k|w|$ (resp., at most $k|w|$).*

Proof. The proof uses the undecidability of the halting problem for 2-counter machines [18]. A close look at the proof of the undecidability of the halting problem for 2-counter machines, where initially one counter has value d_1 and the other counter is zero in [18] reveals that the counters behave in a regular pattern. The 2-counter machine operates in phases in the following way. Let c_1 and c_2 be its counters. The machine's operation can be divided into phases, where each phase starts with one of the counters equal to some positive integer d_i and the other counter equal to 0. During the phase, the positive counter decreases, while the other counter increases. The phase ends with the first counter having value 0 and the other counter having value d_{i+1}. Then in the next phase the modes of the counters are interchanged. Thus, a sequence of configurations corresponding to the phases will be of the form:

$$(q_1, d_1, 0), (q_2, 0, d_2), (q_3, d_3, 0), (q_4, 0, d_4), (q_5, d_5, 0), (q_6, 0, d_6), \ldots, (q_{2m}, 0, d_{2m})$$

where the q_i's are states and $d_1, d_2, d_3, \ldots, d_{2m}$ are positive integers (we assume that the number of steps is odd). Note that the second component of the configuration refers to the value of c_1, while the third component refers to the value of c_2. Without loss of generality, we assume $d_1 = 1$.

Let T be a 2-counter machine that operates as described above. Let $k \geq 1$. Let $L = L_1 \cup L_2$, where

$$L_1 = \{w \mid w \in (0 + 1)^*, w \neq 0^{d_1}1^{d_1}0^{d_2}1^{d_2}0^{d_3}1^{d_3}\ldots0^{d_{2m}}1^{d_{2m}},$$
$$m \geq 1, d_1, d_2, \ldots, d_{2m} \geq 1\}$$

$L_2 = \{w \mid w \in (0+1)^*, w = 0^{d_1}1^{d_1}0^{d_2}1^{d_2}0^{d_3}1^{d_3}\ldots0^{d_{2m}}1^{d_{2m}}$, where $m \geq 1, d_1 = 1, d_2, \ldots, d_{2m} \geq 1, (q_1, d_1, 0), (q_2, 0, d_2), (q_3, d_3, 0), (q_4, 0, d_4), (q_5, d_5, 0), (q_6, 0, d_6), \ldots, (q_{2m}, 0, d_{2m})$ is the halting computation of the 2-counter machine T for some $q_1, \ldots, q_{2m}$, if it halts $\}$.

We construct a 2-ambiguous nondeterministic one-counter machine M which, when given an input $w \in (0 + 1)^*$, nondeterministically selects to process (1) or (2) below:

(1) M operates in such a way that it makes $k - 1$ ε-moves after every symbol it reads and accepts w if it is in L_1. Note that this can be done deterministically.

(2) M simulates the computation of the 2-counter machine T (note that the finite control can check that w is of the form $0^{d_1}1^{e_1}0^{d_2}1^{e_2}0^{d_3}1^{e_3}\ldots0^{d_{2m}}1^{e_{2m}}, m \geq 1, d_1 = 1$; otherwise, M rejects). In the simulation, M assumes that $d_i = e_i$ for $1 \leq i \leq 2m$. It also makes k ε-moves after every symbol it reads. If the simulation succeeds with T halting, M accepts w. Note that it is possible that the simulation succeeds with T halting even when some $d_i \neq e_i$ for some i, and M accepts. In this case, such a w would also be accepted in process (1), so w has two accepting computations. Again, process (2) can be done deterministically.

Clearly, M is 2-ambiguous. If T halts, M will accept a string w of the form $0^{d_1}1^{d_1}0^{d_2}1^{d_2}0^{d_3}1^{d_3}\ldots0^{d_{2m}}1^{d_{2m}}$, which is not in L_1, in exactly $(k+1)|w|$ moves. (Note that there is only one such string, since T is deterministic). If T does not halt, M will accept only strings w in L_1 in exactly $k|w|$ moves.

It follows that M accepts every string w in L in exactly (hence, at most) $k|w|$ moves if and only if T does not halt, which is undecidable. $\square$

The next result shows that determining whether an NPDA accepts in real-time is undecidable, even when it makes only one stack reversal (i.e., when it pops, it can no longer push).

Theorem 5. *For any $k \geq 1$, it is undecidable to determine, given a 2-ambiguous 1-reversal NPDA M, whether every string w in $L(M)$ is accepted by some computation of length exactly $k|w|$ (resp., at most $k|w|$).*

The constructions in the proofs of Theorems 4 and 5 also show:

Corollary 2. *For any $k \geq 1$, it is undecidable to determine, given a 2-ambiguous nondeterministic one-counter machine or a 2-ambiguous 1-reversal NPDA M, whether there is a $d \geq 0$ such that every string w in $L(M)$ is accepted by some computation in exactly $k|w| + d$ moves (resp., in at mos $k|w| + d$ moves).*

Note that Theorems 4 and 5 and Corollary 2 are in contrast to the fact that the problems are decidable when the machines are unambiguous (i.e., 1-ambiguous).

The following shows that Theorem 5 and Corollary 2 also apply to 1-reversal nondeterministic counter machines (i,e., once the counter decrements, it can no longer increment); however, the machine needs unbounded ambiguity:

Proposition 2. *For any $k \geq 1$, it is undecidable to determine, given a 1-reversal nondeterministic one-counter machine M, whether every string w in $L(M)$ is accepted by some computation of length exactly $k|w|$ (resp., at most $k|w|$).*

Open Questions: Can the Proposition 2 be shown when M is both m-ambiguous (for $m \geq 2$) and r-reversal (for $r \geq 1$)? In particular, what about the case when $k = 1$, $m = 2$, and $r = 1$?

5   Time Complexity of NFAs

Throughout this section, we will assume that an NFA has ε moves. We begin by studying the NFA complexity problem with $k = 1$.

Theorem 6. *It is PSPACE-complete to determine, given an NFA M, whether every string w in $L(M)$ can be accepted by a computation of at most $|w|$ moves. The problem is in P (= polynomial time) if the NFA is unambiguous.*

Proof. We will first show that the problem is in PSPACE. Let M be an instance of the problem. It is clear that M is a yes-instance of the problem if and only if every string accepted by M can be accepted without using any ε moves. Let M' be the NFA defined by removing all the *epsilon* moves of M. The problem is to determine if $L(M) = L(M')$. It is well-known [12] that this problem can be solved in PSPACE.

To show PSPACE-hardness, we reduce the following problem [16] (universe problem for union of DFA's, *UDFA for short*) to our problem. The problem involves a collection $\{M_i\}$, $i = 1, 2, ..., m$ of DFAs. The question is to determine if $\bigcup_{i=1}^{m} L(M_i) = \Sigma^*$. We will reduce this problem to our problem as follows: Create an NFA M of size $\Sigma_{i=1}^{n} n_i + 1$ states where $n_i =$ the number of states in M_i as follows. M consists of a copy of each M_i. Let q_1 the start state of M_1 be the start state of M. We add a transition (q_1, a, q) for all $q \in M_j$ for $j = 2, ..., n$ where $q = \delta_j(q_j, a)$ (δ_j is the transition function of M_j) and q_j is the start state of M_j. We also add an accepting state f to M and add an ε transition from q_1 to f. Finally, add a transition from f to itself on all input symbols in Σ, the alphabet over which M_j's are defined. It is clear that M has only one ε transition. It can be shown that reduction is correct, and the problem is PSPACE-hard.

That the problem is in P when the NFA is unambiguous follows from the result in [19] that the equivalence problem for unambiguous NFAs is in P.   □

By suitably modifying the proof of Theorem 6, we can show that the PSPACE-hardness of the claim holds if we replace $|w|$ by $k|w|$ (for a fixed integer $k \geq 2$).

We are currently looking at the problem of whether or not its membership is decidable (or even in PSPACE).

For the proof of the next result, we need the notion of *distance automaton* introduced in [11], where a problem called *limitedness* was studied. Let N denote the set of nonnegative integer numbers. A *distance automaton* is a 5-tuple $< Q, \Sigma, q_0, d, F >$ where Q is the set of states, Σ is the alphabet set, $d : Q \times \Sigma \times Q \to \{0, 1\}$ is the distance function, q_0 is the starting state and F is the set of accepting states.

Given a path $(q_1, a_1, q_2, a_2, ..., a_k, q_{k+1})$, the distance of the path is defined as $d(q_1, a_1, q_2) + d(q_2, a_2, q_3) + ... + d(q_k, a_k, q_{k+1})$. Consider the processing of a string x from state p to state q. Since a distance automaton is nondeterministic, there could be many different paths. We define the distance to be the minimum distance over the different paths.

We extend the distance function d to a function $Q \times \Sigma^* \times Q \to N$ such that $d(p, x, q)$ denotes the distance used by the automaton for going from state p to state q consuming the input string x. The distance of an accepted string x is defined as $d(x) = min_{q \in F} d(p, x, q)$. The distance of a distance automaton M is defined to be $sup_{x \in L(M)} d(x)$. A distance automaton M is said to be *limited* if there is a constant K such that its distance is bounded above by K.

We will need the following result concerning the limitedness problem for NFAs [11] and subsequently improved by [17].

Theorem 7. *Determining, given a distance automaton M, whether it is limited is PSPACE-complete.*

Theorem 8. *It is PSPACE-complete to determine, given an NFA M, whether there exists an integer d such that every string w in $L(M)$ can be accepted by a computation of at most $|w| + d$ moves.*

Proof. Let M an instance of the problem. We will construct a distance automaton M' as follows: For any two states p, q in M, and for $a \in \Sigma$, we determine if q can be reached from p via a path that involves a sequence of ε moves followed by a followed by another sequence of ε moves so that the total path length is at least 2. (i.e., we determine if $q \in \hat{\delta}(p, a)$ and q is not in $\delta(p, a)$ - here $\hat{\delta}$ is the extended δ function [12]). If this is true for the triple (p, a, q), then we assign a transition from p to q on input a with distance $d = 1$. For all the non-ε transitions from p to q on label a, the transition is added with a d value of 0. The resulting distance automaton M' has no ε transitions, and it has a distance function defined. It is easy to see that the above reduction from M to M' can be implemented in polynomial time: compute the ε-closure of each state q by performing a breadth-first search from each state using only the ε transitions as edges. (This step is commonly used, e.g., when removing ε moves from an NFA). Now it is easy to determine which edges of distance 1 should be added to M': An edge (p, q) with distance $= 1$ is added in M' if and only if there are two states r and s such that r is in the ε-closure of p, q is in the ε-closure of s, and there is a transition in M from r to s on a. Finally, add a state q to the accepting state of M' if some accepting state is in the ε-closure of q. It is easy to see that $L(M) = L(M')$ since our construction is essentially the same as the one used for removing ε transitions in an NFA [12].

We will now show that there is a constant d such that any string $w \in L(M)$ can be accepted in $|w| + d$ steps if and only if M' is limited. Suppose M' is limited. Then there is a constant K such that every accepted string w has an accepting path with distance at most K. Consider the corresponding path for w from a starting state to accepting state in M. Every edge of weight 1 in M' (from p to q) involves a sequence of ε transitions from p to r and from s to q. By choosing the shortest path in each case, each path is of length at most $m - 1$ (where m is the number of states in M). Thus each edge of weight 1 uses at most $2(m-1)$ ε transitions. Since the total number of such edges is bounded by K, the total number of ε transitions is at most $2(m-1)K$ and by choosing $d = 2(m-1)K$, we see that every string of length $w \in L(M)$ has an accepting path of length at most $|w| + d$.

Conversely, suppose M is a 'yes' instance of our problem, i.e., there is a constant d such that every string w, there is an accepting path of length at most $|w| + d$. This means it uses at most d ε transitions. Consider the corresponding accepting computation for w in M'. Each ε-transition can be avoided in M' by using an edge of weight one so that the total number of edges of distance one used in an accepting path for w is at most d. Thus, M' has distance value of at most d and hence it is limited. This completes a polynomial time reduction (actually this reduction can be implemented in log-space) from our problem to the limitedness problem.

Since limitedness can be tested in PSPACE according to Theorem 7, it follows that our problem can be solved in PSPACE.

To show that our problem is PSPACE-hard, we will exhibit a reduction in the opposite direction, namely a reduction from limitedness problem to our problem.

Let M be an instance of limitedness problem. We will create an NFA M' using the same set of states (with some additional states as described below): for each edge of distance 0, we will include that edge in M'. For an edge (p, a, q) of distance 1, we add a new node r and edges (p, ε, r) and (r, a, q) in M'. The start state and accepting states of M' are the same as that of M. It is easy to see that $L(M) = L(M')$. Further it is easy to see that this is a polynomial time reduction from limitedness problem to our problem. This shows that our problem is PSPACE-complete.   □

Theorem 9. *Determining, given an unambiguous NFA M, whether there exists an integer d such that every string w in $L(M)$ can be accepted by a computation of at most $|w| + d$ moves is in P.*

The next result is rather unexpected.

Theorem 10. *For any $k \geq 2$, it is undecidable to determine, given an NFA M, whether there is a string w in $L(M)$ which has no accepting computation of length exactly $k|w|$.*

Proof. We prove the case $k = 2$; the generalization for $k \geq 3$ reduces to the case $k = 2$ by padding each input symbol with $k - 2$ dummy symbols. We will use the proof technique introduced in [14] to show the undecidability of the equivalence problem for EFNGSMAs (epsilon-free NGSMs with accepting states) over $\Sigma \times \{1\}$ (i.e., the input alphabet is Σ and the output alphabet is $\{1\}$).

First we note that an EFNGSMA G over $\Sigma \times \{1\}$ can be thought of as an NFA M_G in the following way: If G in state p on input a outputs 1^d and enters state q, then M_G in state p reads a, makes $d-1$ ε-moves (using $d-1$ new states), and enters state q. Thus, G outputting 1^d corresponds to M_G making d moves (the last $d-1$ of which are ε-moves). We call M_G the NFA associated with G.

Let Z be a Turing machine (TM) on an initially blank (one-way) infinite tape with unique initial and halting configurations. Assume that Z makes at least one move. Let Σ be the alphabet used to represent a sequence of IDs. Let $L_Z = \{x \mid x = \#ID_1\# \cdots \#ID_k\#, k \geq 2, ID_1, \ldots, ID_k$ are configurations of Z, ID_1 is the initial configuration on blank tape , ID_k is the unique halting configuration $\}$.

Let $R = \{(x, 1^r) \mid x \in \Sigma^+, x = x_1 x_2 x_3$ for some $x_1, x_2, x_3 \in \Sigma^*, r = |x_1| + 2|x_2| + 3|x_3|\}$.

Clearly, R can be realized by a one-state EFNGSMA. The relation $R(G)$ defined by the EFGSMA G constructed in the proof of Theorem 1 in [14] has the property that $R(G) = R$ if and only if the TM Z does not halt on blank tape.

In that proof, G was constructed as the combination of four EFNGSMAs G_1, G_2, G_3, G_4. Hence, G accepts the union of the transductions accepted by these machines. The machines are constructed so that they realize the following transductions:

1. $R(G_1) = \{(x, 1^r) \mid (x, 1^r) \in R, x \in \Sigma^+ - L_Z\}$
2. $R(G_2) = \{(x, 1^r) \mid (x, 1^r) \in R, r > 2|x|\}$
3. $R(G_3) = \{(x, 1^r) \mid (x, 1^r) \in R, r < 2|x|\}$
4. $R(G_4) = \{(x, 1^r) \mid (x, 1^r) \in R, x = \#ID_1\# \cdots \#ID_k\# \in L_Z$ and either $r \neq 2|x|$ or $r = 2|x|$ and for some $ID_i, 1 \leq i < k, ID_{i+1}$ is not a proper successor of $ID_i\}$

The constructions of G_1, G_2, G_3 are not difficult, but the construction of G_4 is rather intricate (see [14]). Now $R(G) = R(G_1) \cup R(G_2) \cup R(G_3) \cup R(G_4)$. Clearly, there is exactly one string $x \in \Sigma^+$ such that $(x, 1^{2|x|})$ is not in $R(G)$ if and only if the TM Z halts on blank tape (this x corresponds to the unique halting sequence of IDs if the TM halts). If Z halts on blank tape, then $R(G) = R - \{(x, 1^{2|x|})\}$, where x is the unique halting sequence of IDs of Z. Otherwise, $R(G) = R$.

Hence for the NFA M_G associated with G, there is exactly one string x in $L(M_G)$ that is not accepted by M_G in a computation of length exactly $2|x|$ if and only if TM Z halts on blank tape, which is undecidable.   □

Note that Theorem 10 does not hold for unambiguous NFAs, since the problem is decidable, even for NPDAs with reversal-bounded counters (see Corollary 1 and the examples for which the corollary applies in Section 3).

Interestingly, if we are interested in determining if there is a string w which has an accepting computation of length exactly $k|w|$, the problem becomes decidable, even for an NPDA with reversal-bounded counters:

Theorem 11. *For any $k \geq 1$, it is decidable to determine, given an NPDA M augmented with reversal-bounded counters, whether there is a string (resp., unique string) w in $L(M)$ which has an accepting computation of length exactly $k|w|$.*

Next, we consider the problem concerning the number of accepting computations with distinct running times every string can have:

Theorem 12. *It is decidable to determine, given $t \geq 1$ and a finite-crossing two-way NFA M augmented with reversal-bounded counters, whether every string w in $L(M)$ has at most t accepting computations with distinct running times.*

A related question is the following: Given M, is there a t such that every string w in $L(M)$ has at most t accepting computations with distinct running times? We do not know the decidability of this question when M is a finite-crossing two-way NFA augmented with reversal-bounded counters. However, for the special case of NFAs:

Theorem 13. *It is decidable, given an NFA M, whether there exists a $t \geq 1$, such that every string w in $L(M)$ has at most t accepting computations with distinct running times. If such a t exists, the minimal such t can be effectively computed.*

Theorems 12 and 13 do not hold for nondeterministic 1-reversal one-counter machines.

Theorem 14

1. For any $t \geq 1$, it is undecidable to determine, given a nondeterministic 1-reversal one-counter machine M, whether every string w in $L(M)$ has exactly (resp., at most) t accepting computations with distinct running times.
2. It is undecidable to determine, given a nondeterministic 1-reversal one-counter machine, whether there exists a $t \geq 1$ such that every string w in $L(M)$ has exactly (resp., at most) t accepting computations with distinct running times.

6   Run-Time Equivalence of Machines

In this section, we look at run-time equivalence of machines, which is defined as follows:

Definition. Two machines M_1 and M_2 are *run-time equivalent* if for every string w, M_1 has an accepting computation of length t if and only if M_2 has an accepting computation of length t.

Note that if M_1 and M_2 are run-time equivalent, then $L(M_1) = L(M_2)$. However, the converse is not true, in general.

Theorem 15. *It is undecidable to determine, given two NFAs M_1 and M_2, whether they are run-time equivalent.*

However, for unambiguous NFAs, run-time equivalence is decidable:

Theorem 16. *It is decidable to determine, given two unambiguous NFAs M_1 and M_2, whether they are run-time equivalent.*

We do not know if run-time equivalence is decidable for unambiguous NFAs augmented with any form of infinite storage, e.g., with reversal-bounded counters. But, for machines that are known to be equivalent, we have:

Theorem 17. *It is decidable to determine, given two unambiguous finite-crossing two-way NFAs augmented with reversal-bounded counters M_1 and M_2 that are known to be equivalent (i.e., $L(M_1) = L(M_2)$), whether they are run-time equivalent.*

Unfortunately, at present, we do not know if equivalence of unambiguous finite-crossing two-way NFAs with reversal-bounded counters is decidable. In fact, we do not know if equivalence of unambiguous nondeterministic 1-reversal one-counter machines (hence, 1-crossing) is decidable. However, it is known that equivalence of finite-crossing two-way DFAs augmented with reversal-bounded counters is decidable. Hence, from Theorem 17:

Corollary 3. *It is decidable to determine, given two finite-crossing two-way DFAs augmented with reversal-bounded counters M_1 and M_2, whether they are run-time equivalent.*

On the other hand:

Theorem 18. *Run-time equivalence is undecidable for 2-ambiguous nondeterministic one-counter machines (resp., 2-ambiguous 1-reversal NPDAs).*

References

1. Baker, B., Book, R.: Reversal-bounded multipushdown machines. J. Comput. Syst. Sci. 8, 315–332 (1974)
2. Chan, T., Ibarra, O.H.: On the finite-valuedness problem for sequential machines. Theor. Comput. Sci. 23, 95–101 (1983)
3. Cui, C., Dang, Z., Fischer, T.R., Ibarra, O.H.: Similarity in languages and programs (submitted)
4. Chatterjee, K., Doyen, L., Henzinger, T.A.: Quantitative Languages. ACM Transactions on Computational Logic (2010)
5. Ginsburg, S., Greibach, S.A., Harrison, M.A.: One-way stack automata. J. ACM 14, 389–418 (1967)
6. Goldstine, J., Kappes, M., Kintala, C.M.R., Leung, H., Malcher, A., Wotschke, D.: Descriptional complexity of machines with limited resources. Journal of Universal Computer Science 8(2), 193–234 (2002)

7. Greibach, S.A.: A new normal-Form theorem for context-free phrase structure grammars. J. ACM 12, 42–52 (1965)
8. Gurari, E., Ibarra, O.H.: The complexity of decision problems for finite-turn multicounter machines. J. Comput. Syst. Sci. 22, 220–229 (1981)
9. Gurari, E., Ibarra, O.H.: A note on finite-valued and finitely ambiguous transducers. Math. Systems Theory 16, 61–66 (1983)
10. Harju, T., Karhumaki, J.: The equivalence problem of multitape finite automata. Theor. Comput. Sci. 78, 347–355 (1991)
11. Hashiguchi, K.: Limitedness theorem on finite automata with distance functions. J. Comput. Syst. Sci. 24, 233–244 (1982)
12. Hopcroft, J.E., Ullman, J.D.: Introduction to Automata Theory, Languages and Computation. Addison-Wesley Publishing Company (1979)
13. Ibarra, O.H.: Reversal-bounded multicounter machines and their decision problems. J. ACM 25, 116–133 (1978)
14. Ibarra, O.H.: The unsolvability of the equivalence problem for ε-free NGSM's with unary input (output) alphabet and applications. SIAM J. Computing 7, 524–532 (1978)
15. Ibarra, O.H., Jiang, T., Tran, N.Q., Wang, H.: New decidability results concerning two-way counter machines and applications. SIAM J. Comput. 24, 123–137 (1995)
16. Kozen, D.: Lower bounds for natural proof systems. In: IEEE Conf. on Foundations of Computer Science, pp. 254–266 (1977)
17. Leung, H., Podolskiy, V.: The limitedness problem on distance automata: Hashiguchi's method revisited. Theor. Comput. Sci. 310, 147–158 (2004)
18. Minsky, M.: Recursive unsolvability of Post's problem of Tag and other topics in the theory of Turing machines. Ann. of Math. 74, 437–455 (1961)
19. Stearns, R., Hunt, H.: On the equivalence and containment problems for unambiguous regular expressions, grammars and automata. SIAM Journal on Computing 14, 598–611 (1985)
20. Turing, A.M.: On Computable numbers, with an application to the Entscheidungsproblem. Proc. London Math. Soc., s2 42, 230–265 (1937)

Unambiguous Conjunctive Grammars
over a One-Letter Alphabet

Artur Jeż[1,2,*] and Alexander Okhotin[3,**]

[1] Max-Planck-Institut für Informatik, Saarbrücken, Germany
[2] Institute of Computer Science, University of Wrocław, Poland
aje@cs.uni.wroc.pl
[3] Department of Mathematics and Statistics, University of Turku, Finland
alexander.okhotin@utu.fi

Abstract. It is demonstrated that unambiguous conjunctive grammars over a unary alphabet $\Sigma = \{a\}$ have non-trivial expressive power, and that their basic properties are undecidable. The key result is that for every base $k \geqslant 11$ and for every one-way real-time cellular automaton operating over the alphabet of base-k digits $\left\{ \lfloor \frac{k+9}{4} \rfloor, \ldots, \lfloor \frac{k+1}{2} \rfloor \right\}$, the language of all strings a^n with the base-k notation of the form $1w1$, where w is accepted by the automaton, is generated by an unambiguous conjunctive grammar. Another encoding is used to simulate a cellular automaton in a unary language containing almost all strings. These constructions are used to show that for every fixed unambiguous conjunctive language L_0, testing whether a given unambiguous conjunctive grammar generates L_0 is undecidable.

1 Introduction

Conjunctive grammars, introduced by Okhotin [15], are an extension of the context-free grammars, which allows the use of a conjunction operation in any rules, in addition to the implicit disjunction already present in context-free grammars. These grammars maintain the main principle behind the context-free grammars—that of inductive definition of the membership of strings in the language—inherit their parsing techniques and subcubic time complexity [19], and augment their expressive power in a meaningful way. Conjunctive grammars, along the more general Boolean grammars [17], have been a subject of various research [1,7,8,9,12,16,20,21].

Conjunctive grammars over a one-letter alphabet $\Sigma = \{u\}$ were proved non-trivial by Jeż [7], who constructed a grammar for the language $\{a^{4^n} \mid n \geqslant 0\}$, and extended this construction to represent every *automatic set* [2], that is, a unary language with a regular base-k representation. Subsequent work on such grammars revealed their high expressive power and a number of undecidable properties [8]. Testing whether a given string a^n is generated by a grammar

* Supported by NCN grant number 2011/01/D/ST6/07164, 2011–2014.
** Supported by the Academy of Finland under grants 134860 and 257857.

M.-P. Béal and O. Carton (Eds.): DLT 2013, LNCS 7907, pp. 277–288, 2013.
© Springer-Verlag Berlin Heidelberg 2013

G can be done in time $|G| \cdot n(\log n)^3 \cdot 2^{O(\log^* n)}$ [20], and if n is given in binary notation, this problem is EXPTIME-complete already for a fixed grammar G [9]. These results had impact on the study of language equations [13,18], being crucial to understanding their computational completeness over a unary alphabet [10]. They are also related to the complexity results for circuits over sets of numbers [14], as conjunctive grammars over a unary alphabet may be regarded as a generalization of these circuits.

Unambiguous conjunctive grammars [17] are an important subclass of conjunctive grammars defined by analogy with unambiguous context-free grammars, and representing grammars that assign a unique syntactic structure to every well-formed sentence. Little is known about their properties, besides a parsing algorithm with $|G| \cdot O(n^2)$ running time, where n is the length of the input [17]; for a unary alphabet, the running time can be improved to $|G| \cdot n(\log n)^2 \cdot 2^{O(\log^* n)}$ [20]. However, all the known results on the expressive power of conjunctive grammars over a unary alphabet [7,8,9,21] rely upon ambiguous grammars, and it is not even known whether unambiguous grammars can generate anything non-regular.

This paper sets off by presenting the first example of an unambiguous conjunctive grammar that generates a non-regular unary language. This is the same language $\{a^{4^n} \mid n \geqslant 0\}$, yet the grammar generating it, given in Section 3, is more involved than the known ambiguous grammar. Then the paper proceeds with reimplementing, using unambiguous grammars, the main general method for constructing conjunctive grammars over a unary alphabet. This method involves simulating a one-way real-time cellular automaton [6,16,23] over an input alphabet $\Sigma_k = \{0, 1, \ldots, k-1\}$ of base-k digits, by a grammar generating all strings a^n with the base-k representation of n accepted by the cellular automaton. The known construction of such conjunctive grammars [8] essentially uses concatenations of densely populated sets, and hence the resulting grammars are ambiguous. This paper defines a different simulation, under the assumption that the input alphabet of the cellular automaton is not the entire set of base-k digits, but its subset of size around $\frac{k}{4}$. This restriction allows simulating the automaton, so that all concatenations in the grammar remain unambiguous.

The simulation of a cellular automaton presented in Section 4 produces languages that grow exponentially fast, that is, the length of the n-th shortest string in the language is exponential in n. These languages have *density 0*, in the sense that the fraction of strings of length up to ℓ belonging to these languages tends to 0. As the concatenation of any two unary languages of non-zero density is always ambiguous, this limitation of the given construction might appear to be inherent to unambiguous conjunctive grammars. However, the next Section 5 nevertheless succeeds in representing non-regular unary languages of density 1 (that is, containing almost all strings) by an unambiguous conjunctive grammar, and extends the simulation of cellular automata to this kind of unary languages.

These constructions yield undecidability results for unambiguous conjunctive grammars, presented in the last Section 6. For every fixed language L_0 (over an arbitrary alphabet) generated by some unambiguous conjunctive grammar, it is proved that testing whether a given unambiguous conjunctive grammar generates L_0 is undecidable. This is compared to the known decidable properties of the unambiguous case of standard context-free grammars [22].

2   Conjunctive Grammars and Ambiguity

Conjunctive grammars generalize context-free grammars by allowing an explicit conjunction operation in the rules. This is more of a variant of the definition of the context-free grammars than something entirely new, as it leaves the general idea of context-free descriptions intact, and only extends the set of logical connectives used to combine syntactical conditions.

Definition 1 (Okhotin [15]). *A conjunctive grammar is a quadruple $G = (\Sigma, N, P, S)$, in which Σ and N are disjoint finite non-empty sets of terminal and nonterminal symbols respectively; P is a finite set of rules of the form*

$$A \to \alpha_1 \& \ldots \& \alpha_n \quad (with\ A \in N,\ n \geqslant 1\ and\ \alpha_1, \ldots, \alpha_n \in (\Sigma \cup N)^*), \quad (*)$$

while $S \in N$ is a nonterminal designated as the start symbol. A grammar is **linear**, *if each α_i in each rule (*) contains at most one nonterminal symbol.*

A rule (*) informally means that every string generated by each *conjunct* α_i is therefore generated by A. One way of formalizing this understanding is by *language equations*, with the conjunction interpreted as intersection of languages.

Definition 2. *Let $G = (\Sigma, N, P, S)$ be a conjunctive grammar. The* associated *system of language equations is the following system in variables N:*

$$A = \bigcup_{A \to \alpha_1 \& \ldots \& \alpha_n \in P} \bigcap_{i=1}^{n} \alpha_i \quad (for\ all\ A \in N),$$

where each α_i in the equation is a concatenation of variables and constant languages $\{a\}$ representing terminal symbols (or constant $\{\varepsilon\}$ if α_i is the empty string). Let $(\ldots, L_A, \ldots)$ be its least solution (that is, such a solution that every other solution $(\ldots, L'_A, \ldots)$ has $L_A \subseteq L'_A$ for all $A \in N$) and denote $L_G(A) := L_A$ for each $A \in N$. Define $L(G) := L_G(S)$.

As in the case of standard context-free grammars [3], every such system of equations has a least solution expressible through fixpoint iteration, because the right-hand sides of the system are monotone and continuous.

An equivalent definition of conjunctive grammars is given by *term rewriting*, generalizing the more common definition of standard context-free grammars by string rewriting.

Definition 3 ([15]). *Given a conjunctive grammar G, consider terms over concatenation and conjunction with symbols from $\Sigma \cup N$ as atomic terms. The relation $\Longrightarrow$ of immediate derivability on the set of terms is defined as follows:*

- *Using a rule $A \to \alpha_1 \& \ldots \& \alpha_n$, a subterm $A \in N$ of any term $\varphi(A)$ can be rewritten as $\varphi(A) \Longrightarrow \varphi(\alpha_1 \& \ldots \& \alpha_n)$.*
- *A conjunction of several identical strings can be rewritten by one such string: $\varphi(w \& \ldots \& w) \Longrightarrow \varphi(w)$, for every $w \in \Sigma^*$.*

The language generated by a term φ is $L_G(\varphi) = \{w \mid w \in \Sigma^*, \ \varphi \Longrightarrow^* w\}$. The language generated by the grammar is $L(G) = L_G(S) = \{w \mid w \in \Sigma^*, \ S \Longrightarrow^* w\}$.

One can straightforwardly represent any finite intersection of standard context-free languages, such as $\{a^n b^n c^n \mid n \geqslant 0\}$, by a conjunctive grammar. The expressive power of conjunctive grammars actually goes beyond such intersections: for instance, they can represent the language $\{wcw \mid w \in \{a,b\}^*\}$ [15].

This paper concentrates on a subclass of conjunctive grammars defined by analogy with unambiguous context-free grammars. Let a concatenation $L_1 \cdot \ldots \cdot L_k$ be called *unambiguous* if every string $w \in L_1 \cdot \ldots \cdot L_k$ has a unique factorization $w = u_1 \ldots u_k$ with $u_i \in L_i$.

Definition 4 ([17]). *Let $G = (\Sigma, N, P, S)$ be a conjunctive grammar. Then*

  I. *the choice of a rule in G is unambiguous, if different rules for every single nonterminal A generate disjoint languages, that is, for every string w there exists at most one rule $A \to \alpha_1 \& \ldots \& \alpha_m$ with $w \in L_G(\alpha_1) \cap \ldots \cap L_G(\alpha_m)$.*
 II. *concatenation in G is said to be unambiguous, if for every conjunct $\alpha = s_1 \ldots s_\ell$, the concatenation $L_G(s_1) \cdot \ldots \cdot L_G(s_\ell)$ is unambiguous.*

If both conditions are satisfied, the grammar is called unambiguous.

Definition 4 implies that every string has a unique parse tree. The converse is untrue: some grammars define unique parse trees, but condition II does not hold.

3   Representing Powers of Four

Consider the following grammar generating the language $\{a^{4^n} \mid n \geqslant 0\}$, which was the first example of a conjunctive grammar over a unary alphabet representing a non-regular language. Even though much was learned about those grammars since this example, it still remains the smallest and the easiest to understand.

Example 1 (Jeż [7]). The conjunctive grammar

$$A_1 \to A_1 A_3 \& A_2 A_2 \mid a \qquad A_3 \to A_1 A_2 \& A_6 A_6 \mid aaa$$
$$A_2 \to A_1 A_1 \& A_2 A_6 \mid aa \qquad A_6 \to A_1 A_2 \& A_3 A_3$$

with the start symbol A_1 generates the language $L(G) = \{a^{4^n} \mid n \geqslant 0\}$. In particular, $L_G(A_i) = \{a^{i \cdot 4^n} \mid n \geqslant 0\}$ for $i = 1, 2, 3, 6$.

The grammar is best explained in terms of base-4 notation of the lengths of the strings. Let $\Sigma_4 = \{0, 1, 2, 3\}$ be the alphabet of base-4 digits, and for every $w \in \Sigma_4^*$, let $(w)_4$ denote the integer with base-4 notation w. For any $L \subseteq \Sigma_4^*$, denote $a^{(L)_4} = \{a^{(w)_4} \mid w \in L\}$. Then the languages generated by the nonterminals of the above grammar are $a^{(10^*)_4}$, $a^{(20^*)_4}$, $a^{(30^*)_4}$ and $a^{(120^*)_4}$.

Consider the system of language equations corresponding to the grammar: the equation for A_1 is

$$A_1 = (A_1 A_3 \cap A_2 A_2) \cup \{a\},$$

etc. Substituting the given four languages into the intersection $A_1A_3 \cap A_2A_2$ in the first equation, one obtains the following language:

$$a^{(10^*)_4}a^{(30^*)_4} \cap a^{(20^*)_4}a^{(20^*)_4} =$$

$$= \left(a^{(10^+)_4} \cup a^{(10^*30^*)_4} \cup a^{(30^*10^*)_4}\right) \cap \left(a^{(10^+)_4} \cup a^{(20^*20^*)_4}\right) = a^{(10^+)_4}.$$

That is, both concatenations contain some garbage, yet the garbage in the concatenations is disjoint, and is accordingly filtered out by the intersection. Finally, the union with $\{a\}$ yields the language $\{a^{4^n} \mid n \geqslant 0\}$, and thus the first equation turns into an equality. The rest of the equations are verified similarly, and hence the given four languages form a solution. By a standard argument, one can prove that the system has a unique ε-free solution [3, Thm. 2.3].

The grammar in Example 1 is ambiguous, because of the concatenations A_2A_2, A_1A_1, A_6A_6 and A_3A_3: since concatenation of unary strings is commutative, a concatenation of a language with itself is unambiguous only if this language is empty or a singleton. However, it is possible to remake the above grammar without ever using such concatenations, though that requires defining a larger collection of languages. The following grammar becomes the first evidence of non-triviality of unambiguous conjunctive grammars over a unary alphabet.

Example 2. The conjunctive grammar

$$
\begin{aligned}
A_1 &\to A_1A_3 \& A_7A_9 \mid a \mid a^4 & A_7 &\to A_1A_3 \& A_1A_6 \\
A_2 &\to A_1A_7 \& A_2A_6 \mid a^2 & A_9 &\to A_1A_2 \& A_2A_7 \\
A_3 &\to A_1A_2 \& A_3A_9 \mid a^3 & A_{15} &\to A_6A_9 \& A_2A_7 \\
A_6 &\to A_1A_2 \& A_9A_{15} \mid a^6 &
\end{aligned}
$$

is unambiguous and generates the language $\{a^{4^n} \mid n \geqslant 0\}$. Each nonterminal A_i generates the language $L_G(A_i) = \{a^{i \cdot 4^n} \mid n \geqslant 0\}$.

The correctness is established in the same way as in Example 1. For instance, the first equation is checked as

$$a^{(10^*)_4}a^{(30^*)_4} \cap a^{(130^*)_4}a^{(210^*)_4} = \left(a^{(10^+)_4} \cup a^{(10^*30^*)_4} \cup a^{(30^*10^*)_4}\right)\cap$$

$$\cap \left(a^{(10^{\geqslant 2})_4} \cup a^{(2110^*)_4} \cup a^{(2230^*)_4} \cup a^{(130^*210^*)_4} \cup a^{(210^*130^*)_4}\right) = a^{(10^{\geqslant 2})_4}.$$

All concatenations in the grammar are unambiguous. More generally, for every alphabet $\Sigma_k = \{0, 1, \ldots, k-1\}$ of base-k digits, let $(w)_k$ denote the number with the base-k notation w, and $a^{(L)_k} = \{a^{(w)_k} \mid w \in L\}$ for every $L \subseteq \Sigma_k^*$. Then the concatenation of a language $a^{(ij0^*)_k}$ with a language $a^{(i'j'0^*)_k}$ is *in most cases* unambiguous, and none of the concatenations actually used in the grammar are among the few exceptions to this rule.

Lemma 1. *Let $k \geqslant 2$, and consider any two different languages of the form $K = a^{(ij0^*)_k}$ and $L = a^{(i'j'0^*)_k}$, with $i, i' \in \Sigma_k \setminus \{0\}$ and $j, j' \in \Sigma_k$, except those with $i = j = i'$ and $j' = 0$, or vice versa. Then the concatenation KL is unambiguous.*

Using concatenations of this form, Example 2 can be generalized to construct unambiguous grammars for the languages $L_k = \{a^{k^n} \mid n \geqslant 1\}$, with $k \geqslant 9$.

Lemma 2. *For every $k \geqslant 9$, the following conjunctive grammar with the set of nonterminals $N = \{A_{i,j} \mid i, j \in \{0, \ldots, k-1\}, i \neq 0\}$ and with the start symbol $A_{1,0}$ is unambiguous and generates the language $a^{(10^+)_k}$:*

$$A_{1,j} \rightarrow A_{k-1,0}A_{j+1,0} \;\&\; A_{k-2,0}A_{j+2,0} \mid a^{(1j)_k}, \qquad \text{for } j < \tfrac{k}{3} + 2$$

$$A_{i,j} \rightarrow A_{i-1,k-1}A_{j+1,0} \;\&\; A_{i-1,k-2}A_{j+2,0} \mid a^{(ij)_k}, \qquad \text{for } i \geqslant 2,\, j < \tfrac{k}{3} + 2$$

$$A_{i,j} \rightarrow A_{i,j-1}A_{1,0} \;\&\; A_{i,j-2}A_{2,0} \mid a^{(ij)_k}, \qquad \text{for } i \geqslant 1,\, j \geqslant \tfrac{k}{3} + 2$$

In particular, each nonterminal $A_{i,j}$ generates the language $a^{(ij0^*)_k}$.

4   Simulating Trellis Automata

This section extends the known general method for constructing conjunctive grammars over a unary alphabet [8] to the case of unambiguous conjunctive grammars. The overall idea is to simulate a *one-way real-time cellular automaton*, also known as a *trellis automaton*, operating on base-k representations of numbers, by a grammar generating unary representations of the same numbers.

A trellis automaton [4,16], defined as a quintuple $(\Sigma, Q, I, \delta, F)$, processes an input string of length $n \geqslant 1$ using a uniform array of $\frac{n(n+1)}{2}$ nodes, as presented in the figure below. Each node computes a value from a fixed finite set Q. The nodes in the bottom row obtain their values directly from the input symbols using a function $I \colon \Sigma \rightarrow Q$. The rest of the nodes compute the function $\delta \colon Q \times Q \rightarrow Q$ of the values in their predecessors. The string is accepted if and only if the value computed by the topmost node belongs to the set of accepting states $F \subseteq Q$.

Definition 5. *A trellis automaton is a quintuple $M = (\Sigma, Q, I, \delta, F)$, in which:*
- *Σ is the input alphabet,*
- *Q is a finite non-empty set of states,*
- *$I \colon \Sigma \rightarrow Q$ is a function that sets the initial states,*
- *$\delta \colon Q \times Q \rightarrow Q$ is the transition function, and*
- *$F \subseteq Q$ is the set of final states.*

The state computed on a string $w \in \Sigma^+$ is denoted by $\Delta(w)$ and defined inductively as $\Delta(a) = I(a)$ and $\Delta(awb) = \delta(\Delta(aw), \Delta(wb))$, for all $a, b \in \Sigma, w \in \Sigma^*$. Define $L_M(q) = \{w \mid \Delta(w) = q\}$ and $L(M) = \{w \mid \Delta(w) \in F\}$.

Trellis automata are known to be equivalent to linear conjunctive grammars [16], and the family of languages they recognize shall be called *linear conjunctive languages*.

Consider a trellis automaton with the input alphabet $\Sigma_k = \{0, 1, \ldots, k-1\}$ of base-k digits, and assume that it does not accept any strings beginning with

0. Then, every string of digits accepted by the automaton defines a certain non-negative integer, and thus the automaton defines a set of numbers. The goal is to represent the same set of numbers in unary notation by a conjunctive grammar. For conjunctive grammars of the general form, without the unambiguity condition, this is always possible.

Theorem A (Jeż, Okhotin [8]). *For every $k \geqslant 2$ and for every trellis automaton M over the alphabet Σ_k, with $L(M) \cap 0\Sigma_k^* = \varnothing$, there exists a conjunctive grammar generating the language $\{a^{(w)_k} \mid w \in L(M)\}$.*

The grammar simulates the computation of a trellis automaton $M = (\Sigma_k, Q, I, \delta, F)$ using the nonterminal symbols C_q with $q \in Q$, which generate the languages $L(C_q) = \{a^{(1w10^\ell)_k} \mid \Delta(w) = q, \ \ell \geqslant 0\}$, so that each string of digits $w \in \Sigma_k^*$ is represented in unary notation by the strings $a^{(1w1)_k}$, $a^{(1w10)_k}$, $a^{(1w100)_k}$, etc. The definition is inductive on the length of w. As the basis of induction, each C_q should generate all strings of the form $a^{(1j10^\ell)_k}$ with $I(j) = q$ and $\ell \geqslant 0$; this is a language similar to the one in Lemma 2.

The grammar implements a step of induction as follows. A string $a^{(1w10^\ell)_k}$ with $|w| \geqslant 2$ should be generated by C_q if and only if $\Delta(w) = q$, which, according to the definition of a trellis automaton, holds if and only if $w = iuj$ for some $i, j \in \Omega$ and $u \in \Omega^*$ with $q_1 = \Delta(iu)$, $q_2 = \Delta(uj)$ and $q = \delta(q_1, q_2)$. Then, by the induction hypothesis, the nonterminal C_{q_1} generates the string $a^{(1iu10^{\ell+1})_k}$, which is one of the unary encodings of iu, while C_{q_2} generates $a^{(1uj10^\ell)_k}$, an encoding of uj. The rules of the grammar perform a series of concatenations and intersections on these strings, and ultimately generate $a^{(1w10^\ell)_k}$ by C_q.

However, the grammar produced by Theorem A is always ambiguous, and there is no general way of expressing the same languages $L(C_q) \subseteq a^*$ unambiguously, for the following reason. The construction of unambiguous grammars, as in Lemma 2, relies on concatenating *exponentially growing* languages, and the sparsity of such languages in some cases allows their concatenation to be unambiguous. But the languages $L(C_q)$, as defined above, may be denser than that, and their concatenation with any infinite language is always ambiguous.

Thus, the first step towards simulating a trellis automaton by an unambiguous conjunctive grammar is to define a unary encoding of the languages $L_M(q)$ that always grows exponentially, regardless of the form of $L_M(q)$. This is done by choosing the base k to be larger than the cardinality of the input alphabet Ω of M, and assuming that Ω is a subset of the set of all digits.

Theorem 1. *For every trellis automaton M over a d-letter input alphabet Ω, let $c \geqslant \max(5, d+2)$ and assume that $\Omega = \{c, \ldots, c+d-1\}$. Then, for every base $k \geqslant 2c + 2d - 3$, there exists an unambiguous conjunctive grammar generating the language $\{a^{(1w1)_k} \mid w \in L(M)\}$.*

If a base $k \geqslant 11$ is fixed, then, for instance, one can use the values $c = \lfloor \frac{k+9}{4} \rfloor$ and $d = \lfloor \frac{k-1}{4} \rfloor$, which induce the alphabet $\Omega = \{\lfloor \frac{k+9}{4} \rfloor, \ldots, \lfloor \frac{k+1}{2} \rfloor\}$. If the goal is to have an alphabet Ω with $d = 2$ letters, then the smallest values of c and k are $c = 5$ and $k = 11$, so that $\Omega = \{5, 6\}$.

The construction developed in this paper to prove Theorem 1 is generally analogous to the one used in Theorem A; in particular, it adopts a very similar unary representation of strings over Ω. Let $M = (\Omega, Q, I, \delta, F)$ be a trellis automaton. For every state $q \in Q$ and for all $s, t \in \{1, 2\}$, the grammar has a nonterminal $C_q^{s,t}$, which defines the language

$$L(C_q^{s,t}) = \{a^{(swt0^\ell)_k} \mid \ell \geqslant 0, \ \Delta(w) = q\}.$$

In this construction, the digits s and t surrounding the string w may be 1 or 2, whereas Theorem A uses only 1 for that purpose; this is an insignificant technical detail. The crucial difference with Theorem A is that each string w processed by M uses only digits from a small subset of Σ_k, and hence each $C_q^{s,t}$ generates an exponentially growing unary language.

Besides the nonterminals $C_q^{s,t}$, the constructed grammar includes all nonterminals $A_{i,j}$ with $i, j \in \Sigma_k$ and $i \neq 0$, as defined in Lemma 1, which generate the languages $L(A_{i,j}) = a^{(ij0^*)_k}$.

The strings $a^{(swt0^\ell)_k}$ with $\Delta(w) = q$ are generated by the corresponding nonterminals $C_q^{s,t}$ inductively on the length of w. The basis of induction is that $a^{(sjt0^\ell)_k}$ with $j \in \Omega$ and $\ell \geqslant 0$ must be in $L(C_q^{s,t})$, where $I(j) = q$. This means generating the language $a^{(sjt0^*)_k}$, which is achieved by the rules

$$C_q^{s,t} \to A_{1,t} A_{s,j-1} \ \& \ A_{2,t} A_{s,j-2},$$

defined for all $s, t \in \{1, 2\}$, $j \in \Omega$, and $q = I(j)$, similar to those in Example 2.

A string $a^{(swt0^\ell)_k}$ with $|w| \geqslant 2$ is generated by $C_q^{s,t}$ with $q = \Delta(w)$ as follows. Let $w = iuj$, where $i, j \in \Omega$ and $u \in \Omega^*$. In the trellis automaton, $\Delta(iu) = q_1$, $\Delta(uj) = q_2$ and $\delta(q_1, q_2) = q$. In the grammar, the four strings $a^{(s'ujt0^\ell)_k} \in L(C_{q_1}^{s',t})$ and $a^{(siut'0^{\ell+1})_k} \in L(C_{q_2}^{s,t'})$, with $s', t' \in \{1, 2\}$, are already generated, and the goal is to produce the string $a^{(siujt0^\ell)_k}$ by $C_q^{s,t}$. This is done by the rule

$$C_q^{s,t} \to C_{q_2}^{1,t} A_{s,i-1} \ \& \ C_{q_2}^{2,t} A_{s,i-2} \ \& \ C_{q_1}^{s,1} A_{j-1,t} \ \& \ C_{q_1}^{s,2} A_{j-2,t}$$

(with such rules defined for all $s, t \in \{1, 2\}$, $i, j \in \Omega$, $q_1, q_2 \in Q$ and $q = \delta(q_1, q_2)$), which represents the desired string as the following four concatenations:

$$a^{(siujt0^\ell)_k} = a^{(1ujt0^\ell)_k} a^{(s(i-1)0^{|ujt|+\ell})_k} = a^{(2ujt0^\ell)_k} a^{(s(i-2)0^{|ujt|+\ell})_k} =$$

$$= a^{(siu10^{\ell+1})_k} a^{((j-1)t0^\ell)_k} = a^{(siu20^{\ell+1})_k} a^{((j-2)t0^\ell)_k}.$$

The first two conjuncts of this rule are concerned with transforming unary encodings of the string uj to unary encodings of iuj. More precisely, one has to transform the two strings $a^{(1ujt0^\ell)_k}$ and $a^{(2ujt0^\ell)_k}$ with $\Delta(uj) = q_2$, which are generated by $C_{q_2}^{1,t}$ and by $C_{q_2}^{2,t}$, respectively, into the string $a^{(siujt0^\ell)_k}$. This is done by concatenating the string $a^{(1ujt0^\ell)_k}$ to $a^{(s(i-1)0^{|ujt|+\ell})_k} \in L(A_{s,i-1})$, and similarly, $a^{(2ujs'0^\ell)_k}$ is concatenated to $a^{(s(i-2)0^{|ujt|+\ell})_k} \in L(A_{s,i-2})$. It can be proved that the intersection of the first two conjuncts defines the set of all strings $a^{(siujt0^\ell)_k}$ with $\Delta(uj) = q_2$ and with arbitrary $i \in \Omega$.

The last two conjuncts of the rule similarly transform any two strings $a^{(siu10^{\ell+1})_k}$ and $a^{(siu20^{\ell+1})_k}$ into the string $a^{(siujt0^\ell)_k}$. One can prove that, for each $j \in \Omega$, the conjunction $C_{q_1}^{s,1} A_{j-1,t} \& C_{q_1}^{s,2} A_{j-2,t}$ defines the language of all $a^{(siujt0^\ell)_k}$ with $\Delta(iu) = q_1$. Once these four conjuncts are intersected in a single rule, it accordingly generates all $a^{(siujt0^\ell)_k}$ with $\Delta(iuj) = q$, as desired.

Finally, it remains to introduce a new start symbol S, which generates the union of all $C_q^{1,1}$ with $q \in F$, and intersects it with the set $a(a^k)^*$. This is exactly the language $\{a^{(1w1)_k} \mid w \in L(M)\}$.

5   A Density-Preserving Encoding of Trellis Automata

For a language $L \subseteq a^*$, consider the number

$$d(L) = \lim_{n \to \infty} \frac{|L \cap \{\varepsilon, a, a^2, \ldots, a^{n-1}\}|}{n},$$

called the *density* of L [21]. This limit, if it exists, always lies within the bounds $0 \leqslant d(L) \leqslant 1$. Let a language be called *sparse*, if $d(L) = 0$, and *dense*, if $d(L) = 1$.

All unambiguous conjunctive grammars constructed so far generate sparse unary languages (actually, exponentially-growing languages). Using only sparse languages in the constructions is, to some extent, a necessity, because languages are expressed in the grammar by concatenating them to each other, and a concatenation of a non-sparse unary language with any infinite language is bound to be ambiguous [11]. Of course, this does not mean that non-sparse sets cannot be represented at all—for instance, it is easy to modify the grammar in Example 2 to represent the language $\{a^{4^n} \mid n \geqslant 0\} \cup a(aa)^*$ of density $\frac{1}{2}$—but only that, once represented, non-sparse sets cannot be non-trivially concatenated to anything.

This section develops a method of simulating the computation of a trellis automaton in an unambiguous conjunctive grammar generating a unary language of density 1. This result parallels that of Theorem 1, which simulates a trellis automaton in a grammar generating a unary language of density 0. The proof of the new result is actually done on top of the constructions from Theorem 1.

The general idea of the new construction is based on the following representation of a^* as an unambiguous concatenation, due to Enflo et al. [5]:

Example 3. Let $k \geqslant 2$ be any power of two, and consider the languages $L_1, L_2, \ldots, L_{\frac{k}{2}}$, defined by $L_i = a^{(i\{i,0\}^*)_k} \cup \{\varepsilon\}$. Then $L_1 L_2 L_4 L_8 \ldots L_{\frac{k}{2}} = a^*$, and this concatenation is unambiguous; to see this, note that every integer $n \geqslant 0$ is uniquely representable as $n = n_1 + n_2 + n_4 + \ldots + n_{\frac{k}{2}}$ with $n_i \in (i\{0,i\}^*)_k \cup \{0\}$.

Let one of the languages L_i in this concatenation be replaced with a language $L_i' \subseteq L_i$, which encodes the computation of a trellis automaton operating on the two-letter input alphabet $\{0, i\}$, similarly to the encoding in Theorem 1. Then the concatenation $L = L_1 \ldots L_{i-1} L_i' L_{i+1} \ldots L_{k/2}$ is still unambiguous, and the density of the language L is controlled by the given linear conjunctive language, and thus can be set to any desired value. This construction, with $i = 2$, leads to representing the following languages by unambiguous conjunctive grammars.

Theorem 2. *Let L be a linear conjunctive language over a two-letter alphabet $\Gamma = \{e, f\}$, which does not contain any strings beginning with e. Let $k \geqslant 16$ be any power of two and define a homomorphism $h \colon \Sigma_k^* \to \Gamma^*$ by $h(4i) = h(4i+1) = e$ and $h(4i + 2) = h(4i + 3) = f$ for all $i \in \{0, \ldots, \frac{k}{4} - 1\}$. Then the language $\{a^{(w)k} \mid h(w) \in e^* L\}$ is generated by an unambiguous conjunctive grammar. Given a trellis automaton recognizing L, this grammar can be effectively constructed.*

In order to prove the theorem according to the above general idea, there are two claims to be established. First, the necessary constants are representable.

Lemma 3. *For all $k \geqslant 16$ and $t \in \{1, \ldots, \frac{k}{2}\}$, where k is a power of two, there is an unambiguous conjunctive grammar generating the language $a^{(t\{0,t\}^*)k}$.*

Then, a linear conjunctive language is encoded within a subset of L_2 as follows.

Lemma 4. *For every linear conjunctive language $L \subseteq \{0, 2\}^* \setminus 0\{0, 2\}^*$ and for every base $k \geqslant 16$ that is a power of two, there is an unambiguous conjunctive grammar generating the language $\{a^{(w)k} \mid w \in L\}$.*

Consider the language $\widehat{L}$, obtained by renaming the digits in L: 0 to 5 and 2 to 7. If M is a trellis automaton recognizing L, then $\widehat{L}$ is recognized by a similar automaton M'. Applying Theorem 1 with $c = 5$ and $d = 3$ to this language yields a grammar for the unary encoding $a^{(1L(M')1)k}$. Next, this encoding is modified by concatenating it with the constant language $a^{((k-6)^*(k-1))k}$ (for which a grammar can be constructed), so that the digits 5 and 7 are shifted back to 0 and 2. Then the result is intersected with the language $a^{(2\{0,2\}^*)k}$ from Lemma 3, thus filtering out all malformed sums and producing the language $a^{(1L(M)0)k}$.

The full construction is obtained by applying the above transformations to the languages $L^{(0)} = (\{2\}^{-1}L\{0\}^{-1}) \setminus \{\varepsilon\}$ and $L^{(2)} = (\{2\}^{-1}L\{2\}^{-1}) \setminus \{\varepsilon\}$.

6   Decision Problems

Already for standard context-free grammars, many basic decision problems are undecidable, such as testing whether two given grammars generate the same language (the equivalence problem), or even testing whether a given grammar generates the fixed language $\{a, b\}^*$. A few problems are known to be decidable: for instance, one can test in polynomial time whether a given context-free grammar generates the empty set. In contrast, for conjunctive grammars, there is a uniform undecidability result: for every language L_0 generated by some conjunctive grammar, testing whether a given conjunctive grammar generates L_0 is undecidable [8].

Turning to unambiguous subclasses, the decidability status of the equivalence problem for unambiguous context-free grammars is among the major unsolved questions in formal language theory. On the other hand, as proved by Semenov [22], testing whether a given unambiguous context-free grammar generates a given regular language is decidable: this remarkable proof proceeds by reducing the decision problem to a statement of elementary analysis, and then using Tarski's algorithm to solve it.

This section establishes the undecidability of checking whether an unambiguous conjunctive grammar generates a fixed language, for every fixed language. The underlying idea is the same as in the previous results for ambiguous conjunctive grammars [8]: the language of computation histories of a Turing machine (VALC) is represented by a trellis automaton, its alphabet is reinterpreted as an alphabet of digits, so that each computation history is associated to a number, and then the unary notations of these numbers are represented by a conjunctive grammar [8]. However, Theorems 1–2 proved in this paper for the unambiguous case are more restricted than the known constructions of ambiguous conjunctive grammars [8], and the same undecidability methods require a careful re-implementation.

Theorem 3. *For every alphabet Σ and for every language $L_0 \subseteq \Sigma^*$ generated by an unambiguous conjunctive grammar, it is undecidable whether a given unambiguous conjunctive grammar generates L_0.*

The proof proceeds by first establishing the theorem in two special cases, and then inferring the general case out of them. For $L_0 = \varnothing$ (the emptiness problem), the undecidability is proved by encoding the trellis automaton for VALC using Theorem 1. For $L_0 = a^*$, the same automaton can be encoded using Theorem 2. Finally, the problem of equality to $\varnothing$ is reduced to the equality problem for any finite L_0, while equality to a^* reduces to equality to any infinite L_0.

7   Conclusion

The expressive power of unambiguous conjunctive grammars over a unary alphabet has been developed up to the point of simulating a cellular automaton in a "sparse" unary language (Theorem 1), and in a "dense" unary language (Theorem 2). Though these are rather restricted representations, as compared to those constructed earlier for ambiguous conjunctive grammars over the unary alphabet [8], they were sufficient to establish uniform undecidability results for the problem of testing equivalence to a fixed language (Theorem 3). The results of this paper have already been used to investigate the properties of unambiguous conjunctive grammars with two nonterminal symbols [11].

The main research problem for conjunctive grammars, that of finding any non-representable languages, remains open. Though, at the first glance, it seemed that there cannot be any unambiguous conjunctive grammars for unary languages of high density, Theorems 1–2 show that languages of arbitrary density can be represented. Then, what kind of properties of unary languages could rule out their representation by unambiguous conjunctive grammars?

References

1. Aizikowitz, T., Kaminski, M.: LR(0) conjunctive grammars and deterministic synchronized alternating pushdown automata. In: Kulikov, A., Vereshchagin, N. (eds.) CSR 2011. LNCS, vol. 6651, pp. 345–358. Springer, Heidelberg (2011)
2. Allouche, J.-P., Shallit, J.: Automatic Sequences: Theory, Applications, Generalizations. Cambridge University Press (2003)
3. Autebert, J., Berstel, J., Boasson, L.: Context-free languages and pushdown automata. In: Rozenberg, Salomaa (eds.) Handbook of Formal Languages, vol. 1, pp. 111–174. Springer (1997)
4. Culik II, K., Gruska, J., Salomaa, A.: Systolic trellis automata, I and II. International Journal of Computer Mathematics 15, 16, 195–212, 3–22 (1984)
5. Enflo, P., Granville, A., Shallit, J., Yu, S.: On sparse languages L such that $LL = \Sigma^*$. Discrete Applied Mathematics 52, 275–285 (1994)
6. Ibarra, O.H., Kim, S.M.: Characterizations and computational complexity of systolic trellis automata. Theoretical Computer Science 29, 123–153 (1984)
7. Jeż, A.: Conjunctive grammars can generate non-regular unary languages. International Journal of Foundations of Computer Science 19(3), 597–615 (2008)
8. Jeż, A., Okhotin, A.: Conjunctive grammars over a unary alphabet: undecidability and unbounded growth. Theory of Computing Systems 46(1), 27–58 (2010)
9. Jeż, A., Okhotin, A.: Complexity of equations over sets of natural numbers. Theory of Computing Systems 48(2), 319–342 (2011)
10. Jeż, A., Okhotin, A.: On the computational completeness of equations over sets of natural numbers. In: Aceto, L., Damgård, I., Goldberg, L.A., Halldórsson, M.M., Ingólfsdóttir, A., Walukiewicz, I. (eds.) ICALP 2008, Part II. LNCS, vol. 5126, pp. 63–74. Springer, Heidelberg (2008)
11. Jeż, A., Okhotin, A.: On the number of nonterminal symbols in unambiguous conjunctive grammars. In: Kutrib, M., Moreira, N., Reis, R. (eds.) DCFS 2012. LNCS, vol. 7386, pp. 183–195. Springer, Heidelberg (2012)
12. Kountouriotis, V., Nomikos, C., Rondogiannis, P.: Well-founded semantics for Boolean grammars. Information and Computation 207(9), 945–967 (2009)
13. Kunc, M.: What do we know about language equations? In: Harju, T., Karhumäki, J., Lepistö, A. (eds.) DLT 2007. LNCS, vol. 4588, pp. 23–27. Springer, Heidelberg (2007)
14. McKenzie, P., Wagner, K.W.: The complexity of membership problems for circuits over sets of natural numbers. Computational Complexity 16, 211–244 (2007)
15. Okhotin, A.: Conjunctive grammars. Journal of Automata, Languages and Combinatorics 4, 519–535 (2001)
16. Okhotin, A.: On the equivalence of linear conjunctive grammars to trellis automata. Informatique Théorique et Applications 38(1), 69–88 (2004)
17. Okhotin, A.: Unambiguous Boolean grammars. Information and Computation 206, 1234–1247 (2008)
18. Okhotin, A.: Decision problems for language equations. Journal of Computer and System Sciences 76(3-4), 251–266 (2010)
19. Okhotin, A.: Fast parsing for Boolean grammars: a generalization of Valiant's algorithm. In: Gao, Y., Lu, H., Seki, S., Yu, S. (eds.) DLT 2010. LNCS, vol. 6224, pp. 340–351. Springer, Heidelberg (2010)
20. Okhotin, A., Reitwießner, C.: Parsing Boolean grammars over a one-letter alphabet using online convolution. Theoretical Computer Science 457, 149–157 (2012)
21. Okhotin, A., Rondogiannis, P.: On the expressive power of univariate equations over sets of natural numbers. Information and Computation 212, 1–14 (2012)
22. Semenov, A.L.: Algorithmic problems for power series and for context-free grammars. Doklady Akademii Nauk SSSR 212, 50–52 (1973)
23. Terrier, V.: On real-time one-way cellular array. Theoretical Computer Science 141, 331–335 (1995)

Alternative Automata Characterization of Piecewise Testable Languages

Ondřej Klíma and Libor Polák*

Department of Mathematics and Statistics, Masaryk University
Kotlářská 2, 611 37 Brno, Czech Republic
{klima,polak}@math.muni.cz
http://www.math.muni.cz

Abstract. We present a transparent condition on a minimal automaton which is equivalent to piecewise testability of the corresponding regular language. The condition simplifies the original Simon's condition on the minimal automaton in a different way than conditions of Stern and Trahtman. Secondly, we prove that every piecewise testable language L is k-piecewise testable for k equal to the depth of the minimal DFA of L. This result improves all previously known estimates of such k.

Keywords: piecewise testable languages, acyclic automata, locally confluent automata.

1 Introduction

A language L over a non-empty finite alphabet A is called *piecewise testable* if it is a Boolean combination of languages of the form

$$A^*a_1 A^* a_2 A^* \ldots A^* a_\ell A^*, \text{ where } a_1, \ldots, a_\ell \in A, \ \ell \geq 0 . \quad (*)$$

Simon's celebrated theorem [9] states that a regular language L is piecewise testable if and only if the syntactic monoid of L is $\mathcal{J}$-trivial. There exist several proofs of Simon's result based on various methods from algebraic and combinatorial theory of regular languages, e.g. proofs due to Almeida [1], Straubing and Thérien [11], Higgins [2], Klíma [4], Klíma and Polák [5]. For information concerning the role of piecewise testable languages in the theory of star-free languages we refer to Pin's survey paper [6].

In the original paper [9] Simon gave also an alternative characterization of piecewise testable languages by means of a condition for their minimal automata. The proof is included in his PhD thesis [8]. Given a regular language L, with

---

* The authors were supported by the Institute for Theoretical Computer Science (GAP202/12/G061), Czech Science Foundation. The second author also acknowledges the support by ESF Project CZ.1.07/2.3.00/20.0051 Algebraic Methods in Quantum Logic.

M.-P. Béal and O. Carton (Eds.): DLT 2013, LNCS 7907, pp. 289–300, 2013.
© Springer-Verlag Berlin Heidelberg 2013

minimal automaton $\mathcal{A}$, he considered, for an arbitrary subset B of the alphabet A, the subautomaton $\mathcal{A}_B$ of $\mathcal{A}$, where only actions by letters from B are taken into account. Then language L is piecewise testable if and only if $\mathcal{A}$ is acyclic and for every $B \subseteq A$, distinct absorbing states in $\mathcal{A}_B$ belong to distinct connected components of the undirected version of $\mathcal{A}_B$. This characterization was used by Stern [10] to develop a polynomial algorithm testing whether a regular language, given by a minimal deterministic finite automaton, is piecewise testable. Stern's algorithm was improved by Trahtman [12], who lowered the time complexity from $O(n^5)$ to $O(n^2)$, where n is the product of the number of letters of the alphabet and the number of states in the minimal DFA for the language. The idea of these algorithms is that Simon's condition on the automaton does not have to be tested for all subalphabets B, but just for those, which occur as a set of all letters fixing a particular state. We could mention that the proofs of correctness of these algorithms were based on the original Simon's result [9].

In this paper we introduce a new condition on an acyclic automaton which is equivalent to Simon's condition. Using our approach we do not improve Trahtman's algorithm for piecewise testability, because we obtain again a quadratic one. The main advantage is that we will be able to give a new proof which does not use Simon's original proof. Another advantage of this new condition is that it can be formulated in a very transparent way, which could be useful for handmade computations with piecewise testable languages. We call an acyclic automaton $\mathcal{A}$ *locally confluent*, if for each state q and every pair of letters a, b there is a word w over the alphabet $\{a, b\}$ such that $(q \cdot a) \cdot w = (q \cdot b) \cdot w$. Then an alternative statement to Simon's result is the following theorem.

Theorem 1. *A regular language is piecewise testable if and only if its minimal automaton is acyclic and locally confluent.*

In paper [5] the authors pointed out another aspect concerning piecewise testable languages. Several proofs of Simon's result implicitly contain a solution of the following problem: for a given piecewise testable language L, find a number k, such that L is a Boolean combination of languages of the form $(*)$, where $\ell \leq k$, the so-called *k-piecewise testable* language. Simon [9] proved that k could be taken to be equal to $2n - 1$ where n is the maximal length of a $\mathcal{J}$-chain, i.e. the maximal length of a chain of ideals, in the syntactic monoid of L (see the proof of Corollary 1.7 in [7]). A similar estimate was also established in the first author's combinatorial proof of Simon's result [4]: k could be taken as $\ell + r - 2$ where ℓ and r are the maximal lengths of chains of left and right ideals, respectively. In [5] the authors gave an estimate using a new notion of biautomaton: a regular language L is k-piecewise testable whenever its canonical biautomaton is acyclic and has depth k. It was also proved that this estimate is never larger than the mentioned characteristics $2n - 1$ and $\ell + r - 2$ of the syntactic monoid of the language.

Unfortunately, it is not known how the depth of the canonical biautomaton is related to, or bounded by, the size of the minimal automaton, which is the most common description of a regular language. In general, it is known that the size of the canonical biautomaton can be exponential with respect to the size of the minimal automaton [3]. Instead of clarifying the relationship between minimal automata and canonical biautomata for piecewise testable languages, we improve the estimate from [5]. We show an analogous statement, where the depth of the minimal automaton is considered instead of the depth of the canonical biautomaton. Of course, the first characteristic is smaller or equal to the second one.

Theorem 2. *Let L be a regular language whose minimal automaton is acyclic, locally confluent and of the depth k. Then L is k-piecewise testable.*

We should stress that our proof of this theorem does not use Simon's results and it is inspired by the ideas used in the proof of the corresponding result from [5]. It is also a key ingredient of the proof of Theorem 1.

The structure of the paper is as follows. Section 2 overviews notions used in this contribution. The proof of Theorem 2 is contained in Section 3. In Section 4 we show that both Simon's characterizations of piecewise testable languages as well as Theorem 1 follow from Theorem 2. Here we also give an algorithm for testing the piecewise testability which is an alternative to that of Trahtman [12]. Finally, in the last section, we exhibit that the depth of the canonical biautomaton of a piecewise testable language can be arbitrarily larger than the depth of the minimal automaton of the same language. Second example shows that the opposite implication in Theorem 2 is quite far from being true.

2   Preliminaries

2.1   Piecewise Testable Languages

Let A^* be the set of all words over a non-empty finite alphabet A. The empty word is denoted by λ. For $u, v \in A^*$, we write $u \lhd v$ if u is a *subword* of v, which means that $u = a_1 \ldots a_n$, where $a_1, \ldots, a_n \in A$, and there are words $v_0, v_1, \ldots, v_n \in A^*$ such that $v = v_0 a_1 v_1 \ldots a_n v_n$. For a word $u = a_1 \ldots a_n$, $a_1, \ldots, a_n \in A$, we denote by L_u the language of all words which contain u as a subword, i.e.

$$L_u = \{ v \in A^* \mid u \lhd v \} = A^* a_1 A^* \ldots A^* a_n A^* .$$

For such u, we denote by $\mathsf{c}(u) = \{a_1, \ldots, a_n\}$ the *content* of u, and by $|u| = n$ its *length*, and by $u^r = a_n \ldots a_1$ the *reverse* of the word u. The language L over an alphabet A is a subset of A^* and we denote by $L^c = A^* \setminus L$ the *complement*, and by $L^r = \{u^r \mid u \in L\}$ the *reverse* of the language L.

Definition 1. Let $k \geq 0$ be an integer. A language is k-*piecewise testable* if it is a Boolean combination of languages of the form L_u where all u's satisfy $|u| \leq k$. A regular language is *piecewise testable* if it is k-piecewise testable for some k.

Let $k \geq 0$ be an integer. For $v \in A^*$, we denote by

$$\mathsf{Sub}_k(v) = \{\, u \in A^+ \mid u \lhd v,\ |u| \leq k \,\}$$

the set of all subwords of v of length at most k. We define the equivalence relation $\sim_k$ on A^* by the rule: $v \sim_k w$ if and only if $\mathsf{Sub}_k(v) = \mathsf{Sub}_k(w)$. Since $\mathsf{Sub}_k(ua) = \mathsf{Sub}_k(u) \cup \mathsf{Sub}_{k-1}(u) \cdot a$, the fact $u \sim_k v$ implies $ua \sim_k va$ and by a dual argument also $au \sim_k av$. Therefore $\sim_k$ is a congruence on the monoid A^*.

For a regular language $L \subseteq A^*$, we define the relation $\sim_L$ on A^* as follows: for $u, v \in A^*$, we have

$$u \sim_L v \quad \text{if and only if} \quad (\,\forall\, p, q \in A^*\,)\,(\, puq \in L \iff pvq \in L \,).$$

The relation $\sim_L$ is a congruence on A^* and it is called the *syntactic congruence* of L. The corresponding quotient monoid $A^*/\!\!\sim_L$ is called the *syntactic monoid* of L, it is finite and it is isomorphic to the transformation monoid of the minimal automaton of L.

An easy consequence of the definition of piecewise testable languages is the following lemma. A proof can be found e.g. in [8], [4].

Lemma 1. *A language L is k-piecewise testable if and only if $\sim_k \subseteq \sim_L$. Moreover, the last inclusion is equivalent to the fact that L is a union of $\sim_k$-classes.*

A finite monoid is $\mathcal{J}$-trivial if its Green relation $\mathcal{J}$ is the diagonal relation on the monoid. In this paper we use an alternative characterization of this property: a finite monoid M is $\mathcal{J}$-trivial if and only if there is an integer $m \geq 1$ such that for every $a, b \in M$, we have

$$(ba)^m = (ab)^m = b(ab)^m\,.$$

Note that the previous equalities imply also equalities $a(ab)^m = (ab)^m$, $(ab)^m a = (ab)^m$, $(ab)^m b = (ab)^m$ and $a^{m+1} = a^m$.

We recall the original formulation of Simon's result.

Result 1 (Simon [9]). A regular language is piecewise testable if and only if its syntactic monoid is $\mathcal{J}$-trivial.

2.2   Automata for Piecewise Testable Languages

In this paper all considered automata are finite, deterministic and complete. Thus an *automaton* over the alphabet A is a five-tuple $\mathcal{A} = (Q, A, \cdot, i, T)$ where

(i) Q is a nonempty set of *states*,
(ii) $\cdot : Q \times A \to Q$, extended to $\cdot : Q \times A^* \to Q$ by $q \cdot \lambda = q$, $q \cdot (ua) = (q \cdot u) \cdot a$, where $q \in Q$, $u \in A^*$, $a \in A$,
(iii) $i \in Q$ is the *initial* state, and $T \subseteq Q$ is the set of *terminal* states.

The automaton $\mathcal{A}$ accepts $u \in A^*$ if $i \cdot u \in T$ and $\mathcal{A}$ *recognizes* the language $L_{\mathcal{A}} = \{\, u \in A^* \mid i \cdot u \in T \,\}$. We say that a state $q \in Q$ of the automaton $\mathcal{A}$

is *reachable* if there is a word $u \in A^*$ such that $q = i \cdot u$ and q is *absorbing* if, for every $a \in A$, we have $q \cdot a = q$. For an arbitrary state $p \in Q$, we denote $Q_p = \{ p \cdot u \mid u \in A^* \}$ and we put $\mathcal{A}_p = (Q_p, A, \cdot |_{Q_p \times A}, p, T \cap Q_p)$. Hence $\mathcal{A}_p$ is an automaton with all states reachable.

A sequence $(q_0, q_1, \ldots, q_n)$ of states is called a *path* in $\mathcal{A}$ if for each $j \in \{1, \ldots, n\}$ there is $a \in A$ such that $q_j = q_{j-1} \cdot a$. A path $(q_0, q_1, \ldots, q_n)$ is *simple* if the states $q_0, \ldots, q_n$ are pairwise distinct and it is a *cycle* if $n \geq 2$ and $q_n = q_0 \neq q_1$. The automaton $\mathcal{A}$ is called *acyclic* if there is no cycle in $\mathcal{A}$. Note that "loops" are not cycles and for the acyclic automaton $\mathcal{A}$ and its state p, the automaton $\mathcal{A}_p$ is also acyclic. Each simple path in an acyclic automaton with all states reachable can be prolonged to a simple path starting at i. We define the *depth* of an (acyclic) automaton $\mathcal{A}$ with all states reachable as the maximal number n such that there is a simple path $(i, q_1, \ldots, q_n)$ in $\mathcal{A}$.

The languages recognized by acyclic automata were characterized by Eilenberg, who proved that they are exactly languages which have $\mathcal{R}$-trivial syntactic monoids (see [7, Section 4.3]).

Let $\mathcal{A} = (Q, A, \cdot, i, T)$ be an automaton. For a subset $B \subseteq A$, we define the automaton $\mathcal{A}_B = (Q, B, \cdot |_{Q \times B}, i, T)$. We say that a state $q \in Q$ is *B-absorbing* if q is an absorbing state in $\mathcal{A}_B$. We define the undirected graph $G(\mathcal{A}_B) = (Q, E_B)$ where $E_B \subseteq Q \times Q$ is given by:

$$E_B = \{ (q, q \cdot b) \mid q \in Q, b \in B \} \cup \{ (q \cdot b, q) \mid q \in Q, b \in B \}.$$

If we consider the reflexive-transitive closure of the relation E_B on the set Q, we obtain an equivalence relation which is denoted by $\approx_B$. The partition $Q/\approx_B$ is the usual decomposition of an undirected graph on connected components.

Now we are ready to formulate Simon's characterization of piecewise testable languages by a condition on their minimal automata.

Result 2 (Simon [9]). Let L be a regular language and let $\mathcal{A}$ be its minimal automaton. Then L is piecewise testable if and only if
  (i) $\mathcal{A}$ is acyclic, and
  (ii) for each $B \subseteq A$, if $p \approx_B q$ are B-absorbing states, then $p = q$.

3 Locally Confluent Automata and Proof of Theorem 2

We formulate new conditions on acyclic automata.

Definition 2. Let $\mathcal{A} = (Q, A, \cdot, i, T)$ be an automaton and $B \subseteq A$ be a subalphabet. We say that $\mathcal{A}$ is *B-confluent*, if for each state $q \in Q$ and every pair of words $u, v \in B^*$, there is a word $w \in B^*$ such that $(q \cdot u) \cdot w = (q \cdot v) \cdot w$. We say that $\mathcal{A}$ is *confluent* if it is B-confluent for every subalphabet B. We call an acyclic automaton $\mathcal{A}$ *locally confluent*, if for each state $q \in Q$ and every pair of letters $a, b \in A$, there is a word $w \in \{a, b\}^*$ such that $(q \cdot a) \cdot w = (q \cdot b) \cdot w$.[1]

---

[1] An alternative definition could ask for the existence of a pair of words w, w' such that $(q \cdot a) \cdot w = (q \cdot b) \cdot w'$. In this way one obtains an equivalent definition. It is not shown here nor needed here.

Now we present a complete proof of Theorem 2. We show a simple lemma first.

Lemma 2. *Let $\ell \geq 1$ and let $u, v \in A^*$ be such that $\mathsf{Sub}_\ell(u) = \mathsf{Sub}_\ell(v)$. Let a be a letter from $\mathsf{c}(u)$. Then there are uniquely determined words u', u'' and v', v'' such that $u = u'au''$ and $v = v'av''$ and $a \notin \mathsf{c}(u')$, $a \notin \mathsf{c}(v')$. Moreover, for u'' and v'', we have $\mathsf{Sub}_{\ell-1}(u'') = \mathsf{Sub}_{\ell-1}(v'')$.*

Proof. Note that the equality $\mathsf{c}(u) = \mathsf{c}(v)$ follows from the assumption $\mathsf{Sub}_\ell(u) = \mathsf{Sub}_\ell(v)$. We consider the first occurrence of the letter a in u and in v. Thus we have $u = u'au''$ and $v = v'av''$ satisfying $a \notin \mathsf{c}(u')$ and $a \notin \mathsf{c}(v')$. Now if $w \in \mathsf{Sub}_{\ell-1}(u'')$, then $aw \in \mathsf{Sub}_\ell(u) = \mathsf{Sub}_\ell(v)$ from which $w \in \mathsf{Sub}_{\ell-1}(v'')$ follows. This means that $\mathsf{Sub}_{\ell-1}(u'') \subseteq \mathsf{Sub}_{\ell-1}(v'')$ and the opposite inclusion can be proved in the same way. $\square$

The next statement is crucial; namely when using Lemma 1, it yields immediately Theorem 2.

Proposition 1. *Let $\mathcal{A} = (Q, A, \cdot, i, T)$ be an acyclic and locally confluent automaton with all states reachable and with $\mathsf{depth}(\mathcal{A}) = \ell$. Then, for every $u, v \in A^*$ such that $\mathsf{Sub}_\ell(u) = \mathsf{Sub}_\ell(v)$, we have $u \in L_\mathcal{A}$ if and only if $v \in L_\mathcal{A}$.*

Proof. We prove the statement by induction with respect to the depth of the automaton.

For $\ell = 0$, there is nothing to prove, because the assumption $\mathsf{depth}(\mathcal{A}) = 0$ means that $\mathcal{A}$ is a trivial automaton, i.e. $Q = \{i\}$, and hence $L_\mathcal{A} = A^*$ or $L_\mathcal{A} = \emptyset$, depending on the fact whether $i \in T$ or not.

Assume for the remainder of the proof that $\ell \geq 1$ and that the statement holds for all $\ell' < \ell$. And furthermore, assume that the statement is not true for ℓ. We will reach a contradiction by strengthening our assumptions. Let there be a pair of words $u, v \in A^*$ such that

$$\mathsf{Sub}_\ell(u) = \mathsf{Sub}_\ell(v) \text{ and } i \cdot u \in T \text{ and } i \cdot v \notin T.$$

In the state i, we read both words u and v, and we are interested in the positions in the words, where we leave the initial state i. First assume that $i \cdot u = i \in T$, i.e. we do not leave the state i. Recall that the assumption $\mathsf{Sub}_\ell(u) = \mathsf{Sub}_\ell(v)$ implies $\mathsf{c}(u) = \mathsf{c}(v)$. Thus we have $i \cdot v = i \in T$ – a contradiction. From this moment we may assume that $i \cdot u \neq i$ and also $i \cdot v \neq i$.

So we really leave the state i and there are $u', u'' \in A^*$, $a \in A$ such that

$$u = u'au'' \text{ and for each } x \in \mathsf{c}(u'), \text{ we have } i \cdot x = i, \text{ and } i \cdot a \neq i.$$

Similarly, let $v', v'' \in A^*$, $b \in A$ be such that

$$v = v'bv'' \text{ and for each } x \in \mathsf{c}(v'), \text{ we have } i \cdot x = i \text{ and } i \cdot b \neq i.$$

Assume for a moment that $a = b$. We denote $p = i \cdot a = i \cdot u'a = i \cdot v'a$ and we consider the automaton $\mathcal{A}_p$. It is clear that the depth of $\mathcal{A}_p$ is at most $\ell - 1$.

By our assumptions $i \cdot u = p \cdot u'' \in T$ and $i \cdot v = p \cdot v'' \notin T$. By Lemma 2, we have $\mathsf{Sub}_{\ell-1}(u'') = \mathsf{Sub}_{\ell-1}(v'')$. Now we obtain a contradiction to the induction assumption applied on the automaton $\mathcal{A}_p$ and the pair of words u'' and v''. Therefore, we may assume that $a \neq b$.

We will consider the first occurrence of b in u. When we read u in the automaton $\mathcal{A}$, we move from the initial state only when we reach the letter a for the first time. Therefore the first occurrence of b in u is after the first occurrence of a in u. More formally,

$$u = u' a u_0'' b u_1'' \text{ where } a \notin \mathsf{c}(u') \text{ and } b \notin \mathsf{c}(u' a u_0'').$$

and similarly, $v = v' b v_0'' a v_1''$ where $b \notin \mathsf{c}(v')$ and $a \notin \mathsf{c}(v' b v_0'')$. Now, by Lemma 2, we have

$$\mathsf{Sub}_{\ell-1}(u_0'' b u_1'') = \mathsf{Sub}_{\ell-1}(v_1'') \subseteq \mathsf{Sub}_{\ell-1}(v_0'' a v_1'') = \mathsf{Sub}_{\ell-1}(u_1'') \subseteq \mathsf{Sub}_{\ell-1}(u_0'' b u_1''),$$

and thus the previous inclusions hold, in fact, as the equalities. In particular, $\mathsf{Sub}_{\ell-1}(u_0'' b u_1'') = \mathsf{Sub}_{\ell-1}(v_0'' a v_1'')$.

Now assume, for a moment, that $i \cdot a = i \cdot b = p$. We have $i \cdot u = p \cdot u_0'' b u_1'' \in T$ and $i \cdot v = p \cdot v_0'' a v_1'' \notin T$. Again this is a contradiction to the induction assumption applying to the automaton $\mathcal{A}_p$ and the pair of words $u_0'' b u_1''$ and $v_0'' a v_1''$. Altogether we have that $i \cdot a \neq i \cdot b$.

Now we show some consequences of the fact $\mathsf{Sub}_{\ell-1}(u_1'') = \mathsf{Sub}_{\ell-1}(u_0'' b u_1'') = \mathsf{Sub}_{\ell-1}(v_1'') = \mathsf{Sub}_{\ell-1}(v_0'' a v_1'')$. First, since

$$\mathsf{Sub}_{\ell-1}(v_1'') \subseteq \mathsf{Sub}_{\ell-1}(a v_1'') \subseteq \mathsf{Sub}_{\ell-1}(v_0'' a v_1'') = \mathsf{Sub}_{\ell-1}(v_1''),$$

we have $\mathsf{Sub}_{\ell-1}(a v_1'') = \mathsf{Sub}_{\ell-1}(v_1'')$. Similarly, we get $\mathsf{Sub}_{\ell-1}(b u_1'') = \mathsf{Sub}_{\ell-1}(u_1'')$. Let $C \subset A^*$ be a set of all words which are $\sim_{\ell-1}$-related to u_1''. Then from the previous observations, we have

$$u'' = u_0'' b u_1'', \ b u_1'', \ u_1'', \ v'' = v_0'' a v_1'', \ a v_1'', \ v_1'' \in C.$$

We claim that for an arbitrary word $z \in C$ we have $az, bz \in C$. Indeed, since $\sim_{\ell-1}$ is a congruence, $z \sim_{\ell-1} v_1''$ implies $az \sim_{\ell-1} a v_1'' \sim_{\ell-1} v_1''$, i.e. $az \in C$. And in the same way we can deduce $bz \sim_{\ell-1} b u_1'' \sim_{\ell-1} u_1'' \in C$ from $z \sim_{\ell-1} u_1''$. As a consequence of this claim we obtain that

$$\text{for each } z \in C \text{ and } w \in \{a, b\}^*, \text{ we have } wz \in C. \tag{†}$$

We can return to the proof. Our considerations are illustrated in Figure 1. Let $p = i \cdot a$ and $q = i \cdot b$. By the local confluency, there is a word $w \in \{a, b\}^*$ such that $p \cdot w = q \cdot w$. By (†), we have $\mathsf{Sub}_{\ell-1}(wu'') = \mathsf{Sub}_{\ell-1}(u'')$. Using the induction assumption for the automaton $\mathcal{A}_p$ which has depth at most $\ell - 1$, we get $p \cdot wu'' \in T$. In the same way, if we use the induction assumption on automaton $\mathcal{A}_q$, we get $q \cdot wv'' \notin T$. Now we consider the state $r = p \cdot w = q \cdot w$ and we have $r \cdot u'' \in T$, $r \cdot v'' \notin T$ and $\mathsf{Sub}_{\ell-1}(u'') = \mathsf{Sub}_{\ell-1}(v'')$. This is a contradiction with the induction assumption for the automaton $\mathcal{A}_r$ of the depth at most $\ell - 1$ and the pair of words u'' and v''.

We have finished the proof of Proposition 1. $\square$

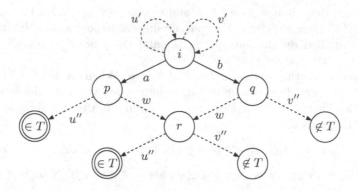

Fig. 1. One possible computation of words u and v in the minimal automaton

4   Consequences of Theorem 2

The following statement implies both Theorem 1 and Result 1.The presence of (iii) is justified by its usage in Proposition 2.

Corollary 1. *For a regular language L, the following conditions are equivalent.*
(i) The language L is piecewise testable.
(ii) The syntactic monoid of L is $\mathcal{J}$-trivial.
(iii) The minimal automaton of L is acyclic and confluent.
(iv) The minimal automaton of L is acyclic and locally confluent.

Proof. "(i) $\Rightarrow$ (ii)" The following argument comes from the original Simon's paper [9, Lemma 7]. If L is k-piecewise testable then $\sim_k \subseteq \sim_L$ by Lemma 1. For arbitrary words $u, v \in A^*$, one can see that $(vu)^k \sim_k (uv)^k \sim_k v(uv)^k$ and therefore $(vu)^k \sim_L (uv)^k \sim_L v(uv)^k$. This implies $(ba)^k = (ab)^k = b(ab)^k$, for a, b being arbitrary elements from the syntactic monoid of L.

"(ii) $\Rightarrow$ (iii)" Let $(q_0, q_1, \ldots, q_n)$ be a cycle in the minimal automaton of L, i.e. $n \geq 2$, $q_n = q_0 \neq q_1$ and for each $j \in \{1, \ldots, n\}$ we have $a_j \in A$ such that $q_j = q_{j-1} \cdot a_j$. Denote $u = a_2 \ldots a_n$. Since the syntactic monoid of L is $\mathcal{J}$-trivial, we have $(a_1 u)^m a_1 \sim_L (a_1 u)^m$ for some integer m. The syntactic monoid is the transformation monoid of the minimal automaton and hence we obtain $q_0 \cdot (a_1 u)^m a_1 = q_0 \cdot (a_1 u)^m$. From $q_0 \cdot a_1 u = q_0$ we get $q_1 = q_0 \cdot a_1 = q_0 \cdot (a_1 u)^m a_1 = q_0 \cdot (a_1 u)^m = q_0$, which is a contradiction. Hence the minimal automaton of L is acyclic.

In a similar manner, we show that this minimal automaton is confluent, let q be a state and $u, v \in B^*$ be words. We need to find a word $w \in B^*$ such that $(q \cdot u) \cdot w = (q \cdot v) \cdot w$. Again from the $\mathcal{J}$-triviality we have $u(uv)^m \sim_L v(uv)^m$ for some integer m. Thus we can take $w = (uv)^m$ and we obtain the required equality $(q \cdot u) \cdot w = (q \cdot v) \cdot w$.

"(iii) $\Rightarrow$ (iv)" It is trivial.

"(iv) $\Rightarrow$ (i)" It follows from Theorem 2. $\square$

Result 2 is also a consequence of Theorem 2 when one applies the following.

Proposition 2. *Let $\mathcal{A} = (Q, A, \cdot, i, T)$ be an acyclic minimal automaton of a language $L \subseteq A^*$ with m states. Then the following conditions are equivalent.*

(i) For each $B \subseteq A$, if $p \approx_B q$ are B-absorbing states, then $p = q$.

(ii) For every $q \in Q$ and every $a, b \in A$, we have $q \cdot a(ab)^m = q \cdot b(ab)^m$.

(iii) The automaton $\mathcal{A}$ is locally confluent.

(iv) The automaton $\mathcal{A}$ is confluent.

Proof. "(i) $\Rightarrow$ (ii)" Let q be an arbitrary state and denote $B = \{a, b\}$. For each $i = 0, \ldots, m$, we consider $q_i = q \cdot a(ab)^i$. Since $\mathcal{A}$ has m states, the sequence $q_0, q_1, \ldots, q_m$ contains some state at least twice. Since $\mathcal{A}$ is acyclic we know that there is $i \in \{0, \ldots, m\}$ such that $q_i \cdot ab = q_i$. Again, from acyclicity of $\mathcal{A}$ we get, that this q_i is B-absorbing. Hence $q \cdot a(ab)^m = q_m = q_i$ is B-absorbing. We can use the same argument and get that the state $q \cdot b(ab)^m$ is also B-absorbing. Now $q \cdot a(ab)^m \approx_B q \approx_B q \cdot b(ab)^m$ and applying condition (i), we get the statement $q \cdot a(ab)^m = q \cdot b(ab)^m$.

"(ii) $\Rightarrow$ (iii)" Put $w = (ab)^m$.

"(iii) $\Rightarrow$ (iv)" By Corollary 1.

"(iv) $\Rightarrow$ (i)" Let p, q be B-absorbing states. From $p \approx_B q$ we can deduce that there is a sequence of states $p = r_0, r_1, \ldots, r_n = q$ such that, for each $i = 1, \ldots, n$ we have $(r_{i-1}, r_i) \in E_B$. We claim that, for each $i = 0, \ldots, n$, there is a word w_i such that $r_i \cdot w_i = p$. We prove this by an induction with respect to i. For $i = 0$, we can take an arbitrary word w_0 from B^*. Let us assume that $r_{i-1} \cdot w_{i-1} = p$. Since $(r_{i-1}, r_i) \in E_B$, there are two possibilities: If there is $b \in B$ such that $r_i \cdot b = r_{i-1}$ then we can put $w_i = bw_{i-1}$ and we have $r_i \cdot bw_{i-1} = r_{i-1} \cdot w_{i-1} = p$. If there is $b \in B$ such that $r_{i-1} \cdot b = r_i$ then, from B-confluency, we get the existence of a word $w_i \in B^*$ such that $p \cdot w_i = r_i \cdot w_i$. Since p is B-absorbing we obtain $r_i \cdot w_i = p \cdot w_i = p$ and we have proved the claim. For $i = n$, we get that there is a word $w_n \in B^*$ such that $q \cdot w_n = p$. Since q is B-absorbing, we have $p = q$. $\qquad\square$

The condition (ii) of Proposition 2 can be used when deciding the piecewise testability of a given language.

Proposition 3. *Let L be an arbitrary regular language with the minimal automaton $\mathcal{A} = (Q, A, \cdot, i, T)$. Let $|Q| = m$ and $|A| = n$. Then one can decide the piecewise testability in time $O(m^2 n^2)$.*

Proof. Suppose that $\mathcal{A}$ is presented by a m/n-type matrix with rows indexed by elements of Q and columns indexed by elements of A having $q \cdot a$ at the position (q, a). We can transform the matrix into the adjacency-list representation (with no repetitions in lists) of the corresponding graph in time $O(mn)$. The graph is acyclic if and only if there in no nontrivial back edge (loops are "trivial") by running Depth First Search on this graph – it takes also $O(mn)$.

Then we calculate, for each $q \in Q$ and each $a, b \in A$, the elements $q, q \cdot a, q \cdot a^2, \ldots, q \cdot a(ab)^m$ and $q, q \cdot b, q \cdot ba, \ldots, q \cdot b(ab)^m$ and we compare $q \cdot a(ab)^m$ and $q \cdot b(ab)^m$. This takes time $O(m^2 n^2)$. $\qquad\square$

5   Examples

Theorem 2 determines, for a given piecewise testable language L, relatively small number k, such that L is k-piecewise testable. Since the minimal automaton of L is a part of the canonical biautomaton of L, this k, the depth of the minimal automaton, is less or equal to the depth of the canonical biautomaton. Therefore Theorem 2 improves the estimate from paper [5] and consequently, overcome all estimates recalled in Section 1. Now, to demonstrate that the new estimate is indeed better than other known ones, we show that there are examples of languages where the difference between the two considered depths could be an arbitrarily large. For that propose we modify the idea used in [5, Example 6].

Example 1. For $m \geq 1$, we denote $A = \{a_1, \ldots, a_m\}$, $B = \{b_1, \ldots, b_m\}$. Let K be the language of all words which contain exactly one occurrence of a letter from the subalphabet A and contain some $a_j b_j$ as a subword. More formally, K is a 2-piecewise testable language given by the following expression

$$K = \bigcap_{j,j'=1}^{m} L^c_{a_j a_{j'}} \ \cap \ \bigcup_{j=1}^{m} L_{a_j b_j} .$$

The minimal automaton of K is described on Figure 2. Here, for each $j = 1, \ldots, m$, we denote $B_j = B \setminus \{b_j\}$. It is easy to see that this automaton has $m + 3$ states and has depth 3.

Note that the minimal automaton of the reverse language K^r is again a part of the canonical biautomaton of K. Of course, K^r is again 2-piecewise testable. To construct the minimal automaton of the reverse K^r of the language K, we must remember all possible letters from the subalphabet B which are contained in the word before a unique occurrence of some letter $a \in A$ is read. Therefore the minimal automaton of K^r is an automaton $\mathcal{A} = (Q, A \cup B, \cdot, i, \{\tau\})$, where $Q = \mathsf{P}(B) \cup \{\tau, \bot\}$, $\mathsf{P}(B)$ is a system of all subsets of the set B, $i = \emptyset \in \mathsf{P}(B)$ and the transition action $\cdot$ is given by the following rules. For $X \in \mathsf{P}(B)$ and $j \in \{1, \ldots, m\}$ we have: if $b_j \in X$ then $X \cdot a_j = \tau$ and $X \cdot b_j = X$; and if $b_j \notin X$ then $X \cdot a_j = \bot$ and $X \cdot b_j = X \cup \{b_j\}$. Moreover, for each $j \in \{1, \ldots, m\}$ we have $\tau \cdot a_j = \bot$, $\tau \cdot b_j = \tau$ and $\bot \cdot a_j = \bot \cdot b_j = \bot$. It is not hard to check that $\mathcal{A}$ is the minimal automaton of K^r, which has $2^m + 2$ states and has depth $m + 2$. Indeed, $(\emptyset, \{b_1\}, \{b_1, b_2\}, \ldots, B, \tau, \bot)$ realizes the longest simple path in $\mathcal{A}$. Thus the depth of the canonical biautomaton of the language K is at least $m + 2$. (One can show that it is $m + 3$.)

One can ask whether a depth of a minimal automaton can be used for a characterization of k-piecewise testable languages. In particular, whether the opposite implication of Theorem 2 can be true. The next example demonstrates that such a statement is not valid because the depth of the minimal automaton could be quite far from the minimal k such that a given piecewise testable language is k-piecewise testable. The example is a modification of Example 7 from paper [5].

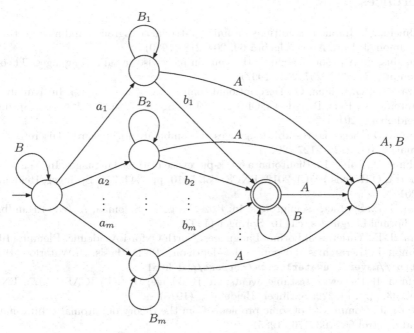

Fig. 2. The minimal automaton of the language K

Example 2. Let $A = \{a, b\}$ and $\ell > 1$ be an integer. We consider the language $L = \{u\}$ consisting of a single word $u = (a^{2\ell}b^{2\ell})^{\ell}$. The minimal automaton of L has the depth equal to the length of the word u, i.e. $4\ell^2$. On the other hand, we claim that the language L is $(4\ell - 1)$-piecewise testable.

First, we see that $r = (ab)^{\ell}$ is a subword of u, and $s = (ab)^{\ell}a$ and $t = b(ab)^{\ell}$ are not subwords of u. Furthermore, for each $i = 1, \ldots, \ell$, we denote $u_i = (ab)^{i-1}a^{2\ell}(ba)^{\ell-i}$ which is a subword of u and $\overline{u}_i = (ab)^{i-1}a^{2\ell+1}(ba)^{\ell-i}$ which is not a subword of u. Similarly, we denote $v_i = (ba)^{i-1}b^{2\ell}(ab)^{\ell-i} \triangleleft u$, and $\overline{v}_i = (ba)^{i-1}b^{2\ell+1}(ab)^{\ell-i} \not\triangleleft u$. Now one can check that

$$L = L_r \cap L_s^{\mathsf{c}} \cap L_t^{\mathsf{c}} \cap \bigcap_{i=1}^{\ell} \left(L_{u_i} \cap L_{\overline{u}_i}^{\mathsf{c}} \cap L_{v_i} \cap L_{\overline{v}_i}^{\mathsf{c}} \right) .$$

All used words $r, s, t, u_1, \overline{u}_1, \ldots, v_{\ell}, \overline{v}_{\ell}$ have length at most $4\ell - 1$ and the claim follows. Note that L is not $(4\ell - 2)$ piecewise testable, because $au \sim_{4\ell-2} u$. The proof of this fact is a bit technical, but one can use Lemma 3 from [9] and the following factorization of the word u:

$$u = (a) \cdot (a) \cdots (a) \cdot (b^{2\ell}a^{\ell}) \cdot (a^{\ell}b^{\ell}) \cdot (b^{\ell}a^{\ell}) \cdots (b^{\ell}a^{\ell}) \cdot (a^{\ell}b^{2\ell}) .$$

Acknowledgment. The authors would like to express their gratitude the anonymous referees whose suggestions considerable improved the transparency of this presentation.

References

1. Almeida, J.: Implicit operations on finite $\mathcal{J}$-trivial semigroups and a conjecture of I. Simon. J. Pure Appl. Algebra 69, 205–218 (1990)
2. Higgins, P.: A proof of Simon's theorem on piecewise testable languages. Theoret. Comput. Sci. 178, 257–264 (1997)
3. Jirásková, G., Klíma, O.: Descriptional complexity of biautomata. In: Kutrib, M., Moreira, N., Reis, R. (eds.) DCFS 2012. LNCS, vol. 7386, pp. 196–208. Springer, Heidelberg (2012)
4. Klíma, O.: Piecewise testable languages via combinatorics on words. Discrete Mathematics 311, 2124–2127 (2011)
5. Klíma, O., Polák, L.: Biautomata for k-piecewise testable languages. In: Yen, H.-C., Ibarra, O.H. (eds.) DLT 2012. LNCS, vol. 7410, pp. 344–355. Springer, Heidelberg (2012)
6. Pin, J.-E.: Syntactic semigroups. In: Rozenberg, G., Salomaa, A. (eds.) Handbook of Formal Languages, ch. 10. Springer (1997)
7. Pin, J.-E.: Varieties of Formal Languages. North Oxford Academic, Plenum (1986)
8. Simon, I.: Hierarchies of events of dot-depth one. Ph.D. thesis. U. Waterloo (1972), http://maveric.uwaterloo.ca/~brzozo/phd.html
9. Simon, I.: Piecewise testable events. In: Brakhage, H. (ed.) ICALP 1975. LNCS, vol. 33, pp. 214–222. Springer, Heidelberg (1975)
10. Stern, J.: Complexity of some problems from the theory of automata. Information and Control 66, 163–176 (1985)
11. Straubing, H., Thérien, D.: Partially ordered finite monoids and a theorem of I. Simon. J. Algebra 119, 393–399 (1988)
12. Trahtman, A.N.: Piecewise and local threshold testability of DFA. In: Freivalds, R. (ed.) FCT 2001. LNCS, vol. 2138, pp. 347–358. Springer, Heidelberg (2001)

Finite Automata with Advice Tapes

Uğur Küçük[1], A.C. Cem Say[1], and Abuzer Yakaryılmaz[2,*]

[1] Boğaziçi University, Istanbul, Turkey
{ugur.kucuk,say}@boun.edu.tr
[2] University of Latvia, Rīga, Latvia
abuzer@lu.lv

Abstract. We define a model of advised computation by finite automata where the advice is provided on a separate tape. We consider several variants of the model where the advice is deterministic or randomized, the input tape head is allowed real-time, one-way, or two-way access, and the automaton is classical or quantum. We prove several separation results among these variants, and establish the relationships between this model and the previously studied ways of providing advice to finite automata.

Keywords: advised computation, finite automata, random advice.

1 Introduction

Advised computation is based on the idea of providing external trusted assistance, depending only on the length of the input, to a computational device in order to extend its capability for solving certain problems [1]. Work on advised finite automaton models started with [2], where the advice string is prefixed to the input tape, and continued with a sequence of papers starting with [3], where the automaton reads the advice in parallel with the input from a separate track.

In this paper, we propose a new architecture for advised finite-state computation which enables the automata to use the advice more flexibly than the setups mentioned above. The idea is simply to let the machine use a separate one-way tape for the advice, thereby enabling it to pause on the input tape while processing the advice, or vice versa. (Examples of finite-state machines with such a separate tape for *untrusted* advice can be seen in [4].) Our model differs from an alternative proposal of Freivalds for advised finite-state automata [5] in the number of allowed advice tapes, and the way in which the advice can be accessed. We consider many variants of our machines, where the advised automaton is classical or quantum, the tapes can be accessed in various alternative modes, and the advice is deterministic or randomized. The power of these variants are compared among themselves, and also with the corresponding instances of the alternative models in the literature.

---

* Yakaryılmaz was partially supported by FP7 FET-Open project QCS.

M.-P. Béal and O. Carton (Eds.): DLT 2013, LNCS 7907, pp. 301–312, 2013.
© Springer-Verlag Berlin Heidelberg 2013

2   Previous Work

Finite automata that take advice were first examined by Damm and Holzer [2]. In their model, the advice string, which depends only on the length of the input, is placed on the input tape so that it precedes the original input. We call such a machine a *finite automaton with advice prefix*. The automaton simply reads the advice first, and then goes on to scan the input. Damm and Holzer studied REG/*const*, which is the class of languages that can be recognized by real-time deterministic finite automata that use constant-length advice, and showed that letting the advice string's length to be an increasing function of the input string's length, say, a polynomial, does not enlarge the class of languages recognized by such automata within this setup. They also used Kolmogorov complexity arguments to prove that every additional bit of advice extends the class of languages that can be recognized by finite automata in this model, that is, REG/$(k-1) \subsetneq$ REG/k, for all $k \geq 1$.

Another model of advised finite automata was examined by Tadaki et al. in [3], and later by T. Yamakami in [6,7,8,9]. This setup enables the automata to process the advice in parallel with the input, by simply placing the advice in a separate track of the input tape. In this manner, an advice string of length n can be provided, and meaningfully utilized, for inputs of length n. This enhances the language recognition power, as can be seen by considering the relative ease of designing such a *finite automaton with advice track* for the language $\{a^n b^n \mid n \in \mathbb{N}\}$, which can not be recognized by any finite automaton with advice prefix. Yamakami studied variants of this model with probabilistic and quantum automata, and randomized advice [7,9], and provided characterizations of the related classes of languages. Note that the track structure in this model both limits the length of the advice by the length of the input, and forces the advice to be scanned synchronously with the input.

R. Freivalds formulates and studies yet another model of advised finite automata in [5,10]. Freivalds' model incorporates one or more separate tapes for the advice to be read from. Both the input and the advice tapes have two-way heads. Unlike the previously mentioned models, the advice string for inputs of length n are supposed to be useful for all shorter inputs as well, and some negative results depend on this additional requirement.

3   Our Model

We model advice as a string provided on a separate read-only tape. As usual, the content of the advice depends only on the length of the input. Formally, the advice to the automaton is determined by an advice function h, which is a mapping from $\mathbb{N}$ to strings in Γ^*, where Γ is the advice alphabet. This function may or may not be computable.

Our advised machine model is then simply a finite automaton with two tapes. The transition function of a (two-way) *deterministic finite automaton with advice tape* (dfat) determines the next move of the machine based on the current internal state, and the symbols scanned by the input and advice tape heads. Each move specifies the next state, and a head movement direction (right, left, or stay-put) for each tape. A tape head that is allowed to move in all these directions is called *two-way*. A head that is not allowed to move left is called *one-way*. We may also require a head to be *real-time*, forcing it to move to the right at every step. As will be shown, playing with these settings changes the computational power of the resulting model. We assume that both the input and the advice strings are delimited by special end-marker symbols, beyond which the automaton does not attempt to move its heads. The machine halts and announces the corresponding decision when it enters one of the two special states q_{accept} and q_{reject}.

Unlike Freivalds [5], we do not allow two-way motion of the advice tape head, as permitting this head to make leftward moves would cause "unfair" accounting of the space complexity of the advised machine.[1]

A language L is said to be recognized by such a dfat M using $O(f(n))$-length advice if there exists an advice function h with the following properties:

- $|h(n)| \in O(f(n))$ for all $n \in \mathbb{N}$, and,
- M eventually halts and accepts when started with the input tape containing a string x of length n, and the advice tape containing $h(n)$, if and only if $x \in L$.

We need a notation for talking about language families corresponding to different settings of the tape access modes, and advice lengths. We will use the template "CLASS/$f(n)$(specification list)" for this purpose. In that template, the name of the complexity class corresponding to the unadvised, two-way version of the automaton in question will appear as the CLASS item. The function description $f(n)$ will denote that the machine uses advice strings of length $O(f(n))$ for inputs of length n. (General descriptors like *poly* and *exp*, for polynomial and exponential bounds, respectively, will also be used.) Any further specifications about, for instance, additionally restricted head movements, will be given in the list within the final parentheses. For example, the class of languages recognized by dfat's with real-time input and one-way advice tapes that use linear amounts of advice will be denoted SPACE(1)/n(rt-input).[2]

We will also be examining randomized advice, as defined by Yamakami [7]. In this case, the advice is randomly selected from a set of alternatives according to a pre-specified probability distribution. Deterministic finite automata which use randomized advice can perform tasks which are impossible with deterministic advice [7]. The use of randomized advice will be indicated by the letter R appearing before the advice length in our class names. We will use an item in the parenthesized specification list to indicate whether bounded or unbounded

---

[1] See Section 5.3.1 of [11] for a discussion of this issue in the context of certificate tape heads.

[2] Although SPACE(1) is well known to equal the regular languages, we avoid the shorter notation REG/n, which was used for the advice track model, and which will turn out to represent a strictly smaller class.

error language recognition is intended, when this is not clear from the core class name.

We define the probabilistic and quantum versions of our advised automata by generalizing the definition for deterministic automata in the standard way, see, for instance, [12]. The transition function of a *probabilistic finite automaton with advice tape* (pfat) specifies not necessarily one, but possibly many choices, associated with selection probabilities, for the next move at every step, with the well-formedness condition that the probabilities of these choices always add up to 1. In the case of *quantum finite automata with advice tapes* (qfat's), each such choice is associated not with a probability, but with an amplitude (a real number in the interval [-1,1]). The presentation of our results on qfat's will not require knowledge of technical details of their definitions such as well-formedness conditions, and we therefore omit these for space constraints, referring the reader to [12]. We should stress that there are many mutually inequivalent quantum finite automaton definitions in the literature, and we use the most powerful one [13,12]. The quantum machines with advice tracks defined in [9] are based on an older model [14], and this difference will be significant in our discussion in Section 6.

The notational convention introduced above is flexible enough to represent the language classes corresponding to the probabilistic and quantum advised machines as well, for instance, $\mathsf{BQSPACE}(1)/n(\mathtt{rt\text{-}input},\mathtt{rt\text{-}advice})$ is the class of languages recognized with bounded error by a qfat using linear-length advice, and real-time input and advice heads.

The model of real-time finite automata with advice tracks [3] is equivalent to our model with a separate advice tape when we set both the input and advice tape heads to be real-time. Therefore, all the results shown for the advice track model are inherited for this setting of our machines. For instance, $\mathsf{SPACE}(1)/n(\mathtt{rt\text{-}input},\mathtt{rt\text{-}advice}) = \mathsf{REG}/n$, where REG/n is defined in [3]. On the other hand, the quantum class $\mathsf{1QFA}/n$ of [9] does *not* equal $\mathsf{BQSPACE}(1)/n(\mathtt{rt\text{-}input},\mathtt{rt\text{-}advice})$, as we will show in Section 6.

Note that we allow only one advice tape in our model. This is justified by the following observation about the great power of one-way finite automata with multiple advice tapes.

Theorem 1. *Every language can be recognized by a finite automaton with a one-way input tape and two one-way advice tapes.*

Proof. Let L be any language on the alphabet Σ. We construct a finite automaton M that recognizes L using a one-way input tape and two one-way advice tapes as follows.

Let $\Gamma = \Sigma \cup \{c_a, c_r\}$ be the advice alphabet, where $\Sigma \cap \{c_a, c_r\} = \emptyset$. For an input of length n, the advice on the first advice tape lists every string in Σ^n in alphabetical order, where every member of L is followed by a c_a, and every nonmember is followed by a c_r. So the content of the first advice tape looks like $w_1 c_1 w_2 c_2 \cdots w_{|\Sigma|^n} c_{|\Sigma|^n}$, where $w_i \in \Sigma^n$, and $c_i \in \{c_a, c_r\}$ for $i \in \{1, \ldots, |\Sigma|^n\}$.

The second advice tape content looks like "$c_a c_r^n c_a c_r^n .. c_a c_r^n c_a$", with $|\Sigma|^n$ repetitions, and will be used by the machine for counting up to $n + 1$ by moving between two consecutive c_a symbols on this tape.

M starts its computation while scanning the first symbols of the input string and w_1 on the first advice tape. It attempts to match the symbols it reads from the input tape and the first advice tape, moving synchronously on both tapes. If the ith input symbol does not match the ith symbol of w_j, M pauses on the input tape, while moving the two advice heads simultaneously until the second advice head reaches the next c_a, thereby placing the first advice tape head on the ith position of w_{j+1}, where $1 \leq i \leq n$, and $1 \leq j < |\Sigma|^n$. M halts when it sees the endmarker on the input tape. When this happens, it accepts the input if the symbol read from the first advice tape is c_a, otherwise, it rejects. $\square$

4 Deterministic Finite Automata with Advice Tapes

It is clear that a machine with advice tape is at least as powerful as a machine of the same type with advice track, which in turn is superior to a corresponding machine with advice prefix, as mentioned in Section 2. We will now show that allowing either one of the input and advice head to pause on their tapes does enlarge the class of recognized languages.

Theorem 2. $\mathsf{REG}/n \subsetneq \mathsf{SPACE}(1)/n(\text{rt-input})$.

Proof. It follows trivially from the definitions of the classes that

$$\mathsf{REG}/n = \mathsf{SPACE}(1)/n(\text{rt-input}, \text{rt-advice}) \subseteq \mathsf{SPACE}(1)/n(\text{rt-input}).$$

Let $|w|_\sigma$ denote the number of occurrences of symbol σ in string w. To show that the above subset relation is proper, we will consider the language $\mathsf{EQUAL}_2 = \{w|\ w \in \{a, b\}^* \text{ and } |w|_a = |w|_b\}$, which is known [3] to lie outside REG/n.

One can construct a finite automaton that recognizes EQUAL_2 with real-time input and one-way access to linear advice as follows. For inputs of odd length, the automaton rejects the input. For inputs of even length, n, the advice function is $h(n) = a^{n/2}$. The automaton moves its advice head one position to the right for each a that it reads on the input. The input is accepted if the number of a's on the two tapes match, and rejected otherwise. $\square$

Tadaki et al. [3] studied one-tape linear-time Turing machines with an advice track, and showed that the class of languages that they can recognize coincides with REG/n. Theorem 2 lets us conclude that simply having a separate head for advice increases the computational power of a real-time dfa, whereas the incorporation of a single two-way head for accessing both advice and a linear amount of read/write memory simultaneously does not.

Theorem 3. $\mathsf{REG}/n \subsetneq \mathsf{SPACE}(1)/n(\text{1w-input}, \text{rt-advice})$.

Proof. Consider the language EQUAL $= \{w | w \in \Sigma^*$, where $\{a, b\} \subseteq \Sigma$, and $|w|_a = |w|_b\}$, which is similar to EQUAL$_2$, but with a possibly bigger alphabet. EQUAL $\notin$ REG/n, as can be shown easily by Theorem 2 of [7]. We will describe a dfat M with one-way input, and real-time access to an advice string that is just a^{2n}, where n is the input length.

M moves the advice head one step to the right for each a that it scans in the input. When it scans a b, it advances the advice head by three steps. For any other input symbol, the advice head is moved two steps. If the advice head attempts to move beyond the advice string, M rejects. When the input tape head reaches the end of the tape, M waits to see if the advice tape head will also have arrived at the end of the advice string after completing the moves indicated by the last input symbol. If this occurs, M accepts, otherwise, it rejects.

Note that the advice head is required to move exactly $|w|_a + 3|w|_b + 2(n - |w|_a - |w|_b)$ steps, which equals $2n$ if and only if the input is a member of EQUAL. $\qquad\square$

As noted earlier, advice lengths that are increasing functions of the input length are not useful in the advice prefix model. Only linear-sized advice has been studied in the context of the advice track model [3,7]. Theorem 4 demonstrates a family of languages for which very small increasing advice length functions are useful in the advice tape model, but not in the advice track model.

Theorem 4. *For every function* $f : \mathbb{N} \to \mathbb{N}$ *such that* $f(n) \in \omega(1)$ *and* $f(n) \in O(\sqrt{n})$, SPACE$(1)/f^2(n)$(1w-input) $\nsubseteq$ REG/n.

Proof. Consider the language L$_f = \{a^k b^m c^k | k \leq f(n), n = k + m + k\}$, for any function f satisfying the properties in the theorem statement.

Theorem 2 of [7] can be used to show L$_f \notin$ REG/n.

One can construct a dfat with one way access to input and advice that recognizes L$_f$ as follows. For inputs of length n, the advice string is of the form $\#\#a\#aa\#aaa\#..\#a^{f(n)}\#$, with length $O(f^2(n))$. During any step, if the automaton detects that the input is not of the form $a^*b^*c^*$, it rejects the input. For each a that it reads from the input tape, the automaton moves the advice tape head to the next $\#$ on the advice tape. (If the advice ends when looking for a $\#$, the input is rejected.) When the input tape head scans the b's, the advice tape head remains idle. Finally, when the input head starts to scan the c's, the automaton compares the number of c's on the input tape with the number of a's that it can scan until the next $\#$ on the advice tape. If these match, the input is accepted; otherwise it is rejected. $\qquad\square$

When restricted to constant size advice, the parallelism and the two-way input access inherent in our model does not make it superior to the advice prefix model. As we show now, one can always read the entire advice before starting to read the input tape without loss of computational power in the constant-length advice case:

Theorem 5. *For every* $k \in \mathbb{N}$, SPACE$(1)/k =$ REG/k.

Proof. The relation $REG/k \subseteq SPACE(1)/k$ is trivial, since an automaton taking constant-length advice in the prefix or track formats can be converted easily to one that reads it from a separate tape. For the other direction, note that a dfat M with two-way input that uses k bits of advice corresponds to a set S of 2^k real-time dfa's *without* advice, each of which can be obtained by hard-wiring a different advice string to the program of M, and converting the resulting two-way dfa to the equivalent real-time machine, which exists by [15]. The advice string's job is just to specify which of these machines will run on the input string. It is then easy to build a dfa with advice prefix which uses the advice to select the appropriate program to run on the input. $\square$

Since our model is equivalent to the advice prefix model for constant-length advice, we inherit the results like Theorem 5 of [2], which states that the longer advice strings one allows, the larger the class of languages we can recognize will be, as long as one makes sure that the advice and input alphabets are identical.

A natural question that arises during the study of advised computation is whether the model under consideration is strong enough to recognize *every* desired language. The combination of two-way input tape head and exponentially long advice can be shown to give this power to finite automata. Let ALL denote the class of all languages on the input alphabet Σ.

Theorem 6. $SPACE(1)/exp(\texttt{rt-advice}) = ALL$.

Proof. The advice string for input length n contains all members of the considered language of length n, separated by substrings consisting of $n + 2$ blank symbols. The automaton compares the input with each of the strings listed on the advice tape. If it is able to match the input to a word on the advice tape, it accepts. If the advice ends without such a match, it rejects. Otherwise, the machine rewinds to the start of the input while consuming blanks from the next string on the advice tape. The advice length is $2^{O(n)}$. $\square$

Whether $SPACE(1)/exp(\texttt{1w-input}) = ALL$ is an open question. In the following, we show a limitation of dfat's with one-way access to input and subexponential-length advice. Let PAL denote the language of even-length palindromes defined on the alphabet $\{a, b\}$.

Theorem 7. $PAL \notin SPACE(1)/poly(\texttt{1w-input}, \texttt{1w-advice})$.

Proof. We assume that a one-way finite automaton M with state set Q recognizes PAL with one-way access to advice $h(n)$ of subexponential length, and show how this assumption leads to a contradiction.

We define a sequence of functions $F_k : \{a, b\}^{(k/2)} \rightarrow Q \times \{1, \ldots, |h(k)|\}$ for $k = 2i$ *and* $i \in \{1, 2, \ldots\}$, which map strings of length $k/2$ to ordered pairs of states of M and positions on advice tape such that $F_k(w) = (q, p)$ if and only if M, started on an input of length k, the first $k/2$ symbols of which match w, reaches the first symbol of the second half of its input when the control state is $q \in Q$, and the advice head position is $p \in \{1, \ldots, |h(k)|\}$.

For sufficiently large values of k, the functions F_k defined in this way can not be one-to-one, since the size of their domain $|\{a,b\}^{(k/2)}| = 2^{(k/2)}$ dominates the size of their range $|Q \times \{1, \ldots, |h(k)|\}| = |Q||h(k)|$ as k grows, since Q is constant in size, and $|h(k)|$ is subexponential.

Let k be large enough so that F_k is not one to one, and let x and y be distinct strings of length $k/2$ which are mapped to the same state, advice position pair by F_k. Let z be any string of length $k/2$. Regardless of whether M starts on xz or yz, the control state and the advice position will be the same when it reaches the first symbol of z. This indicates that the remaining computations will be identical. We therefore conclude that M accepts xz if and only if it accepts yz.

Letting w^R denote the reverse of string w, consider $z = x^R$. Clearly, $xz \in \text{PAL}$, and hence M accepts xz with advice $h(n)$. As we assumed $F_k(x) = F_k(y)$, we know that M also accepts yz, which is a contradiction since $x \neq y$ implies $x^R \neq y^R$, and hence $z \neq y^R$.   □

Since a machine with real-time input does not have time to consume more than a linear amount of advice, we easily have

Corollary 1. *For every function* $f : \mathbb{N} \to \mathbb{N}$, PAL $\notin$ SPACE(1)/$f(n)$(rt-input).

We now show that PAL can be recognized by a two-way dfa with polynomially long advice, which will lead to a separation for machines with one- and two-way input.

Theorem 8. PAL $\in$ SPACE(1)/n^2(2w-input, rt-advice).

Proof. We shall describe how a two-way dfa with real-time access to a quadratic-length advice string can recognize PAL. On an input of length n, the advice tells the automaton to reject if n is odd. For even n, the advice assists the automaton by simply marking the $n/2$ pairs $(i, n - i + 1)$ of positions that should be holding matching symbols on the input string. Consider, for example $h(8) = $ #10000001#01000010#00100100#00011000#. The automaton should just traverse the input from the first symbol to the last while also traversing the part of the advice that lies between two separator symbols (#), and then do the same while going from the last symbol to the first, and so on. At each pass, the automaton should check whether the input symbols whose positions match those of the two 1's on the advice are identical. If this check fails at any pass, the automaton rejects the input, otherwise, it accepts.

The method described above requires a two way automaton with real-time access to an advice of length $n^2/2$. (The separator symbols are for ease of presentation, and are not actually needed for the construction.)   □

We have proven the following separation result.

Theorem 9. SPACE(1)/$poly$(1w-input, 1w-advice)
$\quad\quad\quad\quad\quad$ $\subsetneq$ SPACE(1)/$poly$(2w-input, 1w-advice).

5   Efficient Error Reduction with Randomized Advice

We now turn to randomly selected probabilistic advice given to deterministic machines. Yamakami [7] proved that this setup yields an improvement in language recognition power over REG/n, by demonstrating a deterministic automaton with advice track recognizing the center-marked palindrome language with randomized advice. Considering the amount of randomness involved in the selection of the advice string as a resource, Yamakami's example requires $O(n)$ random bits, since it requires picking a string from a set with $2^{O(n)}$ elements with uniform probability. Furthermore, reducing the error bound of Yamakami's automaton to smaller and smaller values requires extending the advice alphabet to bigger and bigger sizes. In the construction we will present in Theorem 10, the number of random bits does not depend on the input length, and any desired error bound can be achieved without modifying the advice alphabet.

Theorem 10. $SPACE(1)/n(\text{1w-input}, \text{1w-advice})$
$\subsetneq SPACE(1)/Rn(\text{1w-input}, \text{1w-advice}, \text{bounded-error}).$

Proof. We will use the language $EQUAL_3 = \{w|\ w \in \{a, b, c\}^*, |w|_a = |w|_b = |w|_c\}$ to separate the language classes in the theorem statement. Let us begin by showing that $EQUAL_3 \notin SPACE(1)/n(\text{1w-input}, \text{1w-advice})$. The idea is reminiscent of Theorem 7.

Assume that a dfat M with one-way input does recognize $EQUAL_3$ with linear-length advice $h(n)$. We will be considering M at a point in its execution where it has read the first k symbols of a string of length $3k$. Define a sequence of functions $F_n : \{a, b\}^{(n/3)} \to Q \times \{1, \ldots, |h(n)|\}$ for $n = 3k$ and $k \in \{1, 2, \ldots\}$, which map strings of length k to ordered pairs of states of M and positions on advice tape such that $F_n(w) = (q, p)$ if and only if M, when started on an input of length n, the first k symbols of which match w, reaches the $k + 1$'st symbol of its input when the control state is $q \in Q$, and the advice tape position is $p \in \{1, .., |h(n)|\}$. The number of different control state and advice position pairs that M can be in at any step is $O(n)$. On the other hand, note that there are exactly $\binom{k+2}{2}$ different ways of setting the numbers of occurrences of the symbols a, b, and c in the prefix of length k that M has read until this point. This value is $\omega(n)$. Therefore, there should be two distinct strings x and y of length k such that x and y disagree on the number of occurrences of some symbol types, yet still they are mapped to the same control state and advice position pair by F_n. This implies that for any string $z \in \{a, b, c\}^*$, M accepts xz if and only if it accepts yz. This leads to a contradiction, as xz and yz can not have the same number of a's, b's and c's.

To show that $EQUAL_3 \in SPACE(1)/Rn(\text{1w-input}, \text{1w-advice}, \text{bounded-error})$, we will describe a set of advice strings, and show how a randomly selected member of this set can assist a one-way dfat N to recognize $EQUAL_3$ with bounded error. We shall be adapting a technique used by Freivalds in [16].

If the input length n is not divisible by 3, N rejects. If $n = 3k$ for some integer k, the advice is selected with equal probability from a collection of linear-size advice strings $A_i = 1^i \# 1^{ki^2 + ki + k}$ for $i \in \{1, \ldots, s\}$, where s is a constant.

N starts by reading the 1's in the advice string that precede the separator character #, thereby learning the number i. N then starts to scan the input symbols, and moves the advice head 1, i , or i^2 steps to the right for each a, b or c that it reads on the input tape, respectively. The input is accepted if the automaton reaches the ends of the input and advice strings simultaneously, as in the proof of Theorem 3. Otherwise, the input is rejected.

Note that the automaton accepts the input string w if and only if the number of symbols in the advice string that comes after the separator symbol is equal to the total number of moves made by the advice tape head while the input head scans w. N accepts w if and only if $|w|_a + |w|_b i + |w|_c i^2 = k + ki + ki^2$, which trivially holds for $w \in$ EQUAL$_3$ no matter which advice string is selected, since $|w|_a = |w|_b = |w|_c = k$ in that case.

If $w \notin$ EQUAL$_3$, the probability of acceptance is equal to the probability of selecting one of the roots of the quadratic equation $(|w|_c - k)i^2 + (|w|_b - k)i + (|w|_a - k) = 0$ as the value of i. This probability is bounded by $\frac{2}{s}$, and can be pulled down to any desired level by picking a bigger value for s, and reorganizing the automaton accordingly.   □

6   Quantum Finite Automata with Advice Tapes

Yamakami [9] defined the class 1QFA/n as the collection of languages which can be recognized by real-time Kondacs-Watrous quantum finite automata (KWqfa's) with advice tracks. The KWqfa is one of many inequivalent models of quantum finite-state computation that were proposed in the 1990's, and is known to be strictly weaker than classical finite automata in the context of bounded-error language recognition [14]. This weakness carries over to the advised model of [9], with the result that there exist some regular languages that are not members of 1QFA/n. We use a state-of-the-art model of quantum automaton that can simulate its classical counterparts trivially, [13,12] so we have:

Theorem 11. 1QFA/$n \subsetneq$ BQSPACE(1)/n(rt-input, rt-advice).

Whether this properly strong version of qfa can outperform its classical counterparts with advice tapes is an open question. We are able to show a superiority of quantum over classical in the following restricted setup, which may seem silly at first sight: Call an advice tape *empty* if it contains the standard blank tape symbol in all its squares. We say that a machine M receives *empty advice* of length $f(n)$, if it is just allowed to move its advice head on the first $f(n)$ squares of an empty advice tape, where n is the input length. This restriction will be represented by the presence of the designation empty in the specification lists of the relevant complexity classes.

Theorem 12. BPSPACE(1)/n(rt-input, 1w-empty-advice)
              $\subsetneq$ BQSPACE(1)/n(rt-input, 1w-empty-advice).

Proof. An empty advice tape can be seen as an increment-only counter, where each move of the advice tape head corresponds to an incrementation on the counter, with no mechanism for decrementation or zero-testing provided in the programming language. In [17], Yakaryılmaz et al. studied precisely this model. It is obvious that classical automata augmented with such a counter do not gain any additional computational power, so BPSPACE(1)/n(rt-input, 1w-empty-advice) equals the class of regular languages, just like the corresponding class without advice. On the other hand, real-time qfa's augmented with such an increment-only counter were shown to recognize some nonregular languages like EQUAL$_2$ with bounded error in [17].   □

Since increment-only counters are known to increase the computational power of real-time qfa's in the unbounded-error setting as well, [17], we can also state

Theorem 13. PrSPACE(1)/n(rt-input, 1w-empty-advice)
$\subsetneq$ PrQSPACE(1)/n(rt-input, 1w-empty-advice).

7 Open Questions

- Real-time probabilistic automata can be simulated by deterministic automata which receive coin tosses within a randomly selected advice string. It would be interesting to explore the relationship between deterministic automata working with randomized advice, and probabilistic automata working with deterministic advice. For instance, is there a pfat with one-way input and deterministic advice that can recognize EQUAL$_3$ for any desired error bound, as in Theorem 10?
- Are there languages which cannot be recognized with any amount of advice by a dfat with one-way input? Does the answer change for pfat's or qfat's?
- Can qfat's recognize any language which is impossible for pfat's with non-empty advice?

Acknowledgment. We thank Gökalp Demirci for his helpful comments.

References

1. Karp, R., Lipton, R.: Turing machines that take advice. L'Enseignement Mathematique 28, 191–209 (1982)
2. Damm, C., Holzer, M.: Automata that take advice. In: Hájek, P., Wiedermann, J. (eds.) MFCS 1995. LNCS, vol. 969, pp. 149–158. Springer, Heidelberg (1995)
3. Tadaki, K., Yamakami, T., Lin, J.C.H.: Theory of one-tape linear-time Turing machines. Theoretical Computer Science 411(1), 22–43 (2010)
4. Dwork, C., Stockmeyer, L.: Finite state verifiers I: The power of interaction. Journal of the ACM 39(4), 800–828 (1992)
5. Freivalds, R.: Amount of nonconstructivity in deterministic finite automata. Theoretical Computer Science 411(38-39), 3436–3443 (2010)
6. Yamakami, T.: Swapping lemmas for regular and context-free languages with advice. Computing Research Repository abs/0808.4 (2008)

7. Yamakami, T.: The roles of advice to one-tape linear-time Turing machines and finite automata. Int. J. Found. Comput. Sci. 21(6), 941–962 (2010)

8. Yamakami, T.: Immunity and pseudorandomness of context-free languages. Theoretical Computer Science 412(45), 6432–6450 (2011)

9. Yamakami, T.: One-way reversible and quantum finite automata with advice. In: Dediu, A.-H., Martín-Vide, C. (eds.) LATA 2012. LNCS, vol. 7183, pp. 526–537. Springer, Heidelberg (2012)

10. Agadzanyan, R., Freivalds, R.: Finite state transducers with intuition. In: Calude, C.S., Hagiya, M., Morita, K., Rozenberg, G., Timmis, J. (eds.) Unconventional Computation. LNCS, vol. 6079, pp. 11–20. Springer, Heidelberg (2010)

11. Goldreich, O.: Computational Complexity: A Conceptual Perspective. Cambridge University Press (2008)

12. Yakaryılmaz, A., Say, A.C.C.: Unbounded-error quantum computation with small space bounds. Information and Computation 279(6), 873–892 (2011)

13. Hirvensalo, M.: Quantum automata with open time evolution. International Journal of Natural Computing Research 1(1), 70–85 (2010)

14. Kondacs, A., Watrous, J.: On the power of quantum finite state automata. In: Proceedings of the 38th Annual Symposium on Foundations of Computer Science, FOCS 1997, pp. 66–75 (1997)

15. Shepherdson, J.C.: The reduction of two–way automata to one-way automata. IBM Journal of Research and Development 3, 198–200 (1959)

16. Freivalds, R.: Fast probabilistic algorithms. In: Becvar, J. (ed.) MFCS 1979. LNCS, vol. 74, pp. 57–69. Springer, Heidelberg (1979)

17. Yakaryilmaz, A., Freivalds, R., Say, A.C.C., Agadzanyan, R.: Quantum computation with write-only memory. Natural Computing 11(1), 81–94 (2012)

One-Way Multi-Head Finite Automata with Pebbles But No States

Martin Kutrib, Andreas Malcher, and Matthias Wendlandt

Institut für Informatik, Universität Giessen
Arndtstr. 2, 35392 Giessen, Germany
{kutrib,malcher,matthias.wendlandt}@informatik.uni-giessen.de

Abstract. Stateless variants of deterministic one-way multi-head finite automata with pebbles, that is, automata where the heads can drop, sense, and pick up pebbles, are studied. The relation between heads and pebbles is investigated, and a proper double hierarchy concerning these two resources is obtained. Moreover, it is shown that a conversion of an arbitrary automaton to a stateless automaton can always be achieved at the cost of additional heads and/or pebbles. On the other hand, there are languages where one head cannot be traded for any number of additional pebbles and vice versa. Finally, the emptiness problem and related problems are shown to be undecidable even for the 'simplest' model, namely, for stateless one-way finite automata with two heads and one pebble.

1 Introduction

Deterministic finite automata (DFA) and deterministic linear bounded automata (DLBA) describe in a way both extremal cases of a deterministic automaton model with one head working on a bounded input tape: while a DFA may move only from left to right (one-way) and its head can only read from the input tape, a DLBA can move in both directions (two-way) and its head is capable of performing read as well as write operations. It is thus not surprising that the language classes described by DFA and DLBA, that is, the regular languages and the deterministic context-sensitive languages, are far away from each other. So it is naturally of interest to consider automata classes lying in between both ends. One example in this context are deterministic multi-head finite automata which are, basically, DFA provided with a fixed finite number h of heads that can move on the input one-way or two-way. However, the heads can only read the input. Computational complexity aspects of such automata with one-way head motion have been first considered in [19,21]. The latter paper also investigates decidability questions and closure properties, and discusses the question whether $h + 1$ heads are more powerful than h heads. This hierarchy question was affirmatively answered for all $h \geq 1$ in [24] using witness languages over a ternary alphabet. A reduction to a binary alphabet and to witness languages of the form a^*b^* has been obtained in [2,15]. More information on multi-head finite automata may be found in the recent survey [6].

M.-P. Béal and O. Carton (Eds.): DLT 2013, LNCS 7907, pp. 313–324, 2013.
© Springer-Verlag Berlin Heidelberg 2013

One-way multi-head finite automata where each head can also write on the input tape are called multi-head writing finite automata and have been investigated in [1,22]. It is known that three heads are more powerful than two heads and, in the nondeterministic case, every multi-head writing finite automaton with h heads can be simulated with three heads where only one head has to perform write operations. Since writing in connection with multiple heads seems to be very powerful, the following weaker way of writing has been considered in [5]: the automaton has a stock of pebbles (or markers) and every head can write by dropping some pebble to the input tape. Pebbles on the tape are sensed by the heads and can be picked up again and dropped at other positions. In this way, only information with respect to the position and to the type of the pebble can be written. A result discussed in [5] on such deterministic one-way multi-head pebble automata is that such automata with h heads and p individual, that means distinguishable, pebbles can be simulated by automata with h heads and $p + 1$ pebbles that are all of the same type and cannot be distinguished. Another construction gives that unary languages of the form $\{ a^{n^k} \mid n \geq 0 \}$ can be accepted with k heads and $k - 1$ pebbles.

It should be noted that pebble automata have also been considered with a different meaning, namely, as deterministic two-way devices with p individual pebbles and one head which can drop, sense, and pick up pebbles [20]. It is known that such automata with h pebbles can simulate any h-head finite automaton and can themselves be simulated by $(h + 1)$-head finite automata [20,18]. Furthermore, a proper hierarchy on the number of pebbles is shown in [18] where also relations to sensing multi-head finite automata, that is, automata where heads can sense the presence of other heads, are discussed.

A recent issue studied for multi-head finite automata is to consider devices which have one state only, so-called *stateless* automata. This concept has been introduced with a biological motivation in [23]. Despite the strong restriction to one state only, these devices are still very powerful which is demonstrated in [23] by showing that the emptiness problem is undecidable for stateless deterministic one-way three-head finite automata. The best result known so far for one-way devices with two heads is that automata with four states have an undecidable emptiness problem [12]. The two-way case is considered in [9], where the undecidability of the emptiness problem is obtained for stateless automata with two heads. Stateless variants of related models such as, for example, multi-head pushdown automata, multi-counter automata, restarting automata, and two-pushdown automata have been investigated in [3,4,8,13,14].

Here, we investigate stateless deterministic one-way multi-head finite automata with individual pebbles. After some preliminaries in Section 2, we investigate in Section 3 the costs in terms of additional heads and pebbles arising when an arbitrary automaton is reduced to an equivalent stateless automaton. In Section 4, we obtain a double hierarchy concerning heads and pebbles for stateless automata. In detail, it is shown for any number of heads $h \geq 1$ and any number of pebbles $p \geq 0$ that there are, on the one hand, languages acceptable with $h + 1$ heads and p pebbles which are not acceptable with h heads and p pebbles.

On the other hand, it is shown that there are languages acceptable with h heads and $p+1$ pebbles which are not acceptable with h heads and p pebbles. The relation between heads and pebbles is again discussed in Section 5. The main result obtained there is that we exhibit languages which can be accepted by stateless automata with a certain number of heads and pebbles, but which cannot be accepted with one head less although an arbitrary number of pebbles and states may be provided. Finally, we investigate in Section 6 the decidability status of the emptiness problem for stateless one-way two-head DFA with one pebble. Using an encoding of valid computations of the parallel model of one-way cellular automata, it is possible to show the undecidability of the problem. It should be noted that it is currently an open question whether or not the problem is decidable for stateless one-way two-head DFA without pebbles.

2   Preliminaries and Definitions

We write A^* for the set of all words over the finite alphabet A. The empty word is denoted by λ, and $A^+ = A^* \setminus \{\lambda\}$. The reversal of a word w is denoted by w^R and for the length of w we write $|w|$. We use $\subseteq$ for inclusions and $\subset$ for strict inclusions. We write 2^S for the powerset of a set S.

Let $h \geq 1$ and $p \geq 0$ be natural numbers. A one-way h-head finite automaton with p pebbles is a finite automaton having a single read-only input tape whose inscription is the input word in between two endmarkers (we provide two endmarkers in order to have a definition consistent with two-way devices). The h heads of the automaton mutually independently can stay on the current tape square or move to the right but not beyond the endmarkers. In addition, the automaton has a stock of p individual pebbles (or markers) that can be dropped freely by the heads to mark specific positions. More than one pebble may be placed on a single tape square. The pebbles on the tape are noticed by the heads and can be picked up again. A formal definition is:

Definition 1. *A deterministic one-way h-head finite automaton with p pebbles (1DHPA(h,p)) is a system $M = \langle S, A, h, p, \delta, \triangleright, \triangleleft, s_0 \rangle$, where*

1. *S is the finite set of internal states,*
2. *A is the finite set of input symbols,*
3. *$h \geq 1$ is the number of heads,*
4. *$p \geq 0$ is the number of pebbles in the (possibly empty) set of pebbles $P = \{1, 2, \ldots, p\}$,*
5. *$\triangleright \notin A$ is the left and $\triangleleft \notin A$ is the right endmarker,*
6. *$s_0 \subset S$ is the initial state,*
7. *$\delta : S \times 2^P \times ((A \cup \{\triangleright, \triangleleft\}) \times 2^P)^h \to S \times (\{0, 1\} \times 2^P)^h$ is the partial transition function, where 1 means to move a head one square to the right, and 0 means to keep it on the current square. Whenever*

$$\delta(s, P_0, a_1, P_1, a_2, P_2, \ldots, a_h, P_h) = (s', d_1, Q_1, d_2, Q_2, \ldots, d_h, Q_h)$$

is defined, then (i) $d_i = 0$ if $a_i = \triangleleft$, for $1 \leq i \leq h$, (ii) $P_0, P_1, \ldots, P_h$ as well as $Q_1, Q_2, \ldots, Q_h$ are pairwise disjoint, and (iii) $\bigcup_{i=1}^{h} Q_i \subseteq \bigcup_{i=0}^{h} P_i$.

A *configuration* of a 1DHPA(h,p) $M = \langle S, A, h, p, \delta, \triangleright, \triangleleft, s_0 \rangle$ is a quadruple (w, s, τ, μ), where $w \in A^*$ is the input, $s \in S$ denotes the current state, $\tau = (\tau_1, \tau_2, \ldots, \tau_h) \in \{0, 1, \ldots, |w| + 1\}^h$ gives the current head positions, and $\mu : \{0, 1, \ldots, |w| + 1\} \to 2^P$ gives the current distribution of the pebbles. If a position τ_i is 0, then head i scans the symbol $\triangleright$, if it satisfies $1 \leq \tau_i \leq |w|$, then the head scans the τ_ith letter of w, and if it is $|w| + 1$, then the head scans the symbol $\triangleleft$. It is understood that $\mu(i)$ is the set of pebbles on the ith tape square, $\mu(0), \mu(1), \ldots, \mu(|w| + 1)$ are pairwise disjoint, and the automaton currently carries the pebbles $P \setminus (\bigcup_{i=0}^{|w|+1} \mu(i))$. The *initial configuration* for input w is set to $(w, s_0, (0, \ldots, 0), \mu_0)$, where μ_0 denotes the constant mapping $\mu_0(i) = \emptyset$, $0 \leq i \leq |w| + 1$. That is, M starts with carrying all pebbles and having all of its heads on the left endmarker.

During the course of its computation, M runs through a sequence of configurations. One step from a configuration to its successor configuration is denoted by $\vdash$. Let $w = a_1 a_2 \cdots a_n$ be the input, $a_0 = \triangleright$, and $a_{n+1} = \triangleleft$, then we set

$$(w, s, (\tau_1, \tau_2, \ldots, \tau_h), \mu) \vdash (w, s', (\tau_1 + d_1, \tau_2 + d_2, \ldots, \tau_h + d_h), \mu')$$

with $\mu'(\tau_i) = Q_i$, $1 \leq i \leq h$, and $\mu'(j) = \mu(j)$ for $j \notin \{\tau_1, \tau_2, \ldots, \tau_h\}$, if and only if

$$\delta(s, P_0, a_{\tau_1}, \mu(\tau_1), a_{\tau_2}, \mu(\tau_2), \ldots, a_{\tau_h}, \mu(\tau_h)) = (s', d_1, Q_1, d_2, Q_2, \ldots, d_h, Q_h).$$

So, in a computation step, first M picks up all pebbles from the current head positions, thus carrying the pebbles $P_0 \cup \bigcup_{i=1}^h \mu(\tau_i)$, second drops the pebbles Q_i to the head position τ_i, $1 \leq i \leq h$, thus keeping the pebbles $(P_0 \cup \bigcup_{i=1}^h \mu(\tau_i)) \setminus \bigcup_{i=1}^h Q_i$, and third moves its heads and changes the state. Note that by the restrictions of the transition function δ, M cannot drop pebbles which are on tape squares currently not scanned. Moreover, the heads cannot move beyond the right endmarker. As usual we define the reflexive, transitive closure of $\vdash$ by $\vdash^*$, and its transitive closure by $\vdash^+$.

Since in the sequel we also consider stateless 1DHPA(h,p), that is, automata having exactly one state, non-trivial acceptance cannot be defined by accepting states. Instead, we follow the definition in [9] and say that an input is accepted if and only if the computation ends in a loop in which the heads are not moved. A 1DHPA(h,p) blocks and rejects when the transition function is not defined for the current situation. Whenever we consider an accepting computation it is understood that we mean the finite initial part of the computation up to but not including the first loop at the end. The language accepted by a 1DHPA(h,p) M is

$$L(M) = \{ w \in A^* \mid \text{there are } \tau \in \{0, 1, \ldots, |w| + 1\}^h,$$
$$\mu : \{0, 1, \ldots, |w| + 1\} \to 2^P, \text{ and } s \in S \text{ such that}$$
$$(w, s_0, (0, \ldots, 0), \mu_0) \vdash^* (w, s, \tau, \mu) \vdash^+ (w, s, \tau, \mu) \}.$$

In order to clarify our notion we continue with an example. Moreover, the example shows the power gained by pebbles. In particular, a stateless one-way finite automaton having only two heads and one pebble is given that accepts a unary non-semilinear language. So, even automata with a minimum of resources break the border of context-free languages.

Example 2. The non-semilinear language $\{a^{\frac{n^2+n}{2}-1} \mid n \geq 2\}$ is accepted by the stateless 1DHPA(2, 1) $M = \langle \{s_0\}, \{a\}, 2, 1, \delta, \triangleright, \triangleleft, s_0 \rangle$, where the transition function δ is specified as follows:

1. $\delta(s_0, \{1\}, \triangleright, \emptyset, \triangleright, \emptyset) = (s_0, 0, \emptyset, 1, \{1\})$
2. $\delta(s_0, \emptyset, \triangleright, \{1\}, a, \emptyset) = (s_0, 0, \emptyset, 1, \{1\})$
3. $\delta(s_0, \emptyset, \triangleright, \emptyset, a, \emptyset) = (s_0, 1, \emptyset, 1, \emptyset)$
4. $\delta(s_0, \emptyset, a, \{1\}, a, \emptyset) = (s_0, 0, \emptyset, 1, \{1\})$
5. $\delta(s_0, \emptyset, a, \emptyset, a, \emptyset) = (s_0, 1, \emptyset, 1, \emptyset)$
6. $\delta(s_0, \emptyset, a, \{1\}, \triangleleft, \emptyset) = (s_0, 0, \{1\}, 0, \emptyset)$

The basic computation cycles of M are as follows. Both heads have a certain distance to each other and move synchronously to the right. When the left head enters a tape square with the pebble on it, the pebble is picked up and dropped again by the right head. In addition, the left head is stopped for one time step whereby the distance between the heads in increased by one. The computation is accepting if the right head reads the right endmarker exactly when the left head sees the pebble (transition 6).

It is immediately verified that M rejects the input a^1 and accepts the input a^2. So, let the length of the input be at least be three. Initially, by applying transitions 1–3 the second head is placed on the third a, the first head is placed on the first a, and the pebble in dropped on the first a as well. So a situation is reached, where the distance between the heads is two, the right head (head 2) is at position $3 = \frac{n^2+n}{2}$ for $n = 2$, and the pebble is on the tape square of the left head (head 1).

Arguing inductively, we now assume that after some computation cycles, a situation is reached, where the distance between the heads is i, the right head is at position $\frac{i^2+i}{2}$, and the pebble is on the tape square of the left head. Then transition 4 is applied which drops the pebble at position $\frac{i^2+i}{2}$, moves the right head to position $\frac{i^2+i}{2} + 1$, and keeps the left head stationary. So, the distance between the heads is $i+1$ and the distance between head 1 and the pebble is still i. Subsequently, both heads are moved synchronously to the right by transition 5 until the left head sees the pebble, that is, after i steps. Now the distance between the heads is still $i+1$, the right head is at position $\frac{i^2+i}{2}+1+i = \frac{(i+1)^2+(i+1)}{2}$, and the pebble is on the tape square of the left head. Since an input is accepted if and only if in such a situation the right head reads the right endmarker, exactly the inputs of length $\frac{i^2+i}{2} - 1$ are accepted.   □

3   Trading States for Heads and Pebbles

Since, basically, we are interested in stateless automata, this section is devoted to investigate to what extent the number of states can be decreased at the cost of additional heads and/or pebbles, if possible at all. In fact, it turns out that any 1DHPA(h, p) can be made stateless by increasing either the number of pebbles or by increasing the number of heads and adding one pebble. The latter additional pebble cannot be avoided in general.

Given a 1DHPA(h, p) with $m \geq 1$ states, it is shown next how to construct an equivalent stateless version with the same number of heads and $O(\log(m))$ additional pebbles.

Theorem 3. *Let $h \geq 2$, $p \geq 0$, and $m \geq 1$. For any m-state 1DHPA(h, p) an equivalent stateless 1DHPA($h, p + \lceil \log(m) \rceil + 2$) can effectively be constructed.*

Proof. Let $M = \langle S, A, h, p, \delta, \triangleright, \triangleleft, s_0 \rangle$ be an m-state 1DHPA(h, p) with at least two heads. As usual, we denote the set $\{1, 2, \ldots, p\}$ of its pebbles by P and set $\tilde{P} = \{p + 1, p + 2, \ldots, p + \lceil \log(m) \rceil\}$. The equivalent stateless automaton $M' = \langle \{s_0\}, A, h, p + \lceil \log(m) \rceil + 2, \delta, \triangleright, \triangleleft, s_0 \rangle$ uses the $\lceil \log(m) \rceil$ additional pebbles from $\tilde{P}$ to represent the encoding of the states of M. To this end, let $f : S \to 2^{\tilde{P}}$ be an injective function that maps a state to its encoding.

Basically, the simulation of M by M' works in phases, where one phase simulates one transition of M. At the beginning of a phase, the pebbles from P as well as the heads of M' are at the same positions as the pebbles and heads of M. In addition, the pebbles $f(s)$, where s is the current state of M, are at the position of the first head, and M' carries the two extra pebbles $\{p_1, p_2\}$ and the remaining pebbles from $\tilde{P}$.

If M does not move head 1, automaton M' can simulate the transition in a single step. It simply drops the pebbles and moves the heads as M does. In addition, the pebbles $f(s)$ are picked up and pebbles $f(\hat{s})$ are dropped at the position of the first head, where $\hat{s}$ is the successor state of M. Now M' is ready for the next phase.

If M moves head 1, automaton M' simulates the transition in three steps. In a first step, essentially, the transition of M is simulated except for the movement of the first head. That is, M' drops the pebbles and moves all but the first head as M does. In addition, the pebbles $f(s)$ are picked up and pebbles $f(\hat{s})$ are dropped at the position of the first head. In order to memorize the incomplete simulation, M' drops the extra pebble p_1 at the position of the first head. The second step is triggered by the presence of pebble p_1. Now, M' picks up the pebbles $f(\hat{s})$ and p_1 from the position of the first head, moves the first head, and drops the pebbles $f(\hat{s})$ and p_2 to the position of the second head. The third step is triggered by the presence of pebble p_2. It completes the phase by picking up the pebbles $f(\hat{s})$ and p_2 from the position of the second head and drops the pebbles $f(\hat{s})$ at the position of the first head. Now M' is ready for the next phase.

It remains to add a transition to M' which initially, when all heads are on the left endmarker and no pebbles are on the tape, drops the pebbles $f(s_0)$ without moving the heads. Clearly, M' runs into an infinite accepting loop if M does, since the heads are not moved in the loop. Moreover, M' necessarily halts and rejects, when M halts and rejects.   □

The previous theorem cannot be improved to one-head automata.

Proposition 4. *Let $m \geq 2$. Then there is a language accepted by an m-state 1DHPA$(1,0)$ which is not accepted by any stateless 1DHPA$(1,p)$ with an arbitrary number of pebbles $p \geq 0$.*

One-head automata actually cannot utilize their pebbles. Whenever a pebble is dropped and the head moves from the tape square, the pebble is never seen again. So, every 1DHPA$(1,p)$ can be simulated by some 1DHPA$(1,0)$ and accepts a regular language.

Corollary 5. *Let $p \geq 0$ be a number of pebbles. The family of languages accepted by stateless 1DHPA$(1,p)$ is properly included in the family of regular languages.*

Now we turn to trade states for heads. Any m-state 1DHPA(h,p) can be made stateless with additional $O(h \cdot \log(m))$ heads and one additional pebble. In general, this additional pebble cannot be avoided.

Theorem 6. *Let $h \geq 1$, $p \geq 0$, and $m \geq 1$. For any m-state 1DHPA(h,p) an equivalent stateless 1DHPA$(h + h \cdot \lceil \log(m+2) \rceil + 1, p+1)$ can effectively be constructed.*

The previous theorem cannot be improved in the sense that no additional pebble is necessary. In [12] it is shown that, for any prime number m, there is a unary language accepted by some one-head $(m+1)$-state automaton that cannot be accepted by any m-state automaton having an arbitrary number of heads.

Corollary 7. *Let m be a prime number. Then there is a language accepted by an m-state 1DHPA$(1,0)$ which is not accepted by any stateless 1DHPA$(h,0)$ with an arbitrary number of heads $h \geq 1$.*

4   Head and Pebble Double Hierarchy

In this section we show an infinite strict and tight double hierarchy for stateless automata. First we turn to the head hierarchy for any number of pebbles. Let $\ell \geq 1$ be an arbitrary integer. Then we define factors $s(\ell)$ over the alphabet $\{a, b\}$ by $s(\ell) = (ba)^{\ell/2}$ if ℓ is even, and $s(\ell) = (ba)^{(\ell-1)/2}b$ if ℓ is odd. These factors are concatenated to unary words so that the products are bases for witness languages. In particular, for all $k, \ell \geq 1$, we consider languages

$$L \subseteq \{ a^i s(\ell)v \mid 1 \leq i \leq k, v \in \{a,b\}^* \},$$

where L includes at least one word of the form $a^k s(\ell)v$, and call them *languages of type* (k, ℓ). That is, a language of type (k, ℓ) is neither of type $(k-1, \ell)$ nor of type $(k+1, \ell)$. Moreover, L is not empty, and next we give evidence that there are languages of this form accepted by stateless 1DHPA$(h, 0)$ without pebbles.

Example 8. For any $h \geq 1$, language $\{\, a^{h-1}s(h)v \mid v \in \{a,b\}^* \,\}$ of type $(h-1,h)$ is accepted by the stateless 1DHPA$(h,0)$ $M = \langle \{s_0\}, \{a,b\}, h, 0, \delta, \triangleright, \triangleleft, s_0 \rangle$, where the transition function δ is specified as follows. First the heads are moved from the left endmarker one after the other in a row:

$$\delta(s_0, \emptyset, x_1, \emptyset, x_2, \emptyset, \ldots, x_h, \emptyset) = (s_0, d_1, \emptyset, d_2, \emptyset, \ldots, d_h, \emptyset),$$

where $x_1 x_2 \cdots x_h = \triangleright^{h-j} a^j$ and $d_1 d_2 \cdots d_h = 0^{h-(j+1)} 1^{(j+1)}$, for $0 \leq j \leq h-1$. When the last head has left the endmarker, the first one in the row has to read the symbol b. Subsequently the row moves to the right if and only if the prefix a^{h-1} is followed by $s(h-1)$:

$$\delta(s_0, \emptyset, x_1, \emptyset, x_2, \emptyset, \ldots, x_h, \emptyset) = (s_0, 1, \emptyset, 1, \emptyset, \ldots, 1, \emptyset),$$

where $x_1 x_2 \cdots x_h = a^{h-j} s(j)$, for $1 \leq j \leq h-1$. Finally, when the row reads $s(h)$ completely the input is accepted:

$$\delta(s_0, \emptyset, x_1, \emptyset, x_2, \emptyset, \ldots, x_h, \emptyset) = (s_0, 0, \emptyset, 0, \emptyset, \ldots, 0, \emptyset),$$

where $x_1 x_2 \cdots x_h = s(h)$.   □

Theorem 9. *Let $h \geq 1$ and $p \geq 0$. Then there is a language accepted by stateless 1DHPA$(h+1,p)$ but not by any 1DHPA(h,p).*

Proof. For any $h \geq 1$, $p \geq 0$ there are only finitely many stateless 1DHPA(h,p) and, thus, only finitely many stateless 1DHPA(h,p) accepting languages of some type (k,h) (note that here the second component of the type is fixed to h). From these we choose one automaton $M = \langle \{s_0\}, \{a,b\}, h, p, \delta, \triangleright, \triangleleft, s_0 \rangle$ accepting a language whose first component of the type is maximal, say k_{max}. So, $L(M)$ is of type (k_{max}, h). Automaton M needs not to be unique, but it exists. Moreover, no language of type $(k_{max} + i, h)$, for $i \geq 1$, is accepted by any 1DHPA(h,p).

Next, a stateless 1DHPA$(h+1,p)$ $M' = \langle \{s_0\}, \{a,b\}, h+1, p, \delta', \triangleright, \triangleleft, s_0 \rangle$ is constructed from M that accepts a language of type $(k_{max}+1, h)$. In order to construct δ' we modify δ as follows. Let $w \in L(M)$ be a word with prefix $a^{k_{max}}$. First, all transitions not occurring in the accepting computation on w are undefined. In this way the order in which heads leave the left endmarker is made unique. Moreover, the remaining transitions in which a head is moved from the left endmarker are unique with respect to the heads.

The idea is that, basically, M' simulates M on input w. The difference is that all transitions of M moving one or more heads from the left endmarker to the right are simulated by M' in two steps. First, M' simulates the original transition of M. Second, M' moves all heads just moved from the left endmarker to the right once more, whereby all the other heads are kept stationary. In order to distinguish between these two steps the additional head is used. The effect of the construction is that whenever a head leaves the left endmarker it is moved twice, and there are never pebbles on the leftmost a. So, M' accepts a subset of $\{\, au \mid u \in L(M) \,\}$ including aw. Therefore, $L(M')$ is of type $(k_{max}+1, h)$.

More precisely, δ' is constructed as follows. Let head $h + 1$ be the additional head and the remaining h heads be numbered in the order they leave the left endmarker. If more than one head leave at the same time, their order is arbitrary but fixed.

Initially, head $h + 1$ is moved to the right until it reaches the first symbol b:

$$\delta'(s_0, P, \triangleright, \emptyset, \dots, \triangleright, \emptyset, \triangleright, \emptyset) = (s_0, 0, \emptyset, \dots, 0, \emptyset, 1, \emptyset)$$
$$\delta'(s_0, P, \triangleright, \emptyset, \dots, \triangleright, \emptyset, a, \emptyset) = (s_0, 0, \emptyset, \dots, 0, \emptyset, 1, \emptyset).$$

Now, δ is simulated by δ' until head 1 is moved from the left endmarker:

$$\delta'(s_0, P_0, \triangleright, P_1, \dots, \triangleright, P_h, b, \emptyset) = (s_0, d_1, Q_1, d_2, Q_2, \dots, d_h, Q_h, 0, \emptyset) \text{ if}$$
$$\delta(s_0, P_0, \triangleright, P_1, \dots, \triangleright, P_h) = (s_0, d_1, Q_1, d_2, Q_2, \dots, d_h, Q_h).$$

When this happens the situation is unique. That is, head $h + 1$ reads b, head 1 reads an a, and the other heads read $\triangleright$. So, head 1 can be moved once more while at the same time head $h + 1$ is moved from the b to the adjacent a. In this way the two steps of head 1 are distinguished:

$$\delta'(s_0, P_0, a, \emptyset, \triangleright, P_2, \dots, \triangleright, P_h, b, \emptyset) = (s_0, \overset{*}{1}, \emptyset, 0, P_2, \dots, 0, P_h, 1, \emptyset).$$

Subsequently, the situation is again unique, head $h + 1$ reads an a, head 1 reads some symbol $x \in \{a, b, \triangleleft\}$, and the other heads read $\triangleright$. Next, δ is simulated by δ' until head 2 is moved from the left endmarker:

$$\delta'(s_0, P_0, x, P_1, \triangleright, P_2, \dots, \triangleright, P_h, a, \emptyset) = (s_0, d_1, Q_1, d_2, Q_2, \dots, d_h, Q_h, 0, \emptyset) \text{ if}$$
$$\delta(s_0, P_0, x, P_1, \triangleright, P_2, \dots, \triangleright, P_h) = (s_0, d_1, Q_1, d_2, Q_2, \dots, d_h, Q_h).$$

When this happens the situation is again unique. That is, head $h + 1$ reads an a, head 1 reads x, head 2 reads a, and the other heads read $\triangleright$. So, head 2 can be moved once more while at the same time head $h + 1$ is moved from the a to the adjacent b:

$$\delta'(s_0, P_0, x, P_1, a, \emptyset, \triangleright, P_3, \dots, \triangleright, P_h, a, \emptyset) = (s_0, 0, P_1, 1, \emptyset, 0, P_3, \dots, 0, P_h, 1, \emptyset).$$

Subsequently, the situation is again unique. and the simulation continues analogously for the remaining heads. Together with the right endmarker, $s(h)$ provides sufficient symbols to handle all heads by head $h + 1$. Moreover, the construction is straightforwardly generalized to transitions which move more than one head from the left endmarker. □

Now we turn to the pebble hierarchy for any number of heads.

Theorem 10. *Let $h \geq 1$ and $p \geq 0$. Then there is a finite unary language accepted by stateless 1DHPA$(h, p + 1)$ but not by any 1DHPA(h, p).*

5 Heads versus Pebbles

This section is devoted to the comparison of the power of the two resources heads and pebbles for stateless 1DHPA(h, p). The basic questions are whether additional pebbles can compensate for heads or vice versa.

A first answer is already known by Example 2 and a result in [12]. Example 2 gives a stateless one-way two-head finite automaton with one pebble that accepts a non-semilinear unary language. In [12] it is shown that, for any number $h \geq 1$, every unary language accepted by a stateless one-way h-head finite automaton without pebbles is either finite or cofinite and, thus, semilinear. So, in this sense even an arbitrary number of heads cannot compensate for a pebble in general.

Next we turn to answer the converse question, that is, we show that, in general, a head cannot be traded for an arbitrary number of pebbles and states.

Theorem 11. *Let $h \geq 2$. Then there is a $p \geq 0$ and a stateless 1DHPA(h, p) M, so that $L(M)$ is not accepted by any m-state 1DHPA($h - 1, p'$), for any $m \geq 1$ and any $p' \geq 0$.*

Proof (Sketch). As witness language $L^{(h)}$ we will essentially use the language L_b introduced in [24] to establish a proper head hierarchy for classical one-way multi-head DFA. The languages L_b have the form

$$L_b = \{\, w_1 \# w_2 \# \cdots \# w_b \$ w_b \# \cdots \# w_2 \# w_1 \mid w_i \in \{0,1\}^* \,\},$$

and it is shown that, for any $h \geq 1$, language $L_{\binom{h}{2}+1}$ is accepted by some 1DHPA($h + 1, 0$), but not by any 1DHPA($h, 0$). An alternative proof of this result based on Kolmogorov complexity and an incompressibility argument is given in [16], where also general information on Kolmogorov complexity and the incompressibility method can be found.

Here we consider stateless automata with pebbles and set $L^{(h)} = L_b$, where $b = \binom{h-1}{2} + 1$, for $h \geq 2$. Since $L^{(h)}$ is accepted by a 1DHPA($h, 0$), by applying Theorem 3 we obtain that $L^{(h)}$ is accepted by a stateless 1DHPA(h, p) with a certain number of pebbles $p \geq 3$ as well. Next, we turn to show that $L^{(h)}$ cannot be accepted by any 1DHPA($h - 1, p'$) with an arbitrary number of states and an arbitrary number of pebbles $p' \geq 0$.

Let $w \in \{0,1\}^+$ be an arbitrary binary string. The Kolmogorov complexity $C(w)$ of w is defined to be the minimal size of a program describing w. It is well known that there are binary strings w of arbitrary lengths so that $|w| \leq C(w)$.

Let us assume by way of contradiction that language $L^{(h)}$ is accepted by a 1DHPA($h - 1, p'$) M with $p' \geq 0$. We choose some string $w = w_1 w_2 \cdots w_b$, where all subwords $w_1, w_2, \ldots, w_b$ have the same length and whose Kolmogorov complexity meets $C(w) \geq |w|$. So, the input $v = w_1 \# \cdots \# w_b \$ w_b \# \cdots \# w_1$ is accepted by M. Automaton M is said to *check* w_i, if on input v there is a time step at which M has a head on the left copy of w_i and a head on the right copy of

w_i simultaneously. The main ingredient for the proof is the following so-called Matching Lemma, which is shown in [16] for classical one-way multi-head DFA. The lemma says that M must check w_i, for every $1 \leq i \leq b$. Clearly, since M is a one-way device, every pair of heads can check one w_i only. Therefore, there exists some w_i that is not checked if $b > \binom{h-1}{2}$. This contradicts the Matching Lemma and proves that $L^{(h)}$ is not accepted by M. □

Lemma 12 (Matching Lemma). *Let $b \geq 1$, $h \geq 1$, and $p \geq 0$. Then any 1DHPA(h,p) $M = \langle S, \{0, 1, \#, \$\}, h, p, \delta, \triangleright, \triangleleft, s_0 \rangle$ accepting language L_b must check w_i, for every $1 \leq i \leq b$.*

6 Undecidability Results

In this section, we investigate decidability problems for stateless 1DHPA(h, p), where the emptiness problem plays a crucial role. It is known that the emptiness problem is undecidable for stateless two-way DFA with at least two heads [9], and for stateless one-way DFA with at least three heads [23]. Recently, it has been shown that the emptiness problem is also undecidable for one-way DFA having two heads and at least four states [12]. Here we will show that emptiness is undecidable for stateless 1DHPA$(2, 1)$.

To prove this result we use the technique of valid computations which is basically described, for example, in [7]. This technique has also been used in [12], where it is shown that valid computations of deterministic linearly space bounded one-tape, one-head Turing machines can be accepted by one-way DFA with two heads and four states. Currently it is not clear in which way a stateless 1DHPA$(2, 1)$ could accept such sets. So, here we consider the valid computations (VALC) of *one-way cellular automata* which is a parallel computational model (see, for example, [10,11]). The decidability questions for stateless 1DHPA$(2, 1)$ are reduced to those of one-way cellular automata, which have been shown to be undecidable in [17].

Lemma 13. *Let M be an OCA. Then a stateless 1DHPA$(2, 1)$ accepting the language VALC(M) can effectively be constructed.*

Theorem 14. *For $h \geq 2$ and $p \geq 1$, emptiness is undecidable for 1DHPA(h, p).*

Proof. Let M be an OCA. According to Lemma 13 we can effectively construct a 1DHPA$(2, 1)$ M' accepting VALC(M). Clearly, $L(M') = $ VALC(M) is empty if and only if $L(M)$ is empty. Since it shown in [17] that emptiness is undecidable for OCA, the theorem follows. □

Theorem 15. *For $h \geq 2$ and $p \geq 1$, finiteness, infiniteness, inclusion, equivalence, regularity, and context-freeness are undecidable for 1DHPA(h, p).*

References

1. Brandenburg, F.J.: Three write heads are as good as k. Math. Systems Theory 14, 1–12 (1981)
2. Chrobak, M.: Hierarchies of one-way multihead automata languages. Theoret. Comput. Sci. 48, 153–181 (1986)
3. Eğecioğlu, Ö., Ibarra, O.H.: On stateless multicounter machines. In: Ambos-Spies, K., Löwe, B., Merkle, W. (eds.) CiE 2009. LNCS, vol. 5635, pp. 178–187. Springer, Heidelberg (2009)
4. Frisco, P., Ibarra, O.H.: On stateless multihead finite automata and multihead pushdown automata. In: Diekert, V., Nowotka, D. (eds.) DLT 2009. LNCS, vol. 5583, pp. 240–251. Springer, Heidelberg (2009)
5. Gorlick, M.M.: Computation by one-way multihead marker automata. Master's thesis, Department of Computer Science, University of British Columbia (1978)
6. Holzer, M., Kutrib, M., Malcher, A.: Complexity of multi-head finite automata: Origins and directions. Theoret. Comput. Sci. 412, 83–96 (2011)
7. Hopcroft, J.E., Ullman, J.D.: Introduction to Automata Theory, Languages, and Computation. Addison-Wesley (1979)
8. Ibarra, O.H., Eğecioğlu, Ö.: Hierarchies and characterizations of stateless multicounter machines. In: Ngo, H.Q. (ed.) COCOON 2009. LNCS, vol. 5609, pp. 408–417. Springer, Heidelberg (2009)
9. Ibarra, O.H., Karhumäki, J., Okhotin, A.: On stateless multihead automata: Hierarchies and the emptiness problem. Theoret. Comput. Sci. 411, 581–593 (2010)
10. Kutrib, M.: Cellular automata – a computational point of view. In: Bel-Enguix, G., Jiménez-López, M.D., Martín-Vide, C. (eds.) New Developments in Formal Languages and Applications. SCI, vol. 113, pp. 183–227. Springer, Heidelberg (2008)
11. Kutrib, M.: Cellular automata and language theory. In: Encyclopedia of Complexity and System Science, pp. 800–823. Springer (2009)
12. Kutrib, M., Malcher, A., Wendlandt, M.: States and heads do count for unary multi-head finite automata. In: Yen, H.-C., Ibarra, O.H. (eds.) DLT 2012. LNCS, vol. 7410, pp. 214–225. Springer, Heidelberg (2012)
13. Kutrib, M., Messerschmidt, H., Otto, F.: On stateless deterministic restarting automata. Acta Inform. 47, 391–412 (2010)
14. Kutrib, M., Messerschmidt, H., Otto, F.: On stateless two-pushdown automata and restarting automata. Int. J. Found. Comput. Sci. 21, 781–798 (2010)
15. Kutyłowski, M.: One-way multihead finite automata and 2-bounded languages. Math. Systems Theory 23, 107–139 (1990)
16. Li, M., Vitányi, P.M.B.: An Introduction to Kolmogorov Complexity and Its Applications. Springer (1993)
17. Malcher, A.: Descriptional complexity of cellular automata and decidability questions. J. Autom., Lang. Comb. 7, 549–560 (2002)
18. Petersen, H.: Automata with sensing heads. In: Proceedings of the Third Israel Symposium on the Theory of Computing and Systems, pp. 150–157. IEEE Computer Society Press (1995)
19. Rabin, M.O., Scott, D.: Finite automata and their decision problems. IBM J. Res. Dev. 3, 114–125 (1959)
20. Ritchie, R.W., Springsteel, F.N.: Language recognition by marking automata. Inform. Control 20, 313–330 (1972)
21. Rosenberg, A.L.: On multi-head finite automata. IBM J. Res. Dev. 10, 388–394 (1966)
22. Sudborough, I.H.: One-way multihead writing finite automata. Inform. Control 30, 1–20 (1976)
23. Yang, L., Dang, Z., Ibarra, O.H.: On stateless automata and P systems. Int. J. Found. Comput. Sci. 19, 1259–1276 (2008)
24. Yao, A.C., Rivest, R.L.: $k+1$ heads are better than k. J. ACM 25, 337–340 (1978)

Regular Expressions with Binding over Data Words for Querying Graph Databases

Leonid Libkin[1], Tony Tan[2], and Domagoj Vrgoč[1]

[1] University of Edinburgh
libkin@inf.ed.ac.uk, domagoj.vrgoc@ed.ac.uk
[2] Hasselt University and Transnational University of Limburg
tony.tan@uhasselt.be

Abstract. Data words assign to each position a letter from a finite alphabet and a data value from an infinite set. Introduced as an abstraction of paths in XML documents, they recently found applications in querying graph databases as well. Those are actively studied due to applications in such diverse areas as social networks, semantic web, and biological databases. Querying formalisms for graph databases are based on specifying paths conforming to some regular conditions, which led to a study of regular expressions for data words.

Previously studied regular expressions for data words were either rather limited, or had the full expressiveness of register automata, at the expense of a quite unnatural and unintuitive binding mechanism for data values. Our goal is to introduce a natural extension of regular expressions with proper bindings for data values, similar to the notion of freeze quantifiers used in connection with temporal logics over data words, and to study both language-theoretic properties of the resulting class of languages of data words, and their applications in querying graph databases.

1 Introduction

Data words, unlike the usual words over finite alphabet, assign to each position both a letter from a finite alphabet and an element of an infinite set, referred to as a data value. An example of a data word is $\binom{a}{1}\binom{b}{2}\binom{a}{3}\binom{b}{1}$. This is a data word over the finite alphabet $\{a, b\}$, with data elements coming from an infinite domain, in this case, $\mathbb{N}$. Investigations of data words picked up recently due to their importance in the study of XML documents. Those are naturally modeled as ordered unranked trees in which every node has both a label and a datum (these are referred to as data trees). Data words then model paths in data trees, and as such are essential for investigations of many path-based formalisms for XML, for instance, its navigational query language XPath. We refer the reader to [7, 14, 30, 31] for recent surveys.

While the XML data format dominated the data management landscape for a while, primarily in the 2000s, over the past few years the focus started shifting towards the *graph* data model. Graph-structured data appears naturally in a

M.-P. Béal and O. Carton (Eds.): DLT 2013, LNCS 7907, pp. 325–337, 2013.
© Springer-Verlag Berlin Heidelberg 2013

variety of applications, most notably social networks and the Semantic Web (as it underlies the RDF format). Its other applications include biology, network traffic, crime detection, and modeling object-oriented data [13, 21, 24, 26–29]. Such databases are represented as graphs in which nodes are objects and the edge labels specify relationships between them; see [1, 4] for surveys.

Just as in the case of XML, a crucial building block in queries against graph data deals with properties of paths in them. The most basic formalism is that of *regular path queries*, or *RPQs*, which select nodes connected by a path described by a regular language over the labeling alphabet [11]. There are multiple extensions with more complex patterns, backward navigation, regular relations over paths, and non-regular features [3, 5, 6, 8–10]. In real applications we deal with both navigational information and data, so it is essential that we look at properties of paths that also describe how data values change along them. Since such paths (as we shall explain later) are just data words, it becomes necessary to provide expressive and well-behaved mechanisms for describing languages of data words.

One of the most commonly used formalisms for describing the notion of regularity for data words is that of *register automata* [19]. These extend the standard NFAs with registers that can store data values; transitions can compare the currently read data value with values stored in registers.

However, register automata are not convenient for specifying properties – ideally, we want to use regular expressions to define languages. These have been looked at in the context of data words (or words over infinite alphabets), and are based on the idea of using *variables* for binding data values. An initial attempt to define such expressions was made in [20], but it was very limited. Another formalism, called *regular expressions with memory*, was shown to be equivalent to register automata [22, 23]. At the first glance, they appear to be a good formalism: these are expressions like $a \downarrow_x (a[x^=])^*$ saying: read letter a, bind data value to x, and read the rest of the data word checking that all letters are a and the data values are the same as x. This will define data words $\binom{a}{d} \cdots \binom{a}{d}$ for some data value d. This is reminiscent of freeze quantifiers used in connection with the study of data word languages [12].

The serious problem with these expressions, however, is the *binding* of variables. The expression above is fine, but now consider the following expression: $a \downarrow_x (a[x^=]a \downarrow_x)^* a[x^=]$. This expression re-binds variable x inside the scope of another binding, and then crucially, when this happens, the original binding of x is *lost*! Such expressions really mimic the behavior of register automata, which makes them more procedural than declarative. (The above expression defines data words of the form $\binom{a}{d_1}\binom{a}{d_1} \cdots \binom{a}{d_n}\binom{a}{d_n}$.)

Losing the original binding of a variable when reusing it inside its scope goes completely against the usual practice of writing logical expressions, programs, etc., that have bound variables. Nevertheless, this feature was essential for capturing register automata [22]. So natural questions arise:

- Can we define regular expressions for data words that use the acceptable scope/binding policies for variables? Such expressions will be more declarative than procedural, and more appropriate for being used in queries.
- Do these fall short of the full power of register automata?
- What are their basic properties, and what is the complexity of querying graph data with such expressions?

Contributions. Our main contribution is to define a new formalism of *regular expressions with binding*, or REWBs, to study its properties, and to show how it can be used in the context of graph querying. The binding mechanism of REWBs follows the standard scoping rules, and is essentially the same as in LTL extensions with freeze quantifiers [12]. We also look at some subclasses of REWBs based on the types of conditions one can use: in *simple* REWBs, each condition involves at most one variable (all those shown above were such), and in *positive* REWBs, negation and inequality cannot be used in conditions.

We show that the class of languages defined by REWBs is strictly contained in the class of languages defined by register automata. The separating example is rather intricate, and indeed it appears that for most reasonable languages one can think of, if they are definable by register automata, they would be definable by REWBs as well. At the same time, REWBs lower the complexity of some key computational tasks related to languages of data words. For instance, nonemptiness is PSPACE-complete for register automata [12], but we show that it is NP-complete for REWBs (and trivializes for simple and positive REWBs).

We consider the containment and universality problems for REWBs. In general they are undecidable, even for simple REWBs. However, the problem becomes decidable for positive REWBs.

We look at applications of REWBs in querying graph databases. The problem of query evaluation is essentially checking whether the intersection of two languages of data words is nonempty. We use this to show that the complexity of query evaluation is PSPACE-complete (note that it is higher than the complexity of nonemptiness alone); for a fixed REWB, the complexity is tractable.

At the end we also sketch some results concerning a model of data word automaton that uses variables introduced in [16]. We also comment on how these can be combined with register automata to obtain a language subsuming all the previously used ones while still retaining good query evaluation bounds.

Organization. We define data words and data graphs in Section 2. In Section 3 we introduce our notion of regular expression with binding (REWB) and study their nonemptiness and universality problems in Section 4 and Section 5, respectively. In Section 6 we study REWBs as a graph database query language and in Section 7 we consider some possible extensions that could be useful in graph querying. Due to space limitations, complete proofs of all the results are in the appendix.

2 Data Words and Data Graphs

Let Σ be a finite alphabet and $\mathcal{D}$ a countable infinite set of data values. A **data word** is simply a finite string over the alphabet $\Sigma \times \mathcal{D}$. That is, in each position a data word carries a letter from Σ and a data value from $\mathcal{D}$. We will denote data words by $\binom{a_1}{d_1} \ldots \binom{a_n}{d_n}$, where $a_i \in \Sigma$ and $d_i \in \mathcal{D}$.

A **data graph** (over Σ) is pair $G = (V, E)$, where

- V is a finite set of nodes;
- $E \subseteq V \times \Sigma \times \mathcal{D} \times V$ is a set of edges where each edge contains a label from Σ and a data value from $\mathcal{D}$.

We write $V(G)$ and $E(G)$ to denote the set of nodes and edges of G, respectively. An edge e from a node u to a node u' is written in the form $(u, \binom{a}{d}, u')$, where $a \in \Sigma$ and $d \in \mathcal{D}$. We call a the label of the edge e and d the data value of the edge e. We write $\mathcal{D}(G)$ to denote the set of data values in G.

The following is an example of a data graph, with nodes $u_1, \ldots, u_6$ and edges $(u_1, \binom{a}{3}, u_2)$, $(u_3, \binom{b}{1}, u_2)$, $(u_2, \binom{a}{3}, u_5)$, $(u_6, \binom{a}{5}, u_4)$, $(u_2, \binom{a}{1}, u_4)$, $(u_4, \binom{a}{4}, u_3)$ and $(u_5, \binom{c}{7}, u_6)$.

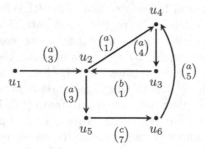

A path from a node v to a node v' in G is a sequence

$$\pi = v_1 \binom{a_1}{d_1} v_2 \binom{a_2}{d_2} v_3 \binom{a_3}{d_3} \cdots v_n \binom{a_n}{d_n} v_{n+1}$$

such that each $(v_i, \binom{a_i}{d_i}, v_{i+1})$ is an edge for each $i \le n$, and $v_1 = v$ and $v_{n+1} = v'$.

A path π defines a data word $w(\pi) = \binom{a_1}{d_1}\binom{a_2}{d_2}\binom{a_3}{d_3} \cdots \binom{a_n}{d_n}$.

Remark. Note that we have chosen a model in which labels and data values appear in edges. Of course other variations are possible, for instance labels appearing in edges and data values in nodes. All of these easily simulate each other, very much in the same way as one can use either labeled transitions systems or Kripke structures as models of temporal or modal logic formulae. In fact both models – with labels in edges and labels in nodes – have been considered in the context of semistructured data and, at least from the point of view of their expressiveness, they are viewed as equivalent. Our choice is dictated by the ease of notation primarily, as it identifies paths with data words.

3   Regular Expressions with Binding

We now define regular expressions with binding for data words. As explained already, expressions with variables for data words were previously defined in [23] but those were really designed to mimic the transitions of register automata, and had very procedural, rather than declarative flavor. Here we define them using proper scoping rules.

Variables will store data values; those will be compared with other variables using conditions. To define them, assume that, for each $k > 0$, we have variables $x_1, \ldots, x_k$. Then the set of conditions C_k is given by the grammar:

$$c := \top \mid \bot \mid x_i^= \mid x_i^{\neq} \mid c \wedge c \mid c \vee c \mid \neg c, \quad 1 \le i \le k.$$

The satisfaction of a condition is defined with respect to a data value $d \in \mathcal{D}$ and a (partial) valuation $\nu : \{x_1, \ldots, x_k\} \to \mathcal{D}$ of variables as follows:

- $d, \nu \models \top$ and $d, \nu \not\models \bot$;
- $d, \nu \models x_i^=$ iff $d = \nu(x_i)$;
- $d, \nu \models x_i^{\neq}$ iff $d \neq \nu(x_i)$;
- the semantics for Boolean connectives $\vee, \wedge$, and $\neg$ is standard.

Next we define regular expressions with binding.

Definition 1. *Let Σ be a finite alphabet and $\{x_1, \ldots, x_k\}$ a finite set of variables. Regular expressions with binding (REWB) over $\Sigma[x_1, \ldots, x_k]$ are defined inductively as follows:*

$$r := \varepsilon \mid a \mid a[c] \mid r + r \mid r \cdot r \mid r^* \mid a \downarrow_{x_i} (r) \qquad (1)$$

where $a \in \Sigma$ and c is a condition in C_k.

A variable x_i is bound if it occurs in the scope of some $\downarrow_{x_i}$ operator and free otherwise. More precisely, free variables of an expression are defined inductively: ε and a have no free variables, in $a[c]$ all variables occurring in c are free, in $r_1 + r_2$ and $r_1 \cdot r_2$ the free variables are those of r_1 and r_2, the free variables of r^* are those of r, and the free variables of $a \downarrow_{x_i} (r)$ are those of r except x_i. We will write $r(x_1, \ldots, x_l)$ if $x_1, \ldots, x_l$ are the free variables in r.

A valuation on the variables $x_1, \ldots, x_k$ is a partial function $\nu : \{x_1, \ldots, x_k\} \mapsto \mathcal{D}$. We denote by $\mathcal{F}(x_1, \ldots, x_k)$ the set of all valuations on $x_1, \ldots, x_k$. For a valuation ν, we write $\nu[x_i \leftarrow d]$ to denote the valuation ν' obtained by fixing $\nu'(x_i) = d$ and $\nu'(x) = \nu(x)$ for all other $x \neq x_i$. Likewise, we write $\nu[\bar{x} \leftarrow \bar{d}]$ for a simultaneous substitution of values from $\bar{d} = (d_1, \ldots, d_l)$ for variables $\bar{x} = (x_1, \ldots, x_l)$. Also notation $\nu(\bar{x}) = \bar{d}$ means that $\nu(x_i) = d_i$ for all $i \le l$.

Semantics. Let $r(\bar{x})$ be an REWB over $\Sigma[x_1, \ldots, x_k]$. A valuation $\nu \in \mathcal{F}(x_1, \ldots, x_k)$ is compatible with r, if $\nu(\bar{x})$ is defined.

A regular expression $r(\bar{x})$ over $\Sigma[x_1, \ldots, x_k]$ and a valuation $\nu \in \mathcal{F}(x_1, \ldots, x_k)$ compatible with r define a language $L(r, \nu)$ of data words as follows.

- If $r = a$ and $a \in \Sigma$, then $L(r, \nu) = \{\binom{a}{d} \mid d \in \mathbb{N}\}$.
- If $r = a[c]$, then $L(r, \nu) = \{\binom{a}{d} \mid d, \nu \models c\}$.
- If $r = r_1 + r_2$, then $L(r, \nu) = L(r_1, \nu) \cup L(r_2, \nu)$.

- If $r = r_1 \cdot r_2$, then $L(r, \nu) = L(r_1, \nu) \cdot L(r_2, \nu)$.
- If $r = r_1^*$, then $L(r, \nu) = L(r_1, \nu)^*$.
- If $r = a \downarrow_{x_i} (r_1)$, then $L(r, \nu) = \bigcup_{d \in \mathcal{D}} \left\{ \binom{a}{d} \right\} \cdot L(r_1, \nu[x_i \leftarrow d])$.

A REWB r defines a language of data words as follows.

$$L(r) = \bigcup_{\nu \text{ compatible with } r} L(r, \nu).$$

In particular, if r is without free variables, then $L(r) = L(r, \emptyset)$. We will call such REWBs *closed*.

Register Automata and Expressions with Memory. As mentioned earlier, *register automata* extend NFAs with the ability to store and compare data values. Formally, an automaton with k registers is $\mathcal{A} = (Q, q_0, F, T)$, where:

- Q is a finite set of states;
- $q_0 \in Q$ is the initial state;
- $F \subseteq Q$ is the set of final states;
- T is a finite set of transitions of the form $(q, a, c) \to (I, q')$, where q, q' are states, a is a label, $I \subseteq \{1, \ldots, k\}$, and c is a condition in $\mathcal{C}_k$.

Intuitively the automaton traverses a data word from left to right, starting in q_0, with all registers empty. If it reads $\binom{a}{d}$ in state q with register configuration $\tau : \{1, \ldots, k\} \to \mathcal{D}$, it may apply a transition $(q, a, c) \to (I, q')$ if $d, \tau \models c$; it then enters state q' and changes contents of registers i, with $i \in I$, to d. For more details on register automata we refer reader to [19, 23].

Expressions introduced in [22] had a similar syntax but rather different semantics. They were built using $a \downarrow_x$, concatenation, union and Kleene star. That is, no binding was introduced with $a \downarrow_x$; rather it directly matched the operation of putting a value in a register. In contrast, we use proper bindings of variables; expression $a \downarrow_x$ appears only in the context $a \downarrow_x (r)$ where it binds x inside the expression r only. This corresponds to the standard binding policies in logic, or in programs.

Example 1. We list several examples of languages expressible with our expressions. In all cases below we have a singleton alphabet $\Sigma = \{a\}$.

- The language that consists of data words where the data value in the first position is different from the others is given by: $a \downarrow_x ((a[x^{\neq}])^*)$.
- The language that consists of data words where the data values in the first and the last position are the same is given by: $a \downarrow_x (a^* \cdot a[x^{=}])$.
- The language that consists of data words where there are two positions with the same data value: $a^* \cdot a \downarrow_x (a^* \cdot a[x^{=}]) \cdot a^*$.

Note that in REWBs in the above example the conditions are very simple: they are either $x^=$ or $x^{\neq}$. We will call such expressions *simple* REWBs.

We shall also consider *positive* REWBs where negation and inequality are disallowed in conditions. That is, all the conditions c are constructed using the following syntax: $c := \top \mid x_i^= \mid c \wedge c \mid c \vee c$, where $1 \leq i \leq k$.

We finish this section by showing that REWBs are strictly weaker than register automata (i.e., proper binding of variables has a cost – albeit small – in terms of expressiveness).

Theorem 1. *The class of languages defined by REWBs is strictly contained in the class of languages accepted by register automata.*

That the class of languages defined by REWBs is contained in the class of languages defined by register automata can be proved by using a similar inductive construction as in [22, Proposition 5.3]. The idea behind the construction of the separating example follows the intuition that defining scope of variables restricts the power of the language, compared to register automata where once stored, the value remains in the register until rewritten. As the proof is rather technical and lengthy, we present it in the appendix.

We note that the separating example is rather intricate, and certainly not a natural language one would think of. In fact, all natural languages definable with register automata that we used here as examples – and many more, especially those suitable for graph querying – are definable by REWBs.

4   The Nonemptiness Problem

We now look at the standard language-theoretic problem of nonemptiness:

NONEMPTINESS FOR REWBS	
Input:	A REWB r over $\Sigma[x_1, \ldots, x_k]$.
Task:	Decide whether $L(r) \neq \emptyset$.

More generally, one can ask if $L(r, \nu) \neq \emptyset$ for a REWB r and a compatible valuation ν.

Recall that for register automata, the nonemptiness problem is PSPACE-complete [12] (and the same bound applied to regular expressions with memory [23]). Introducing proper binding, we lose little expressiveness and yet can lower the complexity.

Theorem 2. *The nonemptiness problem for REWBs is* NP-*complete.*

The proof is in the appendix. Note that for simple and positive REWBs the problem trivializes.

Proposition 1.   – *For every simple REWB r over $\Sigma[x_1, \ldots, x_k]$, and for every valuation ν compatible with r, we have $L(r, \nu) \neq \emptyset$.*
  – *For every positive REWB r over $\Sigma[x_1, \ldots, x_k]$, there is a valuation ν such that $L(r, \nu) \neq \emptyset$.*

5   Containment and Universality

We now turn our attention to language containment. That is we are dealing with the following problem:

CONTAINMENT FOR REWBS
Input: Two REWBs r_1, r_2 over $\Sigma[x_1, \ldots, x_k]$.
Task: Decide whether $L(r_1) \subseteq L(r_2)$.

When r_2 is a fixed expression denoting all data words, this is the universality problem. We show that both are undecidable.

In fact, we show a stronger statement, that *universality* of simple REWBs that use just a single variable is already undecidable.

UNIVERSALITY FOR ONE-VARIABLE REWBS
Input: An REWB r over $\Sigma[x]$.
Task: Decide whether $L(r) = (\Sigma \times \mathcal{D})^*$.

Theorem 3. UNIVERSALITY FOR ONE-VARIABLE REWBS *is undecidable. In particular, containment for REWBs is undecidable too.*

While restriction to simple REWBs does not make the problem decidable, the restriction to positive REWBs does: as is often the case, static analysis tasks become easier without negation.

Theorem 4. *The containment problem for positive REWBs is decidable.*

Proof. It is rather straightforward to show that any positive REWB can be converted into a register automaton without inequality [20]. The decidability of the language containment follows from the fact that the containment problem for register automata without inequality is decidable [32]. □

6   REWBs as a Query Language for Data Graphs

Standard mechanisms for querying graph databases are based on *regular path queries*, or RPQs: those select nodes connected by a path belonging to a given regular language [4, 9–11]. For data graphs, we follow the same idea, but now paths are specified by REWBs, since they contain data. In this section we study the complexity of this querying formalism.

We first explain how the problem of query evaluation can be cast as a problem of checking nonemptiness of language intersection.

Note that a data graph G can be viewed as an automaton, generating data words. That is, given a data graph $G = (V, E)$, and a pair of nodes s, t, we let $\mathcal{L}(G, s, t)$ be $\{w(\pi) \mid \pi \text{ is a path from } s \text{ to } t \text{ in } G\}$; this is a set of data words.

Let $r(\bar{x})$ be a REWB over $\Sigma[x_1, \ldots, x_k]$. For ν compatible with r, we let $\mathcal{L}(G, s, t, r, \nu)$ be $\mathcal{L}(G, s, t) \cap \mathcal{L}(r, \nu)$. Then for a graph $G = (V, E)$, we define the answer to r over G as the set $\mathcal{Q}(r,G)$ of triples $(s,t,\bar{d}) \in V \times V \times \mathcal{D}^k$, such that $\mathcal{L}(G,s,t,r,\nu[\bar{x} \leftarrow \bar{d}])) \neq \emptyset$. In other words, there is a path π in G from s to t such that $w(\pi) \in L(r,\nu)$, where $\nu(\bar{x}) = \bar{d}$.

If r is a closed REWB, we do not need a valuation in the above definition. That is, $\mathcal{Q}(r,G)$ is the set of pairs of nodes (s,t) such that $\mathcal{L}(G,s,t) \cap \mathcal{L}(r) \neq \emptyset$, i.e., there is a path π in G from s to t such that $w(\pi) \in L(r)$.

In what follows we are interested in the query evaluation and query containment problems. For simplicity we will work with closed REWBs only. We start with query evaluation.

Query Evaluation for REWB	
Input:	A data graph G, two nodes $s,t \in V(G)$ and a REWB r.
Task:	Decide whether $(s,t) \in \mathcal{Q}(r,G)$.

Note that in this problem, both the data graph and the query, given by r, are inputs; this is referred to as the *combined complexity* of query evaluation. If the expression r is fixed, we are talking about *data complexity*.

Recall that for the usual graphs (without data), the combined complexity of evaluating RPQs is polynomial, but if conjunctions of RPQs are taken, it goes up to NP (and could be NP-complete, in fact [10, 11]). When we look at data graphs and specify paths with register automata, combined complexity jumps to Pspace-complete [22].

However, we have seen that REWBs are less expressive than register automata, so perhaps a lower NP bound would apply to them? One way to try to do it is to find a polynomial bound on the length of a minimal path witnessing a REWB in a data graph. The next proposition shows that this is impossible, since in some cases the shortest witnessing path will be exponentially long, even if the REWB uses only one variable.

Proposition 2. *Let $\Sigma = \{\$, \mathord{\text{¢}}, a, b\}$ be a finite alphabet. There exists a family of data graphs $\{G_n(s,t)\}_{n>1}$ with two distinguished nodes s and t, and a family of closed REWBs $\{r_n\}_{n>1}$ such that*

- *each $G_n(s,t)$ is of size $O(n)$;*
- *each r_n is a closed REWB over $\Sigma[x]$ of length $O(n)$; and*
- *every data word in $\mathcal{L}(G_n,s,t,r_n)$ is of length $\Omega(2^{\lfloor n/2 \rfloor})$.*

The proof of this is rather involved and can be found in the appendix.

Next we describe the complexity of the query evaluation problem. It turns out that it matches that for register automata.

Theorem 5.   – *The complexity of query evaluation for REWB is* Pspace-*complete.*
- *For each fixed r, the complexity of query evaluation for REWB is in* NLogspace.

In other words, the combined complexity of queries based on REWBs is Pspace-complete, and their data complexity is in NLogspace (and of course it can be NLOGSPACE-complete even for very simple expressions, e.g., Σ^*, which just expresses reachability). Note that the combined complexity is acceptable (it matches, for example, the combined complexity of standard relational query languages such as relational calculus and algebra), and that data complexity is the best possible for a language that can express the reachability problem.

We prove PSPACE membership by showing how to transform REWBs into regular expressions when only finitely many data values are considered. Since the expression in question is of length exponential in the size of the input, standard on-the-fly construction of product with the input graph (viewed as an NFA) gives us the desired bound. Details of this construction, as well as the proof of hardness, can be found in the appendix. The same proof, for a fixed r, gives us the bound for data complexity.

Note that the upper bound follows from the connection with register automata. In order to make our presentation self contained we opted to present a different proof in the appendix.

By examining the proofs of Theorem 5 and Theorem 3 we observe that lower bounds already hold for both simple and positive REWBs. That is we get the following.

Corollary 1. *The following holds for simple REWBs.*

- *Combined complexity of simple (or positive) REWB queries is* PSPACE-*complete.*
- *Data complexity of simple (or positive) REWB queries is* NLOGSPACE-*complete.*

Another important problem in querying graphs is query containment. In general, the query containment problem asks, for two REWBs r_1, r_2 over $\Sigma[x_1, \ldots, x_k]$, whether $\mathcal{Q}(r_1, G) \subseteq \mathcal{Q}(r_2, G)$ for every data graph G. For REWB-based queries we look at, this problem is easily seen to be equivalent to language containment. Using this fact and the results of Section 5 we obtain the following.

Corollary 2. *Query containment is undecidable for REWBs and simple REWBs. It becomes decidable if we restrict our queries to positive REWBs.*

7 Conclusions and Other Models

After conducting an extensive study of their language-theoretic properties and their ability to query graph data we conclude that REWBs can serve as a highly expressive language that still retains good query evaluation properties. Although weaker than register automata and their expression counterpart – regular expressions with memory, REWBs come with a more natural and declarative syntax and have a lower complexity of some language-theoretic properties such as nonemptiness. They also complete a picture of expressions that relate to register automata – a question that often came up in the discussions about the connection of regular expressions with memory (REMs) and register automata [22, 23], as they can be seen as a natural restriction of REMs with proper scoping rules.

As we have seen, both in this paper and in previous work on graph querying, all of the considered formalisms have a combined complexity of query evaluation that is either a low degree polynomial, or PSPACE-complete. A natural question to ask is if there is a formalism whose combined complexity lies between these two classes.

An answer to this can be given using a model of automata that extends NFAs in a similar way that REWBs extend regular expressions – by allowing usage of variables. These automata, called *variable automata*, were introduced in [16] and although originally defined for words over an infinite alphabet, they can easily be modified to handle data words. Intuitively, they can be viewed as NFAs with a guess of data values to be assigned to variables, with the run of the automaton verifying correctness of the guess. An example of a variable automaton recognizing the language of all words where the last data value is different from all others is given in the following image.

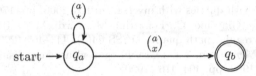

Here we observe that variable automata use two sorts of variables – an ordinary bound variable x that is assigned a unique value, and a special free variable $\star$, whose every occurrence is assigned a value different from the ones assigned to the bound variables.

It can be show that variable automata, used as a graph querying formalism, have NP-complete combined complexity of query evaluation and that their deterministic subclass [16] has coNP query containment. Due to space limitations we defer the technical details of these results to the appendix.

The somewhat synthetic nature of variable automata and their usage of the free variable makes them incomparable with REWBs and register automata, as the example above demonstrates. A natural question then is whether there is a model that encompasses both and still retains the same good query evaluation bounds. It can be shown that by allowing variable automata to use the full power of registers we get a model that subsumes all of the previously studied models and whose combined complexity is no worse that the one of register automata. This approach, albeit in a limited form, was already proposed in e.g. [15]. The details of the construction can be found in the appendix.

Acknowledgement. The second author acknowledges the generous financial support of FWO, under the scheme FWO Pegasus Marie Curie fellowship. The first and the third author were partially supported by the EPSRC grants J015377 and G049165.

References

1. Abiteboul, S., Buneman, P., Suciu, D.: Data on the Web: From Relations to Semistructured Data and XML. Morgan Kauffman (1999)
2. Abiteboul, S., Hull, R., Vianu, V.: Foundations of Databases. Addison-Wesley (1995)
3. Abiteboul, S., Vianu, V.: Regular path queries with constraints. JCSS 58, 428–452 (1999)
4. Angles, R., Gutiérrez, C.: Survey of graph database models. ACM Comput. Surv. 40(1) (2008)
5. Barceló, P., Figueira, D., Libkin, L.: Graph logics with rational relations and the generalized intersection problem. In: LICS (2012)
6. Barceló, P., Libkin, L., Lin, A.W., Wood, P.: Expressive languages for path queries over graph-structured data. ACM TODS 37(4) (2012)
7. Bojanczyk, M.: Automata for Data Words and Data Trees. In: RTA, pp. 1–4 (2010)
8. Calvanese, D., de Giacomo, G., Lenzerini, M., Vardi, M.Y.: Containment of conjunctive regular path queries with inverse. In: KR 2000, pp. 176–185 (2000)
9. Calvanese, D., de Giacomo, G., Lenzerini, M., Vardi, M.Y.: Rewriting of regular expressions and regular path queries. JCSS 64(3), 443–465 (2002)
10. Consens, M.P., Mendelzon, A.O.: GraphLog: a visual formalism for real life recursion. In: PODS 1990, pp. 404–416 (1990)
11. Cruz, I., Mendelzon, A., Wood, P.: A graphical query language supporting recursion. In: SIGMOD 1987, pp. 323–330 (1987)
12. Demri, S., Lazić, R.: LTL with the freeze quantifier and register automata. ACM TOCL 10(3) (2009)
13. Fan, W.: Graph pattern matching revised for social network analysis. In: ICDT 2012, pp. 8–21 (2012)
14. Figueira, D.: Reasoning on words and trees with data. PhD thesis (2010)
15. Figueira, D.: Alternating register automata on finite words and trees. Logical Methods in Computer Science 8(1) (2012)
16. Grumberg, O., Kupferman, O., Sheinvald, S.: Variable automata over infinite alphabets. In: Dediu, A.-H., Fernau, H., Martín-Vide, C. (eds.) LATA 2010. LNCS, vol. 6031, pp. 561–572. Springer, Heidelberg (2010)
17. Grumberg, O., Kupferman, O., Sheinvald, S.: Variable automata over infinite alphabets (2011) (manuscript)
18. Gutierrez, C., Hurtado, C., Mendelzon, A.: Foundations of semantic Web databases. J. Comput. Syst. Sci. 77(3), 520–541 (2011)
19. Kaminski, M., Francez, N.: Finite-memory automata. TCS 134(2), 329–363 (1994)
20. Kaminski, M., Tan, T.: Regular expressions for languages over infinite alphabets. Fundamenta Informaticae 69(3), 301–318 (2006)
21. Leser, U.: A query language for biological networks. Bioinformatics 21(suppl. 2), 33–39 (2005)
22. Libkin, L., Vrgoč, D.: Regular path queries on graphs with data. In: ICDT 2012, pp. 74–85 (2012)
23. Libkin, L., Vrgoč, D.: Regular expressions for data words. In: Bjørner, N., Voronkov, A. (eds.) LPAR-18 2012. LNCS, vol. 7180, pp. 274–288. Springer, Heidelberg (2012)
24. Milo, R., Shen-Orr, S., et al.: Network motifs: simple building blocks of complex networks. Science 298(5594), 824–827 (2002)

25. Neven, F., Schwentick, T., Vianu, V.: Finite state machines for strings over infinite alphabets. ACM TOCL 5(3), 403–435 (2004)
26. Olken, F.: Graph data management for molecular biology. OMICS 7, 75–78 (2003)
27. Pérez, J., Arenas, M., Gutierrez, C.: Semantics and complexity of SPARQL. ACM TODS 34(3), 1–45 (2009)
28. Ronen, R., Shmueli, O.: SoQL: a language for querying and creating data in social networks. In: ICDE 2009, pp. 1595–1602 (2009)
29. San Martín, M., Gutierrez, C.: Representing, querying and transforming social networks with RDF/SPARQL. In: Aroyo, L., et al. (eds.) ESWC 2009. LNCS, vol. 5554, pp. 293–307. Springer, Heidelberg (2009)
30. Schwentick, T.: A Little Bit Infinite? On Adding Data to Finitely Labelled Structures. In: STACS 2008, pp. 17–18 (2008)
31. Segoufin, L.: Automata and logics for words and trees over an infinite alphabet. In: Ésik, Z. (ed.) CSL 2006. LNCS, vol. 4207, pp. 41–57. Springer, Heidelberg (2006)
32. A. Tal. Decidability of Inclusion for Unification Based Automata. M.Sc. thesis (in Hebrew), Technion (1999)

Factorizations and Universal Automaton
of Omega Languages

Vincent Carnino[1] and Sylvain Lombardy[2]

[1] LIGM, Université Paris-Est Marne-la-Vallée
[2] LABRI, Université de Bordeaux - IPB

Abstract. In this paper, we extend the concept of factorization on finite
words to ω-rational languages and show how to compute them. We define
a normal form for Büchi automata and introduce a universal automaton
for Büchi automata in normal form. We prove that, for every ω-rational
language, this Büchi automaton, based on factorization, is canonical and
that it is the smallest automaton that contains the morphic image of
every equivalent Büchi automaton in normal form.

1 Introduction

When considering rational languages on finite words, different kinds of formalism may be studied: automata, semigroups, rational expressions, *etc*. There exist
similar notions for rational languages on infinite words, also called ω-rational languages, which are a rational extension of languages on finite words. Indeed, classical semigroups and rational expressions have been extended to ω-semigroups
and ω-rational expressions respectively. For the automata counterpart, there is
not a unique approach but several ones depending on the acceptance mode:
Büchi automata, Muller automata, Rabin automata, Streett automata, *etc*. In
this paper, we focus on Büchi automata, which are the most intuitive kind of
automata accepting infinite words.

Infinite words are widely used in computer science, especially in order to
describe the behaviour of infinite processes. Some well-known results on finite
words are also transposable to infinite words, like Kleene's theorem which states
that ω-rational languages are exactly languages that are recognized by finite
Büchi automata and, as an extention, by finite ω-semigroups. Yet there are
other properties that are not transposable to infinite words like the existence of
a canonical deterministic minimal automaton: deterministic Büchi automata are
strictly less expressive than nondeterministic ones and, as a consequence, there is
no notion of minimal Büchi automaton. Carton and Michel [1] have proved that
prophetic automata (which are the pertinent notion for "right-left" determinism)
are as expressive as Büchi automata, but there is no unique minimal prophetic
automata for some ω-rational languages.

Yet, the minimal automaton is not the only canonical automaton associated
to a language. In 1971, Conway [2] introduced the notions of factorization and
factor matrix of a rational language. This concept has led to the definition of

M.-P. Béal and O. Carton (Eds.): DLT 2013, LNCS 7907, pp. 338–349, 2013.
© Springer-Verlag Berlin Heidelberg 2013

a new object called the universal automaton of a language [5,9]. It has many significant properties since any automaton that recognizes a language L has a morphic image which is a subautomaton of the universal automaton of L. For example, it may be used to compute a NFA of minimum size [8], or in theoretical proofs for the existence of automata with specific properties (star height [4], reversibility [3], *etc.*).

We extend the concept of factorization on finite words to ω-rational languages. Using these ω-factorizations, we build the universal automaton of an ω-rational language. We prove that, up to a conversion to a normal form, every Büchi automaton has a morphic image in this automaton (universal property), and that this automaton is minimal for this property.

In the first part, we give some basic definitions about languages (on finite and infinite words) and about automata that will be used in the course of this article. We recall some basic notions such as the past and the future of states in both NFA and Büchi automata and some new ones like the normal form of a Büchi automaton. Finally, we recall the definition of ω-semigroups given by [6] and the principle of ω-rational language recognition by ω-semigroup.

In the second part, we define ω-factorizations and pure ω-factorizations which will both be used to define the universal automaton. Then we explain how to compute them using the transition ω-semigroup of a Büchi automaton. For ω-factorizations, we describe another computation based on prophetic automata.

The last part is devoted to the definition of the universal automaton of an ω-rational language $\mathcal{L}$. It involves both pure and standard ω-factorizations as well as positive factorizations on finite words. Finally, we state the main properties of this automaton: it accepts exactly $\mathcal{L}$, has the universal property and is minimal among universal automata for $\mathcal{L}$.

2 Definitions

2.1 Languages and ω-Languages

Classically A^* denotes the free monoid generated by the alphabet A. The length of a word w in A^* is $|w|$, and for $i \in [1; |w|]$, w_i is the i-th letter of w. We denote the empty word by ε and A^+ denotes the free semigroup of non empty words.

The product over A^* naturally extends to languages. The Kleene star L^* of a language L is defined as the union of powers of L, while L^+ is the union of positive powers of L.

The set of infinite words (or ω-words) over A is A^ω; the mixed product of a word u and an ω-word v is the unique ω-word w which can be factorized into a prefix u and a suffix v. A subset of A^ω is called an ω-language. Like the concatenation product, the mixed product naturally extends to a product between languages and ω- languages.

For every word u in A^+, the ω-word u^ω is the infinite repetition of u. If L is a subset of A^+, the ω-language L^ω is the set of infinite concatenations of words of L.

These operations lead to the classical definitions of rational and ω-rational languages.

Definition 1. *The set of rational languages over A is the smallest set of languages which contains every finite language over A and which is closed under union, concatenation and iteration.*

Definition 2. *A set $\mathcal{L}$ of ω-words is an ω-power language if there exists a rational language $K \subseteq A^+$ such that $\mathcal{L} = K^\omega$.*

The set of ω-rational languages over A is the smallest set which contains the empty set, every ω-power language over A, and which is closed under mixed product with rational languages and union.

It is straightforward, that for every ω-rational language $\mathcal{L}$, there exists a finite set of pairs of rational languages $(X_i, Y_i)_{i \in I}$ such that

$$\mathcal{L} = \bigcup_{i \in I} X_i Y_i^\omega. \tag{1}$$

2.2 Automata

An automaton is a 5-tuple $\mathcal{A} = (Q, A, E, I, F)$, where Q is a finite set of states, A is an alphabet, $E \subseteq Q \times A \times Q$ is the set of transitions, $I \subseteq Q$ is the set of initial states and $F \subseteq Q$ is the set of final states. If (p, a, q) is an element of the set of transitions, we denote it $p \xrightarrow{a} q$. As usual, such an automaton can be considered as a labeled graph and we use the graph terminology. A path is a finite sequence of consecutive transitions. An ω-path is an infinite sequence of consecutive transitions. The label of a path is the sequence of the labels of its transitions.

A morphism of automata from $\mathcal{A} = (Q, A, E, I, F)$ into $\mathcal{B} = (R, A, G, J, T)$ is a mapping φ from Q to R such that $\varphi(I) \subseteq J$, $\varphi(F) \subseteq T$, and $\varphi(p) \xrightarrow{a} \varphi(q)$ is in G for every $p \xrightarrow{a} q$ in E.

Definition 3. *Let $\mathcal{A}$ be an automaton. Let p be a state of $\mathcal{A}$. The past of p, $\mathsf{Past}_\mathcal{A}(p)$, is the language of words that label a path from some initial state to p. The set of words that label a path between two states p and q is denoted by $\mathsf{Trans}_\mathcal{A}(p, q)$.*

The semantic of an automaton depends on the acceptance mode. In this paper we only consider two kinds of automata: NFA and Büchi automata; their acceptance mode is described in the two following definitions.

Definition 4. *Let $\mathcal{A}$ be a NFA. A path of $\mathcal{A}$ is accepting if it ends at a final state. The future of a state p, $\mathsf{Fut}_\mathcal{A}(p)$, is the set of words that label accepting paths starting at state p. The (rational) language recognized by $\mathcal{A}$ is the union of futures of the initial states of $\mathcal{A}$.*

Definition 5. *Let $\mathcal{A}$ be a Büchi automaton. An ω-path is accepting if it meets an infinite number of occurrences of final states. The future of a state p, $\mathsf{Fut}_{\mathcal{A}}(p)$, is the set of ω-words that label accepting ω-paths starting at state p. The ω-rational language recognized by $\mathcal{A}$ is the union of futures of the initial states of $\mathcal{A}$.*

We introduce here a normal form for Büchi automata. Indeed, in ω-rational languages, the acceptance of a word is the conjunction of conditions on the finite prefixes of the word and conditions dealing with the infinite behaviour. The normal form we consider here consists in splitting the automaton in two parts: one *transient* part in which finite prefixes are read, and *final* components which process the infinite part.

Definition 6. *A state of Büchi automaton $\mathcal{A}$ is* transient *if it is not accessible from a final state. A strongly connected component (SCC) is* final *if it contains a unique final state, at least one transition, and if every state accessible from this final state is in the same SCC. A Büchi automaton is in* normal form *if*
i) every initial state is transient;
ii) every final state is in a final SCC;
iii) for every non final state q of a final SCC S, every predecessor of q is in S.

Definition 7. *Let $\mathcal{A} = (Q, A, E, I, F)$ be a Büchi automaton. Let G be the set of final states of $\mathcal{A}$ that belong to a non trivial SCC, and let S be the function which maps every state f of G onto the SCC of f. The* normalization *of $\mathcal{A}$ is the automaton $\mathcal{A}_{\mathsf{nf}} = (Q', A, E', I', F')$ defined by :*

- $Q' = Q \cup \{(p, f) \mid f \in G, p \in \mathcal{S}(f)\}$;
- $I' = I$, $F' = \{(f, f) \mid f \in G\}$;
- $E' = E \cup \{p \xrightarrow{a} (f, f) \mid p \xrightarrow{a} f \in E, f \in G\}$
  $\cup \{(p, f) \xrightarrow{a} (q, f) \mid p \xrightarrow{a} q \in E, f \in G, p, q \in \mathcal{S}(f)\}$.

Proposition 1. *Let $\mathcal{A}$ be a Büchi automaton. The normalization $\mathcal{A}_{\mathsf{nf}}$ of $\mathcal{A}$ is an equivalent Büchi automaton in normal form.*

Example 1. Figure 1 shows a Büchi automaton and its normal form.

2.3 Semigroup Recognition

Rational languages can be defined by finite automata and rational expressions, but they can also be characterized by morphisms into finite monoids. A semigroup is a set endowed with an associative product, a monoid is a semigroup with a unit element. If S is a semigroup, S^1 is the monoid obtained by adding a unit element. A morphism of semigroups is a mapping between two semigroups which commutes with the product. A morphism of monoids also preserves the unit element. A language K of A^* is said to be *recognizable* if there exists a finite monoid M and a surjective morphism φ from A^* onto M such that $K = \varphi^{-1}(\varphi(K))$; we then say that K is recognized by (M, φ). The recognizable languages of A^* are exactly the rational languages.

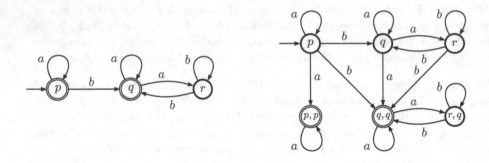

Fig. 1. A Büchi automaton and its normal form

For languages in A^+, there exists a similar notion of recognizability by finite semigroup. Clearly, a language L of A^* is recognized by a finite monoid if and only if $L \cap A^+$ is recognized by a finite semigroup.

This approach has turned out to be very fruitful and there have been many attempts to extend it to ω-rational languages. It appears that the appropriate notion of recognizability of ω-rational languages by a finite algebraic structure requires to embed them in a larger structure, called ω-*semigroup*. The recognizability theorem [6,7] (Theorem 1) requires a few definitions.

Definition 8. *An ω-semigroup is a pair $S = (S_+, S_\omega)$, where S_+ is a semigroup, endowed with:*

- *A mixed product $S_+ \times S_\omega \to S_\omega$ such that for every (u, v, w) in $S_+ \times S_+ \times S_\omega$, $(uv)w = u(vw)$.*
- *A surjective infinite product $S_+^\omega \to S_\omega$ which is compatible with finite associativity and mixed product.*

Definition 9. *Let $S = (S_+, S_\omega)$ and $T = (T_+, T_\omega)$ be two ω-semigroups. A morphism of ω-semigroups is a pair $\varphi = (\varphi_+, \varphi_\omega)$ consisting of a semigroup morphism $\varphi_+ : S_+ \to T_+$ and a mapping $\varphi_\omega : S_\omega \to T_\omega$ that commutes with the infinite product.*

These notions allow to define the recognizability by ω-semigroup.

Definition 10. *Let T be an ω-semigroups and let $\varphi : A^\infty = (A^+, A^\omega) \to T$ be a surjective morphism of ω-semigroups. A set $\mathcal{L}$ of ω-words over A is recognized by (T, φ) if $\mathcal{L} = \varphi_\omega^{-1}(\varphi_\omega(\mathcal{L}))$.*
A set of ω-words over A is recognizable if there exists a finite ω-semigroup T and a surjective morphism φ such that $\mathcal{L}$ is recognized by (T, φ).

Theorem 1. *[6] A set $\mathcal{L}$ of ω-words over A is an ω-rational language if and only if it is recognizable.*

3   Factorizations of Languages

3.1   Definitions and Properties

The factorizations of languages were introduced by Conway in [2]. He has shown that every rational language has a finite number of maximal factorizations and that this property characterizes rational languages.

Definition 11. *Let L be a language over A. A factorization of L is a pair of languages $X = (X_1, X_2)$ such that $X_1 X_2$ is a subset of L. A positive factorization $X = (X_1, X_2)$ is a factorization such that X_1 and X_2 are non empty subsets of A^+. A factorization (resp. positive factorization) X is maximal if for every factorization (resp. positive factorization) X' of L, $X_1 \subseteq X_1'$ and $X_2 \subseteq X_2'$ implies $X = X'$. We denote by $\mathfrak{F}(L)$ the set of maximal factorizations of L and by $\mathfrak{F}_+(L)$ the set of maximal positive factorizations of L.*

We extend this definition to ω-semigroups; in this case, we define two kinds of factorizations: the ω-factorizations which are the straightforward extension of factorizations on finite words and the pure ω-factorizations that involve the infinite iteration.

Definition 12. *Let $S = (S_+, S_\omega)$ be an ω-semigroup and let K be a subset of S_ω. An ω-factorization of K is a pair $X = (X_1, X_2)$, with $X_1 \subseteq S_+^1$ and $X_2 \subseteq S_\omega$ such that $X_1 X_2$ is a subset of K. A pure ω-factorization of K is a pair $Y = (Y_1, Y_2)$, with $Y_1, Y_2 \subseteq S_+$ and $Y_2 \neq \emptyset$, such that $Y_1 Y_2^\omega$ is a subset of K. An ω-factorization (resp. pure ω-factorization) X of K is maximal if for every ω-factorization (resp. pure ω-factorization) Y of K, then $X_1 \subseteq Y_1$ and $X_2 \subseteq Y_2$ implies $X = Y$. We define $\mathfrak{F}(K)$ as the set of maximal ω-factorizations of K and $\mathfrak{F}_p(K)$ as the set of maximal pure ω-factorizations of K.*

If X is a maximal ω-factorization of K then $X_1 = \{x \in S_+^1 \mid x X_2 \subseteq K\}$, and $X_2 = \{y \in S_\omega \mid X_1 y \subseteq K\}$. If Y is a maximal pure ω-factorization of K then $Y_1 = \{x \in S_+ \mid x Y_2^\omega \subseteq K\}$ and $Y_2^+ = Y_2$, but Y_2 is not characterized by Y_1.

Example 2. We consider the language $L_1 = A^*(aa + bb)A^*$ and the ω-regular language $\mathcal{L}_1 = L_1^\omega$. The pairs $(A^*, \mathcal{L}_1)$ and $(\emptyset, A^\omega)$ are the two maximal ω-factorizations of $\mathcal{L}_1$. The pairs $(A^+, L_1 + aA^*a + a)$ and $(A^+, L_1 + bA^*b + b)$ are maximal pure ω-factorizations of $\mathcal{L}_1$.[1]

Proposition 2. *Let $\mathcal{L}$ be an ω-rational language. If Y is in $\mathfrak{F}_p(\mathcal{L})$, then there exists X in $\mathfrak{F}(\mathcal{L})$ such that $Y_1 = X_1 \cap A^+$ and $Y_2^\omega \subseteq X_2$.*

Proposition 3. *Let $\mathcal{L}$ be an ω-rational language over A. Let T be a finite ω-semigroup and $\varphi : A^\infty \to T$ such that $\mathcal{L}$ is recognized by (T, φ). Then, for every X in $\mathfrak{F}(\mathcal{L})$, the set $X_1 \cap A^+$ is recognized by (T_+, φ_+), and X_2 is recognized by (T, φ), and for every Y in $\mathfrak{F}_p(\mathcal{L})$, both Y_1 and Y_2 are recognized by (T_+, φ_+).*

---

[1] Notice that the image of a maximal pure ω-factorizations in the syntactic ω-semigroup is not necessarily a union of *linked pairs* (*cf.* [6]); in $(A^+, L_1 + aA^*a + a)$, the image of a is not an idempotent.

Proof. Let $\mathcal{L}$ be an ω-rational language and let $\varphi : A^\infty \to T$ be a morphism of ω-semigroups that recognizes $\mathcal{L}$, where T is a finite ω-semigroup. Let φ_* be the morphism of monoids from A^* onto T^1_+ which is the natural extention of φ_+.

Let $X \in \mathfrak{F}(\mathcal{L})$, then $\varphi_*(X_1)\varphi_\omega(X_2) \subseteq \varphi_\omega(\mathcal{L})$. Therefore, there exists α in $\mathfrak{F}(\varphi(\mathcal{L}))$ such that $\varphi_*(X_1) \subseteq \alpha_1$ and $\varphi_\omega(X_2) \subseteq \alpha_2$. Since $\mathcal{L} = \varphi_\omega^{-1}(\varphi_\omega(\mathcal{L}))$, then $\varphi_*^{-1}(\alpha_1)\varphi_\omega^{-1}(\alpha_2) \subseteq \mathcal{L}$, thus by maximality of X, it holds $X_1 = \varphi_*^{-1}(\alpha_1)$ and $X_2 = \varphi_\omega^{-1}(\alpha_2)$: the maximal factorizations of $\mathcal{L}$ are recognized by φ.

Let Y be in $\mathfrak{F}_p(\mathcal{L})$. The pair $\gamma = (\varphi_+(Y_1), \varphi_+(Y_2))$ is a pure ω-factorization of $\varphi_\omega(\mathcal{L})$, thus there exists β in $\mathfrak{F}_p(\varphi_\omega(\mathcal{L}))$ such that $\gamma_1 \subseteq \beta_1$ and $\gamma_2 \subseteq \beta_2$. Therefore $Y_1 \subseteq \varphi_+^{-1}(\beta_1)$ and $Y_2 \subseteq \varphi_+^{-1}(\beta_2)$, and, since $\varphi_+^{-1}(\beta_1)(\varphi_+^{-1}(\beta_2))^\omega \subseteq \varphi_\omega^{-1}(\varphi_\omega(\mathcal{L})) = \mathcal{L}$, by maximality of Y, it holds $Y_1 = \varphi_+^{-1}(\beta_1)$ and $Y_2 = \varphi_+^{-1}(\beta_2)$. The maximal pure ω-factorizations of $\mathcal{L}$ are therefore recognized by φ. $\square$

Corollary 1. *Let $\mathcal{L}$ be an ω-rational language over A. The sets $\mathfrak{F}(\mathcal{L})$ and $\mathfrak{F}_p(\mathcal{L})$ are finite and each of their elements is a pair of (ω-)rational languages. Moreover,*

$$\mathcal{L} = \bigcup_{X \in \mathfrak{F}(\mathcal{L})} X_1 X_2 = \bigcup_{Y \in \mathfrak{F}_p(\mathcal{L})} Y_1 Y_2^\omega \qquad (2)$$

Corollary 2. *Let $\mathcal{A}$ be a Büchi automaton and let $\mathcal{L}$ be the ω-rational language accepted by $\mathcal{A}$. The sets $\mathfrak{F}(\mathcal{L})$ and $\mathfrak{F}_p(\mathcal{L})$ are effectively computable.*

Proof. The transition ω-semigroup $S = (S_+, S_\omega)$ of $\mathcal{A}$ is computable (*cf.* [6]), as well as the morphism φ such that $\mathcal{L}$ is recognized by (S, φ). By Proposition 3, every computation of maximal factorizations can be done in S, which is finite. $\square$

This proof induces a bound on the number of maximal ω-factorizations. Since each maximal ω-factorization is characterized by one of its factors, the number of ω-factorizations is at most $\min(2^{|S_+|+1}, 2^{|S_\omega|})$. Likewise, the number of maximal pure ω-factorizations is at most $2^{|S_+|}$.

In contrast with the case of finite words, the finiteness of $\mathfrak{F}(\mathcal{L})$ and $\mathfrak{F}_p(\mathcal{L})$ does not imply that $\mathcal{L}$ is ω-rational.

Example 3. Let $\mathcal{L}_2 = \{\prod_{i \geq 0} a^{f(i)} b \mid f : \mathbb{N} \to \mathbb{N} \text{ and } \forall i, f^{-1}(i) \text{ is finite}\}$; it is not ω-rational and, yet, $\mathfrak{F}(\mathcal{L}_2) = \{(A^*, \mathcal{L}_2), (\emptyset, A^\omega)\}$ and $\mathfrak{F}_p(\mathcal{L}_2) = \{(\emptyset, A^+)\}$.

3.2   Computation of Maximal ω-Factorizations from Prophetic Automata

In this part, we present an alternative computation of the factorizations of an ω-rational language based on a *prophetic* automaton recognizing this language.

Definition 13. *[1] A Büchi automaton $\mathcal{A}$ over A is prophetic if the future of every state is non empty and the futures of states are pairwise disjoint.*

Theorem 2. *[1] Every ω-rational language can be recognized by a prophetic automaton.*

The conversion of a Büchi automaton into a prophetic automaton is effective, but a bit complicated. Nevertheless, some natural ω-rational languages have very simple prophetic automata.

To compute the maximal factorizations from a prophetic automaton, we use the well-known *subset construction*. The aim is not to obtain a deterministic equivalent Büchi automata (which may not exist), but to compute a set of states, that will be used in Proposition 4 to characterize maximal ω-factorizations.

If $\mathcal{A} = (Q, A, E, I, F)$ is a (Büchi) automaton, for each word u in A^*, the set of accessible states by u from a subset K of Q is $\delta_{\mathcal{A}}(K, u) = \{q \mid \exists p \in K$ and $u \in \mathsf{Trans}_{\mathcal{A}}(p, q)\}$.

The *subset construction* of $\mathcal{A}$ is the set $P = \{K \subseteq Q \mid \exists u \in A^*, \ K = \delta_{\mathcal{A}}(I, u)\}$. Notice that $I = \delta_{\mathcal{A}}(I, \varepsilon)$ is always in P and that P can be incrementaly computed, since $\delta_{\mathcal{A}}(I, ua) = \delta_{\mathcal{A}}(\delta_{\mathcal{A}}(I, u), a)$.

Proposition 4. *Let $\mathcal{A}$ be a prophetic automaton with set of states Q recognizing $\mathcal{L} \subseteq A^{\omega}$. Let P be the subset construction of $\mathcal{A}$ and let $P_{\cap}$ be the smallest set containing the element Q, every element of P, and closed under intersection. We set, for every K in $P_{\cap}$, and every X in $\mathfrak{F}(\mathcal{L})$,*

$$\varphi(K) = \left(\bigcap_{p \in K} \mathsf{Past}_{\mathcal{A}}(p), \bigcup_{p \in K} \mathsf{Fut}_{\mathcal{A}}(p)\right), \qquad \psi(X) = \bigcap_{u \in X_1} \delta_{\mathcal{A}}(I, u). \qquad (3)$$

Then, φ is a one-to-one mapping between $P_{\cap}$ and $\mathfrak{F}(\mathcal{L})$ and $\psi = \varphi^{-1}$.

Proof. Let $X \in \mathfrak{F}(\mathcal{L})$ and let $H = \psi(X)$, which is clearly in $P_{\cap}$, since for each word u, the set $\delta_{\mathcal{A}}(I, u)$ is in P (if $X_1 = \emptyset$, then $H = Q$).

We prove now that the factorization $Y = \varphi(H)$ is equal to X. It holds $H = \bigcap_{u \in X_1} \delta_{\mathcal{A}}(I, u) = \{p \in Q \mid X_1 \subseteq \mathsf{Past}_{\mathcal{A}}(p)\}$, thus $X_1 \subseteq \bigcap_{p \in H} \mathsf{Past}_{\mathcal{A}}(p) = Y_1$. For each v in X_2, there exists a unique p in Q such that v is in $\mathsf{Fut}_{\mathcal{A}}(p)$; it holds $X_1 v \subseteq \mathcal{L}$, hence $X_1 \subseteq \mathsf{Past}_{\mathcal{A}}(p)$ and p is in H. Thus, $X_2 \subseteq \bigcup_{p \in H} \mathsf{Fut}_{\mathcal{A}}(p)$, and $X = Y$ by maximality of X.

Conversely, let K be in $P_{\cap}$, and $Y = \varphi(K)$. It holds

$$K = \bigcap \{\delta_{\mathcal{A}}(I, u) \mid u \in A^*, K \subseteq \delta_{\mathcal{A}}(I, u)\}$$

$$= \bigcap \{\delta_{\mathcal{A}}(I, u) \mid u \in \bigcap_{p \in K} \mathsf{Past}_{\mathcal{A}}(p)\} = \bigcap_{u \in Y_1} \delta(I, u). \qquad (4)$$

Let X be in $\mathfrak{F}(\mathcal{L})$ such that $Y_1 \subseteq X_1$ and $Y_2 \subseteq X_2$. Since $Y_1 \subseteq X_1$, it holds $\psi(X) \subseteq K$. For each p in K, there exists v in $\mathsf{Fut}_{\mathcal{A}}(p) \subseteq Y_2 \subseteq X_2$. Thus $X_1 v \subseteq \mathcal{L}$ and p is in $\bigcap_{u \in X_1} \delta_{\mathcal{A}}(I, u) = \psi(X)$. Therefore, $\psi(X) = K$ and $Y = \varphi(\psi(X)) = X$. $\square$

4   Universal Automaton

In this part, we extend the definition of the universal automaton [5] of a language to infinite words. In the case of finite words, the universal automaton of L is the smallest automaton which recognizes L and in which every equivalent automaton has a morphic image.

We shall first describe the definition of the universal automaton of $\mathcal{L}$ and then prove that it is actually the smallest Büchi automaton which recognizes $\mathcal{L}$ and in which every equivalent automaton in normal form has a morphic image.

4.1 Definition of the Universal Automaton

The definition of the universal automaton of an ω-rational language involves ω-factorizations, pure ω-factorizations and positive factorizations.

Definition 14. *Let $\mathcal{L}$ be an ω-rational language on A. For each Y in $\mathfrak{F}_p(\mathcal{L})$, we set $\mathcal{Z}_Y = \{Z \in \mathfrak{F}_+(Y_2) \mid Z \neq (Y_2, Y_2)\}$. The universal automaton $\mathcal{U}_\mathcal{L}$ of $\mathcal{L}$ is a Büchi automaton defined as follows.*

The set of states of $\mathcal{U}_\mathcal{L}$ is the union of $\mathfrak{F}(\mathcal{L})$, $\mathfrak{F}_p(\mathcal{L})$ and $\mathcal{Z}_Y$ for each $Y \in \mathfrak{F}_p(\mathcal{L})$.
The set of final states of $\mathcal{U}_\mathcal{L}$ is $\mathfrak{F}_p(\mathcal{L})$.
The set of initial states of $\mathcal{U}_\mathcal{L}$ is $\{X \in \mathfrak{F}(\mathcal{L}) \mid \varepsilon \in X_1\}$.
The set of transitions of $\mathcal{U}_\mathcal{L}$ is

$$\{X \xrightarrow{a} X' \mid a \in A, \ X, X' \in \mathfrak{F}(\mathcal{L}) \text{ and } X_1 a \subseteq X_1'\}$$
$$\cup \{X \xrightarrow{a} Y \mid a \in A, X \in \mathfrak{F}(\mathcal{L}), Y \in \mathfrak{F}_p(\mathcal{L}), \text{ and } X_1 a Y_2^\omega \subseteq \mathcal{L}\}$$
$$\cup \{K \xrightarrow{a} Y \mid a \in A, \ Y \in \mathfrak{F}_p(\mathcal{L}), K \in \mathcal{Z}_Y \cup \{Y\}, a \in K_2\}$$
$$\cup \{K \xrightarrow{a} Z \mid a \in A, \ \exists Y \in \mathfrak{F}_p(\mathcal{L}), K \in \mathcal{Z}_Y \cup \{Y\}, Z \in \mathcal{Z}_Y \text{ and } aZ_2 \subseteq K_2\}.$$

Example 4. Let $\mathcal{L}_b$ be the language of words with an infinite number of 'b', which is recognized by the automaton of Figure 2 (left). The maximal ω-factorizations and maximal pure ω-factorizations are $\mathfrak{F}(\mathcal{L}_b) = \{(A^*, \mathcal{L}_b), (\emptyset, A^\omega)\}$ and $\mathfrak{F}_p(\mathcal{L}_b) = \{(\emptyset, A^+), (A^+, K_b)\}$, where $K_b = A^*bA^*$. From the right factors of these maximal pure ω-factorizations, we obtain the following sets of maximal positive factorizations: $\mathfrak{F}_+(A^+) = \{(A^+, A^+)\}$ and $\mathfrak{F}_+(K_b) = \{(A^+, K_b), (K_b, A^+)\}$. Notice that $\mathcal{Z}_{(\emptyset, A^+)} = \mathfrak{F}_+(A^+) \setminus \{(A^+, A^+)\} = \emptyset$.

The construction of the universal automaton of $\mathcal{L}_b$ follows; it is shown in Figure 2 (right).

4.2 Basic Properties of the Universal Automaton

The choices concerning the definition of $\mathcal{U}_\mathcal{L}$ could seem arbitrary at first glance. Nevertheless, the following propositions, which seem to naturally follow from the definition of the universal automaton, and that are required for the soundness of this notion, may not be true as soon as the definition of the universal automaton is slightly modified.

The conditions that define the transitions of the universal automaton can be generalized to charaterize the paths in this automaton.

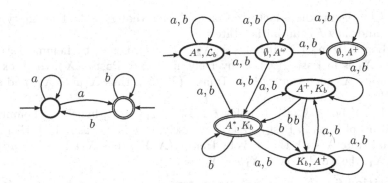

Fig. 2. A prophetic Büchi automaton and the universal automaton recognizing words with an infinite number of 'b'

Lemma 1. *Let $\mathcal{L}$ be an ω-rational language. Let $\mathcal{U}_\mathcal{L}$ be the universal automaton of $\mathcal{L}$ and let $\mathcal{Z}_Y$ be defined as in Definition 14. Let w be a non empty (finite) word. Let X and X' in $\mathfrak{F}(\mathcal{L})$, Y in $\mathfrak{F}_p(\mathcal{L})$, Z in $\mathcal{Z}_Y$, and K in $\mathcal{Z}_Y \cup Y$. It holds:*

a) $w \in \mathsf{Trans}_{\mathcal{U}_\mathcal{L}}(X, X') \Leftrightarrow X_1 w \subseteq X'_1$;   b) $w \in \mathsf{Trans}_{\mathcal{U}_\mathcal{L}}(X, Y) \Leftrightarrow X_1 w Y_2^\omega \subseteq \mathcal{L}$;

c) $w \in \mathsf{Trans}_{\mathcal{U}_\mathcal{L}}(K, Y) \Leftrightarrow w \in K_2$;   d) $w \in \mathsf{Trans}_{\mathcal{U}_\mathcal{L}}(K, Z) \Leftrightarrow w Z_2 \subseteq K_2$.

The future and the past of states of the universal automaton are closely related to the factorizations which define them.

Proposition 5. *Let $\mathcal{L}$ be an ω-rational language and let $\mathcal{U}_\mathcal{L}$ be the universal automaton of $\mathcal{L}$.*
1. For every X in $\mathfrak{F}(\mathcal{L})$,

a) $\mathsf{Past}_{\mathcal{U}_\mathcal{L}}(X) = X_1$ b) $\mathsf{Fut}_{\mathcal{U}_\mathcal{L}}(X) = X_2$.

2. For every Y in $\mathfrak{F}_p(\mathcal{L})$ and for every $Z \in \mathcal{Z}_Y$,

a) $\mathsf{Past}_{\mathcal{U}_\mathcal{L}}(Y) = Y_1$ b) $\mathsf{Trans}_{\mathcal{U}_\mathcal{L}}(Y, Y) = Y_2$ c) $\mathsf{Fut}_{\mathcal{U}_\mathcal{L}}(Y) = Y_2^\omega$
d) $\mathsf{Trans}_{\mathcal{U}_\mathcal{L}}(Y, Z) = Z_1$ e) $\mathsf{Trans}_{\mathcal{U}_\mathcal{L}}(Z, Y) = Z_2$.

Proof. 2 b) and 2 e) are straightforward from Lemma 1 c). 2 c) comes from 2 b) and from the fact that no other final state is accessible from Y.
– 2 d) From Lemma 1 d), $\mathsf{Trans}_{\mathcal{U}_\mathcal{L}}(Y, Z) = \{w \mid w Z_2 \subseteq Y_2\} = Z_1$.
– 1 b) If w is in $\mathsf{Fut}_{\mathcal{U}_\mathcal{L}}(X)$, there exists Y in $\mathfrak{F}_p(\mathcal{L})$ and a factorization of w into xy such that x is in $\mathsf{Trans}_{\mathcal{U}_\mathcal{L}}(X, Y)$ and y is in $\mathsf{Fut}_{\mathcal{U}_\mathcal{L}}(Y)$. Hence $X_1 xy \subseteq X_1 x Y_2^\omega \subseteq \mathcal{L}$ (Lemma 1 b), and $w = xy$ is in X_2. Conversely, if w is in X_2, there exists T in $\mathfrak{F}_p(X_2)$ and a factorization of w into xy, with x in T_1 and y in T_2^ω. Since $X_1 T_1 T_2^\omega \subseteq \mathcal{L}$, there exists Y in $\mathfrak{F}_p(\mathcal{L})$ such that $X_1 T_1 \subseteq Y_1$ and $T_2 \subseteq Y_2$. Therefore $X_1 x Y_2^\omega \subseteq \mathcal{L}$ and y is in Y_2^ω, thus $w = xy$ is in $\mathsf{Trans}_{\mathcal{U}_\mathcal{L}}(X, Y)\mathsf{Fut}_{\mathcal{U}_\mathcal{L}}(Y) \subseteq \mathsf{Fut}_{\mathcal{U}_\mathcal{L}}(X)$.

Let I be the maximal factorization of $\mathcal{L}$ such that $I_2 = \mathcal{L}$. The empty word ε is in I_1 and thus I is an initial state.

– 1 a) Let X be in $\mathfrak{F}(\mathcal{L})$. It holds $X_1 X_2 \subseteq \mathcal{L} = I_2$, hence, by Lemma 1 a), every word of X_1 is in $\mathsf{Past}_{\mathcal{U}_\mathcal{L}}(X)$. Conversely, if w is in $\mathsf{Past}_{\mathcal{U}_\mathcal{L}}(X)$, there exists an initial state X' such that w is in $\mathsf{Trans}_{\mathcal{U}_\mathcal{L}}(X', X)$, thus $X_1' w \subseteq X_1$, and since ε is in X_1', the word w is in X_1.

– 2 a) Let Y be in $\mathfrak{F}_p(\mathcal{L})$. It holds $I_1 Y_1 Y_2^\omega \subseteq I_1 \mathcal{L} = \mathcal{L}$, hence, by Lemma 1 b), every word of Y_1 is in $\mathsf{Past}_{\mathcal{U}_\mathcal{L}}(Y)$. Conversely, if w is in $\mathsf{Past}_{\mathcal{U}_\mathcal{L}}(Y)$, there exists an initial state X such that w is in $\mathsf{Trans}_{\mathcal{U}_\mathcal{L}}(X, Y)$, thus $X_1 w Y_2^\omega \subseteq \mathcal{L}$, and since ε is in X_1, the word w is in $X_1 w \subseteq Y_1$. $\qquad\square$

Proposition 6. *The universal automaton of an ω-language is a finite Büchi automaton in normal form.*

In the universal automaton, the states corresponding to ω-factorizations are transient states, while every pure ω-factorization Y is a final state and elements of $\mathcal{Z}_Y$ are the other states of the final SCC of Y.

From Proposition 5, it is straightforward that:

Proposition 7. *The universal automaton of $\mathcal{L}$ recognizes $\mathcal{L}$.*

The universal automaton of $\mathcal{L}$ is canonical w.r.t. $\mathcal{L}$. We state now that it is indeed universal for $\mathcal{L}$, *i.e.* it contains the morphic image of any Büchi automaton in normal form that recognizes $\mathcal{L}$.

Proposition 8. *(Universality) Let $\mathcal{A}$ be a Büchi automaton in normal form that recognizes $\mathcal{L}$. Then, there exists a morphism φ from $\mathcal{A}$ into $\mathcal{U}_\mathcal{L}$. Moreover, this morphism can be chosen such that transient states of $\mathcal{A}$ map onto transient states of $\mathcal{U}_\mathcal{L}$.*

Depending on the nature of the state, the mapping φ is defined as follows :

1. If p is a transient state, let $X_2 = \{v \mid \mathsf{Past}_\mathcal{A}(p)v \subseteq \mathcal{L}\}$ and $X_1 = \{u \mid uX_2 \subseteq \mathcal{L}\}$. We set $\varphi(p) = (X_1, X_2)$.
2. If p is a final state. Let $Y_1 = \{u \mid u\mathsf{Trans}_\mathcal{A}(p, p)^\omega \subseteq \mathcal{L}\}$ and $Y_2 \in max\{T \mid Y_1 T^\omega \subseteq \mathcal{L} \text{ and } \mathsf{Trans}_\mathcal{A}(p, p) \subseteq T\}$. We set $\varphi(p) = (Y_1, Y_2)$.
3. If p belongs to a SCC containing a final state q distinct from p. Let $Y = \varphi(q)$, $Z_1 = \{u \mid u\mathsf{Trans}_\mathcal{A}(p, q) \subseteq Y_2\}$ and $Z_2 = \{v \mid Z_1 v \subseteq Y_2\}$. If $(Z_1, Z_2) = (Y_2, Y_2)$, then let $K = Y$, otherwise $K = Z$. We set $\varphi(p) = K$.

Every automaton in normal form that recognizes the ω-language $\mathcal{L}$ and that fulfils the universal property contains the universal automaton. The maximality of factorizations implies that the merging of distinct states leads to accept more ω-words.

Proposition 9. *Let $\mathcal{V}$ be an automaton in normal form recognizing $\mathcal{L}$ such that there exists a morphism φ from $\mathcal{U}_\mathcal{L}$ into $\mathcal{V}$. Then φ is injective.*

Propositions 7,8 and 9 put together give the main result of this paper.

Theorem 3. *For every ω-rational language $\mathcal{L}$, the universal automaton of $\mathcal{L}$ is the smallest Büchi automaton in normal form that recognizes $\mathcal{L}$ in which every equivalent Büchi automaton in normal form has a morphic image.*

5 Conclusion

This paper introduces factorizations of ω-rational languages. They lead to the definition of the universal automaton of such a language. This automaton is effectively computable, since the maximal ω-factorizations are computable either from a finite ω-semigroup that recognizes the language or from a prophetic automaton. The maximal pure ω-factorizations are also computable from the same ω-semigroup, but it remains open whether they can be efficiently computed from some automaton accepting the language.

Like for rational languages on finite words, the universal automaton may be useful, in spite of its size, which is at most exponential in the size of the syntactic ω-semigroup, in the proofs of existence of Büchi automata with specific properties. On finite words, the universal automaton allows for instance to prove that a reversible rational language can be recognized by a NFA which is both reversible and star-height minimal [4].

Moreover, since every automaton has a morphic image in the universal automaton, it can be a tool for the construction of automata with a small number of states. In the case of Büchi automata, since the universal automaton is in normal form, the computation of a Büchi automaton with a minimal number of states is not as straightforward as in the case of NFA. It is probably a question which deserves more studies.

References

1. Carton, O., Michel, M.: Unambiguous Büchi automata. Theoret. Comput. Sci. 297, 37–81 (2003)
2. Conway, J.H.: Regular algebra and finite machines. Mathematics series. Chapman and Hall, London (1971)
3. Lombardy, S.: On the construction of reversible automata for reversible languages. In: Widmayer, P., Triguero, F., Morales, R., Hennessy, M., Eidenbenz, S., Conejo, R. (eds.) ICALP 2002. LNCS, vol. 2380, pp. 170–182. Springer, Heidelberg (2002)
4. Lombardy, S., Sakarovitch, J.: Star height of reversible languages and universal automata. In: Rajsbaum, S. (ed.) LATIN 2002. LNCS, vol. 2286, pp. 76–90. Springer, Heidelberg (2002)
5. Lombardy, S., Sakarovitch, J.: The universal automaton. In Logic and Automata. Texts in Logic and Games, vol. 2, pp. 457–504. Amsterdam University Press (2008)
6. Perrin, D., Pin, J.-É.: Semigroups and automata on infinite words. In: Semigroups, Formal Languages and Groups, pp. 49–72. Kluwer Academic Publishers (1995)
7. Perrin, D., Pin, J.-É.: Infinite Words. Pure and Applied Mathematics, vol. 141. Elsevier (2004)
8. Polák, L.: Minimalizations of nfa using the universal automaton. Int. J. Found. Comput. Sci. 16(5), 999–1010 (2005)
9. Sakarovitch, J. Elements of Automata Theory. Cambridge University Press (2009)

Deciding Determinism
of Unary Languages Is coNP-Complete*

Ping Lu[1,2,3], Feifei Peng[4], and Haiming Chen[1]

[1] State Key Laboratory of Computer Science,
Institute of Software, Chinese Academy of Sciences, Beijing 100190, China
{luping,chm}@ios.ac.cn
[2] Graduate University of Chinese Academy of Sciences
[3] University of Chinese Academy of Sciences
[4] China Agricultural University, Beijing 100083, China

Abstract. In this paper, we give the complexity of deciding determinism of unary languages. First, we derive a set of arithmetic progressions from an expression E over a unary alphabet, and give the relations between numbers in these arithmetic progressions and words in $L(E)$. Next, we define a problem related to arithmetic progressions and investigate the complexity of this problem. Finally, by reduction from this problem we show that deciding determinism of unary languages is **coNP**-complete.

1 Introduction

The XML schema languages, e.g., DTD and XML Schema, require that the content model should be deterministic, which ensures fast parsing documents [20,1]. Intuitively, determinism means that a symbol in the input word should be matched to a unique position in the regular expression without looking ahead in the word [21,4].

However, this determinism is defined in a semantic way, without a known simple syntax definition [1]. It is not easy for users to understand such kind of expressions. Lots of work [1,4,3,11,14,5,15] studied properties of deterministic expressions. But most of these work merely handled determinism of expressions and only little progress has been made about determinism of languages.

For standard regular expressions, Brüggemann-Klein and Wood [4] showed that the problem, whether an expression denotes a deterministic language, is decidable. Recently Bex et al. [1] proved that the problem is **PSPACE**-hard, but it is unclear whether the problem is in **PSPACE**. The problem becomes much harder when we consider expressions with counting. It is not known to be decidable whether a language can be defined by a deterministic expression with counting. In [9], Gelade et al. showed that for unary languages, deterministic expressions with counting are expressively equivalent to standard deterministic

* Work supported by the National Natural Science Foundation of China under Grant No. 61070038.

M.-P. Béal and O. Carton (Eds.): DLT 2013, LNCS 7907, pp. 350–361, 2013.
© Springer-Verlag Berlin Heidelberg 2013

expressions. Hence considering determinism of standard expressions over a unary alphabet can give a lower bound for the problem. This is our starting point.

Covering systems were introduced by Paul Erdős [7,2]. This is an interesting topic in mathematics and there are many unsolved problems about covering systems [12]. Here, we are only concerned with the problem whether a set of arithmetic progressions forms a covering system. This problem has been shown to be **coNP**-complete [19,8].

Unary languages are actually sets of numbers. Then for an expression E, we derive a set of arithmetic progressions and show the relations between these arithmetic progressions and $L(E)$. After that, we give the complexity of deciding determinism of unary languages by reduction from covering systems.

The rest of the paper is organized as follows. Section 2 gives some basic definitions and some facts from the number theory, which we will use later. We associate a set of arithmetic progressions with a given regular expression in Section 3. Section 4 shows the complexity of deciding determinism of unary languages. Section 5 gives the conclusion and the future work.

2 Preliminaries

Let Σ be an alphabet of symbols. A regular expression over Σ is recursively defined as follows: $\emptyset$, ε and $a \in \Sigma$ are regular expressions; For any two regular expressions E_1 and E_2, the union $E_1 + E_2$, the concatenation $E_1 E_2$ and the star E_1^* are regular expressions. For a regular expression E, we denote $L(E)$ as the language specified by E and $|E|$ as the size of E, which is the number of symbols occurrence in E.

We mark each symbol a in E with a different integer i such that each marked symbol a_i occurs only once in the marked expression. For example $a_1^* a_2$ is a marking of $a^* a$. The marking of E is denoted by $\overline{E}$. We use $E^\natural$ to denote the result of dropping subscripts from the marked symbols. These notations are extended for words and sets of symbols in the obvious way.

Deterministic regular expressions are defined as follows.

Definition 1 ([4]). *An expression E is deterministic if and only if, for all words $uxv, uyw \in L(\overline{E})$ where $|x| = |y| = 1$, if $x \neq y$ then $x^\natural \neq y^\natural$. A regular language is deterministic if it is denoted by some deterministic expression.*

For example, $a^* a$ is not deterministic, since $a_2, a_1 a_2 \in L(a_1^* a_2)$. A natural characterization of deterministic regular expressions is that: E is deterministic if and only if the Glushkov automaton of E is deterministic [3]. Deterministic regular expressions denote a proper subclass of regular languages [4].

The following notations are basic mathematical operators [10]: $\lfloor x \rfloor = \max\{ n \mid n \leq x, n \in \mathbb{Z} \}$; $x \bmod y = x - y\lfloor \frac{x}{y} \rfloor$, for $y \neq 0$; $x \equiv y \pmod{p} \Leftrightarrow x \bmod p = y \bmod p$; $m|n \Leftrightarrow m > 0$ and $n = mx$ for some integer x; $gcd(x_1, x_2, \ldots, x_n) = \max\{k|(k|x_1) \wedge (k|x_2) \wedge \ldots (k|x_n)\}$; $lcm(x_1, x_2, \ldots, x_n) = \min\{k|k > 0 \wedge (x_1|k) \wedge (x_2|k) \wedge \ldots (x_n|k)\}$. Notice that $gcd(0, 0, \ldots, 0)$ is undefined and $lcm(x_1, x_2, \ldots, x_n)$ is also undefined when one of the parameters is not larger than 0. In this paper, we denote $gcd(0, 0, \ldots, 0) = 0$ and $lcm(x_1, x_2, \ldots, x_n) = 0$ when one of the parameters is 0.

Here, we give some facts, which we will use later.

Lemma 1 ([22]). *Given two integers $a, b > 0$, each number of the form $ax+by$, with $x, y \geq 0$, is a multiple of $gcd(a, b)$. Furthermore, the largest multiple of $gcd(a, b)$ that cannot be represented as $ax+by$, with $x, y \geq 0$, is $lcm(a, b) - (a+b)$.*

From Lemma 1, we can obtain more generalized results as follows.

Corollary 1. *Given two integers $a, b > 0$, each number of the form $ax + by$, with $x \geq X \geq 0$ and $y \geq Y \geq 0$, is a multiple of $gcd(a, b)$. Furthermore, the largest multiple of $gcd(a, b)$ that cannot be represented as $ax + by$, with $x \geq X$ and $y \geq Y$, is $aX + bY + lcm(a, b) - (a + b)$.*

Corollary 2. *Given n ($n \geq 2$) integers $a_1 > 0, a_2 > 0, \ldots, a_n > 0$, each number of the form $a_1 x_1 + a_2 x_2 \ldots + a_n x_n$, with $x_1 \geq X_1 \geq 0$, $x_2 \geq X_2 \geq 0$, $\ldots$, and $x_n \geq X_n \geq 0$, is a multiple of $gcd(a_1, a_2, \ldots, a_n)$. Furthermore, all multiples of $gcd(a_1, a_2, \ldots, a_n)$ no less than $\sum_{i=1}^{n} a_i X_i + n \prod_{i=1}^{n} a_i$ can be represented as $a_1 x_1 + a_2 x_2 \ldots + a_n x_n$, with $x_1 \geq X_1 \geq 0$, $x_2 \geq X_2 \geq 0$, $\ldots$, and $x_n \geq X_n \geq 0$.*

In this paper, we primarily discuss unary languages. For a regular language L over the alphabet $\{a\}$, there is a correspondence between words in L and their lengths. For convenience when we say the word n, we mean the word a^n.

3   The Arithmetic Progressions of Unary Languages

In this section, we handle the following problem: Given an expression E, how to construct a set $\mathcal{P} = \{\langle l_1, s_1 \rangle, \langle l_2, s_2 \rangle, \ldots, \langle l_n, s_n \rangle\}$ of arithmetic progressions such that any word $w \in L(E)$ satisfies $|w| \equiv s_i \pmod{l_i}$ for some i ($1 \leq i \leq n$), and $\mathcal{P}$ reflects some structural properties of E.

To this end, we define the following function for an expression E.

Definition 2. *The function $\mathcal{S}(E)$ is defined as*

$\mathcal{S}(\varepsilon) = \{\langle 0, 0 \rangle\}$
$\mathcal{S}(a) = \{\langle 0, 1 \rangle\} \quad a \in \Sigma$
$\mathcal{S}(E_1 + E_2) = \mathcal{S}(E_1) \cup \mathcal{S}(E_2)$
$\mathcal{S}(E_1 E_2) = \{\langle gcd(l_i, l_j), s_i + s_j \rangle | \langle l_i, s_i \rangle \in \mathcal{S}(E_1) \wedge \langle l_j, s_j \rangle \in \mathcal{S}(E_2)\}$
$\mathcal{S}(E_1^*) = \begin{cases} \{\langle 0, 0 \rangle\} & \text{if } \mathcal{S}(E_1) = \{\langle 0, 0 \rangle\}, \\ \{\langle l, 0 \rangle | l = \max\{x | \forall l_i \forall s_i (\langle l_i, s_i \rangle \in \mathcal{S}(E_1) \wedge (x | l_i) \wedge (x | s_i))\}\} & \text{otherwise;} \end{cases}$

The intuition behind the construction of $\mathcal{S}(E)$ is Lemma 1. The cases ε, a, and $E_1 + E_2$ are obvious. For the case $E_1 E_2$, let $\langle l_1, s_1 \rangle \in \mathcal{S}(E_1)$ and $\langle l_2, s_2 \rangle \in \mathcal{S}(E_2)$. Then numbers in $\mathcal{S}(E)$ should be in the form $k_1 l_1 + s_1 + k_2 l_2 + s_2$, where $k_1, k_2 \in \mathbb{N}$. From Lemma 1, these numbers can be written as $k \cdot gcd(l_1, l_2) + s_1 + s_2$, where $k \in \mathbb{N}$. Moreover, we have added some extra numbers in $\mathcal{S}(E)$, but the number of new natural numbers is finite. The case E_1^* is similar.

Example 1. Let $E = (aaa + aa)^* + (aaa)^*((aa)^*aaa + (aaa)^*aa)$. The process of computing S is shown in Table 1. E_1 in the table stands for subexpressions of E. At last, $S(E) = \{\langle 1,0\rangle, \langle 1,3\rangle, \langle 3,2\rangle\}$. It is easy to see that $S(E)$ contains all natural numbers. However, $a \notin L(E)$ and a is the only word, which is not in $L(E)$.

Table 1. The process of computing S

E_1	$S(E_1)$	E_1	$S(E_1)$	E_1	$S(E_1)$	E_1	$S(E_1)$
a	$\langle 0,1\rangle$	$aa + aaa$	$\langle 0,2\rangle$ $\langle 0,3\rangle$	$(aa + aaa)^*$	$\langle 1,0\rangle$	$(aa)^*aaa + (aaa)^*aa$	$\langle 2,3\rangle$ $\langle 3,2\rangle$
aa	$\langle 0,2\rangle$	$(aa)^*$	$\langle 2,0\rangle$	$(aa)^*aaa$	$\langle 2,3\rangle$	$(aaa)^*((aa)^*aaa + (aaa)^*aa)$	$\langle 1,3\rangle$ $\langle 3,2\rangle$
aaa	$\langle 0,3\rangle$	$(aaa)^*$	$\langle 3,0\rangle$	$(aaa)^*aa$	$\langle 3,2\rangle$	$(aaa + aa)^* +$ $(aaa)^*((aa)^*aaa + (aaa)^*aa)$	$\langle 1,0\rangle$ $\langle 1,3\rangle$ $\langle 3,2\rangle$

At first, we have the following property about the tuples in $S(E)$.

Proposition 1. *Let E be an expression. For any $\langle l, s\rangle \in S(E)$ there exists an L ($L \geq 0$) such that* (1) : *$L + t \cdot l + s \in L(E)$ for any integer t ($t \geq L$);* (2) : *if $l \neq 0$, then $l | L$.* (3) : *if $l = 0$, then $L = 0$.*

Proof. We prove it by induction on the structure of E. For simplicity, we denote $\mathcal{Q}(l, s, L, E)$ as the conditions in the proposition. That is $\mathcal{Q}(l, s, L, E) = true$ if and only if (1), (2), and (3) hold for the parameters l, s, L and E.

The cases $E = \varepsilon$ or a, $a \in \Sigma$ are obvious, since $L = 0$ satisfy the conditions.

$E = E_1 + E_2$: Suppose $\langle l, s\rangle \in S(E)$. From the definition of S, we have $\langle l, s\rangle \in S(E_1)$ or $\langle l, s\rangle \in S(E_2)$. If $\langle l, s\rangle \in S(E_1)$, then by the inductive hypothesis there is an L_1 ($L_1 \geq 0$) such that $\mathcal{Q}(l, s, L_1, E_1) = true$. Then $L = L_1$ satisfies $\mathcal{Q}(l, s, L, E) = true$. The case $\langle l, s\rangle \in S(E_2)$ can be proved in a similar way.

$E = E_1 E_2$: Suppose $\langle l, s\rangle \in S(E)$. From the definition of S, there are $\langle l_1, s_1\rangle \in S(E_1)$ and $\langle l_2, s_2\rangle \in S(E_2)$ such that $l = gcd(l_1, l_2)$ and $s = s_1 + s_2$. By the inductive hypothesis there are L_1 ($L_1 \geq 0$) and L_2 ($L_2 \geq 0$) such that $\mathcal{Q}(l_1, s_1, L_1, E_1) = true$ and $\mathcal{Q}(l_2, s_2, L_2, E_2) = true$. Let $L = L_1 + L_2 + l_1 L_1 + l_2 L_2 + lcm(l_1, l_2) - l_1 - l_2 + gcd(l_1, l_2)$. Then from Corollary 1, for any integer t ($t \geq L$) there are $k_1 > L_1$ and $k_2 \geq L_2$ such that:

$L + t \cdot l + s$
$= L_1 + L_2 + l_1 L_1 + l_2 L_2 + lcm(l_1, l_2) - l_1 - l_2 + gcd(l_1, l_2) + t \cdot gcd(l_1, l_2) + s_1 + s_2$
$= L_1 + L_2 + k_1 l_1 + k_2 l_2 + s_1 + s_2$
$= L_1 + k_1 l_1 + s_1 + L_2 + k_2 l_2 + s_2$

From the inductive hypothesis and $E = E_1 E_2$, we have $L + t \cdot l + s \in L(E)$. If $l \neq 0$, then $l_1 \neq 0$ or $l_2 \neq 0$. Suppose $l_1 \neq 0$ and $l_2 = 0$. By the inductive hypothesis, $l_1 | L_1$ and $L_2 = 0$. Hence $L = L_1 + l_1 L_1$. Since $l = gcd(l_1, l_2)$, $l | L$.

The other cases can be proved in a similar way. If $l = 0$, then $l_1 = 0$ and $l_2 = 0$. Therefore $L_1 = 0$ and $L_2 = 0$ from the inductive hypothesis. Hence $L = 0$.

$E = E_1^*$: Suppose $\langle l, 0 \rangle \in \mathcal{S}(E)$ and $\mathcal{S}(E_1) = \{\langle l_1, s_1 \rangle, \langle l_2, s_2 \rangle, \ldots, \langle l_n, s_n \rangle\}$. Then $l = gcd(l_1, l_2, \ldots, l_n, s_1, s_2, \ldots, s_n)$. Because $\mathcal{S}(E_1) = \{\langle l_1, s_1 \rangle, \langle l_2, s_2 \rangle, \ldots, \langle l_n, s_n \rangle\}$, by the inductive hypothesis there are $L_1 \geq 0, L_2 \geq 0, \ldots, L_n \geq 0$ such that $\mathcal{Q}(l_1, s_1, L_1, E_1) = true$, $\mathcal{Q}(l_2, s_2, L_2, E_1) = true$, $\ldots$, $\mathcal{Q}(l_n, s_n, L_n, E_1) = true$. Therefore for any integer $t_1 \geq L_1$, $t_2 \geq L_2, \ldots$, $t_n \geq L_n$ we have $L_1 + t_1 l_1 + s_1 \in L(E_1)$, $L_2 + t_2 l_2 + s_2 \in L(E_1)$, $\ldots$, and $L_n + t_n l_n + s_n \in L(E_1)$. Let $L = \sum_{i=1}^{n} 2(L_i + l_i L_i + s_i) + 2n \prod_{i=1}^{n} (L_i + l_i L_i + s_i) l_i$ and $g = gcd(l_1, l_2, \ldots, l_n, L_1 + l_1 L_1 + s_1, L_2 + l_2 L_2 + s_2, \ldots, L_n + l_n L_n + s_n)$. If $g = 0$, then $l_1 = 0$, $l_2 = 0, \ldots$, $l_n = 0$, $L_1 + l_1 L_1 + s_1 = 0$, $L_2 + l_2 L_2 + s_2 = 0, \ldots$, $L_n + l_n L_n + s_n = 0$, $l = 0$ and $L = 0$. It is easy to see that the statements (1), (2) and (3) hold. Otherwise, suppose $g \neq 0$. Then $g | L$. From Corollary 2, we know that for any integer t $(t \geq L)$, $\sum_{i=1}^{n} 2(L_i + l_i L_i + s_i) + 2n \prod_{i=1}^{n} (L_i + l_i L_i + s_i) l_i + t \cdot g$ can be represented as $\sum_{i=1}^{n} (L_i + l_i L_i + s_i) x_i + \sum_{i=1}^{n} l_i y_i$, with $x_i \geq 2$, $y_i \geq 0$ $(1 \leq i \leq n)$. By the inductive hypothesis, for any $1 \leq i \leq n$, $gcd(l_i, L_i + l_i L_i + s_i) = gcd(l_i, s_i)$. Then $gcd(l_1, l_2, \ldots, l_n, L_1 + l_1 L_1 + s_1, L_2 + l_2 L_2 + s_2, \ldots, L_n + l_n L_n + s_n) = gcd(l_1, l_2, \ldots, l_n, s_1, s_2, \ldots, s_n)$. Therefore for any integer t $(t \geq L)$,

$L + t \cdot gcd(l_1, l_2, \ldots, l_n, s_1, s_2, \ldots, s_n)$
$= \sum_{i=1}^{n} 2(L_i + l_i L_i + s_i) + 2n \prod_{i=1}^{n} (L_i + l_i L_i + s_i) l_i + t \cdot g$
$= \sum_{i=1}^{n} (L_i + l_i L_i + s_i) x_i + \sum_{i=1}^{n} l_i y_i$
$= \sum_{i=1}^{n} [(L_i + l_i L_i + s_i) x_i + l_i y_i]$

For all i $(1 \leq i \leq n)$, since $x_i \geq 2$, $(L_i + l_i L_i + s_i) x_i + l_i y_i = (L_i + l_i L_i + s_i)(x_i - 1) + L_i + l_i (L_i + y_i) + s_i$. By the inductive hypothesis, we have $L_i + l_i L_i + s_i \in L(E_1)$ and $L_i + l_i (L_i + y_i) + s_i \in L(E_1)$. Hence $(L_i + l_i L_i + s_i) x_i + l_i y_i \in L(E)$. Therefore $\sum_{i=1}^{n} [(L_i + l_i L_i + s_i) x_i + l_i y_i] \in L(E)$. That is $L + t \cdot gcd(l_1, l_2, \ldots, l_n, s_1, s_2, \ldots, s_n) \in L(E)$. If $l \neq 0$, then $l | l_i$ and $l | s_i$ for any $0 \leq i \leq n$. Hence $l | L$ by the inductive hypothesis. If $l = 0$, then $l_i = 0$ and $s_i = 0$. Therefore $L_i = 0$. Hence $L = 0$.   □

This proposition means that for any $\langle l, s \rangle \in \mathcal{S}(E)$ there exists an L such that any word w, satisfying $|w| = L + t \cdot l + s$ for some integer t $(t \geq L)$, is in $L(E)$.

On the other hand, for any word w in $L(E)$ there is a tuple $\langle l, s \rangle$ in $\mathcal{S}(E)$ such that w satisfies $|w| = t \cdot l + s$ for some integer t $(t \in \mathbb{Z})$. This statement can be ensured by the following proposition.

Proposition 2. *Let E be an expression. For all $w \in L(E)$, there exists $\langle l, s \rangle \in \mathcal{S}(E)$ such that if $l \neq 0$, then $|w| \equiv s \pmod{l}$, or if $l = 0$, then $|w| = s$.*

Proof. We prove it by induction on the structure of E. For simplicity, we denote $\mathcal{R}(l, s, w)$ as the conditions in the proposition. That is $\mathcal{R}(l, s, w) = true$ if and only if the following statement holds: if $l \neq 0$, then $|w| \equiv s$ (mod l), or if $l = 0$, then $|w| = s$. The cases $E = \varepsilon$ or a, $a \in \Sigma$ are obvious.

$E = E_1 + E_2$: For all $w \in L(E)$, we know that $w \in L(E_1)$ or $w \in L(E_2)$. If $w \in L(E_1)$, then by the inductive hypothesis there exists $\langle l_1, s_1 \rangle \in \mathcal{S}(E_1)$ such that $\mathcal{R}(l_1, s_1, w) = true$. Because $\mathcal{S}(E) = \mathcal{S}(E_1) \cup \mathcal{S}(E_2)$, there exists $\langle l_1, s_1 \rangle \in \mathcal{S}(E)$ such that $\mathcal{R}(l_1, s_1, w) = true$. The case $w \in L(E_2)$ can be proved in a similar way.

$E = E_1 E_2$: For all $w \in L(E)$, there are $w_1 \in L(E_1)$ and $w_2 \in L(E_2)$ such that $w = w_1 w_2$. By the inductive hypothesis there exists $\langle l_1, s_1 \rangle \in \mathcal{S}(E_1)$ and $\langle l_2, s_2 \rangle \in \mathcal{S}(E_2)$ such that $\mathcal{R}(l_1, s_1, w_1) = true$ and $\mathcal{R}(l_2, s_2, w_2) = true$. Therefore there are $k_1, k_2 \in \mathbb{Z}$ such that $|w_1| = k_1 l_1 + s_1$ and $|w_2| = k_2 l_2 + s_2$. From the definition of $\mathcal{S}$, we know that $\langle gcd(l_1, l_2), s_1 + s_2 \rangle \in \mathcal{S}(E)$. Hence

$$|w| = |w_1 w_2| = k_1 l_1 + s_1 + k_2 l_2 + s_2$$
$$= k_1 k_1' gcd(l_1, l_2) + s_1 + k_2 k_2' gcd(l_1, l_2) + s_2$$
$$= (k_1 k_1' + k_2 k_2') gcd(l_1, l_2) + s_1 + s_2$$

That is $\mathcal{R}(gcd(l_1, l_2), s_1 + s_2, w) = true$.

$E = E_1^*$: Suppose $\langle l, 0 \rangle \in \mathcal{S}(E)$. If $w = \epsilon$, then clearly $\mathcal{R}(l, 0, \epsilon) = true$. For all $w \in L(E)$ and $w \neq \epsilon$, there are $w_1, w_2, \ldots, w_n \in L(E_1)$ such that $w = w_1 w_2 \ldots w_n$. By the inductive hypothesis there exists $\langle l_1, s_1 \rangle \in \mathcal{S}(E_1)$, $\langle l_2, s_2 \rangle \in \mathcal{S}(E_2), \ldots$, and $\langle l_n, s_n \rangle \in \mathcal{S}(E_1)$ satisfying $\mathcal{R}(l_1, s_1, w_1) = true$, $\mathcal{R}(l_2, s_2, w_2) = true, \ldots$, and $\mathcal{R}(l_n, s_n, w_n) = true$. From the definition of $\mathcal{S}$, $w \in L(E)$ and $w \neq \epsilon$, it is easy to prove that $l \neq 0$. Then for any $\langle l', s' \rangle \in \mathcal{S}(E_1)$, $l | l'$ and $l | s'$. Therefore there are $k_{11}, k_{12}, k_{21}, k_{22}, \ldots, k_{n1}, k_{n2} \in \mathbb{Z}$ such that $|w_1| = k_{11} l + k_{12} l$, $|w_2| = k_{21} l + k_{22} l, \ldots$, and $|w_n| = k_{n1} l + k_{n2} l$. Hence

$$|w| = |w_1 w_2 \ldots w_n| = k_{11} l + k_{12} l + k_{21} l + k_{22} l + \ldots + k_{n1} l + k_{n2} l$$
$$= (k_{11} + k_{12} + k_{21} + k_{22} + \ldots + k_{n1} + k_{n2}) l$$

That is $\mathcal{R}(l, 0, w) = true$.   □

The following lemma is straightforward.

Lemma 2. *Let E be an expression. The following statements hold:* (1) *For all $\langle 0, s \rangle \in \mathcal{S}(E)$ we have $s \in L(E)$;* (2) *$l = 0$ for all $\langle l, s \rangle \in \mathcal{S}(E)$ iff $L(E)$ is finite;* (3) *If there is a tuple $\langle 1, s \rangle \in \mathcal{S}(E)$, then there exists an L $(L \geq 0)$ such that $w \in L(E)$ for any w $(|w| > L)$.*

From the above properties, we build the relations between words in $L(E)$ and tuples in $\mathcal{S}(E)$. Then we can study properties of $L(E)$ by investigating attributes of $\mathcal{S}(E)$.

Now we analyze the time used to compute $\mathcal{S}(E)$. Given an expression E, we compute $\mathcal{S}(E)$ in the following way: We first construct the syntax tree of E, after that we use a bottom-up traversal to compute $\mathcal{S}$ for each node. It is known that for two m-bit numbers, the greatest common divisor can be computed in $O(m^2)$ time [6]. In our computation, each number has $O(\log |E|)$ bits. Then computing $\mathcal{S}$ for each node takes $O(|E|^2 \log^2 |E|)$ time. Therefore the total time to compute $\mathcal{S}(E)$ is $O(|E|^3 \log^2 |E|)$.

Given an NFA $\mathcal{N}$, Sawa [16] also gave an algorithm to construct a set of arithmetic progressions such that the union of these arithmetic progressions is the language accepted by $\mathcal{N}$. The algorithm runs in $O(n^2(n+m))$ time, where n is the number of states in $\mathcal{N}$ and m is the number of transitions in $\mathcal{N}$. The advantage of our method is that it works merely on original expressions and reaches some kind of the lower bound of the algorithm in [16], since there is an expression E_n such that $|E_n| = n$ and any NFA describing $L(E_n)$ has $\Omega(n \cdot (\log n)^2)$ transitions [17,13]. But the price is that we add words in the language. However, we will see later that adding such words does not affect determinism of the language.

4 Determinism of Unary Languages

In the previous section, we derived a set of arithmetic progressions from a given expression E. We will show how to use these arithmetic progressions to check determinism of $L(E)$ in this section.

4.1 Decision Problems for *Covering Systems*

A *covering system* CS is a set of ordered pairs $\{\langle l_1, s_1\rangle, \langle l_2, s_2\rangle, \ldots, \langle l_n, s_n\rangle\}$, with $0 \le s_i < l_i$ $(1 \le i \le n)$ and $\sum_{i=1}^{n}(l_i + s_i) \le p(n)$ for some polynomial function p, such that any integer x satisfies $x \equiv s_i \pmod{l_i}$ for some i $(1 \le i \le n)$ [2]. For example, the set of pairs $\{\langle 2, 0\rangle, \langle 4, 1\rangle, \langle 4, 3\rangle\}$ forms a *covering system*. Since any integer i satisfies one of the following conditions: $i \equiv 0 \pmod 2$; $i \equiv 1 \pmod 4$; $i \equiv 3 \pmod 4$.

The *covering problem* (**CP**) is the following problem: Whether a given set of ordered pairs $\{\langle l_1, s_1\rangle, \langle l_2, s_2\rangle, \ldots, \langle l_n, s_n\rangle\}$, with $0 \le s_i < l_i$ $(1 \le i \le n)$ and $\sum_{i=1}^{n}(l_i + s_i) \le p(n)$ for some polynomial function p, forms a *covering system*? The complexity of **CP** is shown in the following theorem.

Theorem 1 ([19],[8]). **CP** *is* coNP-*complete*[1].

Similarly, an equal difference covering system (**EDCS**) is a set $\mathcal{P}$ of pairs, $\{\langle l_1, s_1\rangle, \langle l_2, s_2\rangle, \ldots, \langle l_n, s_n\rangle\}$, with $0 \le s_i < l_i$ $(1 \le i \le n)$ and $\sum_{i=1}^{n}(l_i + s_i) \le p(n)$ for some polynomial function p, such that there exist two integers (y, x) $(0 \le x < y)$ satisfying the following condition: For any integer k, $x \equiv k \pmod y$ if and only if $k \equiv s_i \pmod{l_i}$ for some i $(1 \le i \le n)$. We define (y, x) as the answer to $\mathcal{P}$. Let $\mathcal{P} = \{\langle 4, 1\rangle, \langle 4, 3\rangle\}$. It is straightforward to see that $\mathcal{P}$ is an **EDCS**, but is not a **CS**. The answer to $\mathcal{P}$ is $(2, 1)$. Intuitively, the union of the

---

[1] The polynomial bound is not necessary for this theorem [8]. However, we concentrate on unary languages in this paper, and we need this condition for the definition of **CP**. For this restricted case, the theorem also holds [19].

numbers represented by an **EDCS** forms an arithmetic progression, while the union of the numbers represented by a **CS** contains all integers.

The equal difference covering problem (**EDCP**) is defined as follows: Whether a given set of ordered pairs $\{\langle l_1, s_1\rangle, \langle l_2, s_2\rangle, \ldots, \langle l_n, s_n\rangle\}$, with $0 \le s_i < l_i$ ($1 \le i \le n$) and $\sum_{i=1}^{n}(l_i + s_i) \le p(n)$ for some polynomial function p, forms an **EDCS**?

The union of arithmetic progressions is used in the study of the evenly spaced integer topology [18]. However, the complexity of **EDCP**, as far as we know, has not been given.

We have the following properties of an **EDCS**.

Lemma 3. *Suppose $\mathcal{P}$ is an* **EDCS**. *The answer to $\mathcal{P}$ is unique.*

For a set $\mathcal{P}$ of ordered pairs $\{\langle l_1, s_1\rangle, \langle l_2, s_2\rangle, \ldots, \langle l_n, s_n\rangle\}$, we denote $L = gcd(l_1, l_2, \ldots, l_n)$ and suppose $0 < p_1 < p_2 < p_3 \ldots < p_m$ are the distinct divisors of L.

Lemma 4. *Suppose $\mathcal{P}$ is an* **EDCS** *and the answer to $\mathcal{P}$ is (y, x). Then there is an integer k such that $y = p_k$, $k = \max\{l | \forall i \forall j (\langle l_i, s_i\rangle \in \mathcal{P} \wedge \langle l_j, s_j\rangle \in \mathcal{P} \wedge s_i \equiv s_j \pmod{p_l})\}$ and $x = (s_1 \bmod p_k)$.*

Hence given a set $\mathcal{P}$ of ordered pairs, if we know $\mathcal{P}$ is an **EDCS**, we can find the answer to $\mathcal{P}$ from the tuples in $\mathcal{P}$. Since $\sum_{i=1}^{n}(l_i + s_i) \le p(n)$, this computation is polynomial-time computable. It is easy to see that the converse of the lemma does not hold. Consider the following set of ordered pairs: $\{\langle 3, 0\rangle, \langle 4, 0\rangle\}$. $y = 1$ and $x = 0$ satisfy all the conditions, but obviously this set is not an **EDCS**.

Bickel et al. [2] gave a method to construct a *covering system* from a set $\mathcal{P}$ of ordered pairs, where the union of numbers represented by the pairs contains an arithmetic progression. Inspired by this idea, we can prove that given an **EDCS**, we can construct a *covering system*, and vice versa.

Theorem 2. **EDCP** *is* coNP-*complete.*

Proof. At first, we prove that the problem is **coNP**-hard. This can be proved by reduction from **CP**. Given a set $\mathcal{P}$ of ordered pairs $\{\langle l_1, s_1\rangle, \langle l_2, s_2\rangle, \ldots, \langle l_n, s_n\rangle\}$, we construct the set $\mathcal{P}_1 = \{\langle 3l_1, 3s_1\rangle, \langle 3l_2, 3s_2\rangle, \ldots, \langle 3l_n, 3s_n\rangle\}$. We claim that $\mathcal{P}$ is a **CS** if and only if $\mathcal{P}_1$ is an **EDCS** and the answer to $\mathcal{P}_1$ is $(3, 0)$.

Suppose $\mathcal{P}$ is a **CS**. For any integer k such that $k \equiv 0 \pmod 3$, let $k = 3k_1$. Because $\mathcal{P}$ is a **CS**, there is a tuple $\langle l, s\rangle \in \mathcal{P}$ satisfying $k_1 \equiv s \pmod l$. Hence $k_1 = lt + s$ for some integer t ($t \in \mathbb{Z}$). Therefore $3k_1 = 3lt + 3s$ and $k = 3lt + 3s$. That is $k \equiv 3s \pmod{3l}$, and obviously $\langle 3l, 3s\rangle \in \mathcal{P}_1$. If $k \equiv 3s \pmod{3l}$ for some tuple $\langle 3l, 3s\rangle \in \mathcal{P}_1$, then $k = 3lt + 3s$ for some integer t ($t \in \mathbb{Z}$). Hence $3 | k$. That is $k \equiv 0 \pmod 3$. Therefore $\mathcal{P}_1$ is an **EDCS** and $(3, 0)$ is the answer to $\mathcal{P}_1$.

Suppose $\mathcal{P}_1$ is an **EDCS** and $(3, 0)$ is the answer to $\mathcal{P}_1$. For any integer k, since $3k \equiv 0 \pmod 3$, there is a tuple $\langle 3l, 3s\rangle \in \mathcal{P}_1$ satisfying $3k \equiv 3s \pmod{3l}$. Hence $3k = 3lt + 3s$ for some integer t ($t \in \mathbb{Z}$). Therefore $k = lt + s$. That is $k \equiv s \pmod l$. Since $\langle l, s\rangle \in \mathcal{P}$, we can conclude that $\mathcal{P}$ is a **CS**.

From Theorem 1 we conclude that **EDCP** is **coNP**-hard.

Next, we show that **EDCP** is in **coNP**. This can be proved by reduction to **CP**. Suppose $\mathcal{P}$ is the set of ordered pairs $\{\langle l_1, s_1 \rangle, \langle l_2, s_2 \rangle, \ldots, \langle l_n, s_n \rangle\}$. Then let $p = gcd(l_1, l_2, \ldots, l_n)$. From the definition of **EDCP**, we know that $p > 0$. Suppose $0 < p_1 < p_2 < \ldots < p_m$ are the distinct divisors of p. We look for an integer k such that $k = \max\{l | \forall i \forall j (\langle l_i, s_i \rangle \in \mathcal{P} \wedge \langle l_j, s_j \rangle \in \mathcal{P} \wedge s_i \equiv s_j \pmod{p_l})\}$. If there is not such k, then from Lemma 4 we know that $\mathcal{P}$ is not an **EDCS**. Hence suppose there is a k satisfying the condition. Denote $s = (s_1 \bmod p_k)$. We construct the following set $\mathcal{Q}$ of ordered pairs, $\{\langle \frac{l_1}{p_k}, \frac{s_1-s}{p_k} \rangle, \langle \frac{l_2}{p_k}, \frac{s_2-s}{p_k} \rangle, \ldots, \langle \frac{l_n}{p_k}, \frac{s_n-s}{p_k} \rangle\}$. We have the following relationship between $\mathcal{P}$ and $\mathcal{Q}$.

Claim. $\mathcal{Q}$ is a **CS** iff $\mathcal{P}$ is an **EDCS**.

Proof. Suppose $\mathcal{Q}$ is a **CS**. Then let $x = s$ and $y = p_k$. For any integer j, such that $j \equiv s_i \pmod{l_i}$ for some i ($1 \leq i \leq n$), there is j_1 such that $j = j_1 l_i + s_i = j_1 j' y + j'' y + s = (j_1 j' + j'') y + x$. Hence $x \equiv j \pmod{y}$. For any integer j such that $x \equiv j \pmod{y}$, there is j_1 satisfying $j = j_1 y + x$. Since $\mathcal{Q}$ is a **CS**, there is an integer j_2 such that $j_1 = j_2 \frac{l_i}{p_k} + \frac{s_i-s}{p_k}$ for some i ($1 \leq i \leq n$). Then $j_1 p_k = j_2 l_i + s_i - s$, so $j = j_1 y + x = j_2 l_i + s_i - s + x = j_2 l_i + s_i$. Therefore $\mathcal{P}$ is an **EDCS**.

On the other hand, suppose $\mathcal{P}$ is an **EDCS**. From Lemma 4, we know that (p_k, s) is the answer to $\mathcal{P}$. For any integer j', let $J = j' p_k + s$. Since $s \equiv J \pmod{p_k}$, $J \equiv s_i \pmod{l_i}$ for some i ($1 \leq i \leq n$). Hence there is an integer j_1 such that $J = j_1 l_i + s_i = j_1 p_k \frac{l_i}{p_k} + p_k \frac{s_i-s}{p_k} + s = (j_1 \frac{l_i}{p_k} + \frac{s_i-s}{p_k}) p_k + s = j' p_k + s$. Since $p_k \neq 0$, $j' = j_1 \frac{l_i}{p_k} + \frac{s_i-s}{p_k}$. Hence $\mathcal{Q}$ is a **CS**. $\qquad\square$

Then to decide whether $\mathcal{P}$ is an **EDCS**, we only need to check whether $\mathcal{Q}$ is a **CS**. Since **CP** is in **coNP** and the computations of p, p_k and s take polynomial time, **EDCP** is in **coNP**.

We conclude that **EDCP** is **coNP**-complete. $\qquad\square$

4.2   The Complexity of Determinism of Unary Languages

In this section, we will discuss the complexity of deciding determinism of unary languages.

For an expression E, suppose $\mathcal{S}(E) = \{\langle l_1, s_1 \rangle, \langle l_2, s_2 \rangle, \ldots, \langle l_n, s_n \rangle\}$. Define $L(\mathcal{S}(E)) = \bigcup_{\langle l, s \rangle \in \mathcal{S}(E)} \{kl + s | k \in \mathbb{Z} \wedge kl + s \geq 0\}$.

To build the relations between $L(E)$ and $L(\mathcal{S}(E))$, we need the following lemma.

Lemma 5 ([1]). *For every deterministic language L, the following statements hold: (1) If string $w \in L$, then the language $L \setminus \{w\}$ is deterministic; (2) If string $w \notin L$, then the language $L \cup \{w\}$ is deterministic.*

Then we can simplify $\mathcal{S}(E)$ in the following ways: (1) Delete all tuples $\langle 0, s \rangle$; (2) For $\langle l, s \rangle$, where $l \leq s$, we replace $\langle l, s \rangle$ with $\langle l, s \bmod l \rangle$. This process just deletes or adds a finite number of words to $L(\mathcal{S}(E))$. From Lemma 5, it does not change determinism of the language. From now on, when we say $S(E)$, we mean the simplified one. After the simplification, any tuple $\langle l, s \rangle \in S(E)$ satisfies $0 \leq s < l$. Moreover, from the definition of S, it is easy to see that $\sum_{i=1}^{|S(E)|} (l_i + s_i) \leq 2|E|^2$, where $\langle l_i, s_i \rangle \in S(E)$.

From the construction of $S(E)$, the relations between $L(E)$ and $L(S(E))$ can be characterized as follows.

Theorem 3. *Let E be an expression. There exists a number $M \in \mathbb{N}$ such that $|L(S(E)) \setminus L(E)| + |L(E) \setminus L(S(E))| \leq M$.*

Corollary 3. *Given an expression E, $L(S(E))$ is deterministic if and only if $L(E)$ is deterministic.*

Then to decide determinism of $L(E)$, we can check determinism of $L(S(E))$.

For a unary language, the corresponding minimal DFA consists of a chain of states or a chain followed by a cycle [9]. Then to check determinism we need the following characterization of determinism of unary languages.

Lemma 6 ([9]). *Let $\Sigma = \{a\}$, and L be a regular language, then L is a deterministic language if and only if L is finite or the cycle of the minimal DFA of L has at most one final state.*

From the definition of $L(S(E))$ and Lemma 6, we can easily see that to check determinism of $L(S(E))$ we only need to check whether $S(E)$ is an **EDCS**.

Theorem 4. *Suppose $L(S(E))$ is infinite. $L(S(E))$ is deterministic if and only if $S(E)$ is an **EDCS**.*

Proof. ($\Rightarrow$) Since $L(S(E))$ is deterministic, the cycle of the minimal DFA of $L(S(E))$ has at most one final state. Let the size of the cycle be p and $n = |S(E)|$. Denote the start state as q_0 and the only final state in the cycle as q_1. Suppose w is the shortest word such that $\delta(q_0, w) = q_1$. Because $l_i > 0$ $(1 \leq i \leq n)$, there is an integer k' $(k' > 0)$ such that $M = k' \prod_{i=1}^{n} l_i$ and $M > w$. Since $L(S(E))$ is infinite, $p \neq 0$. For any $\langle l_i, s_i \rangle$ $(1 \leq i \leq n)$, since q_1 is the only final state in the cycle, there is an integer k such that $kl_i + s_i > w$, $\delta(q_0, kl_i + s_i) = q_1$ and $\delta(q_0, kl_i + l_i + s_i) = q_1$. Then $p|l_i$, $p|M$, and there is an integer k_1 such that $w + k_1 p = kl_i + s_i$. Hence $w \equiv s_i \pmod{p}$. Let $s = (w \bmod p)$. We prove that (p, s) is the answer to $S(E)$. For any integer k satisfying $s \equiv k \pmod{p}$, there is an integer k'_1 such that $k'_1 > 0$ and $k + k'_1 M > w$ and $(k + k'_1 M) \equiv s \pmod{p}$. So there is an integer k'' satisfying $k + k'_1 M = w + k'' p$. That is $k + k'_1 M \in L(S(E))$. Hence $(k + k'_1 M) \equiv s_i \pmod{l_i}$ for some i $(1 \leq i \leq n)$. Because $l_i | M$, $k \equiv s_i \pmod{l_i}$. For any integer k satisfying $k \equiv s_i \pmod{l_i}$ for some i $(1 \leq i \leq n)$, there is an integer k_1 such that $k = k_1 l_i + s_i$. So

$$k \equiv k_1 l_i + s_i \pmod{p}$$
$$\equiv s_i \pmod{p}$$

$$\equiv w \pmod{p}$$
$$\equiv s \pmod{p}$$

Hence (p, s) is the answer to $\mathcal{S}(E)$. Therefore $\mathcal{S}(E)$ is an **EDCS**.

($\Leftarrow$) Since $\mathcal{S}(E)$ is an **EDCS**, suppose (p, s) is the answer to $\mathcal{S}(E)$. If $L(\mathcal{S}(E))$ is not deterministic, then there are two final states p_1 and p_2 in the cycle of the minimal DFA of $L(\mathcal{S}(E))$. Since p_1, p_2 are not equivalent, there is a word k such that $\delta(q_1, k) \in L(\mathcal{S}(E)) \wedge \delta(q_2, k) \notin L(\mathcal{S}(E))$ or $\delta(q_1, k) \notin L(\mathcal{S}(E)) \wedge \delta(q_2, k) \in L(\mathcal{S}(E))$. Suppose the case $\delta(q_1, k) \in L(\mathcal{S}(E)) \wedge \delta(q_2, k) \notin L(\mathcal{S}(E))$ holds. Denote the start state as q_0. Then there are words k_1 and k_2 such that $\delta(q_0, k_1) = q_1$ and $\delta(q_0, k_2) = q_2$. Since q_1 and q_2 are final states, $k_1 \in L(\mathcal{S}(E))$ and $k_2 \in L(\mathcal{S}(E))$. From $\mathcal{S}(E)$ is an **EDCS**, we have $k_1 \equiv s \pmod{p}$ and $k_2 \equiv s \pmod{p}$. Because $\delta(q_1, k) \in L(\mathcal{S}(E))$, $k + k_1 \equiv s \pmod{p}$. Hence $p | k$. However, from $\delta(q_2, k) \notin L(\mathcal{S}(E))$, we have $k + k_2 \not\equiv s \pmod{p}$. So $p \nmid k$, which is a contradiction. The case $\delta(q_1, k) \notin L(\mathcal{S}(E)) \wedge \delta(q_2, k) \in L(\mathcal{S}(E))$ can be proved in a similar way. Hence $L(\mathcal{S}(E))$ is deterministic. $\square$

From Theorem 2 and Theorem 4, we can obtain the main result of the paper.

Theorem 5. *Given a regular expression E over a unary alphabet, the problem of deciding whether $L(E)$ is deterministic is* **coNP**-*complete.*

For any expression $E = E_1^*$, we have $|\mathcal{S}(E_1^*)| = 1$ from the definition of $\mathcal{S}$. Then $\mathcal{S}(E_1^*)$ is an **EDCS**. Hence we can easily obtain the following theorem.

Theorem 6 ([15]). *Let L be any language over a unary alphabet. Then L^* is deterministic.*

5   Conclusion and Future Work

In this paper, we give the complexity of deciding determinism of regular languages over a unary alphabet. By studying unary languages, we can conclude that the problem, whether a language can be defined by a deterministic expression with counting, is **coNP**-hard.

There are a few problems for future research. It is easy to see that we have only handled standard regular expressions. What is the complexity when the input is an expression with counting? To solve this problem in the way we used in this paper, we have to handle the following problems: (1) For a word w and an expression E with counting over a unary alphabet, can the membership problem be tested in time $O(\log^k |w|)$ for some integer $k > 0$? (2) For an expression E with counting over a unary alphabet, how to define $\mathcal{S}(E)$? The hardness of these problems mainly comes from the fact that we do not have a good tool to handle expressions with counting.

Acknowledgments. We thank Wim Martens for sending us the full version of the paper [15].

References

1. Bex, G.J., Gelade, W., Martens, W., Neven, F.: Simplifying XML schema: effortless handling of nondeterministic regular expressions. In: SIGMOD 2009, pp. 731–743 (2009)
2. Bickel, K., Firrisa, M., Ortiz, J., Pueschel, K.: Constructions of Coverings of the Integers: Exploring an Erdős problem. Summer Math Institute, Cornell University (2008)
3. Brüggemann-Klein, A.: Regular expressions into finite automata. Theoretical Computer Science 120(2), 197–213 (1993)
4. Brüggemann-Klein, A., Wood, D.: One-unambiguous regular languages. Information and Computation 142(2), 182–206 (1998)
5. Chen, H., Lu, P.: Assisting the Design of XML Schema: Diagnosing Nondeterministic Content Models. In: Du, X., Fan, W., Wang, J., Peng, Z., Sharaf, M.A. (eds.) APWeb 2011. LNCS, vol. 6612, pp. 301–312. Springer, Heidelberg (2011)
6. Cormen, T.H., Leiserson, C.E., Rivest, R.L., Stein, C.: Introduction to Algorithms, 2nd edn. The MIT Press, Cambridge (2001)
7. Erdős, P.: On integers of the form $2^k + p$ and some related problems. Summa Brasil. Math. 2, 113–123 (1950)
8. Garrey, M.R., Johnson, D.S.: Computers and Intractability: A Guide to the Theory of NP-completeness. Freeman (1979)
9. Gelade, W., Gyssens, M., Martens, W.: Regular expressions with counting: weak versus strong determinism. SIAM J. Comput. 41(1), 160–190 (2012)
10. Graham, R.L., Knuth, D.E., Patashnik, O.: Concrete Mathematics: a foundation for computer science, 2nd edn. Addison-Wesley (1994)
11. Groz, B., Maneth, S., Staworko, S.: Deterministic regular expressions in linear time. In: PODS 2012, pp. 49–60 (2012)
12. Guy, R.K.: Unsolved problems in Number Theory, 3rd edn. Problem Books in Math. Springer, New York (2004)
13. Holzer, M., Kutrib, M.: The complexity of regular(-like) expressions. In: Gao, Y., Lu, H., Seki, S., Yu, S. (eds.) DLT 2010. LNCS, vol. 6224, pp. 16–30. Springer, Heidelberg (2010)
14. Kilpeläinen, P.: Checking determinism of XML Schema content models in optimal time. Informat. Systems 36(3), 596–617 (2011)
15. Losemann, K., Martens, W., Niewerth, M.: Descriptional complexity of deterministic regular expressions. In: Rovan, B., Sassone, V., Widmayer, P. (eds.) MFCS 2012. LNCS, vol. 7464, pp. 643–654. Springer, Heidelberg (2012)
16. Sawa, Z.: Efficient construction of semilinear representations of languages accepted by unary NFA. In: Kučera, A., Potapov, I. (eds.) RP 2010. LNCS, vol. 6227, pp. 176–182. Springer, Heidelberg (2010)
17. Schnitger, G.: Regular expressions and NFAs without ε-transitions. In: Durand, B., Thomas, W. (eds.) STACS 2006. LNCS, vol. 3884, pp. 432–443. Springer, Heidelberg (2006)
18. Steen, L.A., Seebach, J.A.: Counterexamples in topology, 2nd edn. Springer, New York (1978)
19. Stockmeyer, L.J., Meyer, A.R.: Word problems requiring exponential time: preliminary report. In: STOC 1973, pp. 1–9 (1973)
20. van der Vlist, E.: XML Schema. O'Reilly (2002)
21. World Wide Web Consortium, http://www.w3.org/wiki/UniqueParticleAttribution
22. Yu, S., Zhuang, Q., Salomaa, K.: The state complexities of some basic operations on regular langauges. Theoretical Computer Science 125(2), 315–328 (1994)

Ultimate Periodicity of b-Recognisable Sets: A Quasilinear Procedure

Victor Marsault* and Jacques Sakarovitch

Telecom-ParisTech and CNRS, 46 rue Barrault 75013 Paris, France
victor.marsault@telecom-paristech.fr

Abstract. It is decidable if a set of numbers, whose representation in a base b is a regular language, is ultimately periodic. This was established by Honkala in 1986.

We give here a structural description of minimal automata that accept an ultimately periodic set of numbers. We then show that it can be verified in linear time if a given minimal automaton meets this description.

This yields a $O(n \log(n))$ procedure for deciding whether a general deterministic automaton accepts an ultimately periodic set of numbers.

1 Introduction

Given a fixed positive integer b, called the *base*, every positive integer n is represented (in base b) by a *word* over the digit alphabet $A_b = \{0, 1, \ldots, b-1\}$ which does not start with a 0. Hence, *sets* of numbers are represented by *languages* of $A_b{}^*$. Depending on the base, a given set of integers may be represented by a simple or complex language: the set of powers of 2 is represented by the rational language 10^* in base 2; whereas in base 3, it can only be represented by a context-sensitive language, much harder to describe.

A set of numbers is said to be *b-recognisable* if it is represented by a recognisable, or rational, or regular, language over $A_b{}^*$. On the other hand, a set of numbers is *recognisable* if it is, via the identification of $\mathbb{N}$ with a^* ($n \leftrightarrow a^n$), a recognisable, or rational, or regular, language of the free monoid a^*. A set of numbers is recognisable if, and only if it is *ultimately periodic* (UP) and we use the latter terminology in the sequel as it is both meaningful and more distinguishable from b-recognisable. It is common knowledge that every UP-set of numbers is b-recognisable for every b, and the above example shows that a b-recognisable set for some b is not necessarily UP, nor c-recognisable for all c. It is an exercice to show that if b and c are *multiplicatively dependent* integers (that is, there exist integers k and l such that $b^k = c^l$), then every b-recognisable set is a c-recognisable set as well (*cf.* [9] for instance). A converse of these two properties is the theorem of Cobham: *a set of numbers which is both b- and c-recognisable, for multiplicatively independent b and c, is UP*, established in 1969 [5], a strong and deep result whose proof is difficult (*cf.* [4]).

After Cobham's theorem, the next natural (and last) question left open on b-recognisable sets of numbers was the decidability of ultimate periodicity. It was positively solved in 1986:

---

* Corresponding author.

M.-P. Béal and O. Carton (Eds.): DLT 2013, LNCS 7907, pp. 362–373, 2013.
© Springer-Verlag Berlin Heidelberg 2013

Theorem 1 (Honkala [11]). *It is decidable whether an automaton over A_b^* accepts a UP-set of numbers.*

The complexity of the decision procedure was not an issue in the original work. Neither were the properties or the structure of automata accepting UP-set of numbers. Given an automaton $\mathcal{A}$ over A_b^*, bounds are computed on the parameters of a potential UP-set of numbers accepted by $\mathcal{A}$. The property is then decidable as it is possible to enumerate all automata that accept sets with smaller parameters and check whether any of them is equivalent to $\mathcal{A}$.

As explained below, subsequent works on automata and number representations brought some answers regarding the complexity of the decision procedure, explicitly or implicitly. The present paper addresses specifically this problem and yields the following statement.

Theorem 2. *It is decidable* in linear time *whether a minimal DFA $\mathcal{A}$ over A_b^* accepts a UP-set of numbers.*

As it is often the case, this complexity result is obtained as the consequence of a structural characterisation. Indeed, we describe here a set of structural properties for an automaton: the shape of its strongly connected components (SCC's) and that of its graph of SCC's, that we gather under the name of UP-criterion. Theorem 2 then splits into two results:

Theorem 3. *A minimal DFA $\mathcal{A}$ over A_b^* accepts a UP-set of numbers if, and only if, it satisfies the UP-criterion.*

Theorem 4. *It is decidable in linear time whether a minimal DFA $\mathcal{A}$ over A_b^* satisfies the UP-criterion.*

As for Cobham's theorem (*cf.* [4,7]), new insights on the problem tackled here are obtained when stating it in a higher dimensional space. Let $\mathbb{N}^d$ be the additive monoid of d-tuples of integers. Every d-tuple of integers may be represented in base b by a d-tuple of words of A_b^* of *the same length*, as shorter words can be padded by 0's without changing the corresponding value. Such d-tuples can be read by (finite) automata over $(A_b^d)^*$ — automata reading on d synchronised tapes — and a subset of $\mathbb{N}^d$ is b-recognisable if the set of the b-representations of its elements is accepted by such an automaton.

On the other hand, recognisable and rational sets of $\mathbb{N}^d$ are defined in the classical way but they do not coincide as $\mathbb{N}^d$ *is not a free monoid*. A subset of $\mathbb{N}^d$ is *recognisable* if is saturated by a congruence of finite index, and the family of recognisable sets is denoted by Rec $\mathbb{N}^d$. A subset of $\mathbb{N}^d$ is *rational* if is denoted by a rational expression, and the family of rational sets is denoted by Rat $\mathbb{N}^d$. Rational sets of $\mathbb{N}^d$ have been characterised by Ginsburg and Spanier as sets definable in the *Presburger arithmetic* $\langle \mathbb{N}, + \rangle$ ([10]), hence the name *Presburger definable* that is most often used in the literature.

It is also common knowledge that every rational set of $\mathbb{N}^d$ is b-recognisable for every b, and the example in dimension 1 is enough to show that a b-recognisable set is not necessarily rational. The generalisation of Cobham's theorem: *a subset of* $\mathbb{N}^d$ *which is both b- and c-recognisable, for multiplicatively independent b and c, is rational,* is due to Semenov (*cf.* [4,7]). The generalisation of Honkala's theorem went as smoothly.

Theorem 5 (Muchnik [15]). *It is decidable whether a b-recognisable subset of* $\mathbb{N}^d$ *is rational.*

Theorem 6 (Leroux [13]). *It is decidable* in polynomial time *whether a minimal DFA* $\mathcal{A}$ *over* $(A_b{}^d)^*$ *accepts a rational subset of* $\mathbb{N}^d$.

The algorithm underlying Theorem 5 is triply exponential whereas the one described in [13], based on sophisticated geometric constructions, is quadratic — an impressive improvement — but not easy to explain.

There exists another way to devise a proof for Honkala's theorem which yields another extension. In [10], Ginsburg and Spanier also proved that there exists a formula in Presburger arithmetic deciding whether a given subset of $\mathbb{N}^d$ is recognisable. In dimension 1, it means that being a UP-set of numbers is expressible in Presburger arithmetic. In [2], it was then noted that since addition in base p is realised by a finite automaton, every Presburger formula is realised by a finite automaton as well. Hence a decision procedure that establishes Theorem 1.

Generalisation of base p by non-standard numeration systems then gives an extension of Theorem 1, best expressed in terms of abstract numeration systems. Given a totally ordered alphabet A, any *rational* language L of A^* defines an *abstract numeration system* (ANS) $\mathcal{S}_L$ in which the integer n is represented by the n+1-th word of L in the radix ordering of A^* (*cf.* [12]). A set of integers whose representations in the ANS $\mathcal{S}_L$ form a rational language is called $\mathcal{S}_L$-recognisable and it is known that every UP-set of numbers is $\mathcal{S}_L$-recognisable for every ANS $\mathcal{S}_L$ ([12]). The next statement then follows.

Theorem 7. *If* $\mathcal{S}_L$ *is an abstract numeration system in which addition is realised by a finite automaton, then it is decidable whether a* $\mathcal{S}_L$-*recognisable set of numbers is UP.*

For instance, Theorem 7 implies that ultimate periodicity is decidable for sets of numbers represented by rational sets in a Pisot base system [8]. The algorithm underlying Theorem 7 is exponential (if the set of numbers is given by a DFA) and thus (much) less efficient than Leroux's constructions for integer base systems. On the other hand, it applies to a much larger family of numeration systems. All this was mentioned for the sake of completeness, and the present paper does not follow this pattern.

Theorem 6, restricted to dimension 1, readily yields a quadratic procedure for Honkala's theorem. The improvement from quadratic to quasilinear complexity achieved in this article is not a natural simplification of Leroux's construction for the case of dimension 1. Although the UP-criterion bears similarities with some features of Leroux's construction, it is not derived from [13], nor is the proof of quasilinear complexity.

The paper is organised as follows. In Section 2, we treat the special case of determining whether a given minimal group automaton accepts an ultimately periodic set of numbers. We describe canonical automata, which we call Pascal automata, that accept such sets. We then show how to decide in linear time whether a given minimal group automaton is the quotient of some Pascal automaton.

Section 3 introduces the UP-criterion and sketches both its completeness and correctness. An automaton satisfying the UP-criterion is a directed acyclic graph (DAG) 'ending' with at most two layers of non-trivial strongly connected components (SCC's). If the root is seen at the top, the upper (non-trivial) SCC's are circuits of 0's and the lower ones are quotients of Pascal automata. It is easy, and of linear complexity to verify that an automaton has this overall structure.

Proofs have been consistently removed in order to comply with space constraints. A full version of this work is available on arXiv [14].

2   The Pascal Automaton

2.1   Preliminaries

On Automata. We consider only finite deterministic finite automata, denoted by $\mathcal{A} = \langle Q, A, \delta, i, T \rangle$, where Q is the set of *states*, i the *initial state* and T the set of *final states*; A is the *alphabet*, A^* is the *free monoid* generated by A and the *empty word* is denoted by ε; $\delta : Q \times A \to Q$ is the *transition function*.

As usual, δ is extended to a function $Q \times A^* \to Q$ by $\delta(q, \varepsilon) = q$ and $\delta(q, ua) = \delta(\delta(q, u), a)$; and $\delta(q, u)$ will also be denoted by $q \cdot u$. When δ is a total function, $\mathcal{A}$ is said to be *complete*. In the sequel, we only consider automata that are *accessible*, that is, in which every state is reachable from i.

A word u of A^* is *accepted* by $\mathcal{A}$ if $i \cdot u$ is in T. The set of words accepted by $\mathcal{A}$ is called *the* language of $\mathcal{A}$, and is denoted by $L(\mathcal{A})$.

Let $\mathcal{A} = \langle Q, A, \delta, i, T \rangle$ and $\mathcal{B} = \langle R, A, \eta, j, S \rangle$ be two deterministic automata. A map $\varphi : Q \to R$ is an *automaton morphism*, written $\varphi : \mathcal{A} \to \mathcal{B}$ if $\varphi(i) = j$, $\varphi(T) \subseteq S$, and for all q in Q and a in A, such that $\delta(q, a)$ is defined, then $\eta(\varphi(q), a)$ is defined, and $\varphi(\delta(q, a)) = \eta(\varphi(q), a)$. We call φ a *covering* if the following two conditions hold: i) $\varphi(T) = S$ and ii) for all q in Q and a in A, if $\eta(\varphi(q), a)$ is defined, then so is $\delta(q, a)$. In this case, $L(\mathcal{A}) = L(\mathcal{B})$, and $\mathcal{B}$ is called *a quotient* of $\mathcal{A}$. Note that if $\mathcal{A}$ is complete, every morphism satisfies (ii).

Every complete deterministic automaton $\mathcal{A}$ has a *minimal quotient* which is *the minimal automaton* accepting $L(\mathcal{A})$. This automaton is unique up to isomorphism and can be computed from $\mathcal{A}$ in $O(n \log(n))$ time, where n is the number of states of $\mathcal{A}$ (*cf.* [1]).

Given a deterministic automaton $\mathcal{A}$, every word u induces an application $(q \mapsto q \cdot u)$ over the state set. These applications form a finite monoid, called the *transition monoid* of $\mathcal{A}$. When this monoid happens to be a group (meaning that the action of every letter is a permutation over the states), $\mathcal{A}$ is called a *group automaton*.

On Numbers. The base b is fixed throughout the paper (it will be a *parameter* of the algorithms, not an input) and so is the digit alphabet A_b. As a consequence, the number of transitions of any deterministic automaton over A_b^* is linear in its number of states. Verifying that an automaton is deterministic (resp. a group automaton) can then be done in linear time.

For our purpose, it is far more convenient to write the integers *least significant digits first* (LSDF), and to keep the automata reading *from left to right* (as in Leroux's work [13]). The *value* of a word $u = a_0 a_1 \cdots a_n$ of A_b^*, denoted by $\overline{u}$, is then $\overline{u} = \sum_{i=0}^{n}(a_i b^i)$ and may be obtained by the recursive formula:

$$\overline{ua} = \overline{u} + a\,b^{|u|} \tag{1}$$

Conversely, every integer n has a unique canonical representation in base b that does not *end* with 0, and is denoted by $\langle n \rangle$. A word of A_b^* has value n if, and only if, it is of the form $\langle n \rangle 0^k$.

By abuse of language, we may talk about the *set of numbers* accepted by an automaton. An integer n is then accepted if there exists a word of value n accepted by the automaton.

A set $E \subseteq \mathbb{N}$ is *periodic*, of *period* q, if there exists $S \subseteq \{0, 1, \ldots, q-1\}$ such that $E = \{n \in \mathbb{N} \mid \exists r \in S \quad n \equiv r\,[q]\}$. Any periodic set E has a *smallest* period p and a corresponding set of *residues* R: the set E is then denoted by E_p^R. The set of numbers in E_p^R and larger than an integer m is denoted by $E_{p,m}^R$.

2.2 Definition of a Pascal Automaton

We begin with the construction of an automaton $\mathcal{P}_p^R$ that accepts the set E_p^R, in the case where

$$p \text{ is coprime with } b.$$

We call any such automaton a *Pascal automaton*.[1] If p is coprime with b, there exists a (smallest positive) integer ψ such that:

$$b^\psi \equiv 1\,[p] \qquad \text{and thus} \qquad \forall x \in \mathbb{N} \qquad b^x \equiv b^{x \bmod \psi}\,[p] .$$

Therefore, from Equation (1), knowing $\overline{u} \bmod p$ and $|u| \bmod \psi$ is enough to compute $\overline{ua} \bmod p$.

Hence the definition of $\mathcal{P}_p^R = \langle\, \mathbb{Z}/p\mathbb{Z} \times \mathbb{Z}/\psi\mathbb{Z}, A_b, \eta, (0,0), R \times \mathbb{Z}/\psi\mathbb{Z}\,\rangle$, where

$$\forall (s,t) \in \mathbb{Z}/p\mathbb{Z} \times \mathbb{Z}/\psi\mathbb{Z},\ \forall a \in A_b \qquad \eta((s,t), a) = (s,t) \cdot a = (s + a b^t, t+1) \tag{2}$$

By induction on $|u|$, it follows that $(0,0) \cdot u = (\overline{u} \bmod p, |u| \bmod \psi)$ for every u in A_b^* and consequently that E_p^R is the set of number accepted by $\mathcal{P}_p^R$.

Example 1. Fig. 1 shows $\mathcal{P}_3^2$, the Pascal automaton accepting integers written in binary and congruent to 2 modulo 3. For clarity, the labels are omitted; transitions labelled by 1 are drawn with thick lines and those labelled by 0 with thin lines.

---

[1] As early as 1654, Pascal describes a computing process that generalises the casting out nines and that determines if an integer n, written *in any base* b, is divisible by an integer p (see [16, Prologue]).

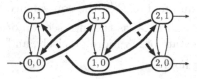

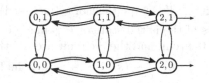

Fig. 1. The Pascal automaton $\mathcal{P}_3^2$ **Fig. 2.** The modified Pascal automaton $\mathcal{P'}_3^2$

2.3 Recognition of Quotients of Pascal Automata

The tricky part of achieving a linear complexity for Theorem 4 is contained in the following statement:

Theorem 8. *It is decidable in linear time whether a minimal DFA $\mathcal{A}$ over A_b is the quotient of a Pascal automaton.*

Simplifications. Since $\mathcal{P}_p^R$ is a group automaton, all its quotients are group automata.

The permutation on $\mathbb{Z}/p\mathbb{Z} \times \mathbb{Z}/\psi\mathbb{Z}$ realised by $0^{(\psi-1)}$ is the inverse of the one realised by 0 and we call it the action of the "digit" 0^{-1}. Let g be a new letter whose action on $\mathbb{Z}/p\mathbb{Z} \times \mathbb{Z}/\psi\mathbb{Z}$ is the one of 10^{-1}. It follows from (2) that for every a in A_b — where a is understood both as a *digit* and as a *number* — the action of a on $\mathbb{Z}/p\mathbb{Z} \times \mathbb{Z}/\psi\mathbb{Z}$ (in $\mathcal{P}_p^R$) is equal to the one of $g^a 0$. The same relation holds in any group automaton $\mathcal{A}$ over A_b^* that is a quotient of a Pascal automaton, and this condition is tested in linear time.

Let $B = \{0, g\}$ be a new alphabet. Any group automaton $\mathcal{A} = \langle Q, A_b, \delta, i, T \rangle$ may be transformed into an automaton $\mathcal{A}' = \langle Q, B, \delta', i, T \rangle$ where, for every q in Q, $\delta'(q, 0) = \delta(q, 0)$ and $\delta'(q, g) = \delta(q, 10^{-1})$. Fig. 2 shows $\mathcal{P'}_3^2$ where transitions labelled by 0 are drawn with thin lines and those labelled by g with double lines.[2]

Analysis: Computation of the Parameters. From now on, and for the rest of the section, $\mathcal{A} = \langle Q, A_b, \delta, i, T \rangle$ is a group automaton which has been consistently transformed into an automaton $\mathcal{A}' = \langle Q, B, \delta', i, T \rangle$. If $\mathcal{A}'$ is a quotient of a Pascal automaton $\mathcal{P'}_p^R$, then the parameters p and R may be computed (or 'read') in $\mathcal{A}'$; this is the consequence of the following statement.

Proposition 1. *Let $\varphi \colon \mathcal{P'}_p^R \to \mathcal{A}'$ be a covering. Then, for every (x, y) and (x', y') in $\mathbb{Z}/p\mathbb{Z} \times \mathbb{Z}/\psi\mathbb{Z}$, if $x \neq x'$ and $\varphi(x, y) = \varphi(x', y')$, then $y \neq y'$.*

Corollary 1. *If $\mathcal{A}' = \langle Q, B, \delta', i, T \rangle$ is a quotient of a modified Pascal automaton $\mathcal{P'}_p^R$, then p is the length of the g-circuit in $\mathcal{A}'$ which contains i and $R = \{r \mid i \cdot g^r \in T\}$.*

---

[2] The transformation highlights that the transition monoid of $\mathcal{P}_p^R$ (and thus of $\mathcal{P'}_p^R$) is the *semi-direct product* $\mathbb{Z}/p\mathbb{Z} \rtimes \mathbb{Z}/\psi\mathbb{Z}$.

Next, if $\mathcal{A}'$ is a quotient of a (modified) Pascal automaton $\mathcal{P}'^R_p$, the equivalence class of the initial state of $\mathcal{P}'^R_p$ may be 'read' as well in $\mathcal{A}'$ as the intersection of the 0-circuit and the g-circuit around the initial state of $\mathcal{A}'$. More precisely, and since $(0,0) \xrightarrow[\mathcal{P}'^R_p]{g^s} (s,0) \xrightarrow[\mathcal{P}'^R_p]{0^t} (s,t)$, the following holds.

Proposition 2. Let $\varphi: \mathcal{P}'^R_p \to \mathcal{A}'$ be a covering. For all s in $\mathbb{Z}/p\mathbb{Z}$ and t in $\mathbb{Z}/\psi\mathbb{Z}$, $\varphi(s,t) = \varphi(0,0)$ if, and only if, $i \cdot g^s = i \cdot 0^{-t}$.

From this proposition follows that, given $\mathcal{A}'$, it is easy to compute the class of $(0,0)$ modulo φ if $\mathcal{A}'$ is indeed a quotient of a (modified) Pascal automaton by φ. Starting from i, one first marks the states on the g-circuit C. Then, starting from i again, one follows the 0^{-1}-transitions: the *first time* C is crossed yields t. This parameter is *characteristic* of φ, as explained now.

Let (s,t) be an element of the semidirect product $G_p = \mathbb{Z}/p\mathbb{Z} \times \mathbb{Z}/\psi\mathbb{Z}$ and $\tau_{(s,t)}$ the permutation on G_p induced by the multiplication *on the left* by (s,t):

$$\tau_{(s,t)}((x,y)) = (s,t)(x,y) = (x\,b^t + s, y + t) . \tag{3}$$

The same element (s,t) defines a permutation $\sigma_{(s,t)}$ on $\mathbb{Z}/p\mathbb{Z}$ as well:

$$\forall x \in \mathbb{Z}/p\mathbb{Z} \qquad \sigma_{(s,t)}(x) = x\,b^t + s . \tag{4}$$

Given a permutation σ over a set S, *the orbit of an element s of S under σ* is the set $\{\sigma^i(s) \mid i \in \mathbb{N}\}$. *An orbit of σ* is one of these sets.

Proposition 3. Let $\varphi: \mathcal{P}'^R_p \to \mathcal{A}'$ be a covering and let (s,t) be the state φ-equivalent to $(0,0)$ with the smallest second component. Then, every φ-class is an orbit of $\tau_{(s,t)}$ (in G_p) and R is an union of orbits of $\sigma_{(s,t)}$ (in $\mathbb{Z}/p\mathbb{Z}$).

Synthesis: Verification That a Given Automaton Is a Quotient of a Pascal Automaton.

Given $\mathcal{A}' = \langle Q, B, \delta', i, T \rangle$, let p, R and (s,t) computed as explained above. It is easily checked that R is an union of orbits of $\sigma_{(s,t)}$ and that $\|Q\| = pt$. The last step is the verification that $\mathcal{A}'$ is indeed (isomorphic to) the quotient of $\mathcal{P}'^R_p$ by the morphism φ defined by (s,t).

A corollary of Proposition 3 (and of the multiplication law in G_p) is that every class modulo φ contains one, and exactly one, element whose second component is smaller than t. From this observation follows that the multiplication by the generators $0 = (0,1)$ and $g = (1,0)$ in the quotient of $\mathcal{P}'^R_p$ by φ may be described on the set of representatives $Q_\varphi = \{(x,z) \mid x \in \mathbb{Z}/p\mathbb{Z}, z \in \mathbb{Z}/t\mathbb{Z}\}$ (beware that z is in $\mathbb{Z}/t\mathbb{Z}$ and not in $\mathbb{Z}/\psi\mathbb{Z}$) by the following formulas:

$$\forall (x,z) \in Q_\varphi$$

$$(x,z) \cdot 0 = (x,z)(0,1) = \begin{cases} (x, z+1) & \text{if} \quad z < t-1 \\ \tau^{-1}_{(s,t)}(x, z+1) = (\frac{x-s}{b^t}, 0) & \text{if} \quad z = t-1 \end{cases}$$

$$(x,z) \cdot g = (x,z)(1,0) = (x + b^z, z) .$$

Hence $\mathcal{A}'$ is the quotient of $\mathcal{P}'^R_p$ by φ if one can mark Q according to these rules, starting from i with the mark $(0,0)$, without conflicts and in such a way that two distinct states have distincts marks. Such a marking is realised by a simple traversal of $\mathcal{A}'$, thus in linear time, and this concludes the proof of Theorem 8.

Remark 1. Theorem 8 states that one can decide in linear time whether a *given* automaton $\mathcal{A}$ is a quotient of a Pascal automaton, and in particular $\mathcal{A}$ has a fixed initial state that plays a crucial role in the verification process.

The following proposition shows that the property (being a quotient of a Pascal automaton) is actually independent of the state chosen to be initial. If it holds for $\mathcal{A}$, it also holds for any automaton derived from $\mathcal{A}$ by changing the initial state. This is a general property that will be used in the general verification process described in the next section.

Proposition 4. *If an automaton* $\mathcal{A} = \langle Q, A_b, \delta, i, T \rangle$ *is the quotient of* $\mathcal{P}^R_p$, *then for every state q in Q,* $\mathcal{A}_q = \langle Q, A_b, \delta, q, T \rangle$ *is the quotient of* $\mathcal{P}^S_p$ *for some set S.*

3   The UP-Criterion

Let $\mathcal{A} = \langle Q, A, E, I, T \rangle$ be an automaton, σ the strong connectivity equivalence relation on Q, and γ the surjective map from Q onto Q/σ. The *condensation* $\mathcal{C}_\mathcal{A}$ of $\mathcal{A}$ is the directed acyclic graph with loops (V, E) such that V is the image of Q by γ; and the edge (x, y) is in E if there exists a transition $q \xrightarrow{a} s$ in $\mathcal{A}$, for some q in $\gamma^{-1}(x)$, s in $\gamma^{-1}(y)$ and a in A. The condensation of $\mathcal{A}$ can be computed in linear time by Tarjan's algorithm (*cf.* [6]).

We say that an SCC C of an automaton $\mathcal{A}$ is *embeddable* in another SCC D of $\mathcal{A}$ if there exists an injective function $f : C \to D$ such that, for all q in C and a in A: if $q \cdot a$ is in C then $f(q \cdot a) = (f(q) \cdot a)$, and if $q \cdot a$ is not in C, then $f(q) \cdot a = q \cdot a$.

Definition 1 (The UP-criterion). *Let $\mathcal{A}$ be a complete deterministic automaton and $\mathcal{C}_\mathcal{A}$ its condensation. We say that $\mathcal{A}$ satisfies the UP-criterion (or equivalently that $\mathcal{A}$ is a UP-automaton) if the following five conditions hold.*

UP-0. *The successor by 0 of a final (resp. non-final) state of $\mathcal{A}$ is final (resp. non-final).*

UP-1. *Every non-trivial SCC of $\mathcal{A}$ that contains an internal transition labelled by a digit different from 0 is mapped by γ to a leaf of $\mathcal{C}_\mathcal{A}$. Such and SCC is called a Type 1 SCC.*

UP-2. *Every non-trivial SCC of $\mathcal{A}$ which is not of Type 1:*
i) is a simple circuit labelled by 0 (or 0-circuit);
ii) is mapped by γ to a vertex of $\mathcal{C}_\mathcal{A}$ which has a unique successor, and this successor is a leaf.
Such an SCC is called a Type 2 SCC.

UP-3. *Every Type 1 SCC is the quotient of a Pascal automaton* $\mathcal{P}_p^R$, *for some* R *and* p.

UP-4. *Every Type 2 SCC is embeddable in the unique Type 1 SCC associated with it by (UP-2).*

It should be noted that (UP-0) is not a specific condition, it is more of a precondition (hence its numbering 0) to ensure that either all representations of an integer are accepted, or none them are. Moreover, (UP-1) and (UP-2) (together with the completeness of $\mathcal{A}$) imply the converse of (UP-1), namely that every SCC mapped by γ to a leaf of $\mathcal{C}_\mathcal{A}$ is a Type 1 SCC.

Example 2. Fig. 3 shows a simple but complete example of a UP-automaton. The framed subautomata are the minimisation of Pascal automata $\mathcal{P}_3^{\{1,2\}}$ on the top and $\mathcal{P}_5^{\{1,2,3,4\}}$ on the bottom. The two others non-trivial SCC's, $\{B_2, C_2\}$ and $\{D_2\}$, are reduced to 0-circuits. Each of them has successors in only one Pascal automaton.

The dotted lines highlight (UP-4). The circuit (B_2, C_2) is embeddable in the Pascal automaton $\{A, B, C\}$ with the map $B_2 \mapsto B$ and $C_2 \mapsto C$. A similar observation can be made for the circuit (D_2).

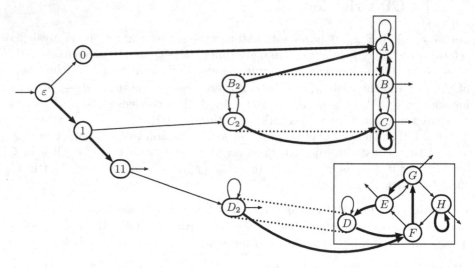

Fig. 3. A complete example of the UP-criterion

Completeness and correctness of the UP-criterion are established as follows.

1. Every UP-set of numbers is accepted by a UP-automaton;
2. The UP-criterion is stable by quotient;
3. Every UP-automaton accepts a UP-set of numbers.

The first two steps ensure completeness for minimal automata (as every b-recognisable set of numbers is accepted by a *unique minimal automaton*), the third one plays for correctness.

3.1   Every UP-Set of Numbers Is Accepted by a UP-Automaton

Proposition 5. *For every integers m and p and for every set R of residues there exists a UP-automaton accepting $E_{p,m}^R$.*

When the Period Divides a Power of the Base. Let E_p^R be a periodic set of numbers such that $p \mid b^j$ for some j. An automaton accepting E_p^R is obtained by a generalisation of the method for recognising if an integer written in base 10 is a multiple of 5, namely checking if its unit digit is 0 or 5: from (1) follows:

Lemma 1. *Let d be an integer such that $d \mid b^j$ (and $d \nmid b^{j-1}$) and u in A_b^* of length j. Then, w in A_b^* is such that $\overline{w} \equiv \overline{u} \, [d]$ if, and only if, $w = uv$ for a certain v.*

The Case of Periodic Sets of Numbers. Let E_p^R be a periodic set of numbers. In contrast with Sect. 2.2, p and b are not supposed to be coprime anymore. Given a integer p, there exist k and d such that $p = kd$, k and b are coprime, and $d \mid b^j$ for a certain j. The Chinese remainder theorem, a simplified version of which is given below, allows to break the condition: 'being congruent to r modulo p' into two simpler conditions.

Theorem 9 (Chinese remainder theorem). *Let k and d be two coprime integers. Let r_k, r_d be two integers. There exists a unique integer $r < kd$ such that $r \equiv r_k \, [k]$ and $r \equiv r_d \, [d]$.*

Moreover, for every n such that $n \equiv r_k \, [k]$ and $n \equiv r_d \, [d]$, we have $n \equiv r \, [kd]$.

Let us assume for now that R is a singleton $\{r\}$, with r in $\{0, 1, \ldots, p-1\}$ and define $r_d = (r \bmod d)$ and $r_k = (r \bmod k)$. Theorem 9 implies:

$$\forall n \in \mathbb{N} \qquad n \equiv r \, [p] \qquad \Longleftrightarrow \qquad n \equiv r_k \, [k] \quad \text{and} \quad n \equiv r_d \, [d] \ . \qquad (5)$$

The Pascal automaton $\mathcal{P}_k^{r_k}$ accepts the integers satisfying $n \equiv r_k \, [k]$ and an automaton accepting the integers satisfying $n \equiv r_d \, [d]$ can be defined from Lemma 1. The *product* of the two automata accepts the integers satisfying both equations of the right-hand side of (5) and this is a UP-automaton. Figure 4 shows the automaton accepting integers congruent to 5 modulo 12 in base 2.

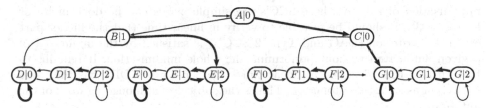

Fig. 4. Automaton accepting integers congruent to 5 modulo 12 in base 2

The case where R is not a singleton is laboured but essentially the same. We denote by $\mathcal{B}_p^R$ the automaton accepting E_p^R.

The Case of Arbitrary UP-Sets of Numbers. Let us denote by $\mathcal{D}_m$ the automaton accepting words whose value is greater than m. It consists in a complete b-tree T_m of depth $\lceil log_b(m) \rceil$ plus a final sink state. Every state may be labelled by the value of the word reaching it and it is final if its label is greater than m. Additionally, every leaf of T_m loops onto itself by reading a 0 and reaches the sink state by reading any other digit. Every $\mathcal{D}_m$ is obviously a UP-automaton.

An arbitrary UP-set of numbers $E_{p,m}^R$ is accepted by the product $\mathcal{B}_p^R \times \mathcal{D}_m$, denoted by $\mathcal{B}_{p,m}^R$. The very special form of $\mathcal{D}_m$ makes it immediate that this product is a UP-automaton, and this complete the proof of Proposition 5.

3.2   The UP-Criterion Is Stable by Quotient

Proposition 6. *If $\mathcal{A}$ is a UP-automaton, then every quotient of $\mathcal{A}$ is also a UP-automaton.*

The UP-criterion relies on properties of SCC's that are stable by quotient. The proof of Proposition 6 then consists essentially of proving that SCC's are mapped into SSC's by the quotient.

3.3   Every UP-Automaton Accepts a UP-Set of Numbers

Let $\mathcal{A}$ be a UP-automaton and $\mathcal{C}_\mathcal{A}$ its condensation. We call *branch* of $\mathcal{C}_\mathcal{A}$ any path going from the root to a leaf using no loops. There is finitely many of them. The inverse image by γ of a branch of $\mathcal{C}_\mathcal{A}$ define a subautomaton of $\mathcal{A}$. Since a finite union of UP-sets of numbers is still UP, it is sufficient to prove the following statement.

Proposition 7. *Let $\mathcal{A}$ be a UP-automaton and $\mathcal{C}_\mathcal{A}$ its condensation. The inverse image by γ of a branch of $\mathcal{C}_\mathcal{A}$ accepts a UP-set of numbers.*

4   Conclusion and Future Work

This work almost closes the complexity question raised by the Honkala's original paper [11]. The simplicity of the arguments in the proof should not hide that the difficulty was to make the proofs simple. Two questions remain: getting rid, in Theorem 2 of the minimality condition; or of the condition of determinism.

We are rather optimistic for a positive answer to the first one. Since the minimisation of a DFA whose SCC's are simple cycles can be done in linear time (*cf.* [3]), it should be possible to verify in linear time that the higher part of the UP-criterion (DAG and Type 2 SCC's) is satisfied by the minimised of a given automaton without performing the whole minimisation. It remains to find an algorithm deciding in linear time whether a given DFA has the same behaviour as a Pascal automaton. This is the subject of still ongoing work of the authors.

On the other hand, defining a similar UP-criterion for nondeterministic automata seems to be much more difficult. The criterion relies on the form and relations between SCC's, and the determinisation process is prone to destroy them.

References

1. Aho, A.V., Hopcroft, J.E., Ullman, J.D.: The Design and Analysis of Computer Algorithms. Addison-Wesley (1974)
2. Allouche, J.-P., Rampersad, N., Shallit, J.: Periodicity, repetitions, and orbits of an automatic sequence. Theoret. Comput. Sci. 410, 2795–2803 (2009)
3. Almeida, J., Zeitoun, M.: Description and analysis of a bottom-up DFA minimization algorithm. Inf. Process. Lett. 107(2), 52–59 (2008)
4. Bruyère, V., Hansel, G., Michaux, C., Villemaire, R.: Logic and p-recognizable sets of integers. Bull. Belg. Soc. Math. 1, 191–238 (1994); Corrigendum. Bull. Belg. Soc. Math. 1, 577 (1994)
5. Cobham, A.: On the base-dependance of the sets of numbers recognizable by finite automata. Math. Systems Theory 3, 186–192 (1969)
6. Cormen, T.H., Leiserson, C.E., Rivest, R.L., Stein, C.: Introduction to Algorithms, 3rd edn. MIT Press (2009)
7. Durand, F., Rigo, M.: On Cobham's theorem, HAL-00605375. Pin, J.-E. (ed.) To Appear in AutoMathA Handbook. E.M.S. (2011)
8. Frougny, C.: Representation of numbers and finite automata. Math. Systems Theory 25, 37–60 (1992)
9. Frougny, C., Sakarovitch, J.: Number representation and finite automata. In: Berthé, V., Rigo, M. (eds.) Combinatorics, Automata and Number Theory. Encyclopedia of Mathematics and its Applications, vol. 135, pp. 34–107. Cambridge Univ. Press (2010)
10. Ginsburg, S., Spanier, E.H.: Semigroups, Presburger formulas and languages. Pacif. J. Math. 16, 285–296 (1966)
11. Honkala, J.: A decision method for the recognizability of sets defined by number systems. RAIRO Theor. Informatics and Appl. 20, 395–403 (1986)
12. Lecomte, P., Rigo, M.: Abstract numeration systems. In: Berthé, V., Rigo, M. (eds.) Combinatorics, Automata and Number Theory. Encyclopedia of Mathematics and its Applications, vol. 135, pp. 108–162. Cambridge Univ. Press (2010)
13. Leroux, J.: A polynomial time Presburger criterion and synthesis for number decision diagrams. In: Logic in Computer Science 2005 (LICS 2005), pp. 147–156. IEEE Comp. Soc. Press (2005); New version at arXiv:cs/0612037v1
14. Marsault, V., Sakarovitch, J.: Ultimate periodicity of b-recognisable sets: a quasilinear procedure, http://arxiv.org/abs/1301.2691
15. Muchnik, A.: The definable criterion for definability in Presburger arithmetic and its applications. Theoret. Computer Sci. 290, 1433–1444 (1991)
16. Sakarovitch, J.: Elements of Automata Theory. Cambridge University Press (2009); Corrected English translation of Éléments de théorie des automates. Vuibert (2003)

3-Abelian Cubes
Are Avoidable on Binary Alphabets

Robert Mercaş[1,*] and Aleksi Saarela[2,**]

[1] Otto-von-Guericke-Universität Magdeburg, Fakultät für Informatik,
PSF 4120, D-39016 Magdeburg, Germany
robertmercas@gmail.com
[2] Department of Mathematics and Statistics, University of Turku,
FI-20014 Turku, Finland
amsaar@utu.fi

Abstract. A k-abelian cube is a word uvw, where u, v, w have the same factors of length at most k with the same multiplicities. Previously it has been known that k-abelian cubes are avoidable over a binary alphabet for $k \geq 5$. Here it is proved that this holds for $k \geq 3$.

Keywords: combinatorics on words, k-abelian equivalence, repetition-freeness.

1   Introduction

The study of repetition-free infinite words (or even the whole area of combinatorics on words) was begun by Axel Thue [15,16]. He proved that using three letters one can construct an infinite word that does not contain a square, that is a factor of the form uu where u is a non-empty word. Further, using two letters one can construct an infinite word that does not contain a cube, that is a factor of the form uuu where u is a non-empty word, or even an overlap, that is a factor of the form $auaua$ where u is a word and a is a letter. Due to their initial obscure publication, these results have been rediscovered several times.

The problem of repetition-freeness has been studied from many points of view. One is to consider fractional powers. This leads to the concept of repetition threshold and the famous Dejean's conjecture, which was proved in many parts. For example, an infinite number of cases were settled in [3], while the last remaining cases were settled independently in [4] and [14]. Another example is the repetition-freeness of partial words. It was shown that there exist infinite ternary words with an infinite number of holes whose factors are not matching any squares (overlaps) of words of length greater than one [12,2]. For the abelian case an alphabet with as low as 5 letters is enough in order to construct an infinite word with an infinite number of holes with factors that do not match an abelian square of any word of length greater than two [1].

---

* Supported by the Alexander von Humboldt Foundation.
** Supported by the Academy of Finland under grant 257857.

M.-P. Béal and O. Carton (Eds.): DLT 2013, LNCS 7907, pp. 374–383, 2013.
© Springer-Verlag Berlin Heidelberg 2013

In this paper abelian repetition-freeness is an important concept. An abelian square is a non-empty word uv, where u and v have the same number of occurrences of each symbol. Abelian cubes and nth powers are defined in a similar way. Erdős [6] raised the question whether abelian squares can be avoided, i.e., whether there exist infinite words over a given alphabet that do not contain two consecutive permutations of the same factor. It is easily seen that abelian squares cannot be avoided over a three-letter alphabet: Each word of length eight over three letters contains an abelian square. Dekking [5] proved that over a binary alphabet there exists a word that avoids abelian fourth powers, and over a ternary alphabet there exists a word that avoids abelian cubes. The problem of whether abelian squares can be avoided over a four-letter alphabet was open for a long time. In [11], using an interesting combination of computer checking and mathematical reasoning, Keränen proved that abelian squares are avoidable on four letters.

Recently, several questions have been studied from the point of view of k-abelian equivalence. For a positive integer k, two words are said to be k-abelian equivalent if they have the same number of occurrences of every factor of length at most k. These equivalence relations provide a bridge between abelian equivalence and equality, because 1-abelian equivalence is the same as abelian equivalence, and as k grows, k-abelian equivalence becomes more and more like equality. The topic of this paper is k-abelian repetition-freeness, but there has also been research on other topics related to k-abelian equivalence [9,10].

In [9], the authors show that 2-abelian squares are avoidable only on a four letter alphabet. For $k \geq 3$, the question of avoiding k-abelian squares is open, the minimal alphabet size being either three or four. Computer experiments would suggest that there are 3-abelian square-free ternary words, but it is known that there are no pure morphic k-abelian square-free ternary words for any k [7].

It was conjectured in [9] that for avoiding k-abelian cubes a binary alphabet suffices whenever $k \geq 2$ since computer generated words of length 100000 having the property have been found. This was proved for $k \geq 8$ in [8] and for $k \geq 5$ in [13].

In this work it is proved that 3-abelian cubes can be avoided on a binary alphabet. The methods used are somewhat similar to those used in [8] and [13]: A word is constructed by mapping an abelian cube-free ternary word by a morphism. However, there are some crucial differences. Most importantly, the morphisms used in this paper are not uniform, and this makes many parts of the proofs different and more difficult. The method used in this article is fairly general, but using it requires an extensive case analysis, which can only be carried out with the help of a computer. The 2-abelian case remains open.

2   Preliminaries

We denote by Σ a finite set of symbols called *alphabet*. For $n \geq 0$, the n-letter alphabet $\{0, \ldots, n-1\}$ will be denoted by Σ_n. A *word* w represents a concatenation of letters from Σ. By ε we denote the *empty word*. We denote by $|w|$ the *length* of w and by $|w|_u$ the number of occurrences of u in w. For a factorization $w = uxv$, we say that x is a *factor* of w, and whenever u is empty x is a *prefix* and, respectively, when v is empty x is a *suffix* of w. The prefix of w of length k will be denoted by $\mathrm{pref}_k(w)$ and the suffix of length k by $\mathrm{suff}_k(w)$.

The *powers* of a word w are defined recursively, $w^0 = \varepsilon$ and $w^n = ww^{n-1}$ for $n > 0$. We say that w is an nth power if there exists a word u such that $w = u^n$. Second powers are called *squares* and third powers *cubes*.

Words u and v are *abelian equivalent* if $|u|_a = |v|_a$ for all letters $a \in \Sigma$. For a word $u \in \Sigma_n^*$, let $P_u = (|u|_0, \ldots, |u|_{n-1})$ be the *Parikh vector* of u. Words $u, v \in \Sigma_n^*$ are abelian equivalent if and only if $P_u = P_v$.

Two words u and v are *k-abelian equivalent* if $|u|_t = |v|_t$ for every word t of length at most k. Obviously, 1-abelian equivalence is the same as abelian equivalence, and words of length less than $k - 1$ (or, in fact, words of length less than $2k$) are k-abelian equivalent only if they are equal. For words u and v of length at least $k - 1$, another equivalent definition can be given: u and v are k-abelian equivalent if $|u|_t = |v|_t$ for every word t of length k, $\mathrm{pref}_{k-1}(u) = \mathrm{pref}_{k-1}(v)$ and $\mathrm{suff}_{k-1}(u) = \mathrm{suff}_{k-1}(v)$. This latter definition is actually the one used in the proofs of this article.

A *k-abelian nth power* is a word $u_1 u_2 \ldots u_n$, where $u_1, u_2 \ldots, u_n$ are k-abelian equivalent. For $k = 1$ this gives the definition of an *abelian nth power*.

A mapping $f : A^* \to B^*$ is a *morphism* if $f(xy) = f(x)f(y)$ for any words $x, y \in A^*$, and is completely determined by the images $f(a)$ for all $a \in A$.

If no non-empty square is a factor of a word w, then it is said that w is *square-free*, or that w *avoids squares*. If there exists an infinite square-free word over an alphabet Σ, then it is said that *squares are avoidable on Σ*. Of course the only thing that matters here is the size of Σ. Similar definitions can be given for cubes and higher powers, as well as for k-abelian powers.

Unlike ordinary cubes, abelian cubes are not avoidable on a binary alphabet, and unlike ordinary squares, abelian squares are not avoidable on a ternary alphabet. However, Dekking showed in [5] that two letters are sufficient for avoiding abelian fourth powers, and three letters suffice for avoiding abelian cubes. An extension of the latter result is stated in the following theorem. It is proved that the word of Dekking avoids also many other factors in addition to abelian cubes.

Theorem 1. *Let $w = \sigma^\omega(0)$ be a fixed point of the morphism $\sigma : \Sigma_3^* \to \Sigma_3^*$ defined by*

$$\sigma(0) = 0012, \qquad \sigma(1) = 112, \qquad \sigma(2) = 022.$$

Then w is abelian cube-free and contains no factor $apbqcrd$ where a, b, c, d are letters and one of the following conditions is satisfied:

1. *$abcd = 0112$ and $P_p = P_q = P_r$,*
2. *$abcd = 0210$ and $P_p = P_q - (1, -1, 1) = P_r - (0, -1, 1)$,*
3. *$abcd = 0211$ and $P_p = P_q - (1, -1, 1) = P_r - (1, -2, 1)$,*
4. *$abcd = 0220$ and $P_p = P_q - (1, -1, 1) = P_r - (0, 0, 0)$,*
5. *$abcd = 0221$ and $P_p = P_q - (1, -1, 1) = P_r - (1, -1, 0)$,*

6. $abcd = 1001$ and $P_p = P_q = P_r$,
7. $abcd = 1002$ and $P_p = P_q = P_r$,

Proof. The word w was shown to be abelian cube-free in [5]. Similar ideas can be used to show that w avoids the factors $apbqcrd$. Case 1 was proved in [13]. Case 2 is proved here. Cases 3–6 are similar to the first two, so their proofs are omitted. Case 7 is more difficult, so it is proved here.

Let $f : \Sigma^* \to \mathbb{Z}_7$ be the morphism defined by

$$f(0) = 1, \qquad f(1) = 2, \qquad f(2) = 3$$

(here $\mathbb{Z}_7$ is the additive group of integers modulo 7). Then $f(\sigma(x)) = 0$ for all $x \in \Sigma$. If $apbqcrd$ is a factor of w, then there are u, s, t such that $\sigma(u) = sapbqcrdt$ and u is a factor of w. Consider the values

$$f(s), f(sa), f(sap), f(sapb), f(sapbq), f(sapbqc), f(sapbqcr), f(sapbqcrd). \quad (1)$$

These elements are of the form $f(\sigma(u')v') = f(v')$, where v' is a prefix of one of $0012, 112, 022$. The possible values for $f(v')$ are 0, 1, 2 and 4.

Consider Case 2. Let $abcd = 0210$. If $P_p = P_q - (1, -1, 1) = P_r - (0, -1, 1)$, then $f(p) = f(q) - 2 = f(r) - 1$. If we denote $i = f(s)$, $j = f(p)$, then the values for (1) are

$$i, i+1, i+j+1, i+j+4, i+2j+6, i+2j+1, i+3j+2, i+3j+3.$$

For all values of i and j, one is not 0, 1, 2 or 4. This is a contradiction.

Consider Case 7. Let $abcd = 1002$. Let $apbqcrd$ be the shortest factor of w satisfying the conditions of Case 7. Then $P_p = P_q = P_r$ and $f(p) = f(q) = f(r)$. If we denote $i = f(s)$, $j = f(p)$, then the values for (1) are

$$i, i+2, i+j+2, i+j+3, i+2j+3, i+2j+4, i+3j+4, i+3j.$$

It must be $i = 0$ and $j = 6$, because otherwise one of the values is not 0, 1, 2 or 4. There are letters a', b', c', d' and words $s', p', q', r', t', s_2, p_1, p_2, q_1, q_2, r_1, r_2, t_1$ such that

$$
\begin{aligned}
u &= s'a'p'b'q'c'r'd't' & s &= \sigma(s')s_2 \\
s_2 1 p_1 &= \sigma(a') & p &= p_1\sigma(p')p_2 \\
p_2 0 q_1 &= \sigma(b') & q &= q_1\sigma(q')q_2 \\
q_2 0 r_1 &= \sigma(c') & r &= r_1\sigma(r')r_2 \\
r_2 2 t_1 &= \sigma(d') & t &= t_1\sigma(t'),
\end{aligned}
$$

i.e. the situation is like in the following diagram:

s	1	p	0	q	0	r	2	t
	s_2	p_1		p_2	q_1		q_2	r_1		r_2	t_1	
$\sigma(s')$	$\sigma(a')$	$\sigma(p')$	$\sigma(b')$	$\sigma(q')$	$\sigma(c')$	$\sigma(r')$	$\sigma(d')$	$\sigma(t')$

Because $i = 0$, $s_2 = \varepsilon$. Then $\sigma(a')$ begins with 1, so $a' = 1$ and $p_1 = 12$. Thus $p = 12\sigma(p')p_2$. It must be $f(p_2) = f(p) - f(\sigma(p')) - f(12) = j - 0 - 5 = 1$, so $p_2 = 0$. Then $\sigma(b')$ begins with 00, so $b' = 0$ and $q_1 = 12$. Like above, it can be concluded that $q = 12\sigma(q')0$, and similarly also $r = 12\sigma(r')0$. But then $1p'0q'0r'2$ is a factor of w. If

$$M = \begin{pmatrix} |\sigma(0)|_0 & |\sigma(1)|_0 & |\sigma(2)|_0 \\ |\sigma(0)|_1 & |\sigma(1)|_1 & |\sigma(2)|_1 \\ |\sigma(0)|_2 & |\sigma(1)|_2 & |\sigma(2)|_2 \end{pmatrix} = \begin{pmatrix} 2 & 1 & 1 \\ 0 & 2 & 1 \\ 1 & 0 & 2 \end{pmatrix}$$

and Parikh vectors are interpreted as column vectors, then

$$MP_{p'} = P_{\sigma(p')}, \qquad MP_{q'} = P_{\sigma(q')}, \qquad MP_{r'} = P_{\sigma(r')}.$$

Because M is invertible and $\sigma(p'), \sigma(q'), \sigma(r')$ are abelian equivalent, also p', q', r' are abelian equivalent. Because $1p'0q'0r'2$ is shorter than $1p0q0r2$, this contradicts the minimality of $1p0q0r2$. □

If abelian cubes are avoidable on some alphabet, then so are k-abelian cubes. This means that k-abelian cubes are avoidable on a ternary alphabet for all k. But for which k are they avoidable on a binary alphabet? In [8] it was proved that this holds for $k \geq 8$, and conjectured that it holds for $k \geq 2$. In [13] it was proved that this holds for $k \geq 5$. In this article it is proved that this holds for $k \geq 3$. The case when $k = 2$ remains open.

3   3-Abelian Cube-Freeness

Let $w \in \Sigma_m^\omega$. The following remarks will be used in the case where $m = 3$, $n = 2$, w is abelian cube-free and $k = 4$ or $k = 3$, but they hold also more generally.

For a word $v \in \Sigma_n^*$, let $Q_v = (|v|_{t_0}, \ldots, |v|_{t_{N-1}})$, where $t_0, \ldots, t_{N-1}$ are the words of Σ_n^k in lexicographic order. When doing matrix calculations, all vectors P_u and Q_v will be interpreted as column vectors.

Let $h : \Sigma_m^* \to \Sigma_n^*$ be a morphism. It needs to be assumed that h satisfies three conditions:

- There is a word $s \in \Sigma_m^{k-1}$ that is a prefix of $h(a)$ for every $a \in \Sigma_m$.
- The matrix M whose columns are $Q_{h(0)s}, \ldots, Q_{h(m-1)s}$ has rank m.
- There are no k-abelian equivalent words v_1, v_2, v_3 of length less than

$$2 \max \{h(a) \mid a \in \Sigma_m\}$$

such that $v_1v_2v_3$ is a factor of $h(w)$.

Let M^+ be the Moore-Penrose pseudoinverse of M. The only properties of M^+ needed in this article are that it exists and can be efficiently computed, and that since the columns of M are linearly independent, M^+M is the $m \times m$ identity matrix. For any word $u \in \Sigma^*$, $Q_{h(u)s} = MP_u$.

Lemma 2. *If the word $h(w)$ has a factor $v_1 v_2 v_3$, where v_1, v_2, v_3 are k-abelian equivalent, then there are letters $a_0, a_1, a_2, a_3, b_2, b_3 \in \Sigma_m$, words $u_1, u_2, u_3 \in \Sigma_m^*$ and indices*

$$
\begin{aligned}
i_0 &\in \{0, \ldots, |h(a_0)| - 1\}, \\
i_1 &\in \{k - 1, \ldots, |h(a_1)| + k - 2\}, \\
i_2 &\in \{k - 1, \ldots, |h(a_2)| + k - 2\}, \\
i_3 &\in \{k - 1, \ldots, |h(a_3)| + k - 2\}
\end{aligned}
\tag{2}
$$

such that $a_0 u_1 a_1 b_2 u_2 a_2 b_3 u_3 a_3$ is a factor of w and $v_i = x_i h(u_i) y_i$ for $i \in \{1, 2, 3\}$, where

$$
\begin{array}{ll}
x_1 = \mathrm{suff}_{|h(a_0)| - i_0}(h(a_0)) & y_1 = \mathrm{pref}_{i_1}(h(a_1 b_2)), \\
x_2 = \mathrm{suff}_{|h(a_1 b_2)| - i_1}(h(a_1 b_2)) & y_2 = \mathrm{pref}_{i_2}(h(a_2 b_3)), \\
x_3 = \mathrm{suff}_{|h(a_2 b_3)| - i_2}(h(a_2 b_3)) & y_3 = \mathrm{pref}_{i_3}(h(a_3)s).
\end{array}
\tag{3}
$$

Proof. It was assumed that $h(w)$ does not contain short k-abelian cubes, and a longer k-abelian cube $v_1 v_2 v_3$ must be of the form specified in the claim. $\square$

Because s is a prefix of $h(u_i)$ and y_i, it follows that $Q_{v_i} = Q_{x_i s} + Q_{h(u_i)s} + Q_{y_i}$.

The idea is to iterate over all values of $a_0, a_1, a_2, a_3, b_2, b_3$ and i_0, i_1, i_2, i_3 and in each case try to deduce that one of the following holds:

- There are no u_1, u_2, u_3 such that the words $v_i = x_i h(u_i) y_i$ are k-abelian equivalent.
- If $v_i = x_i h(u_i) y_i$ are k-abelian equivalent, then $a_0 u_1 a_1 b_2 u_2 a_2 b_3 u_3 a_3$ contains an abelian cube or a factor of the form mentioned in Theorem 1.

If we succeed, then there are words w such that $h(w)$ is k-abelian cube-free. The following lemmas will be useful.

Lemma 3. *Let $a_0, a_1, a_2, a_3, b_2, b_3 \in \Sigma_m$, indices i_0, i_1, i_2, i_3 be as in (2) and words $x_1, x_2, x_3, y_1, y_2, y_3$ be as in (3). Let the following condition be satisfied:*

$$
\begin{aligned}
&\mathrm{pref}_{k-1}(x_1 s), \mathrm{pref}_{k-1}(x_2), \mathrm{pref}_{k-1}(x_3) \text{ are not equal or} \\
&\mathrm{suff}_{k-1}(y_1), \mathrm{suff}_{k-1}(y_2), \mathrm{suff}_{k-1}(y_3) \text{ are not equal.}
\end{aligned}
\tag{C1}
$$

Then there are no u_1, u_2, u_3 such that the three words $v_i = x_i h(u_i) y_i$ would be k-abelian equivalent.

Proof. If the prefixes or suffixes of v_1, v_2, v_3 of length $k - 1$ are not equal, then v_1, v_2, v_3 cannot be k-abelian equivalent. $\square$

Lemma 4. *Let $a_0, a_1, a_2, a_3, b_2, b_3 \in \Sigma_m$, indices i_0, i_1, i_2, i_3 be as in (2) and words $x_1, x_2, x_3, y_1, y_2, y_3$ be as in (3). Let $R_i = Q_{x_i s} + Q_{y_i}$ for $i \in \{1, 2, 3\}$. Let the following condition be satisfied:*

$$
\begin{aligned}
&M^+(R_1 - R_i) \text{ is not an integer vector or} \\
&MM^+(R_1 - R_i) + R_i \text{ are not equal for } i \in \{1, 2, 3\}.
\end{aligned}
\tag{C2}
$$

Then there are no u_1, u_2, u_3 such that the three words $v_i = x_i h(u_i) y_i$ would be k-abelian equivalent.

Proof. If $v_i = x_i h(u_i) y_i$, then $Q_{v_i} = Q_{h(u_i)s} + R_i = MP_{u_i} + R_i$. If $Q_{v_1} = Q_{v_2} = Q_{v_3}$, then $P_{u_i} - P_{u_1} = M^+(R_1 - R_i)$. This must be an integer vector. The vectors $Q_{v_i} - MP_{u_1} = MM^+(R_1 - R_i) + R_i$ must be equal for $i \in \{1, 2, 3\}$. □

Lemma 5. *Let $a_0, a_1, a_2, a_3, b_2, b_3 \in \Sigma_m$, indices i_0, i_1, i_2, i_3 be as in (2) and words $x_1, x_2, x_3, y_1, y_2, y_3$ be as in (3). Let $R_i = Q_{x_i s} + Q_{y_i}$ for $i \in \{1, 2, 3\}$. Let the following condition be satisfied:*

For $i \in \{0, 1, 2, 3\}$ there are $c_i, d_i \in \{a_i, \varepsilon\}$ such that $c_i d_i = a_i$ and

$$
\begin{aligned}
&M^+(R_1 - R_1) + P_{d_0 c_1} \\
&= M^+(R_1 - R_2) + P_{d_1 b_2 c_2} \\
&= M^+(R_1 - R_3) + P_{d_2 b_3 c_3}.
\end{aligned}
\tag{C3}
$$

If $a_0 u_1 a_1 b_2 u_2 a_2 b_3 u_3 a_3$ is abelian cube-free, then the three words $v_i = x_i h(u_i) y_i$ cannot be k-abelian equivalent.

Proof. Like in the proof of Lemma 4, the k-abelian equivalence of v_1, v_2, v_3 implies $P_{u_i} - P_{u_1} = M^+(R_1 - R_i)$. From this and (C3) it follows that

$$
P_{u_1} + P_{d_0 c_1} = P_{u_2} + P_{d_1 b_2 c_2} = P_{u_3} + P_{d_2 b_3 c_3},
$$

so $d_0 u_1 c_1, d_1 b_2 u_2 c_2, d_2 b_3 u_3 c_3$ are abelian equivalent. This contradicts the abelian cube-freeness of $a_0 u_1 a_1 b_2 u_2 a_2 b_3 u_3 a_3$. □

Lemma 6. *Let $a_0, a_1, a_2, a_3, b_2, b_3 \in \Sigma_m$, indices i_0, i_1, i_2, i_3 be as in (2) and words $x_1, x_2, x_3, y_1, y_2, y_3$ be as in (3). Let $R_i = Q_{x_i s} + Q_{y_i}$ for $i \in \{1, 2, 3\}$ and $S_i = M^+(R_1 - R_i) + P_{b_i}$ for $i \in \{2, 3\}$. Let the following condition be satisfied:*

$$
\begin{array}{lll}
(0 = S_2 = S_3 & \text{and} & a_0 a_1 a_2 a_3 = 0112) \text{ or} \\
(0 = S_2 - (1, -1, 1) = S_3 - (0, -1, 1) & \text{and} & a_0 a_1 a_2 a_3 = 0210) \text{ or} \\
(0 = S_2 - (1, -1, 1) = S_3 - (1, -2, 1) & \text{and} & a_0 a_1 a_2 a_3 = 0211) \text{ or} \\
(0 = S_2 - (1, -1, 1) = S_3 - (0, 0, 0) & \text{and} & a_0 a_1 a_2 a_3 = 0220) \text{ or (C4)} \\
(0 = S_2 - (1, -1, 1) = S_3 - (1, -1, 0) & \text{and} & a_0 a_1 a_2 a_3 = 0221) \text{ or} \\
(0 = S_2 = S_3 & \text{and} & a_0 a_1 a_2 a_3 = 1001) \text{ or} \\
(0 = S_2 = S_3 & \text{and} & a_0 a_1 a_2 a_3 = 1002).
\end{array}
$$

If $a_0 u_1 a_1 b_2 u_2 a_2 b_3 u_3 a_3$ is not of the form $apbqcrd$ specified in Theorem 1, then the three words $v_i = x_i h(u_i) y_i$ cannot be k-abelian equivalent.

Proof. Like in the proof of Lemma 4, the k-abelian equivalence of v_1, v_2, v_3 implies $P_{u_i} - P_{u_1} = M^+(R_1 - R_i)$. From this and the first row of (C4) it follows that

$$
P_{u_1} = P_{u_2} + P_{b_2} = P_{u_3} + P_{b_3},
$$

so $u_1, b_2 u_2, b_3 u_3$ are abelian equivalent, which is a contradiction. The other rows lead to a contradiction in a similar way. □

We can iterate over all values of $a_0, a_1, a_2, a_3, b_2, b_3$ and i_0, i_1, i_2, i_3. If in all cases one of Conditions C1, C2, C3 is true, then h maps all abelian cube-free words to k-abelian cube-free words. If in all cases one of Conditions C1, C2, C3, C4 is true, then h maps the word of Theorem 1 to a k-abelian cube-free word. In this way we obtain Theorems 7 and 8.

Concerning the actual implementation of the above algorithm, there are some optimizations that can be made. First, if i_1 and i_2 are such that b_1 and b_2 do not affect the definition of $x_1, x_2, x_3, y_1, y_2, y_3$ in (3), then instead of iterating over all values of b_1 and b_2, they can be combined with u_2 and u_3. Second, in most of the cases Condition C1 is true, and these cases can be handled easily. In the following theorems, there are a couple of thousand nontrivial cases, i.e. cases where Condition C1 is false. A Python file used for proving Theorems 7 and 8 is available on the Internet[1].

Theorem 7. *The morphism defined by*

$$0 \mapsto 10110100110, \qquad 1 \mapsto 101101001001, \qquad 2 \mapsto 1011001100100,$$

maps every abelian cube-free ternary word to a 4-abelian cube-free word.

Proof. The morphism satisfies all conditions stated at the beginning of this section:

- The images of 0, 1 and 2 have the common prefix 101.
- The rows of M corresponding to the factors 0010, 0101 and 1100 are $(0, 1, 2)$, $(1, 0, 1)$ and $(0, 0, 2)$, respectively. These are linearly independent, so the rank of M is 3.
- It can be checked that the image of any abelian cube-free word does not contain 4-abelian cubes of words shorter than 26.

Thus it suffices to check all cases as in the algorithm described above. Observe that here Condition C4 is not needed.   □

Theorem 8. *The morphism defined by*

$$0 \mapsto 01010, \qquad 1 \mapsto 0110010, \qquad 2 \mapsto 0110110,$$

maps the word w of Theorem 1 to a 3-abelian cube-free word.

Proof. The morphism satisfies all conditions stated at the beginning of this section:

- The images of 0, 1 and 2 have the common prefix 01.
- The rows of M corresponding to the factors 010, 011 and 101 are $(2, 1, 0)$, $(0, 1, 2)$ and $(1, 0, 1)$, respectively. These are linearly independent, so the rank of M is 3.
- It can be checked that the image of w does not contain 3-abelian cubes of words shorter than 14.

Thus it suffices to check all cases as in the algorithm described above.   □

---

[1] http://users.utu.fi/amsaar/en/code.htm

We end this work with some remarks regarding how the search of these morphisms was performed. A first observation is that in order to avoid short cubes and given the fact that we want the obtained images to have the same prefix of length $k - 1$, we can only look at morphisms obtained by concatenation of elements from the set $\{ab, aab, abb, aabb\}$. Moreover, when investigating infinite words obtained by application of some morphism to the Dekking word, we note that not only all the images but also their concatenation with themselves must be k-abelian cube-free. Hence, one can generate all words up to some length, say 30, and check for which of these both them and their squares occur among factors. Next, using some backtracking one can check if any triple made of these words would in fact be fit for application on the Dekking word. One final observation is that in order to ensure that any of these triples constitute good candidates, one must check the k-abelian cube-freeness property for factors up to length 10,000, as it happened that the first occurrence of a 3-abelian cube of length over 1,000 started after position 7,000 of the generated word.

References

1. Blanchet-Sadri, F., Kim, J.I., Mercaş, R., Severa, W., Simmons, S., Xu, D.: Avoiding abelian squares in partial words. Journal of Combinatorial Theory. Series A 119(1), 257–270 (2012)
2. Blanchet-Sadri, F., Mercaş, R., Scott, G.: A generalization of Thue freeness for partial words. Theoretical Computer Science 410(8-10), 793–800 (2009)
3. Carpi, A.: On Dejean's conjecture over large alphabets. Theoretical Computer Science 385(1-3), 137–151 (2007)
4. Currie, J., Rampersad, N.: A proof of Dejean's conjecture. Mathematics of Computation 80, 1063–1070 (2011)
5. Dekking, F.M.: Strongly nonrepetitive sequences and progression-free sets. Journal of Combinatorial Theory. Series A 27(2), 181–185 (1979)
6. Erdős, P.: Some unsolved problems. Magyar Tudományos Akadémia Matematikai Kutató Intézete 6, 221–254 (1961)
7. Huova, M., Karhumäki, J.: On the unavoidability of k-abelian squares in pure morphic words. Journal of Integer Sequences 16(2) (2013)
8. Huova, M., Karhumäki, J., Saarela, A.: Problems in between words and abelian words: k-abelian avoidability. Theoretical Computer Science 454, 172–177 (2012)
9. Huova, M., Karhumäki, J., Saarela, A., Saari, K.: Local squares, periodicity and finite automata. In: Calude, C.S., Rozenberg, G., Salomaa, A. (eds.) Rainbow of Computer Science. LNCS, vol. 6570, pp. 90–101. Springer, Heidelberg (2011)
10. Karhumäki, J., Puzynina, S., Saarela, A.: Fine and Wilf's theorem for k-abelian periods. In: Yen, H.-C., Ibarra, O.H. (eds.) DLT 2012. LNCS, vol. 7410, pp. 296–307. Springer, Heidelberg (2012)
11. Keränen, V.: Abelian squares are avoidable on 4 letters. In: Proceedings of the 19th International Colloquium on Automata, Languages and Programming, pp. 41–52 (1992)
12. Manea, F., Mercaş, R.: Freeness of partial words. Theoretical Computer Science 389(1-2), 265–277 (2007)
13. Mercas, R., Saarela, A.: 5-abelian cubes are avoidable on binary alphabets. In: Proceedings of the 14th Mons Days of Theoretical Computer Science (2012)

14. Rao, M.: Last cases of Dejean's conjecture. Theoretical Computer Science 412(27), 3010–3018 (2011)
15. Thue, A.: Über unendliche Zeichenreihen. Norske Vid. Selsk. Skr. I, Mat. Nat. Kl. Christiania 7, 1–22 (1906); Reprinted in Selected Mathematical Papers of Axel Thue. Nagell, T. (ed.) Universitetsforlaget, Oslo, Norway, pp. 139–158 (1977)
16. Thue, A.: Über die gegenseitige Lage gleicher Teile gewisser Zeichenreihen. Norske Vid. Selsk. Skr. I, Mat. Nat. Kl. Christiania 1, 1–67 (1912); Reprinted in Selected Mathematical Papers of Axel Thue. Nagell, T. (ed.) Universitetsforlaget, Oslo, Norway, pp. 413–478 (1977)

Repetition Avoidance in Circular Factors

Hamoon Mousavi and Jeffrey Shallit

School of Computer Science, University of Waterloo, Waterloo, ON N2L 3G1 Canada
{sh2mousa,shallit}@uwaterloo.ca

Abstract. We consider the following novel variation on a classical avoidance problem from combinatorics on words: instead of avoiding repetitions in all factors of a word, we avoid repetitions in all factors where
each individual factor is considered as a "circular word", i.e., the end of
the word wraps around to the beginning. We determine the best possible
avoidance exponent for alphabet size 2 and 3, and provide a lower bound
for larger alphabets.

1   Introduction

Repetition in words is an active research topic and has been studied for over a
hundred years. For example, Axel Thue [10,11] constructed an infinite word over
a three-letter alphabet that contains no squares (i.e., no nonempty word of the
form xx), and another infinite word over a two-letter alphabet that contains no
cubes (i.e., no nonempty word of the form xxx).

In 1972, Dejean refined these results by considering fractional powers. An α-
power for a rational number $\alpha \geq 1$ is a word of the form $w = x^{\lfloor \alpha \rfloor} x'$, where x'
is a (possibly empty) prefix of x and $|w| = \alpha|x|$. The word w is a *repetition*,
with a *period* x and an *exponent* α. Among all possible exponents, we let $\exp(w)$
denote the largest one, corresponding to the shortest period. For example, the
word `alfalfa` has shortest period `alf` and exponent $\frac{7}{3}$. The *critical exponent*
of a word w is the supremum, over all factors f of w, of $\exp(f)$. We write it as
$\exp(w)$.

For a real number α, an α^+-*power* is a β-power where $\beta > \alpha$. For example
$ababa = (ab)^{\frac{5}{2}}$ is a 2^+-power. A word w is

- α^+-*power-free*, if none of the factors of w is an α^+-power;
- α-*power-free* if, in addition to being α^+-power-free, the word w has no factor
  that is an α-power.

We also say that w *avoids* α^+-powers (resp., avoids α-powers). Dejean asked,
what is the smallest real number r for which there exist infinite r^+-power-free
words over an alphabet of size k? This quantity is called the *repetition threshold*
[2], and is denoted by $\mathrm{RT}(k)$. From results of Thue we know that $\mathrm{RT}(2) = 2$.
Dejean [5] proved $\mathrm{RT}(3) = \frac{7}{4}$, and conjectured that

$$\mathrm{RT}(k) = \begin{cases} \frac{7}{5}, & \text{if } k = 4; \\ \frac{k}{k-1}, & \text{if } k > 4. \end{cases}$$

M.-P. Béal and O. Carton (Eds.): DLT 2013, LNCS 7907, pp. 384–395, 2013.
© Springer-Verlag Berlin Heidelberg 2013

This conjecture received much attention in the last forty years, and its proof was recently completed by Currie and Rampersad [4] and Rao [9], independently, based on work of Carpi [3] and others.

We consider the following novel variation on Dejean, which we call "circular α-power avoidance". We consider each finite factor x of a word w, but interpret such a factor as a "circular" word, where the end of the word wraps around to the beginning. Then we consider each factor f of this interpretation of x; for w to be circularly α-power-free, each such f must be α-power-free. For example, consider the English word $w = $ dividing with factor $x = $ dividi. The circular shifts of x are

$$\text{dividi}, \text{ividid}, \text{vididi}, \text{ididiv}, \text{didivi}, \text{idivid},$$

and (for example) the word ididiv contains a factor idididi that is a $\frac{5}{2}$-power. In fact, w is circularly cubefree and circularly $(\frac{5}{2})^+$-power-free.

To make this more precise, we recall the notion of conjugacy. Two words x, y are *conjugate* if one is a cyclic shift of the other; that is, if there exist words u, v such that $x = uv$ and $y = vu$.

Definition 1. *Let w be a finite or infinite word. The largest circular α-power in a word w is defined to be the supremum of $\exp(f)$ over all factors f of conjugates of factors of w. We write it as $\mathrm{cexp}(w)$.*

Although Definition 1 characterizes the subject of this paper, we could have used a different definition, based on the following.

Proposition 2. *Let w be a finite word or infinite word. The following are equivalent:*

(a) s is a factor of a conjugate of a factor of w;
(b) s is a prefix of a conjugate of a factor of w;
(c) s is a suffix of a conjugate of a factor of w;
(d) $s = vt$ for some factor tuv of w.

Proof. (a) $\Longrightarrow$ (b): Suppose $s = y''x'$, where xy is a factor of w and $x = x'x''$ and $y = y'y''$. Another conjugate of xy is then $y''x'x''y'$ with prefix $y''x'$.

(b) $\Longrightarrow$ (c): Such a prefix s is either of the form y' or yx', where xy be a factor of w and $x = x'x''$ and $y = y'y''$. Considering the conjugate $y''xy'$ of yx, we get a suffix y', and consider the conjugate $x''yx'$ we get a suffix yx'.

(c) $\Longrightarrow$ (d): Such a suffix s is either of the form $s = x''$ or $s = y''x$, where xy be a factor of w and $x = x'x''$ and $y = y'y''$. In the former case, let $t = x''$, $u = v = \epsilon$. In the latter case, let $t = x$, $u = y'$, and $v = y''$.

(d) $\Longrightarrow$ (a): Let tuv be a factor of w. Then vtu is a conjugate of tuv, and vt is a factor of it. $\square$

Let $\Sigma_k = \{0, 1, \ldots, k-1\}$. Define RTC($k$), the *repetition threshold for circular factors*, to be the smallest real number r for which there exist infinite circularly r^+-power-free words in Σ_k. Clearly we have

$$\mathrm{RTC}(k) \geq \mathrm{RT}(k).$$

In this paper we prove that $\mathrm{RTC}(2) = 4$ and $\mathrm{RTC}(3) = \frac{13}{4}$. For larger alphabets, we conjecture that

$$\mathrm{RTC}(k) = \begin{cases} \frac{5}{2}, & \text{if } k = 4; \\ \frac{105}{46}, & \text{if } k = 5; \\ \frac{2k-1}{k-1}, & \text{if } k \geq 6. \end{cases}$$

Finally, we point out that the quantities we study here are *not* closely related to the notion of *avoidance in circular words*, studied previously in [1,6,7].

2   Notation

For a finite alphabet Σ, let Σ^* denote the set of finite words over Σ. Let Σ^ω denote the set of right infinite words over Σ, and let $\Sigma^\infty = \Sigma^\omega \cup \Sigma^*$. Let $w = a_0 a_1 \cdots \in \Sigma^\infty$ be a word. Let $w[i] = a_i$, and let $w[i..j] = a_i \cdots a_j$. By convention we have $w[i] = \epsilon$ for $i < 0$ and $w[i..j] = \epsilon$ for $i > j$. Note that if x is a period of w and $|x| = p$, then $w[i+p] = w[i]$ for $0 \leq i < |w| - p$.

For a word x, let $\mathrm{pref}(x)$ and $\mathrm{suff}(x)$, respectively, denote the set of prefixes and suffixes of x. For words x, y, let $x \preceq y$ denote that x is a factor of y. Let $x \preceq_p y$ (resp., $x \preceq_s y$) denote that x is a prefix (resp., suffix) of y.

A morphism $h : \Sigma^* \to \Phi^*$ is said to be *q-uniform* if $|h(a)| = q$ for all $a \in \Sigma$. A morphism is uniform if it is q-uniform for some q. The fixed point of a morphism $h : \Sigma^* \to \Phi^*$ starting with $a \in \Sigma$, if it exists, is denoted by $h^\omega(a)$.

In the next section, we prove some preliminary results. We get some bounds for $\mathrm{RTC}(k)$, and in particular, we prove that $\mathrm{RTC}(2) = 2\,\mathrm{RT}(2) = 4$. In Section 4, we study the three-letter alphabet, and we prove that $\mathrm{RTC}(3) = \frac{13}{4}$. Finally, in Section 5, we give another interpretation for repetition threshold for circular factors.

3   Binary Alphabet

First of all, we prove a bound on $\mathrm{RTC}(k)$.

Theorem 3. $1 + \mathrm{RT}(k) \leq \mathrm{RTC}(k) \leq 2\,\mathrm{RT}(k)$.

Proof. Let $r = \mathrm{RT}(k)$. We first prove that $\mathrm{RTC}(k) \leq 2r$. Let $w \in \Sigma_k^\omega$ be an r^+-power-free word. We prove that w is circularly $(2r)^+$-power-free. Suppose that $xty \preceq w$, such that yx is $(2r)^+$-power. Now either y or x is an r^+-power. This implies that w contains an r^+-power, a contradiction.

Now we prove that $1 + r \leq \mathrm{RTC}(k)$. Let l be the length of the longest r-power-free word over Σ_k, and let $w \in \Sigma_k^\omega$. Considering the factors of length $n = l + 1$ of w, we know some factor f must occur infinitely often. This f contains an r-power: z^r. Therefore $z^r t z$ is a factor of w. Therefore w contains a circular $(1 + r)$-power. This proves that $1 + r \leq \mathrm{RTC}(k)$. $\square$

Note that since $\mathrm{RT}(k) > 1$, we have $\mathrm{RTC}(k) > 2$.

Lemma 4. $\mathrm{RTC}(2) \geq 4$.

Proof. Let $w \in \Sigma_2^\omega$ be an arbitrary word. It suffices to prove that w contains circular 4-powers. There are two cases: either 00 or 11 appears infinitely often, or w ends with a suffix of the form $(01)^\omega$. In the latter case, obviously there are circular 4-powers; in the former there are words of the form $aayaa$ for $a \in \Sigma_2$ and $y \in \Sigma_2^*$ and hence circular 4-powers. $\qquad\square$

Theorem 5. $\mathrm{RTC}(2) = 4$.

Proof. A direct consequence of Theorem 3 and Lemma 4. $\qquad\square$

The Thue-Morse word is an example of a binary word that avoids circular 4^+-powers.

4   Ternary Alphabet

Our goal in this section is to show that $\mathrm{RTC}(3) = \frac{13}{4}$. For this purpose, we frequently use the notion of synchronizing morphism, which was introduced in Ilie et al. [8].

Definition 6. *A morphism* $h : \Sigma^* \to \Gamma^*$ *is said to be* synchronizing *if for all* $a, b, c \in \Sigma$ *and* $s, r \in \Gamma^*$, *if* $h(ab) = rh(c)s$, *then either* $r = \epsilon$ *and* $a = c$ *or* $s = \epsilon$ *and* $b = c$.

Definition 7. *A synchronizing morphism* $h : \Sigma^* \to \Gamma^*$ *is said to be* strongly synchronizing *if for all* $a, b, c \in \Sigma$, *if* $h(c) \in \mathrm{pref}(h(a)) \, \mathrm{suff}(h(b))$, *then either* $c = a$ *or* $c = b$.

The following technical lemma is applied several times throughout the paper.

Lemma 8. *Let* $h : \Sigma^* \to \Gamma^*$ *be a synchronizing q-uniform morphism. Let* $n > 1$ *be an integer, and let* $w \in \Sigma^*$. *If* $z^n \preceq_p h(w)$ *and* $|z| \geq q$, *then* $u^n \preceq_p w$ *for some* u. *Furthermore* $|z| \equiv 0 \pmod{q}$.

Proof. Let $z = h(u)z'$, where $|z'| < q$ and $u \in \Sigma^*$. Note that $u \neq \epsilon$, since $|z| \geq q$. Clearly, we have $z'h(u[0]) \preceq_p h(w[|u|..|u|+1])$. Since h is synchronizing, the only possibility is that $z' = \epsilon$, so $|z| \equiv 0 \pmod q$. Now we can write $z^n = h(u^n) \preceq_p h(w)$. Therefore $u^n \preceq_p w$. $\qquad\square$

The next lemma states that if the fixed point of a strongly synchronizing morphism (SSM) avoids small n'th powers, where n is an integer, it avoids n'th powers of all lengths.

Lemma 9. *Let* $h : \Sigma^* \to \Sigma^*$ *be a strongly synchronizing q-uniform morphism. Let* $n > 1$ *be an integer. If* $h^\omega(0)$ *avoids factors of the form* z^n, *where* $|z^n| < 2nq$, *then* $h^\omega(0)$ *avoids n'th powers.*

Proof. Let $w = a_0 a_1 a_2 \cdots = h^\omega(0)$. Suppose w has n'th powers of length greater than or equal to $2nq$. Let z be the shortest such word, i.e., $|z^n| \geq 2nq$ and $z^n \preceq w$. We can write

$$z^n = xh(w[i..j])y,$$
$$x \preceq_s h(a_{i-1}),$$
$$y \preceq_p h(a_{j+1}),$$
$$|x|, |y| < q,$$

for some integers $i, j \geq 0$. If $x = y = \epsilon$, then using Lemma 8, since $|z| \geq q$, the word $w[i..j]$ contains an n'th power. Therefore w contains an n'th power of length smaller than $|z^n|$, a contradiction. Now suppose that $xy \neq \epsilon$. Since $|z| \geq \frac{2nq}{n} = 2q$, and $|xh(w[i])|, |h(w[j])y| < 2q$, we can write

$$xh(w[i]) \preceq_p z,$$
$$h(w[j])y \preceq_s z.$$

Therefore $h(w[j])yxh(w[i]) \preceq z^2 \preceq z^n$. Since h is synchronizing

$$h(w[j])yxh(w[i]) \preceq h(w[i..j]).$$

Hence $yx = h(a)$ for some $a \in \Sigma$. Since h is an SSM, we have either $a = a_{i-1}$ or $a = a_{j+1}$. Without loss of generality, suppose that $a = a_{i-1}$. Then we can write $h(w[i-1..j]) = yxh(w[i..j])$. The word $yxh(w[i..j])$ is an n'th power, since it is a conjugate of $xh(w[i..j])y$. So we can write

$$h(w[i-1..j]) = z_1^n$$

where z_1 is a conjugate of z. Note that $|z_1| = |z| \geq 2q$. Now using Lemma 8, the word $w[i-1..j]$ contains an n'th power, and hence w contains an n'th power of length smaller than $|z^n|$, a contradiction. $\qquad\square$

The following lemma states that, for an SSM h and a well-chosen word w, all circular $(\frac{13}{4})^+$-powers in $h(w)$ are small.

Lemma 10. *Let $h : \Sigma^* \to \Gamma^*$ be a strongly synchronizing q-uniform morphism. Let $w = a_0 a_1 a_2 \cdots \in \Sigma^\omega$ be a circularly cubefree word. In addition, suppose that w is squarefree. If $x_1 t x_2 \preceq h(w)$ for some words t, x_1, x_2, and $x_2 x_1$ is a $(13/4)^+$-power, then $|x_2 x_1| < 22q$.*

Proof. The proof is by contradiction. Suppose there are words t, x_1, x_2, and z in Γ^* and a rational number $\alpha > \frac{13}{4}$ such that

$$x_1 t x_2 \preceq h(w)$$

$$|x_2 x_1| \geq 22q$$

$$x_2 x_1 = z^\alpha.$$

Suppose $|z| < q$. Let k be the smallest integer for which $|z^k| \geq q$. Then $|z^k| < 2q$, because otherwise $|z^{k-1}| \geq q$, a contradiction. We can write $x_2x_1 = (z^k)^\beta$, where $\beta = \frac{|x_2x_1|}{|z^k|} > \frac{22q}{2q} > \frac{13}{4}$. Therefore we can assume that $|z| \geq q$, since otherwise we can always replace z with z^k, and α with β.

There are three cases to consider.

(a) Suppose that x_1 and x_2 are both long enough, so that each contains an image of a word under h. More formally, suppose that

$$x_1 = y_1 h(w[i_1..j_1])y_2, \tag{1}$$
$$x_2 = y_3 h(w[i_2..j_2])y_4, \tag{2}$$
$$i_1 \leq j_1, i_2 \leq j_2,$$
$$y_1 \preceq_s h(a_{i_1-1}),$$
$$y_2 \preceq_p h(a_{j_1+1}),$$
$$y_3 \preceq_s h(a_{i_2-1}),$$
$$y_4 \preceq_p h(a_{j_2+1}),$$
$$|y_1|, |y_2|, |y_3|, \text{ and } |y_4| < q, \text{ and}$$
$$y_2 t y_3 = h(w[j_1 + 1..i_2 - 1]).$$

Let $v_1 = w[i_1..j_1]$ and $v_2 = w[i_2..j_2]$. See Fig 1.

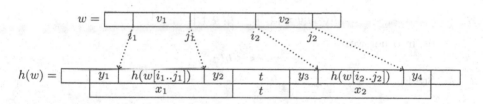

Fig. 1. $x_1 t x_2$ is a factor of $h(w)$

There are two cases to consider.

(1) Suppose that $y_4 y_1 = \epsilon$. Let $v = w[i_2..j_2]w[i_1..j_1]$.
The word $h(v)y_2$ is a factor of $y_3 h(v)y_2 = z^\alpha$ of length $\geq 22q - q = 21q$, and so

$$h(v)y_2 = z_1^\beta,$$

where z_1 is a conjugate of z, and $\beta > \frac{21}{22}\alpha > 3$. Therefore we can write

$$z_1^3 \preceq_p h(v)y_2 \preceq_p h(vw[j_1 + 1]).$$

Note that $|z_1| = |z| \geq q$, so using Lemma 8, we can write $|z_1| \equiv 0 \pmod{q}$. Therefore

$$z_1^3 \preceq_p h(v).$$

Using Lemma 8 again, the word v contains a cube, which means that the word w contains a circular cube, a contradiction.

(2) Suppose that $y_4 y_1 \neq \epsilon$. We show how to get two new factors $x_1' = h(v_1')y_2'$ and $x_2' = y_3'h(v_2')$, with v_1', v_2' nonempty, such that $x_2'x_1' = x_2 x_1$. Then we use case (1) above to get a contradiction.

Let $s = h(w[j_2])y_4 y_1 h(w[i_1])$, and let m be the smallest integer for which $|z^m| \geq |s|$. Note that if $|z| < |s|$, then

$$|z^m| < 2|s| < 8q. \tag{3}$$

We show that at least one of the following inequalities holds:

$$|h(v_1)| \geq q + |z^m|,$$
$$|h(v_2)| \geq q + |z^m|.$$

Suppose that both inequalities fail. Then using (1) and (2) we can write

$$|x_2 x_1| < 2q + 2|z^m| + |y_1 y_2 y_3 y_4| < 6q + 2|z^m|. \tag{4}$$

If $|z| < |s|$, then by (3) and (4) one gets $|x_2 x_1| < 22q$, contradicting our assumption. Otherwise $|z| \geq |s|$, and hence $m = 1$. Then

$$|x_2 x_1| = \alpha|z| < 2q + 2|z| + |y_1 y_2 y_3 y_4| < 6q + 2|z|,$$

and hence $|z| < 6q$. So $|x_2 x_1| < 6q + 2|z| < 18q$, contradicting our assumption. Without loss of generality, suppose that $|h(v_1)| \geq q + |z^m|$. Using the fact that z is a period of $x_2 x_1$, we can write

$$h(v_1)[q + |z^m| - |s|..q + |z^m| - 1] = s,$$

or, in other words,

$$s \preceq h(v_1).$$

See Fig 2.

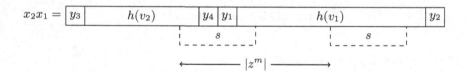

$$x_2 x_1 = \boxed{\; y_3 \;|\; h(v_2) \;|\; y_4 \;|\; y_1 \;|\; h(v_1) \;|\; y_2 \;}$$

Fig. 2. $h(v_1)$ contains a copy of s

Using the fact that h is synchronizing, we get that $y_4 y_1 = h(a)$ for some $a \in \Sigma$. Since h is an SSM, we have either $a = a_{i_1-1}$ or $a = a_{j_2+1}$. Without loss of generality, suppose that $a = a_{j_2+1}$. Now look at the following factors of $h(w)$, which can be obtained from x_1 and x_2 by extending x_2 to the right and shrinking x_1 from the left:

$$x_1' = h(w[i_1..j_1])y_2$$
$$x_2' = y_3 h(w[i_2..j_2 + 1]).$$

See Fig 3.

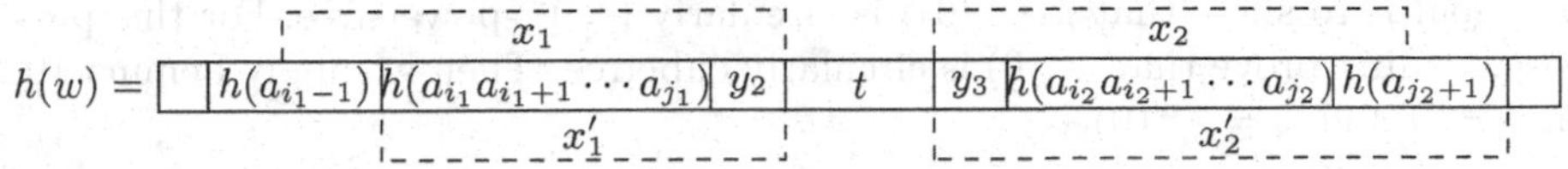

Fig. 3. x_1' and x_2' are obtained from x_1 and x_2

We can see that

$$x_2'x_1' = y_3h(w[i_2..j_2+1])h(w[i_1..j_1])y_2 = y_3h(w[i_2..j_2])y_4y_1h(w[i_1..j_1])y_2 = x_2x_1.$$

Now using case (1) we get a contradiction.

(b) Suppose that x_2 is too short to contain an image of a word under h. More formally, we can write

$$x_1 = y_1h(v)y_2 \text{ where } |x_2| < 2q \text{ and } |y_1|, |y_2| < q$$

for some words $y_1, y_2 \in \Gamma^*$ and a word $v \preceq w$. Then $h(v)$ is a factor of $x_2x_1 = z^\alpha$ of length $\geq 22q - 4q = 18q$, and so

$$h(v) = z_1^\beta,$$

where z_1 is a conjugate of z, and $\beta \geq \frac{18}{22}\alpha > 2$. Note that $|z_1| = |z| \geq q$, so using Lemma 8, the word v contains a square, a contradiction.

(c) Suppose that x_1 is not long enough to contain an image of a word under h. An argument similar to (b) applies here to get a contradiction.

$\square$

The following 15-uniform morphism is an example of an SSM:

$$\mu(0) = 012102120102012$$
$$\mu(1) = 201020121012021$$
$$\mu(2) = 012102010212010$$
$$\mu(3) = 201210212021012$$
$$\mu(4) = 102120121012021$$
$$\mu(5) = 102010212021012.$$

Another example of an SSM is the 4-uniform morphism $\psi : \Sigma_6^* \to \Sigma_6^*$ as follows:

$$\psi(0) = 0435$$
$$\psi(1) = 2341$$
$$\psi(2) = 3542$$
$$\psi(3) = 3540$$
$$\psi(4) = 4134$$
$$\psi(5) = 4105.$$

Our goal is to show that $\mu(\psi^\omega(0))$ is circularly $(\frac{13}{4})^+$-power-free. For this purpose, we first prove that $\psi^\omega(0)$ is circularly cubefree. Then we apply Lemma 10, for $h = \mu$ and $w = \psi^\omega(0)$.

Lemma 11. *The fixed point $\psi^\omega(0)$ is squarefree.*

Proof. Suppose that $\psi^\omega(0)$ contains a square. Using Lemma 9, there is a square $zz \preceq \psi^\omega(0)$ such that $|zz| < 16$. Using a computer program, we checked all factors of length smaller than 16 in $\psi^\omega(0)$, and none of them is a square. This is a contradiction. $\qquad\square$

Lemma 12. *The fixed point $\psi^\omega(0)$ is circularly cubefree.*

Proof. By contradiction. Let $w = a_0 a_1 a_2 \cdots = \psi^\omega(0)$. Suppose $x_1 t x_2 \preceq w$, and $x_2 x_1 = z^3$ for some words t, x_1, x_2, z, where

$$x_1 = y_1 \psi(w[i_1..j_1])y_2,$$
$$x_2 = y_3 \psi(w[i_2..j_2])y_4,$$
$$y_1 \preceq_s \psi(a_{i_1-1}),$$
$$y_2 \preceq_p \psi(a_{j_1+1}),$$
$$y_3 \preceq_s \psi(a_{i_2-1}),$$
$$y_4 \preceq_p \psi(a_{j_2+1}),$$
$$|y_1|, |y_2|, |y_3|, \text{ and } |y_4| < 4,$$
$$y_2 t y_3 = \psi(w[j_1+1..i_2-1]),$$

for proper choices of the integers i_1, i_2, j_1, j_2. Let $v_1 = w[i_1..j_1]$ and $v_2 = w[i_2..j_2]$.

Using a computer program, we searched for circular cubes of length not greater than 66, and it turns out that there is no such circular cube in w. Thus we can assume that $|x_2 x_1| > 66$ so $|z| > 22$. Moreover suppose that $x_2 x_1$ has the smallest possible length.

There are two cases to consider.

(a) Suppose that $y_4 y_1 = \epsilon$. If $y_2 y_3 = \epsilon$, then $\psi(v_2 v_1) = z^3$. Using Lemma 8, we get that $v_2 v_1$ contains a cube. Hence w contains a smaller circular cube, a contradiction.

Suppose that $y_2 y_3 \neq \epsilon$. Since $|y_3 \psi(w[i_2])|, |\psi(w[j_1])y_2| < 8$ and $|z| > 22$, we can write

$$y_3 \psi(w[i_2]) \preceq_p z,$$
$$\psi(w[j_1])y_2 \preceq_s z.$$

Therefore $\psi(w[j_1])y_2 y_3 \psi(w[i_2]) \preceq z^3$, and since ψ is synchronizing

$$\psi(w[j_1])y_2 y_3 \psi(w[i_2]) \preceq \psi(v_2 v_1).$$

Hence $y_2 y_3 = \psi(b)$ for some $b \in \Sigma_6$. Since ψ is an SSM, we have either $b = a_{i_2-1}$, or $b = a_{j_1+1}$. Without loss of generality, suppose that $b = a_{i_2-1}$. So we can write

$$\psi(w[i_2-1..j_2]w[i_1..j_1]) = y_2 y_3 \psi(w[i_2..j_2]w[i_1..j_1]).$$

The word $y_2 y_3 \psi(v_2 v_1)$ is a cube, since it is a conjugate of $y_3 \psi(v_2 v_1) y_2$. So we can write

$$\psi(w[i_2 - 1..j_2]w[i_1..j_1]) = z_1^3$$

where z_1 is a conjugate of z. Then using Lemma 8, the word $w[i_2 - 1..j_2]$ $w[i_1..j_1]$ contains a cube. Note that since $y_2 y_3 \neq \epsilon$ we have $j_1 < i_2 - 1$. Hence $w[i_2 - 1..j_2]w[i_1..j_1]$ is a circular cube of w, a contradiction.

(b) Suppose that $y_4 y_1 \neq \epsilon$. We show how to get two new factors $x_1' = h(v_1')y_2'$ and $x_2' = y_3' h(v_2')$ of w, for nonempty words v_1', v_2', such that $x_2' x_1' = x_2 x_1$. Then we use case (a) above to get a contradiction.

The word w is squarefree due to Lemma 11. Therefore $|x_1|, |x_2| > |z| > \frac{66}{3}$ and hence $|v_1|, |v_2| > 0$. One can observe that either $|\psi(v_1)| \geq 4 + |z|$ or $|\psi(v_2)| \geq 4 + |z|$. Without loss of generality, suppose that $|\psi(v_1)| \geq 4 + |z|$. Let $s = w[j_2]y_4 y_1 w[i_1]$. Now, using the fact that z is a period of $x_2 x_1$, we can write

$$\psi(v_1)[4 + |z| - |s|..4 + |z| - 1] = s,$$

or, in other words,

$$s \preceq \psi(v_1).$$

Using the fact that ψ is synchronizing, we get that $y_4 y_1 = \psi(a)$ for some $a \in \Sigma_6$. Since ψ is an SSM, we have either $a = a_{i_1 - 1}$, or $a = a_{j_2 + 1}$. Without loss of generality, suppose that $a = a_{j_2 + 1}$. Now look at the following factors of w, which can be obtained from x_1 and x_2 by extending x_2 to the right and shrinking x_1 from the left

$$x_1' = \psi(w[i_1..j_1])y_2$$
$$x_2' = y_3 \psi(w[i_2..j_2 + 1]).$$

We can write

$$x_2' x_1' = y_3 \psi(w[i_2..j_2 + 1])\psi(w[i_1..j_1])y_2 = y_3 \psi(v_2)y_4 y_1 \psi(v_1)y_2 = x_2 x_1 = z^3.$$

So using case (a) we get a contradiction. □

Theorem 13. RTC(3) $= \frac{13}{4}$.

Proof. First let us show that RTC(3) $\geq \frac{13}{4}$.

Suppose there exists an infinite word w that avoids circular α-powers, for $\alpha < 4$. We now argue that for every integer C, there exists an infinite word w' that avoids both squares of length $\leq C$ and circular α-powers. Note that none of the factors of w looks like $xxyxx$, since w avoids circular 4-powers. Therefore, every square in w occurs only finitely many times. Therefore w' can be obtained by removing a sufficiently long prefix of w.

Computer search verifies that the longest circularly $\frac{13}{4}$-power-free word over a 3-letter alphabet that avoids squares xx where $|xx| < 147$ has length 147. Therefore the above argument for $C = 147$ shows that circular $\frac{13}{4}$-powers are unavoidable over a 3-letter alphabet.

Now to prove $\mathrm{RTC}(3) = \frac{13}{4}$, it is sufficient to give an example of an infinite word that avoids circular $(\frac{13}{4})^+$-powers. We claim that $\mu(\psi^\omega(0))$ is such an example. We know that $\psi^\omega(0)$ is circularly cubefree. Therefore we can use Lemma 10 for $w = \psi^\omega(0)$ and $h = \mu$. So if $xty \preceq \mu(\psi^\omega(0))$, and yx is a $(\frac{13}{4})^+$-power, then $|yx| < 22 \times 15$. Now there are finitely many possibilities for x and y. Using a computer program, we checked that none of them leads to a $(\frac{13}{4})^+$-power. This completes the proof.   □

5   Another Interpretation

We could, instead, consider the supremum of $\exp(p)$ over all products of i factors of w. Call this quantity $\mathrm{pexp}_i(w)$.

Proposition 14. *If w is a recurrent infinite word, then $\mathrm{pexp}_2(w) = \mathrm{cexp}(w)$.*

Proof. Let s be a product of two factors of w, say $s = xy$. Let y occur for the first time at position i of w. Since w is recurrent, x occurs somewhere after position $i + |y|$ in w. So there exists z such that yzx is a factor of w. Then xy is a factor of a conjugate of a factor of w.

On the other hand, from Proposition 2, we know that if s is a conjugate of a factor of w, then $s = vt$ where tuv is a factor of w. Then s is the product of two factors of w.   □

We can now study the repetition threshold for i-term products, $\mathrm{RT}_i(k)$, which is the infimum of $\mathrm{pexp}_i(w)$ over all words $w \in \Sigma_k^\omega$. Note that

$$\mathrm{RT}_2(k) \geq \mathrm{RTC}(k).$$

The two recurrent words, the Thue-Morse word and $\mu(\psi^\omega(0))$, introduced in Section 4, are circularly $\mathrm{RTC}(2)^+$-power-free and circularly $\mathrm{RTC}(3)^+$-power-free, respectively. Using Proposition 14, we get that $\mathrm{RT}_2(k) = \mathrm{RTC}(k)$ for $k = 2, 3$.

Theorem 15. *For $i \geq 1$ we have $\mathrm{RT}_i(2) = 2i$.*

Proof. From Thue we know there exists an infinite 2^+-power-free word. If some product of factors $x_1 x_2 \cdots x_i$ contains a $(2i)^+$-power, then some factor contains a 2^+-power, a contradiction. So $\mathrm{RT}_i(2) \leq 2i$.

For the lower bound, fix $i \geq 2$, and let $w \in \Sigma_2^\omega$ be an arbitrary word. Either 00 or 11 appears infinitely often, or w ends in a suffix of the form $(01)^\omega$. In the latter case we get arbitrarily high powers, and the former case there is a product of i factors with exponent $2i$.   □

It would be interesting to obtain more values of $\mathrm{RT}_i(k)$. We propose the following conjectures which are supported by numerical evidence:

$$\mathrm{RT}_2(4) = \mathrm{RTC}(4) = \frac{5}{2}\,,$$

$$\mathrm{RT}_2(5) = \mathrm{RTC}(5) = \frac{105}{46}\,,\text{ and}$$

$$\mathrm{RT}_2(k) = \mathrm{RTC}(k) = 1 + \mathrm{RT}(k) = \frac{2k-1}{k-1}\text{ for }k \geq 6.$$

We know that the values given above are lower bounds for $\mathrm{RTC}(k)$.

Acknowledgments. We thank the referees for their careful reading of this paper.

References

1. Aberkane, A., Currie, J.D.: There exist binary circular $5/2^+$ power free words of every length. Electronic J. Combin. 11(1) (2004) Paper #R10, http://www1.combinatorics.org/Volume_11/Abstracts/v11i1r10.html
2. Brandenburg, F.-J.: Uniformly growing k-th power-free homomorphisms. Theoret. Comput. Sci. 23, 69–82 (1983)
3. Carpi, A.: On Dejean's conjecture over large alphabets. Theoret. Comput. Sci. 385, 137–151 (2007)
4. Currie, J., Rampersad, N.: A proof of Dejean's conjecture. Math. Comp. 80, 1063–1070 (2011)
5. Dejean, F.: Sur un théorème de Thue. J. Combin. Theory. Ser. A 13, 90–99 (1972)
6. Gorbunova, I.A.: Repetition threshold for circular words. Electronic J. Combin. 19(4) Paper #11, http://www.combinatorics.org/ojs/index.php/eljc/article/view/v19i4p11
7. Harju, T., Nowotka, D.: Cyclically repetition-free words on small alphabets. Inform. Process Lett. 110, 591–595 (2010)
8. Ilie, L., Ochem, P., Shallit, J.: A generalization of repetition threshold. Theoret. Comput. Sci. 345, 359–369 (2005)
9. Rao, M.: Last cases of Dejean's conjecture. Theoret. Comput. Sci. 412, 3010–3018 (2011)
10. Thue, A.: Über unendliche Zeichenreihen. Norske vid. Selsk. Skr. Mat. Nat. Kl. 7, 1–22 (1906), Reprinted in Selected Mathematical Papers of Axel Thue. Nagell, T. (ed.) Universitetsforlaget, Oslo, pp. 139–158 (1977)
11. Thue, A.: Über die gegenseitige Lage gleicher Teile gewisser Zeichenreihen. Norske vid. Selsk. Skr. Mat. Nat. Kl. 1, 1–67 (1912); Reprinted in Selected Mathematical Papers of Axel Thue. Nagell, T. (ed.) Universitetsforlaget, Oslo, pp. 413–478 (1977)

Operator Precedence ω-Languages

Federica Panella[1], Matteo Pradella[1], Violetta Lonati[2], and Dino Mandrioli[1]

[1] DEIB - Politecnico di Milano, via Ponzio 34/5, Milano, Italy
{federica.panella,matteo.pradella,dino.mandrioli}@polimi.it
[2] DI - Università degli Studi di Milano, via Comelico 39/41, Milano, Italy
lonati@di.unimi.it

Abstract. Recent literature extended the analysis of ω-languages from the regular ones to various classes of languages with "visible syntax structure", such as visibly pushdown languages (VPLs). Operator precedence languages (OPLs), instead, were originally defined to support deterministic parsing and exhibit interesting relations with these classes of languages: OPLs strictly include VPLs, enjoy all relevant closure properties and have been characterized by a suitable automata family and a logic notation. We introduce here operator precedence ω-languages (ωOPLs), investigating various acceptance criteria and their closure properties. Whereas some properties are natural extensions of those holding for regular languages, others require novel investigation techniques. Application-oriented examples show the gain in expressiveness and verifiability offered by ωOPLs w.r.t. smaller classes.

Keywords: ω-languages, Operator precedence languages, Push-down automata, Closure properties, Infinite-state model checking.

1 Introduction

Languages of infinite strings, i.e. ω-languages, have been introduced to model nonterminating processes; thus they are becoming more and more relevant nowadays when most applications are "ever-running", often in a distributed environment. The pioneering work by Büchi and others investigated their main algebraic properties in the context of finite state machines, pointing out commonalities and differences w.r.t. the finite length counterpart [4,17].

More recent literature, mainly under the motivation of widening the application of model checking techniques to larger language families, extended this analysis to various classes of languages with "visible structure", i.e., languages whose syntax structure is immediately visible in their strings: parenthesis languages, tree languages, visibly pushdown languages (VPLs) [1] are examples of such classes.

Operator precedence languages, instead, were defined by Floyd in the 1960s, and still are in use [9], with the original motivation of supporting deterministic parsing, which is trivial for visible structure languages but is crucial for general context-free languages such as programming languages [8], where structure is often left implicit (e.g. in arithmetic expressions). Recently, these seemingly unrelated classes of languages have been shown to share most major features; precisely OPLs strictly include VPLs and enjoy all the same closure properties [7]. This observation motivated characterizing

M.-P. Béal and O. Carton (Eds.): DLT 2013, LNCS 7907, pp. 396–408, 2013.
© Springer-Verlag Berlin Heidelberg 2013

OPLs in terms of a suitable automata family [10] and in terms of a logic notation [11], which was missing in previous literature.

In this paper we further the investigation of OPLs properties to the case of infinite strings, i.e., we introduce and study operator precedence ω-languages (ωOPLs). We prove closure and decidability properties that are a fundamental condition enabling infinite-state model checking. Also, we present a few simple application-oriented examples that show the considerable gain in expressiveness and verifiability offered by ω-OPLs w.r.t. previous classes.

We follow traditional lines of research in theory on ω-languages considering various acceptance criteria, their mutual expressiveness relations, and their closure properties, yet departing from the classical path for a number of critical and new issues. Not surprisingly, some properties are natural extensions of those holding for, say, regular languages or VPLs, whereas others required different and novel investigation techniques essentially due to the more general managing of the stack.

Due to space limitations, herein we focus on the newest and most interesting aspects. Also, we limit the technicalities of formal arguments to a minimum, relying instead on intuition and examples. The reader can find more results and all details in the technical report [14]. The paper is organized as follows. The next section provides basic concepts on operator precedence languages of finite-length words and on operator precedence automata able to recognize them. Section 3 defines operator precedence automata which can deal with infinite strings, analyzing various classical acceptance conditions for ω-abstract machines. Section 4 proves the closure properties they enjoy w.r.t typical operations on ω-languages and shows also that the emptiness problem is decidable for these formalisms. Finally, Section 5 draws some conclusions.

2 Preliminaries

Operator precedence languages [7,8] have been characterized in terms of both a generative formalism (operator precedence grammars, OPGs) and an equivalent operational one (operator precedence automata, OPAs, named Floyd automata or FAs in [10]), but in this paper we consider the latter, as it is better suited to model and verify nonterminating computations.

Let Σ be an alphabet. The empty string is denoted ε. Between the symbols of the alphabet three types of operator precedence (OP) binary relations can hold: *yields* precedence, *equal* in precedence and *takes* precedence, denoted $\lessdot$, $\doteq$ and $\gtrdot$ respectively. Notice that $\doteq$ is not necessarily an equivalence relation, and $\lessdot$ and $\gtrdot$ are not necessarily strict partial orders. We use a special symbol # not in Σ to mark the beginning and the end of any string. This is consistent with the typical operator parsing technique that requires the lookback and lookahead of one character to determine the next action to perform [9]. The initial # can only yield precedence, and other symbols can only take precedence on the ending #.

Definition 1. *An* operator precedence matrix *(OPM) M over an alphabet Σ is a $|\Sigma \cup \{\#\}| \times |\Sigma \cup \{\#\}|$ array that with each ordered pair (a, b) associates the set M_{ab} of OP* *relations holding between a and b. M is conflict-free iff* $\forall a, b \in \Sigma, |M_{ab}| \leq 1$*. We call* (Σ, M) *an operator precedence alphabet if M is a conflict-free OPM on* Σ*.*

Between two OPMs M_1 and M_2, we define set inclusion and union:

$$M_1 \subseteq M_2 \text{ if } \forall a, b : (M_1)_{ab} \subseteq (M_2)_{ab}, \qquad M = M_1 \cup M_2 \text{ if } \forall a, b : M_{ab} = (M_1)_{ab} \cup (M_2)_{ab}$$

If $M_{ab} = \{\circ\}$, with $\circ \in \{\lessdot, \doteq, \gtrdot\}$, we write $a \circ b$. For $u, v \in \Sigma^*$ we write $u \circ v$ if $u = xa$ and $v = by$ with $a \circ b$. Two matrices are *compatible* if their union is conflict-free. A matrix is *complete* if it contains no empty case.

In the following we assume that M is $\doteq$-*acyclic*, which means that $c_1 \doteq c_2 \doteq \cdots \doteq c_k \doteq c_1$ does not hold for any $c_1, c_2, \ldots, c_k \in \Sigma, k \geq 1$. See [14] for a discussion on this hypothesis. Let also (Σ, M) be an OP alphabet.

Definition 2. A nondeterministic operator precedence automaton *(OPA) is a tuple* $\mathcal{A} = \langle \Sigma, M, Q, I, F, \delta \rangle$ *where:*

- (Σ, M) *is an operator precedence alphabet,*
- Q *is a set of states (disjoint from* Σ*),*
- $I \subseteq Q$ *is a set of initial states,*
- $F \subseteq Q$ *is a set of final states,*
- $\delta : Q \times (\Sigma \cup Q) \to 2^Q$ *is the transition function.*

The transition function can be seen as the union of two disjoint functions:

$$\delta_{push} : Q \times \Sigma \to 2^Q \qquad \delta_{flush} : Q \times Q \to 2^Q$$

An OPA can be represented by a graph with Q as the set of vertices and $\Sigma \cup Q$ as the set of edge labels: there is an edge from state q to state p labeled by $a \in \Sigma$ if and only if $p \in \delta_{push}(q, a)$, and there is an edge from state q to state p labeled by $r \in Q$ if and only if $p \in \delta_{flush}(q, r)$. To distinguish flush transitions from push transitions we denote the former ones by a double arrow.

To define the semantics of the automaton, we introduce some notation. We use letters $p, q, p_i, q_i, \ldots$ for states in Q and we set $\Sigma' = \{a' \mid a \in \Sigma\}$; symbols in Σ' are called *marked* symbols.

Let Γ be $(\Sigma \cup \Sigma' \cup \{\#\}) \times Q$; we denote symbols in Γ as $[a \; q]$, $[a' \; q]$, or $[\# \; q]$, respectively. We set $symbol([a \; q]) = symbol([a' \; q]) = a$, $symbol([\# \; q]) = \#$, and $state([a \; q]) = state([a' \; q]) = state([\# \; q]) = q$. Given a string $\beta = B_1 B_2 \ldots B_n$ with $B_i \in \Gamma$, we set $state(\beta) = state(B_n)$.

A *configuration* is any pair $C = \langle \beta, w \rangle$, where $\beta = B_1 B_2 \ldots B_n \in \Gamma^*$, $symbol(B_1) = \#$, and $w = a_1 a_2 \ldots a_m \in \Sigma^* \#$. A configuration represents both the contents β of the stack and the part of input w still to process.

A computation (run) of the automaton is a finite sequence of moves $C \vdash C_1$; there are three kinds of moves, depending on the precedence relation between $symbol(B_n)$ and a_1:

push move: if $symbol(B_n) \doteq a_1$ then $C_1 = \langle \beta[a_1 \; q], a_2 \ldots a_m \rangle$, with $q \in \delta_{push}(state(\beta), a_1)$;

mark move: if $symbol(B_n) \lessdot a_1$ then $C_1 = \langle \beta[a_1' \; q], a_2 \ldots a_m \rangle$, with $q \in \delta_{push}(state(\beta), a_1)$;

flush move: if $symbol(B_n) \gtrdot a_1$ then let i the greatest index such that $symbol(B_i) \in \Sigma'$ (such index always exists). Then $C_1 = \langle B_1 B_2 \ldots B_{i-2}[symbol(B_{i-1})\ q]\ ,\ a_1 a_2 \ldots a_m \rangle$, with $q \in \delta_{flush}(state(B_n), state(B_{i-1}))$.

Push and mark moves both push the input symbol on the top of the stack, together with the new state computed by δ_{push}; such moves differ only in the marking of the symbol on top of the stack. The flush move is more complex: the symbols on the top of the stack are removed until the first marked symbol (*included*), and the state of the next symbol below them in the stack is updated by δ_{flush} according to the pair of states that delimit the portion of the stack to be removed; notice that in this move the input symbol is not consumed and it remains available for the following move.

Finally, we say that a configuration $[\#\ q_I]$ is *starting* if $q_I \in I$ and a configuration $[\#\ q_F]$ is *accepting* if $q_F \in F$. The language accepted by the automaton is defined as:

$$L(\mathcal{A}) = \left\{ x \mid \langle [\#\ q_I]\ ,\ x\# \rangle \overset{*}{\vdash} \langle [\#\ q_F]\ ,\ \# \rangle, q_I \in I, q_F \in F \right\}.$$

An OPA is *deterministic* when I is a singleton and $\delta_{push}(q, a)$ and $\delta_{flush}(q, p)$ have at most one element, for every $q, p \in Q$ and $a \in \Sigma$.

An *operator precedence transducer* is defined in the natural way.

Example 1. As an introductory example, consider a language of queries on a database expressed in relational algebra. We consider a subset of classical operators (union, intersection, selection σ, projection π and natural join $\bowtie$). Just like mathematical operators, the relational operators have precedences between them: unary operators σ and π have highest priority, next highest is the "*multiplicative*" operator $\bowtie$, lowest are the "*additive*" operators $\cup$ and $\cap$. Denote as T the set of tables of the database and, for the sake of simplicity, let E be a set of conditions for the unary operators. The OPA depicted in Figure 1 accepts the language of queries without parentheses on the alphabet $\Sigma = T \cup \{\bowtie, \cup, \cap\} \cup \{\sigma, \pi\} \times E$, where we use letters $A, B, R \ldots$ for elements in T and we write σ_{expr} for a pair $(\sigma, expr)$ of selection with condition $expr$ (similarly for projection π_{expr}). The same figure also shows an accepting computation on input $A \cup B \bowtie C \bowtie \pi_{expr} D$.

Notice that the sentences of this language show the same structure as arithmetic expressions with prioritized operators and without parentheses, which cannot be represented by VPAs due to the particular shape of their OPM [7].

Definition 3. *A simple chain is a word* $a_0 a_1 a_2 \ldots a_n a_{n+1}$, *written as* $\langle {}^{a_0} a_1 a_2 \ldots a_n{}^{a_{n+1}} \rangle$, *such that:* $a_0 \in \Sigma \cup \{\#\}$, $a_i \in \Sigma$ *for every* $i : 1 \leq i \leq n + 1$, $M_{a_0 a_{n+1}} \neq \emptyset$, *and* $a_0 \lessdot a_1 \doteq a_2 \ldots a_{n-1} \doteq a_n \gtrdot a_{n+1}$.

A composed chain is a word $a_0 x_0 a_1 x_1 u_2 \ldots a_n x_n a_{n+1}$, *where* $\langle {}^{a_0} a_1 a_2 \ldots a_n{}^{a_{n+1}} \rangle$ *is a simple chain, and* $x_i \in \Sigma^*$ *is the empty word or is such that* $\langle {}^{a_i} x_i{}^{a_{i+1}} \rangle$ *is a chain (simple or composed), for every* $i : 0 \leq i \leq n$. *Such a composed chain will be written as* $\langle {}^{a_0} x_0 a_1 x_1 a_2 \ldots a_n x_n{}^{a_{n+1}} \rangle$.

A word w *over* (Σ, M) *is compatible with* M *iff for each pair of consecutive letters* c, d *in* w $M_{cd} \neq \emptyset$, *and for each factor* x *of* $\#w\#$ *such that* $x = a_0 x_0 a_1 x_1 a_2 \ldots a_n x_n a_{n+1}$ *where* $a_0 \lessdot a_1 \doteq a_2 \ldots a_{n-1} \doteq a_n \gtrdot a_{n+1}$ *and* $x_i \in \Sigma^*$ *is the empty word or is such that* $\langle {}^{a_i} x_i{}^{a_{i+1}} \rangle$ *is a chain (simple or composed) for every* $0 \leq i \leq n$, $M_{a_0 a_{n+1}} \neq \emptyset$.

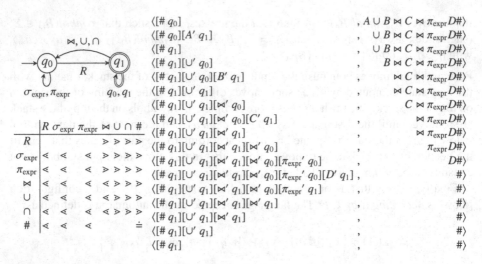

Fig. 1. Automaton, precedence matrix and example of computation for language of Example 1

Definition 4. *Let $\mathcal{A}$ be an operator precedence automaton. A support for the simple chain $\langle {}^{a_0} a_1 a_2 \ldots a_n {}^{a_{n+1}} \rangle$ is any path in $\mathcal{A}$ of the form*

$$\xrightarrow{a_0} q_0 \xrightarrow{a_1} q_1 \longrightarrow \ldots \longrightarrow q_{n-1} \xrightarrow{a_n} q_n \xRightarrow{q_0} q_{n+1} \qquad (1)$$

The label of the last (and only) flush is exactly q_0, i.e. the first state of the path; this flush is executed because of relation $a_n \gtrdot a_{n+1}$.

A support for the composed chain $\langle {}^{a_0} x_0 a_1 x_1 a_2 \ldots a_n x_n {}^{a_{n+1}} \rangle$ is any path in $\mathcal{A}$ of the form

$$\xrightarrow{a_0} q_0 \overset{x_0}{\leadsto} q'_0 \xrightarrow{a_1} q_1 \overset{x_1}{\leadsto} q'_1 \xrightarrow{a_2} \ldots \xrightarrow{a_n} q_n \overset{x_n}{\leadsto} q'_n \xRightarrow{q'_0} q_{n+1} \qquad (2)$$

where, for every $i : 0 \leq i \leq n$:

- *if $x_i \neq \varepsilon$, then $\xrightarrow{a_i} q_i \overset{x_i}{\leadsto} q'_i$ is a support for the chain $\langle {}^{a_i} x_i {}^{a_{i+1}} \rangle$, i.e., it can be decomposed as $\xrightarrow{a_i} q_i \overset{x_i}{\leadsto} q''_i \xRightarrow{q_i} q'_i$.*
- *if $x_i = \varepsilon$, then $q'_i = q_i$.*

Notice that the label of the last flush is q'_0.

The chains fully determine the structure of the parsing of any automaton on a word compatible with M, and hence the structure of the syntax tree of the word. Indeed, if the automaton performs the computation $\langle [a\ q_0]\ ,\ xb \rangle \overset{*}{\vdash} \langle [a\ q]\ ,\ b \rangle$ on a factor axb, then $\langle {}^a x^b \rangle$ is necessarily a chain over (Σ, M) and there exists a support like (2) with $x = x_0 a_1 \ldots a_n x_n$ and $q_{n+1} = q$.

3   Operator Precedence ω-Languages and Automata

Traditionally, ω-automata have been classified on the basis of the acceptance condition they are equipped with. All acceptance conditions refer to the occurrence of states which are visited in a computation of the automaton, and they generally impose constraints on those states that are encountered infinitely (or also finitely) often during a run. Classical notions of acceptance (introduced by Büchi [4], Muller [12], Rabin [15], Streett [16]) can be naturally adapted to ω-automata for operator precedence languages and can be characterized according to a peculiar acceptance component of the automaton on ω-words. Here we focus mainly on nondeterministic Büchi-operator precedence ω-automata with acceptance by final state; other models are briefly presented and compared in Section 3.1.

As usual, we denote by Σ^ω the set of infinite-length words over Σ. Thus, the symbol # occurs only at the beginning of an ω-word. Given a precedence alphabet (Σ, M), the definition of an ω-word compatible with the OPM M and the notion of syntax tree of an infinite-length word are the natural extension of these concepts for finite strings. We also use the notation "$\exists^\omega i$" as a shorthand for "there exist infinitely many i".

Definition 5. *A nondeterministic Büchi-operator precedence ω-automaton (ωOPBA) is given by a tuple $\mathcal{A} = \langle \Sigma, M, Q, I, F, \delta \rangle$, where Σ, Q, I, F, δ are defined as for OPAs; the operator precedence matrix M is restricted to be a $|\Sigma \cup \{\#\}| \times |\Sigma|$ array, since ω-words are not terminated by #. Configurations and (infinite) runs are defined as for OP automata on finite-length words.*

Let $\mathcal{S}$ be a run of the automaton on a given word $x \in \Sigma^\omega$. Define $In(\mathcal{S}) = \{q \in Q \mid \exists^\omega i \langle \beta_i, x_i \rangle \in \mathcal{S}$ with $state(\beta_i) = q\}$ as the set of states that occur infinitely often at the top of the stack of configurations in $\mathcal{S}$. A run $\mathcal{S}$ of an ωOPBA on an infinite word $x \in \Sigma^\omega$ is successful iff there exists a state $q_f \in F$ such that $q_f \in In(\mathcal{S})$. $\mathcal{A}$ accepts $x \in \Sigma^\omega$ iff there is a successful run of $\mathcal{A}$ on x.

As in the finite-length case, the class of languages accepted by ωBVPAs (*nondeterministic Büchi visibly pushdown ω-automata*) is a proper subset of that accepted by ωOPBAs. Indeed, classical families of automata, like Visibly Pushdown Automata [1], imply several restrictions that hinder them from being able to deal with the concept of precedence among symbols and make them unsuitable to model several interesting aspects often exhibited by real-world systems in various contexts.

To mention a few examples, a natural field of application of ωOPLs is the representation of processes or tasks which are assigned a priority to fulfill their requirements, which is a common paradigm in the area of operating systems, e.g. the processing of interrupts from peripherals with different priorities, or in the context of Web services, where servers provide services to users with different privileges. ωOPLs can also model the run-time behavior of database systems, e.g. for modeling sequences of user's transactions with possible rollbacks, and revision control systems (such as subversion or git). Examples of such systems are more extensively presented in [14].

3.1 Other Automata Models for Operator Precedence ω-Languages

There are several possibilities to define other classes of OP ω-languages. We may introduce a variant of ωOPBA (called ωOPBEA) which recognizes a word if the automaton traverses final states with an empty stack infinitely often, and we may as well consider automata with acceptance conditions other than Büchi's, as e.g. Muller operator precedence ω-automata (ωOPMAs). Furthermore, *deterministic ωOPA and OP* *ω-transducers* can be specified in the natural way as for operator precedence automata on finite-length words.

In general, the relationships among languages recognized by the different classes of operator precedence ω-automata and visibly pushdown ω-languages are summarized in Figure 2, where ωDOPBA and ωDOPMA denote the classes of deterministic ωOPBAs and deterministic ωOPMAs respectively. The detailed proofs of the strict containment relations holding among the classes in Figure 2 are presented in [13, Chapter 4]. The proofs regarding the relationships between those classes which are not comparable are described in [14]. In the sequel we will consider only the most expressive class of ωOPAs, i.e. ωOPBA.

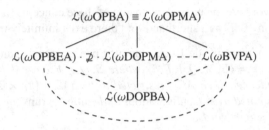

Fig. 2. Containment relations for ωOPLs. Solid lines denote strict inclusions; dashed lines link classes which are not comparable.

4   Closure Properties and Emptiness Problem

In this section we focus on the most interesting closure properties of ωOPAs, which are summarized in Table 1, where they are compared with the properties enjoyed by VPAs on infinite-length words. All operations are assumed to be applied to, and, when closure subsists, to produce, languages with compatible OPMs.

Table 1. Closure properties of families of ω-languages. ($L_1 \cdot L_2$ denotes the concatenation of a language of finite-length words L_1 and an ω-language L_2).

	$\mathcal{L}(\omega\text{DOPBA})$	$\mathcal{L}(\omega\text{DOPMA})$	$\mathcal{L}(\omega\text{OPBA}){\equiv}\mathcal{L}(\omega\text{OPMA})$	$\mathcal{L}(\omega\text{BVPA})$
Intersection	Yes	Yes	Yes	Yes
Union	Yes	Yes	Yes	Yes
Complement	No	Yes	Yes	Yes
$L_1 \cdot L_2$	No	No	Yes	Yes

The main family ωOPBA is closed under Boolean operations and under concatenation with a language of finite words accepted by an OPA. Furthermore, the emptiness problem is decidable for ωOPAs in polynomial time because they can be interpreted as pushdown automata on infinite-length words: e.g. [5] shows an algorithm that decides the alternation-free modal μ-calculus for context-free processes, with linear complexity in the size of the system's representation; thus the emptiness problem for the intersection of the language recognized by a pushdown process and the language of a given property in this logic is decidable.

Closure under intersection and union hold for ωOPBAs as for classical ω-regular languages: these closure properties, together with those enjoyed by ωDOPBAs and ωDOPMAs, can be proved in a similar way as for classical families of ω-automata, and their proofs can be found in [13, Chapter 5] and [14, Section 4]. Closure under complementation and concatenation for ωOPBAs, instead, required novel investigation techniques.

Closure under Concatenation

Unlike other families of languages, closure under concatenation has been proved for finite-length word OPLs by using their generating grammars with some difficulty [6], essentially due to the peculiar structure of their syntax trees. In the case of OPAs, and of infinite-length words, difficulties are further exacerbated by the fact that an OPA relies on the end-marker # to empty the stack before accepting a string; on the contrary, when parsing the concatenation of two OPL strings, the stack *cannot* always be emptied after reading the former one; for instance, consider a language $L_1 \subseteq \Sigma^*$ with an OPM where $a \lessdot a$ and $b \lessdot a$: a word in L_1 ending with a b concatenated with a^ω compels the OPA to let the stack indefinitely grow with no chance for any flush move after the reading of the L_1 word.

To overtake this difficulty we use a new approach which heavily exploits nondeterminism; remember in fact that, similarly to regular languages and VPLs, ωDOPBAs are strictly less powerful than ωOPBAs (see Figure 2). The basic idea consists in *guessing* the end of the first word and deciding whether it could be accepted by the original OPA recognizing L_1 *without emptying the stack*. This is a nontrivial job which requires storing suitable information in the stack at any mark move as it will be explained shortly.

To achieve our goal we first introduce a variant of the semantics of the transition relation and of the acceptance condition for OPAs: a string x is accepted if the automaton reaches a final state right at the end of the parsing of the whole word, and it does not perform any flush move determined by the ending delimiter # to empty the stack; thus it stops just after having put the last symbol of x on the stack. Precisely, the semantics of the transition relation differs from the definition of classical OPAs in that, once a configuration with the endmarker as lookahead is reached, the computation cannot evolve in any subsequent configuration, i.e. a flush move $C \vdash C_1$ with $C = \langle B_1 B_2 \ldots B_n\, ,\ y\# \rangle$ and $symbol(B_n) \gtrdot y\#$ is performed only if $y \neq \varepsilon$. The language accepted by this variant of the automaton (denoted as $\widetilde{L}$) is the set of words:

$$\widetilde{L}(\mathcal{A}) = \{x \mid \langle [\# \, q_I]\, ,\ x\# \rangle \overset{*}{\vdash} \langle \gamma[a \, q_F]\, ,\ \# \rangle, q_I \in I, q_F \in F, \gamma \in \Gamma^*, a \in \Sigma \cup \{\#\}\}$$

We emphasize that, unlike normal acceptance by final state of a pushdown automaton, which can perform a number of ε-moves after reaching the end of a string and accept if just one of the visited states is final, this type of automaton cannot perform any (flush) move after reaching the endmarker through the last look-ahead.

Nevertheless, the variant and the classical definition of OPA are equivalent: the following lemma shows the first direction of inclusion between the two formalisms.

Statement 1 in [14], although not necessary to prove closure under concatenation of $\mathcal{L}(\omega OPBA)$, completes the proof of equivalence between traditional and variant OPAs.

Lemma 1. *Let $\mathcal{A}_1$ be a nondeterministic OPA defined on an OP alphabet (Σ, M) with s states. Then there exists a nondeterministic OPA $\mathcal{A}_2$ with the same precedence matrix as $\mathcal{A}_1$ and $O(|\Sigma|s^2)$ states such that $L(\mathcal{A}_1) = \widetilde{L}(\mathcal{A}_2)$.*

Sketch of the proof. Consider a word of finite length w which is preceded by a delimiter # but which is not ended with such a symbol. Define a chain in a word w as *maximal* if it does not belong to a larger composed chain. In a word of finite length preceded and ended by # only the outer chain $\langle^\# w^\# \rangle$ is maximal.

The *body* of a chain $\langle^a w^b \rangle$, simple or composed, is the word w. A word w which is preceded but not ended by a delimiter # can be factored in a unique way as a sequence of bodies of maximal chains w_i and letters a_i as $\# \, w = \# \, w_1 a_1 w_2 a_2 \ldots w_n a_n$ where $\langle^{a_{i-1}} w_i {}^{a_i} \rangle$ are maximal chains and each w_i can be possibly missing, with $a_0 = \#$ and $\forall i : 1 \leq i \leq n-1 \; a_i \lessdot a_{i+1}$ or $a_i \doteq a_{i+1}$. During the parsing of word w, the symbols of the string are put on the stack and, whenever a chain is recognized, the letters of its body are flushed away. Hence, after the parsing of #w the stack contains only the symbols $\# \, a_1 \, a_2 \ldots a_n$, which we call *pending letters*, and is structured as a sequence

$$\# \lessdot a_{i_1} = a_1 \doteq a_2 \doteq \ldots \lessdot a_{i_2} \doteq a_{i_2+1} \doteq \ldots \lessdot a_{i_3} \doteq a_{i_3+1} \doteq \ldots \lessdot a_{i_k} \doteq a_{i_k+1} \doteq \ldots \doteq a_n$$

of k *open chains*, i.e., sequences of symbols $b_0 \lessdot b_1 \doteq b_2 \doteq \ldots \doteq b_m$, for $m \geq 1$. At the end of the computation a classical OPA performs a series of flush moves due to the presence of the final symbol #. These moves progressively empty the stack, removing one by one the open chains.

A nondeterministic automaton that, unlike classical OPAs, does not resort to the last # for the recognition, guesses nondeterministically the ending point of each open chain on the stack and guesses how, in an accepting run, the states in these points of the stack would be updated if the final flush moves were progressively performed. The automaton must behave as if, at the same time, it simulates two steps of the accepting run of a classical OPA: a move during the parsing of the string and a step during the final flush transitions which will later on empty the stack, leading to a final state. To this aim, the states of a classical OPA are augmented with an additional component to store the necessary information. If the forward path consisting of moves during the parsing of the string and the backward path of flush moves guessed by the automaton can consistently meet and be rejoined when the parsing of the input string stops, then combined they constitute an accepting run of the classical OPA.

A variant OPA $\mathcal{A}_2$ equivalent to a given OPA $\mathcal{A}_1$ thus may be defined so that, after reading each prefix of a word, it reaches a final state whenever, if the word were completed in that point with #, $\mathcal{A}_1$ could reach an accepting state with a sequence of flush moves. In this way, $\mathcal{A}_2$ can guess in advance which words may eventually lead to an accepting state of $\mathcal{A}_1$, without having to wait until reading the delimiter # and to perform final flush moves. To illustrate, we use the following example.

Example 2. Consider Figure 1. If we take the input word of this computation without the ending marker #, then the sequence of pending letters on the stack, after the automaton puts on the stack the last symbol D, is $\# \lessdot \cup \lessdot \Join \; \lessdot \; \Join \lessdot \pi_{expr} \lessdot D$. There are five open chains with starting symbols $\cup$, $\bowtie$, $\bowtie$, π_{expr}, D, hence the computation ends with five consecutive flush moves determined by the delimiter #. The following figure shows the configuration just before looking ahead at the symbol #. The states (depicted within boxes) at the end of the open chains are those placeholders that an equivalent variant OPA should guess in order to find in advance the last flush moves $q_1 = \boxed{q_1} \overset{q_0}{\Longrightarrow} \boxed{q_1} \overset{q_0}{\Longrightarrow} \boxed{q_1} \overset{q_1}{\Longrightarrow} \boxed{q_1} \overset{q_1}{\Longrightarrow} \boxed{q_1} \overset{q_1}{\Longrightarrow} \boxed{q_1 \in F_1}$ of the accepting run.

$$\langle [\# \; q_1] \quad [\cup' \; q_1] \quad [\bowtie' \; q_1] \quad [\bowtie' \; q_0] \quad [\pi_{expr} \; q_0] \quad [D' \; q_1] \; , \qquad \# \rangle$$

$$\boxed{q_1 \in F_1} \qquad \boxed{q_1} \qquad \boxed{q_1} \qquad \boxed{q_1} \qquad \boxed{q_1} \qquad \boxed{q_1}$$

The corresponding configuration of the variant OPA, with the augmented states, would be:

$$\langle [\# \; q_1, \boxed{q_1}] \quad [\cup' \; q_1, \boxed{q_1}] \quad [\bowtie' \; q_1, \boxed{q_1}] \quad [\bowtie' \; q_0, \boxed{q_1}] \quad [\pi_{expr} \; q_0, \boxed{q_1}] \quad [D' \; q_1, \boxed{q_1}] \; , \quad \# \rangle$$

The formal definition of the variant automaton and the proof of its equivalence with a classical OPA are presented in [14].

We are now ready for the main result.

Theorem 1. *Let $L_1 \subseteq \Sigma^*$ be a language of finite words recognized by an OPA with OPM M_1 and s_1 states. Let $L_2 \subseteq \Sigma^\omega$ be an ω-language recognized by a nondeterministic ωOPBA with OPM M_2 compatible with M_1 and s_2 states. Then the concatenation $L_1 \cdot L_2$ is also recognized by a ωOPBA with OPM $M_3 \supseteq M_1 \cup M_2$ and $O(|\Sigma|(s_1^2 + s_2^2))$ states.*

Sketch of the proof. Intuitively, a nondeterministic ωOPBA recognizing $L_1 \cdot L_2$ first simulates the variant automaton recognizing L_1, guesses the end of the L_1 word, and leaves a suitable "marker" on top of the stack before beginning the simulation of the second ωOPBA. In this process, the only nontrivial technical aspect is the fact that the second phase cannot leave unaffected the part of the stack that is left as a "legacy" by the first phase; thus, some flush moves must "invade" the lower part of the stack and the two phases cannot be completely independent, somewhat mimicking the construction of the OP grammar generating the concatenation of two OPLs [7]. □

Closure under Complementation

Theorem 2. *Let M be a conflict-free precedence matrix on an alphabet Σ. Denote by $L_M \subseteq \Sigma^\omega$ the ω-language comprising all infinite words $x \in \Sigma^\omega$ compatible with M.*

Let L be an ω-language on Σ that can be recognized by a nondeterministic ωOPBA with precedence matrix M and s states. Then the complement of L w.r.t L_M is recognized by an ωOPBA with the same precedence matrix M and $2^{O(s^2)}$ states.

Sketch of the proof. The proof follows to some extent the structure of the corresponding proof for Büchi VPAs [1], but it exhibits some relevant technical aspects which distinctly characterize it; in particular, we need to introduce an ad-hoc factorization of ω-words due to the more complex management of the stack performed by ωOPAs.

Let $\mathcal{A} = \langle \Sigma, M, Q, I, F, \delta \rangle$ be a nondeterministic ωOPBA with $|Q| = s$. Without loss of generality $\mathcal{A}$ can be considered complete with respect to the transition function δ, i.e. such that there is a run of $\mathcal{A}$ on every ω-word on Σ compatible with M.

In general, a sentence on Σ^ω can be factored in a unique way so as to distinguish the subfactors of the string that can be recognized without resorting to the stack of the automaton and those subwords for which the use of the stack is necessary.

More precisely, an ω-word $w \in \Sigma^\omega$ can be factored as a sequence of chains and pending letters $w = w_1 w_2 w_3 \ldots$ where either $w_i = a_i \in \Sigma$ is a pending letter or $w_i = a_{i1} a_{i2} \ldots a_{in}$ is a finite sequence of letters such that $\langle {}^{l_i} w_i {}^{first_{i+1}} \rangle$ is a chain, where l_i denotes the last pending letter preceding w_i in the word and $first_{i+1}$ denotes the first letter of word w_{i+1}. Let also, by convention, $a_0 = \#$ be the first pending letter.

Notice that such factorization is not unique, since a string w_i can be nested into a larger chain having the same preceding pending letter. The factorization is unique, however, if we additionally require that w_i has no prefix which is a chain.

As an example, for the word $w = \underbrace{\text{<}a \text{<} c \text{>}}b \underbrace{\text{<}a\text{>}} \underbrace{d\text{>}} b \ldots$, with precedence relations in the OPM $a \gtrdot b$ and $b \lessdot d$, the unique factorization is $w = w_1 b w_3 w_4 b \ldots$, where b is a pending letter and $\langle {}^{\#}ac^b \rangle$, $\langle {}^{b}a^d \rangle$, $\langle {}^{b}d^b \rangle$ are chains.

Define a *semisupport* for a chain $\langle {}^{a_0} x^{a_{n+1}} \rangle$ (simple or composed) as any path in $\mathcal{A}$ which is a support for the chain (Equations 1 and 2), where however the initial state of the path is not restricted to be the state reached after reading symbol a_0.

Let $x \in \Sigma^*$ be such that $\langle {}^{a} x^{b} \rangle$ is a chain for some a, b and let $T(x)$ be the set of all triples $(q, p, f) \in Q \times Q \times \{0, 1\}$ such that there exists a semisupport $q \overset{x}{\rightsquigarrow} p$ in $\mathcal{A}$, and $f = 1$ iff the semisupport contains a state in F. Also let $\mathcal{T}$ be the set of all such $T(x)$, i.e., $\mathcal{T}$ contains set of triples identifying all semisupports for some chain, and set $PR = \Sigma \cup \mathcal{T}$. The *pseudorun* for w in $\mathcal{A}$ is the ω-word $w' = y_1 y_2 y_3 \ldots \in PR^\omega$ where $y_i = a_i$ if $w_i = a_i$, otherwise $y_i = T(w_i)$. For the example above, then, $w' = T(ac) b T(a) T(d) b \ldots$.

Deferring to [14] further details of our proof, which from this point on resembles [1] with the necessary adaptions, we can define a nondeterministic Büchi finite-state automaton $\mathcal{A}_R$ over alphabet PR which has $O(s)$ states and accepts a pseudorun iff the corresponding words on Σ belong to $L(\mathcal{A})$. Consider then a deterministic Streett automaton $\mathcal{B}_R$ that accepts the complement of $L(\mathcal{A}_R)$ on the alphabet PR and, receiving pseudoruns as input words, accepts only words in $L_M \backslash L(\mathcal{A})$. The automaton $\mathcal{B}_R$ has $2^{O(s \log s)}$ states and $O(s)$ accepting constraints [17]. We can build a nondeterministic transducer ωOPBA $\mathcal{B}$ that on reading w generates online the pseudorun w', which will be given as input to $\mathcal{B}_R$. The final automaton, that recognizes the complement of $L = L(\mathcal{A})$ w.r.t L_M, is the ωOPBA representing the product of $\mathcal{B}_R$ (converted to a Büchi automaton), which has $2^{O(s \log s)}$ states, and $\mathcal{B}$, with $2^{O(s^2)}$ states; thus it has $2^{O(s^2)}$ states. $\square$

5   Conclusions and Further Research

We presented a formalism for infinite-state model checking based on operator precedence languages, continuing to explore the paths in the lode of operator precedence languages started up by Robert Floyd a long time ago. We introduced various classes of automata able to recognize operator precedence languages of infinite-length words whose expressive power outperforms classical models for infinite-state systems as Visibly Pushdown ω-languages, allowing to represent more complex systems in several practical contexts. We proved the closure properties of ωOPLs under Boolean operations that, along with the decidability of the emptiness problem, are fundamental for the application of such formalism to model checking.

Our results open further directions of research. A first interesting topic deals with the characterization of ωOPLs in terms of suitable monadic second order logical formulas, that has already been studied for operator precedence languages of finite-length strings [11]. This would further strengthen applicability of model checking techniques. The next step of investigation will regard the actual design and study of complexity issues of algorithms for model checking of expressive logics on these pushdown models. We expect that the peculiar features of Floyd languages, as their "locality principle" which makes them suitable for parallel and incremental parsing [2,3] and their expressivity, might be interestingly exploited to devise efficient and attractive software model-checking procedures and approaches.

References

1. Alur, R., Madhusudan, P.: Adding nesting structure to words. Journ. ACM 56(3) (2009)
2. Barenghi, A., Crespi Reghizzi, S., Mandrioli, D., Pradella, M.: Parallel parsing of operator precedence grammars. Information Processing Letters (2013), doi:10.1016/j.ipl.2013.01.008
3. Barenghi, A., Viviani, E., Crespi Reghizzi, S., Mandrioli, D., Pradella, M.: PAPAGENO: a parallel parser generator for operator precedence grammars. In: Czarnecki, K., Hedin, G. (eds.) SLE 2012. LNCS, vol. 7745, pp. 264–274. Springer, Heidelberg (2013)
4. Büchi, J.R.: Weak Second-Order Arithmetic and Finite Automata. Mathematical Logic Quarterly 6(1-6), 66–92 (1960)
5. Burkart, O., Steffen, B.: Model checking for context-free processes. In: Cleaveland, W.R. (ed.) CONCUR 1992. LNCS, vol. 630, pp. 123–137. Springer, Heidelberg (1992)
6. Crespi Reghizzi, S., Mandrioli, D.: Operator Precedence and the Visibly Pushdown Property. In: Dediu, A.-H., Fernau, H., Martín-Vide, C. (eds.) LATA 2010. LNCS, vol. 6031, pp. 214–226. Springer, Heidelberg (2010)
7. Crespi Reghizzi, S., Mandrioli, D.: Operator Precedence and the Visibly Pushdown Property. Journal of Computer and System Science 78(6), 1837–1867 (2012)
8. Floyd, R.W.: Syntactic Analysis and Operator Precedence. Journ. ACM 10(3), 316–333 (1963)
9. Grune, D., Jacobs, C.J.: Parsing techniques: a practical guide. Springer, New York (2008)
10. Lonati, V., Mandrioli, D., Pradella, M.: Precedence Automata and Languages. In: Kulikov, A., Vereshchagin, N. (eds.) CSR 2011. LNCS, vol. 6651, pp. 291–304. Springer, Heidelberg (2011)
11. Lonati, V., Mandrioli, D., Pradella, M.: Logic Characterization of Invisibly Structured Languages: the Case of Floyd Languages. In: van Emde Boas, P., Groen, F.C.A., Italiano, G.F., Nawrocki, J., Sack, H. (eds.) SOFSEM 2013. LNCS, vol. 7741, pp. 307–318. Springer, Heidelberg (2013)
12. Muller, D.E.: Infinite sequences and finite machines. In: Proceedings of the Fourth Annual Symposium on Switching Circuit Theory and Logical Design, SWCT 1963, pp. 3–16. IEEE Computer Society, Washington, DC (1963)
13. Panella, F.: Floyd languages for infinite words. Master's thesis, Politecnico di Milano (2011), http://home.dei.polimi.it/panella

14. Panella, F., Pradella, M., Lonati, V., Mandrioli, D.: Operator precedence ω-languages. CoRR abs/1301.2476 (2013), http://arxiv.org/abs/1301.2476
15. Rabin, M.: Automata on infinite objects and Church's problem. Regional conference series in mathematics. Published for the Conference Board of the Mathematical Sciences by the American Mathematical Society (1972)
16. Streett, R.S.: Propositional dynamic logic of looping and converse is elementarily decidable. Information and Control 54(1-2), 121–141 (1982)
17. Thomas, W.: Automata on infinite objects. In: Handbook of Theoretical Computer Science, vol. B, pp. 133–191. MIT Press, Cambridge (1990)

New Results on Deterministic Sgraffito Automata*

Daniel Průša[1], František Mráz[2], and Friedrich Otto[3]

[1] Czech Technical University, Faculty of Electrical Engineering
Karlovo náměstí 13, 121 35 Prague 2, Czech Republic
prusapa1@cmp.felk.cvut.cz
[2] Charles University, Faculty of Mathematics and Physics
Malostranské nám. 25, 118 25 Prague 1, Czech Republic
frantisek.mraz@mff.cuni.cz
[3] Fachbereich Elektrotechnik/Informatik, Universität Kassel
D-34109 Kassel, Germany
otto@theory.informatik.uni-kassel.de

Abstract. The deterministic *sgraffito automaton* is a two-dimensional computing device that allows a clear and simple design of important computations. The family of picture languages it accepts has many nice closure properties, but when restricted to one-row inputs (that is, strings), this family collapses to the class of regular languages. Here we compare the deterministic sgraffito automaton to some other two-dimensional models: the two-dimensional deterministic forgetting automaton, the four-way alternating automaton and the sudoku-deterministically recognizable picture languages. In addition, we prove that deterministic sgraffito automata accept some unary picture languages that are outside the class REC of recognizable picture languages.

Keywords: picture languages, sgraffito automaton, recognizable picture languages.

1 Introduction

The two-dimensional *sgraffito automaton* (2SA) was introduced recently as a device for accepting picture languages [12]. It is a bounded two-dimensional Turing machine that in each step replaces the currently scanned symbol by a symbol of smaller weight. Hence, it can be seen as a two-dimensional variant of the Hennie machine [6], which visits each of its tape positions just a bounded

---

* The first author was supported by the Grant Agency of the Czech Republic under the project P103/10/0783, the second author under the projects P103/10/0783 and P202/10/1333. The work presented here was done while the third author was visiting at Charles University in Prague. He gratefully acknowledges the hospitality of the Faculty of Mathematics and Physics.

M.-P. Béal and O. Carton (Eds.): DLT 2013, LNCS 7907, pp. 409–419, 2013.
© Springer-Verlag Berlin Heidelberg 2013

number of times independent of the size of the input. As Hennie machines recognize regular languages only, the 2SA is equivalent to the finite-state acceptor when restricted to one-dimensional inputs. This is seen as a basic requirement on models of automata that define a family of picture languages that is to correspond to the ground level of a Chomsky-like hierarchy for picture languages. A well established example of such a class is the family of recognizable picture languages (REC) [5]. However, as REC also contains some rather complicated languages (e.g., some NP-complete languages), various models that recognize deterministic ground level classes have been proposed recently. These include the family DREC [1], the sudoku-deterministically recognizable languages [3], and the deterministic Wang automata [10,11].

Here we study the computational power of the deterministic sgraffito automaton (2DSA) by comparing it to various other models. The 2DSA allows an easy and clear design of computations, and it is quite powerful. As shown in [12], the family of picture languages $\mathcal{L}(\text{2DSA})$ that are accepted by 2DSA properly includes the class DREC, and it has the same closure properties. Here we compare the 2DSA to the following two-dimensional models: the two-dimensional deterministic forgetting automaton [8], the four-way alternating automaton [9], and the sudoku-deterministically recognizable picture languages [3]. In addition, we show that the 2DSA accepts some unary picture languages that are not in REC, which implies that the 2DSA is more powerful than DREC and incomparable to REC even when only inputs over a one-letter alphabet are considered.

2   The Sgraffito Automaton

Here we use the common notation and terms on pictures and picture languages (see, e.g., [5]). Let Σ be a finite alphabet, and let $P \in \Sigma^{*,*}$ be a *picture* over Σ, that is, P is a two-dimensional array of symbols from Σ. If P is of size $m \times n$, then we write $P \in \Sigma^{m,n}$. We introduce a set of five special markers (*sentinels*) $\mathcal{S} = \{\vdash, \dashv, \top, \bot, \#\}$, and we assume that $\Sigma \cap \mathcal{S} = \emptyset$ for any alphabet Σ considered. In order to enable an automaton to detect the border of P easily, we define the *boundary picture* $\widehat{P}$ over $\Sigma \cup \mathcal{S}$ of size $(m+2) \times (n+2)$. It is illustrated by the scheme in Figure 1.

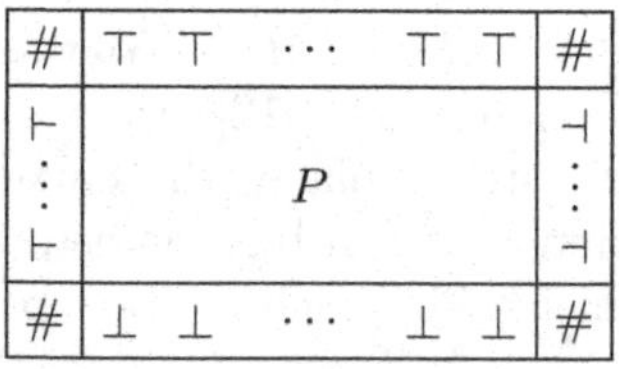

Fig. 1. The boundary picture $\widehat{P}$

Let $\mathcal{H} = \{R, L, D, U, Z\}$ be the set of possible *head movements*, where the first four elements denote directions (right, left, down, up) and Z represents no movement.

Definition 1. A two-dimensional sgraffito automaton (2SA) *is given by a 7-tuple* $\mathcal{A} = (Q, \Sigma, \Gamma, \delta, q_0, Q_F, \mu)$, *where*

- Q *is a finite non-empty set of states,*
- Σ *is an input alphabet,*
- Γ *is a working alphabet containing* Σ,
- $q_0 \in Q$ *is the initial state,*
- $Q_F \subseteq Q$ *is a set of final states,*
- $\delta : (Q \smallsetminus Q_F) \times (\Gamma \cup \mathcal{S}) \to 2^{Q \times (\Gamma \cup \mathcal{S}) \times \mathcal{H}}$ *is a transition relation, and*
- $\mu : \Gamma \to \mathbb{N}$ *is a weight function.*

In addition, the following two properties are satisfied:

1. $\mathcal{A}$ *is* bounded, *that is, whenever it scans a symbol from* $\mathcal{S}$, *then it immediately moves to the nearest field of* P *without changing this symbol,*
2. $\mathcal{A}$ *is* weight-reducing, *that is, for all* $q, q' \in Q$, $d \in \mathcal{H}$, *and* $a, a' \in \Gamma$, *if* $(q', a', d) \in \delta(q, a)$, *then* $\mu(a') < \mu(a)$.

Finally, $\mathcal{A}$ *is* deterministic (a 2DSA), *if* $|\delta(q, a)| \leq 1$ *for all* $q \in Q$ *and* $a \in \Gamma \cup \mathcal{S}$.

The notions of configuration and computation are defined as usual. In the initial configuration on input P, the tape contains $\widehat{P}$, $\mathcal{A}$ is in state q_0, and its head scans the top-left corner of P. If P is the empty picture, then the head initially scans the bottom-right corner of $\widehat{P}$ which contains the sentinel #. The automaton $\mathcal{A}$ accepts P iff there is a computation of $\mathcal{A}$ on input P that finishes in a state from Q_F.

When designing a sgraffito automaton for a picture language, it suffices to describe a bounded two-dimensional Turing machine that visits each tape cell only a constant number of times (i.e., a two-dimensional Hennie machine). In [12] it was shown that any such machine can be transformed into an equivalent sgraffito automaton (preserving determinism). This fact will be utilized in our constructive proofs below.

3   Comparing the 2DSA to the Forgetting Automaton

The two-dimensional deterministic forgetting automaton (2DFA) was introduced by Jiřička and Král in [8]. It is a bounded two-dimensional Turing machine that is allowed to rewrite by only using a special symbol @. In comparison to the 2DSA, there is no bound on the number of visits of any tape field. Here we study the question of whether the larger number of possible rewrites is an advantage for the 2DSA over the 2DFA. We will see, however, that this is not the case. On the contrary, we show that the 2DFA is more powerful than the 2DSA.

Jiřička and Král described a technique of how a 2DFA can store information on its tape while still remembering its original content. Let $P \in \Sigma^{*,*}$ be an input, let M be the set of tape fields storing P, and let $B \subseteq M$ be a rectangular subarea of M. Divide B into two parts, say H and G, where H consists of the $|\Sigma|$ first fields of B, and $G = B \setminus H$. For $a \in \Sigma$, let $G(a)$ denote the subset of those fields of G that initially contain the symbol a. Choose $a \in \Sigma$ such that $|G(a)| \geq |G|/|\Sigma|$. The idea is to use $G(a)$ to record information by erasing some of its fields, while memorizing which a has been chosen in order to be able to reconstruct the original content. Suppose a is the i-th symbol of Σ. Erase the i-th field of H to indicate this, and represent the erased symbol in unary using the first $|\Sigma|$ fields of $G(a)$. All the other fields of $G(a)$ can each store one bit of information: it is either erased or not erased. Hence, the total number of available bits is

$$|G(a)| - |\Sigma| \geq \frac{|B| - |\Sigma|}{|\Sigma|} - |\Sigma|.$$

Theorem 1. $\mathcal{L}(\text{2DSA})$ *is a proper subset of* $\mathcal{L}(\text{2DFA})$.

Proof. We first show how a 2DSA $\mathcal{A} = (Q, \Sigma, \Gamma, \delta, q_0, Q_F, \mu)$ can be simulated by a 2DFA $\mathcal{F}$. Let P be an input picture, let M be the set of fields storing P, and let n be an integer, dependent on $\mathcal{A}$, whose value will be determined later.

We split M into blocks of size $n \times n$, except for those blocks neighbouring the right border or the bottom – their size could be up to $2n - 1$. If the height or the width of P is smaller than n, then the blocks will be only as high or as wide as the picture. In case that both, the height and the width of P are smaller than n, $\mathcal{F}$ can decide whether to accept or to reject P simply by table look-up. Figure 2 summarizes the non-trivial variants.

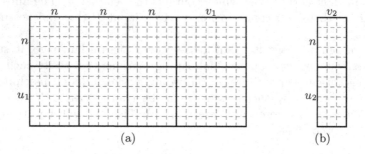

Fig. 2. Decomposition of M into blocks of fields, when (a) both dimensions of M are at least n, or (b) the width v_2 is smaller than n

Note that $\mathcal{F}$ can detect the borders between blocks by counting modulo n for both, the number of movements in the vertical and the horizontal directions.

The construction is now done as follows. $\mathcal{F}$ simulates $\mathcal{A}$ in the top-left block. All symbols in the block are scanned and memorized in states first, then $\mathcal{F}$ determines whether $\mathcal{A}$ finishes its computation inside the block or whether it enters one of its neighbouring blocks (B). In the latter case $\mathcal{F}$ writes to B, in which state and in which position B was entered. Then it again simulates $\mathcal{A}$ within this block and so on. The information on entry points to a block (ordered by time) is sufficient to reconstruct its content. There are at most $8n$ positions on the border of a block, thus a position can be encoded in binary using $\lceil \log_2 8n \rceil$ bits, while a state is encoded using $\lceil \log_2 |Q| \rceil$ bits. Any (border) field in M is entered at most $|\Gamma|$ times. This implies that $\mathcal{O}(n \log n)$ bits are needed. We have seen above that the capacity of each block is $\Theta(n^2)$ bits. Thus, a suitable value of n fulfilling the memory requirement can be found. When the height or the width of P is only $m < n$, then all blocks are entered across the borders of length m, which means that only $\mathcal{O}(m \log m)$ bits are needed, while the capacity of the block is $\mathcal{O}(mn)$. Thus, also in this case a corresponding value for n exists.

Finally, it is known that 2DFAs working over strings accept the deterministic context-free languages [7]. This proves that the above inclusion is proper, as 2DSAs only accept the regular string languages.   □

4   Simulations of Other Models by 2DSA

The simulations we will present in this section are based on representing a computation as a directed graph and traversing this graph by a 2DSA in the depth first manner (DFS) [4]. Therefore, we start with a description of the general principles shared by our constructions.

Let $G = (V, E)$ be a directed graph that satisfies the following conditions:

1. $V \subseteq \{1, \ldots, m\} \times \{1, \ldots, n\} \times U$ for some integers m, n and a finite set U.
2. For every edge $((i_1, j_1, u_1), (i_2, j_2, u_2))$ in E, $|i_1 - i_2| + |j_1 - j_2| \leq 1$.

The graph G can be represented in a picture P of size $m \times n$, where the field at a position (i, j) records the vertices of the form (i, j, u) $(u \in U)$ in V and the outgoing edges of these vertices. Since the edges only go to the vertices represented in the field itself and in its neighbouring fields, it is only necessary to represent $\mathcal{O}(|U|)$ many vertices and edges in each tape field.

Assume that a 2DSA has created a representation of G and moved its head back to the initial position. To traverse G, it assigns a status to vertices as well as to edges. Initially, each vertex has status *fresh*. When visiting a vertex v during DFS for the first time, its status is changed to *open*, and when DFS backtracks to v, since the whole subtree rooted at v has been searched, then its status is set to *closed*. Analogously, each edge e has initially the status *unexplored*. This changes when e is being traversed. If it leads to a *fresh* vertex, its status is set to *discovery*, otherwise its status is set to *back*. The edges with status *discovery* will form DFS-trees at the end of the search. The search is now implemented as follows.

1. If there is no vertex with status *fresh* at the field scanned, go to step 3.
2. While there a vertex v with status *fresh* (possibly fulfilling some additional requirement), mark it as *open* and start DFS, which will return to v at the end.
3. If not all the fields have been scanned, move the head to the next field in the row or to the first field of the next row when the right border is reached. Continue with step 1.

In our constructions U will correspond, for example, to the set of states of an automaton to be simulated. The requirement mentioned in step 2 may select, for example, source nodes only (vertices without incoming edges). The whole process visits and rewrites each tape field $\mathcal{O}(|U|)$ many times. Hence, it can be realized by a 2DSA.

4.1   The Four-Way Alternating Automaton

In [12], it is shown that every nondeterministic four-way automaton (4FA) [2] can be simulated by a 2DSA. Here we strengthen this result by showing that it is even possible to simulate four-way alternating automata (4AFA) [9]. A *four-way alternating automaton* is given by a 6-tuple $\mathcal{A} = (Q(\exists), Q(\forall), \Sigma, \delta, q_0, Q_F)$, where the set of states is split into $Q(\exists)$ (existential states) and $Q(\forall)$ (universal states). The set of final states Q_F is a subset of $Q(\exists) \cup Q(\forall)$, and the transition relation δ assigns a finite subset of $(Q(\exists) \cup Q(\forall)) \times \mathcal{H}$ to each pair of the form $(q, a) \in (Q(\exists) \cup Q(\forall)) \times (\Sigma \cup \mathcal{S})$.

Theorem 2. $\mathcal{L}(\text{4AFA})$ *is a proper subset of* $\mathcal{L}(\text{2DSA})$.

Proof. Let $\mathcal{A} = (Q(\exists), Q(\forall), \Sigma, \delta, q_0, Q_F)$ be a 4AFA, and let $P \in \Sigma^{m,n}$ be an input picture. We define a directed graph $G = (V, E)$ representing all possible transitions of $\mathcal{A}$ as follows:

- $V = \{1, \ldots, m\} \times \{1, \ldots, n\} \times (Q(\exists) \cup Q(\forall))$;
- $((i_1, j_1, q_1), (i_2, j_2, q_2))$ is an edge in E iff $(q_2, d) \in \delta(q_1, P(i_1, j_1))$, where (i_2, j_2) is the coordinate reached from (i_1, j_1) by moving into the direction given by d.

Special care is required for transitions that visit the borders of $\widehat{P}$. In these cases we represent the composition of two consecutive transitions by a single edge (the head must not finish on the border).

Let $\rho : V \to \{0, 1\}$ be the function that expresses the result (0 – reject, 1 – accept) of the computation of $\mathcal{A}$ over P if started at position (i, j) in state q. The values of ρ can be computed recurrently by the following rules:

1. if $q \in Q_F$, then $\rho(i, j, q) = 1$ for all admissible values of i and j;
2. if $q \in Q(\exists) \smallsetminus Q_F$, then $\rho(i, j, q) = 1$ iff there is an edge from (i, j, q) to some (i', j', q') such that $\rho(i', j', q') = 1$;
3. if $q \in Q(\forall) \smallsetminus Q_F$, then $\rho(i, j, q) = 1$ iff for each edge from (i, j, q) to some (i', j', q'), $\rho(i', j', q') = 1$.

$\mathcal{A}$ accepts P iff $\rho(1, 1, q_0) = 1$. We design a computation that computes ρ for each vertex in V, applying the above rules. The values of ρ are initially set to 1 only for the vertices in the set V_I that is defined as follows: $v = (i, j, q) \in V_I$ iff $q \in Q_F$ or $q \in Q(\forall)$ and there is no outgoing edge from v. Then DFS is performed on the reversion of G. It is started at the vertices in V_I. Moreover, it continues through outgoing edges of a vertex v only if $\rho(v)$ has been set to 1.

Since $\mathcal{L}(\text{4AFA})$ is not closed under complement [9], while $\mathcal{L}(\text{2DSA})$ is [12], it follows that the above inclusion is proper. □

4.2 Sudoku-Deterministically Recognizable Picture Languages

The *sudoku-deterministically recognizable picture languages* (SDREC) were introduced by Borchert and Reinhardt in [3]. They are defined via domino tiling systems [5] and the sudoku-deterministic process. A domino tiling system $\mathcal{T} = (\Sigma, \Gamma, \Delta, \pi)$ is specified by a set of dominoes Δ (that is, pictures over $\Gamma \cup \mathcal{S}$ of sizes 1×2 and 2×1) and a projection $\pi : \Gamma \to \Sigma$. These systems can be used to define picture languages in REC that are projections of local languages whose tiles all match dominoes in Δ.

The sudoku-deterministic process follows a different approach. Given a picture $P \in \Sigma^{m,n}$, it is initialized by the picture S_P of the same size in which each position (i,j) stores the set $S_P(i,j) := \pi^{-1}(P(i,j)) \in 2^\Gamma$. These preimages are then reduced iteratively in the same manner as a Sudoku-puzzle is solved. In one step, all those symbols that do not conform locally to the set of allowed dominoes are discarded from a given position. Formally, for $S, S' \in (2^\Gamma)^{m,n}$, S is reduced to S' if

$$S'(i,j) = \{\, x \in S(i,j) \mid \exists y_1, y_2, y_3, y_4 \in \Gamma \cup \mathcal{S} : y_1 \in \widehat{S}(i+1,j), y_2 \in \widehat{S}(i-1,j),$$

$$y_3 \in \widehat{S}(i,j+1), y_4 \in \widehat{S}(i,j-1), \text{and } \boxed{x|y_1}, \boxed{y_2|x}, \boxed{\begin{array}{c} x \\ \hline y_3 \end{array}}, \boxed{\begin{array}{c} y_4 \\ \hline x \end{array}} \in \Delta \,\}$$

for all positions (i,j). The accepted picture language is the set of all pictures P for which there exists a way to transform the initialized picture S_P in finitely many steps into a picture in which every position consists of exactly one element and which cannot be transformed further.

Theorem 3. SDREC *is contained in* $\mathcal{L}(\text{2DSA})$.

Proof. Let L be a language in SDREC, and let $\mathcal{T} = (\Sigma, \Gamma, \Delta, \pi)$ be a domino tiling system for L. To design a 2DSA $\mathcal{A}$ accepting L, we benefit from the known fact that the result of a sudoku-deterministic process does not depend on the order in which particular symbols are discarded. Let $P \in \Sigma^{m,n}$ be an input picture. Then $\mathcal{A}$ simulates the sudoku-deterministic process as follows:

1. For each position (i,j), $\mathcal{A}$ initializes $S(i,j)$ by the set $S(i,j) := \pi^{-1}(P(i,j))$.
2. $\mathcal{A}$ moves its head row by row. Every position (i,j) visited is marked as 'reached'. $\mathcal{A}$ checks whether some elements in $S(i,j)$ can be discarded by the sudoku-deterministic process – this is decided based on the content of the neighbouring fields.
3. If at least one element is discarded, then $\mathcal{A}$ searches all those neighbouring fields that have already been marked as 'reached'. It checks whether further symbols can be discarded at these fields, and then it repeats this process for the corresponding neighbours. Actually, this process can be realized by an extended version of the DFS algorithm. $\mathcal{A}$ visits neighbouring fields marked as 'reached' using the depth-first principle, but when returning from some field (i',j') to its predecessor (i,j) in the depth-first tree, $\mathcal{A}$ checks again

whether some element at this field must be discarded. If so, then the corresponding elements are discarded, and the DFS algorithm is restarted at this field, again visiting the (already visited) neighbours of the field (i,j) again. After the extended DFS finishes, $\mathcal{A}$ returns to the last field marked as 'reached', and it continues with the row by row movement.

4. When the simulation has been completed, then $\mathcal{A}$ revisits all fields in a sequential manner and accepts iff $|S(i,j)| = 1$ for all positions (i,j).

The simulation guarantees that all symbols that are discarded by the sudoku-deterministic process are also discarded by $\mathcal{A}$. Finally, let us determine how many times a field f is visited during steps 2 and 3. When it is visited for the first time, then it is being marked as 'reached'. When the DFS is launched at f, then f is visited at most $8 \cdot |\Gamma|$ many times, as f has at most 4 neighbours, and each time a symbol is being discarded from one of these neighbours, two more visits to f may become necessary. And finally, f is visited at most $4 \cdot |\Gamma|$ times when the DFS is launched at other fields – a DFS contributes a visit to f only when a symbol is discarded from a neighbouring field, and this happens in any field at most $|\Gamma|$ many times. Hence, $\mathcal{A}$ visits every field only a bounded number of times. □

Remark 1. Borchert and Reinhardt proved that $\mathcal{L}(4\mathsf{AFA}) \subseteq \mathsf{SDREC}$. This means that Theorem 2 is a consequence of Theorem 3. However, the direct proof of Theorem 2 we presented above is simpler, and it nicely demonstrates the power of the 2DSA.

5   Recognition of Unary Picture Languages

Giammaresi and Restivo studied the problem of which functions can be represented by recognizable picture languages. Let $\Sigma = \{\square\}$ denote a one-letter alphabet. A function $f : \mathbb{N} \to \mathbb{N}$ is called *representable* if the language $L(f) = \{\square^{n,f(n)} \mid n \in \mathbb{N}\}$ belongs to REC. A representable function cannot grow faster than an exponential function [5]. Here we show that using 2DSA, functions can be represented that grow faster than any exponential function.

Proposition 1. *The language* $L_1 = \{\square^{n,n!} \mid n \in \mathbb{N}\}$ *is accepted by a 2DSA.*

Proof. We first describe a 2DSA $\mathcal{A}$ that accepts pictures of size $n \times (n \cdot n!)$ over Σ. Let $\mathcal{P}(m)$ be the set of all permutations on $M = \{1,\ldots,m\}$. Each element of $\mathcal{P}(m)$ can be represented as a sequence $(a_1,\ldots,a_m)$, where the a_i's are the different numbers from M. For two permutations $a = (a_1,\ldots,a_m)$ and $b = (b_1,\ldots,b_m)$, we write $a < b$ iff there is an index j such that $a_j < b_j$ and $a_i = b_i$ for all $i = 1,\ldots,j-1$. We encode a permutation $(a_1,\ldots,a_m)$ by a square picture of size $m \times m$ over $\{\square, \blacksquare\}$ as follows: for all i, row i is composed of white pixels except at position a_i, which contains a black pixel. An example is given in Figure 3(a).

Given an input picture $P \in \Sigma^{m,n}$, $\mathcal{A}$ tries to write the sequence of all permutations of M in ascending order into this picture (see Figure 3(b)). The critical

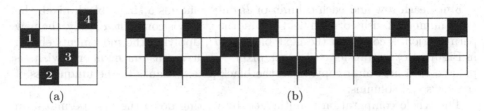

(a)   (b)

Fig. 3. (a) The permutation $(4,1,3,2)$ represented as a black and white picture. (b) Representations of all permutations on $\{1,2,3\}$ written side by side.

part of this design is a procedure that computes the next permutation from a given permutation $(a_1, \ldots, a_m)$. This is realized by the following algorithm:

1. find the greatest index j such that $a_{j-1} < a_j$;
2. among $a_j, \ldots, a_m$, find the smallest element a_k that is larger than a_{j-1};
3. switch elements a_{j-1} and a_k;
4. reverse the order of the elements $a_j, \ldots, a_m$.

This algorithm can be adopted to work over a picture R that represents $(a_1, \ldots, a_m)$. W.l.o.g. we can assume that $\mathcal{A}$ can also see the symbols in the neighbouring fields. It traverses R row by row from the bottom to the top, finds j and marks the j-th row. After that it traverses R from the left to the right column by column, starting at column $a_{j-1} + 1$, and finds k. The situation is illustrated in Figure 4(a). A swap of elements a_{j-1} and a_k follows, see the general pattern in Figure 4(b). It suffices to mark the rows and the columns in which the corresponding black pixels are located, to erase the pixels, and to write them at the new positions (note that the markers are always discarded when they are no longer needed). The whole process only visits the related marked fields. In the next step the order of the elements $a_j, \ldots, a_m$ is reversed. This is done by swapping a_j with a_m, a_{j+1} with a_{m-1}, and so forth.

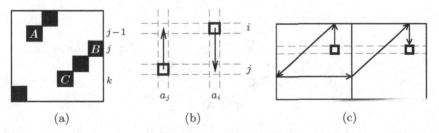

(a)   (b)   (c)

Fig. 4. (a) Finding the indices j and k. B is the lowest one which is positioned to the right of the black pixel in the previous row (which is A). C has to lie in rows below (and including) B. Its column is the first one to the right of A's column that contains a black pixel below row $j-1$. (b) The swap of the i-th and j-th element of a permutation. (c) Copy one element of a permutation to the next block. The sequence consists of vertical, horizontal and diagonal movements. The related row is marked.

Since each row and each column of R only contains a single black pixel, the total number of visits of each field during the swaps is constant. Finally, the new permutation is copied to the next block in P, applying the movements shown in Figure 4(c) to transfer each black pixel. Again, a constant number of visits is paid to each field after copying all the black pixels, thanks to the uniqueness of the rows and columns.

The whole computation is initialized by writing down the first permutation $(1, \ldots, m)$. If the area of P is exactly covered by all permutations, then its size is $m \times (m \cdot m!)$, and $\mathcal{A}$ accepts. A minor modification is needed to accept pictures of size $m \times m!$ – extend pictures representing permutations by one (dummy) row and column (that is, permutations on $M' = \{1, \ldots, m-1\}$ are generated).   □

Since it is known that the number of different crossing sequences of a 2DSA between two neighbouring columns of height n is $2^{\mathcal{O}(n \log n)}$ [12], we have found a representable function with the fastest possible growth, which shows that the given bound is tight. Also we obtain the following consequence.

Corollary 1. DREC $\cap \{\square\}^{*,*}$ *is a proper subset of* $\mathcal{L}(\text{2DSA}) \cap \{\square\}^{*,*}$.

6   Conclusions

We have shown that the two-dimensional deterministic sgraffito automaton has a great potential to simulate other models, thanks to its ability to perform the depth first search in a graph represented on the tape. Also we have seen that it accepts some unary picture languages not in REC.

Jiřička and Král proved that two-dimensional deterministic forgetting automata are more powerful than nondeterministic four-way automata [8]. We have strengthened their result by using a weaker automaton to recognize a larger family, as is demonstrated by the relationships shown in the diagram in Figure 5.

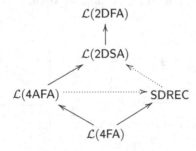

Fig. 5. Hierarchy of Classes of Picture Languages. Each arrow represents a proper inclusion, while a dotted arrow represents an inclusion that is not known to be proper.

It remains open whether SDREC is a proper subclass of $\mathcal{L}(\text{2DSA})$. It seems that closure properties of SDREC have not yet been studied much. It would be worth to know them. If SDREC is not closed under complement, then it would be a proper subclass of $\mathcal{L}(\mathsf{2DSA})$. On the other hand, if SDREC is closed under complement, then $\mathcal{L}(\mathsf{4AFA})$ would be a proper subset of SDREC (an open question in [3]).

References

1. Anselmo, M., Giammarresi, D., Madonia, M.: From determinism to non-determinism in recognizable two-dimensional languages. In: Harju, T., Karhumäki, J., Lepistö, A. (eds.) DLT 2007. LNCS, vol. 4588, pp. 36–47. Springer, Heidelberg (2007)
2. Blum, M., Hewitt, C.: Automata on a 2-dimensional tape. In: Proceedings of the 8th Annual Symposium on Switching and Automata Theory, SWAT, FOCS 1967, pp. 155–160. IEEE Computer Society, Washington, DC (1967)
3. Borchert, B., Reinhardt, K.: Deterministically and sudoku-deterministically recognizable picture languages. In: Loos, R., Fazekas, S.Z., Martín-Vide, C. (eds.) LATA. Research Group on Mathematical Linguistics, vol. 35/07, pp. 175–186. Universitat Rovira i Virgili, Tarragona (2007)
4. Even, S., Even, G.: Graph Algorithms. Computer Software Engineering Series. Cambridge University Press (2011)
5. Giammarresi, D., Restivo, A.: Two-dimensional languages. In: Rozenberg, G., Salomaa, A. (eds.) Handbook of Formal Languages, vol. 3, pp. 215–267. Springer, New York (1997)
6. Hennie, F.C.: One-tape, off-line Turing machine computations. Information and Control 8(6), 553–578 (1965)
7. Jančar, P., Mráz, F., Plátek, M.: Characterization of context-free languages by erasing automata. In: Havel, I.M., Koubek, V. (eds.) MFCS 1992. LNCS, vol. 629, pp. 307–314. Springer, Heidelberg (1992)
8. Jiřička, P., Král, J.: Deterministic forgetting planar automata are more powerful than non-deterministic finite-state planar automata. In: Rozenberg, G., Thomas, W. (eds.) Developments in Language Theory, pp. 71–80. World Scientific, Singapore (1999)
9. Kari, J., Moore, C.: New results on alternating and non-deterministic two-dimensional finite-state automata. In: Ferreira, A., Reichel, H. (eds.) STACS 2001. LNCS, vol. 2010, pp. 396–406. Springer, Heidelberg (2001)
10. Lonati, V., Pradella, M.: Picture recognizability with automata based on Wang tiles. In: van Leeuwen, J., Muscholl, A., Peleg, D., Pokorný, J., Rumpe, B. (eds.) SOFSEM 2010. LNCS, vol. 5901, pp. 576–587. Springer, Heidelberg (2010)
11. Lonati, V., Pradella, M.: Towards more expressive 2D deterministic automata. In: Bouchou-Markhoff, B., Caron, P., Champarnaud, J.-M., Maurel, D. (eds.) CIAA 2011. LNCS, vol. 6807, pp. 225–237. Springer, Heidelberg (2011)
12. Průša, D., Mráz, F.: Two-dimensional sgraffito automata. In: Yen, H.-C., Ibarra, O.H. (eds.) DLT 2012. LNCS, vol. 7410, pp. 251–262. Springer, Heidelberg (2012)

On the Number of Abelian Bordered Words

Narad Rampersad[1], Michel Rigo[2], and Pavel Salimov[2,3,*]

[1] Dept. of Math. and Statistics,
University of Winnipeg, 515 Portage Ave. Winnipeg, MB, R3B 2E9, Canada
narad.rampersad@gmail.com
[2] Dept. of Math., University of Liège,
Grande traverse 12 (B37), B-4000 Liège, Belgium
M.Rigo@ulg.ac.be
[3] Sobolev Institute of Math.,
4 Acad. Koptyug avenue, 630090 Novosibirsk, Russia

Abstract. In the literature, many bijections between (labeled) Motzkin paths and various other combinatorial objects are studied. We consider abelian (un)bordered words and show the connection with irreducible symmetric Motzkin paths and paths in $\mathbb{Z}$ not returning to the origin. This study can be extended to abelian unbordered words over an arbitrary alphabet and we derive expressions to compute the number of these words. In particular, over a 3-letter alphabet, the connection with paths in the triangular lattice is made. Finally, we study the lengths of the abelian unbordered factors occurring in the Thue–Morse word.

1 Introduction

A finite word is *bordered* if it has a proper prefix that is also a suffix of the whole word. Otherwise, the word is said to be *unbordered*. Such properties have been investigated for a long time in combinatorics on words. For instance, the famous Duval's conjecture about the relationship between the length of a word and the maximum length of its unbordered factors has been solved in [8]. A classic result by Ehrenfeucht and Silberger [5] states that if an infinite word has only finitely many unbordered factors, then it is ultimately periodic.

Let us denote the *Parikh vector* of the word u over A by $\Psi(u)$: i.e., $\Psi(u)$ is the element of $\mathbb{N}^A$ representing the number of occurrences of each letter in u. Two words u and v are *abelian equivalent*, if $\Psi(u) = \Psi(v)$. The notions of (un)bordered words are naturally extended to their abelian analogues by replacing equality with abelian equivalence. Such an extension is considered, for example, in [9].

This paper is organized as follows. Below, we make precise the basic definitions. In Section 2, we show that abelian unbordered words over a two letter alphabet are in one-to-one correspondence with particular Motzkin paths, i.e., lattice paths of $\mathbb{N}^2$ that begin at the origin, never pass below the x-axis, and

---

* The author is supported by the Russian President's grant no. MK-4075.2012.1, the Russian Foundation for Basic Research grant no. 12-01-00089 and by a University of Liège post-doctoral grant.

M.-P. Béal and O. Carton (Eds.): DLT 2013, LNCS 7907, pp. 420–432, 2013.
© Springer-Verlag Berlin Heidelberg 2013

use only up diagonal, down diagonal and horizontal steps. In Section 3, abelian unbordered words over a two letter alphabet are shown to be in one-to-one correspondence with n-step walks in $\mathbb{Z}$ starting from the origin but not returning to it. In particular, the number of these n-step walks is well-known and is given by the sequence A063886 in Sloane's Encyclopedia [12]. In Section 4, we extend the size of the alphabet and are still able to relate abelian unbordered words with specific paths and then derive a recursion formula to get the number of such words of length n. Interestingly, for a three letter alphabet, the connection is made with paths in the so-called triangular lattice. Finally, in Section 5, we consider the abelian unbordered factors occurring in abelian periodic automatic sequences (such as the Thue–Morse word). In this last section, we make use of Büchi's theorem and the formalism of first order logic as it was recently used in combinatorics on words, see for instance [3,6].

Definition 1. *A word $u \in A^*$ is* abelian bordered *if there exist $v, v', x, y \in A^+$ such that $u = vx = yv'$ and $\Psi(v) = \Psi(v')$. In that case v is an* abelian border *of u. Otherwise, u is said to be* abelian unbordered.

It is easy to see that if u is abelian bordered, it has an abelian border of length at most $|u|/2$.

Note that a word u over $\{a, b\}$ is abelian unbordered if and only if its complement $\overline{u}$, where all a's are replaced with b's and all b's with a's, is also abelian unbordered. If a word is bordered, then it is trivially abelian bordered. But, in general, the converse does not hold. For instance, $aabbabab$ is abelian bordered but not bordered.

Example 1. We consider the first few abelian unbordered words over $\{a, b\}$ that start with a: $a, ab, aab, abb, aaab, aabb, abbb, aaaab, aaabb, aabbb, abbbb, aabab, ababb$. The first few values for the number of abelian unbordered words of length $n \geq 0$ over $\{a, b\}$ are: 2, 2, 4, 6, 12, 20, 40, 70, 140, 252, 504, 924, 1848, 3432, 6864,... These values match Sloane's sequence A063886.

Remark 1. The language of abelian bordered words is not context-free. Indeed, if we intersect the language of abelian bordered words over $\{a, b\}$ with the regular language $a^+b^+a^+b^+$, then we get the language $\{a^i b^j a^k b^\ell : k \geq i \text{ and } j \geq \ell\}$. Using the pumping lemma, it is easy to show that this language is not context-free.

2 Connection with Motzkin Words

The following is an immediate consequence of the definition "abelian unbordered".

Lemma 1. *Let $n \geq 1$. A word $u_1 \cdots u_n c v_n \cdots v_1$, where for all $i \in \{1, \ldots, n\}$, $u_i \in A$, $v_i \in A$, and $c \in \{\varepsilon\} \cup A$, is abelian unbordered if and only if, for all $i \in \{1, \ldots, n\}$, $\Psi(u_1 \cdots u_i) \neq \Psi(v_i \cdots v_1)$.*

Let us fix the alphabet $A = \{a, b\}$. If $x = x_1 \cdots x_n$ and $y = y_1 \cdots y_n$ are words of length n over A, we define $\begin{pmatrix} x \\ y \end{pmatrix} \in (A \times A)^*$ by

$$\begin{pmatrix} x \\ y \end{pmatrix} := \begin{pmatrix} x_1 \\ y_1 \end{pmatrix} \cdots \begin{pmatrix} x_n \\ y_n \end{pmatrix}.$$

We also define the *projection map*

$$\pi_1 : (A \times A)^* \to A, \begin{pmatrix} x \\ y \end{pmatrix} \mapsto x.$$

We write x^R to denote the *reversal* of x; that is, $x^R = x_n \cdots x_1$. We now define the map

$$m : A^* \to (A \times A)^*, \ u \mapsto \begin{pmatrix} u \\ u^R \end{pmatrix}.$$

Let $P \subseteq (A \times A)^*$ be the context-free language $P = m(A^*) = \left\{ \begin{pmatrix} u \\ u^R \end{pmatrix} \mid u \in A^* \right\}$. Lemma 1 can be restated as follows.

Lemma 2. *A word $u \in A^+$ is abelian bordered if and only if there exists a nonempty proper prefix p of $m(u)$ such that the numbers of occurrences of $\begin{pmatrix} a \\ b \end{pmatrix}$ and $\begin{pmatrix} b \\ a \end{pmatrix}$ in p are the same.*

Definition 2. *A* Grand Motzkin path *of length n is a lattice path of $\mathbb{N}^2$ running from $(0,0)$ to $(n,0)$, whose permitted steps are the up diagonal step $(1,1)$, the down diagonal step $(1,-1)$ and the horizontal step $(1,0)$, called* rise, fall *and* level step, *respectively.*

A Motzkin path *is a Grand Motzkin path that never passes below the x-axis.*

An irreducible *(or* elevated) *Motzkin path is a Motzkin path that does not touch the x-axis except for the origin and the final destination [1].*

If the level steps are labeled by k colors (here colors will be letters from the alphabet A) we obtain a k-colored Motzkin path [11]. A k-colored Motzkin path is described by a word over the alphabet $\{R, F, L_1, \ldots, L_k\}$ and the context-free language of the k-colored Motzkin paths is denoted by $\mathcal{M}_k$. In particular, a Motzkin path described by a word over $\{R, F\}$ is a Dyck path.

Let $h : (A \times A)^* \to \{R, F, L_a, L_b\}^*$ be the coding

$$h : \begin{pmatrix} a \\ b \end{pmatrix} \mapsto R, \ \begin{pmatrix} b \\ a \end{pmatrix} \mapsto F, \ \begin{pmatrix} a \\ a \end{pmatrix} \mapsto L_a, \ \begin{pmatrix} b \\ b \end{pmatrix} \mapsto L_b.$$

Note that if p belongs to P, then $h(p)$ is a *symmetric Grand Motzkin path* having a symmetry axis $x = n/2$. Let $\iota : \{R, F, L_a, L_b\}^* \to \{R, F, L_a, L_b\}^*$ defined by $\iota(R) = F$, $\iota(F) = R$, $\iota(L_a) = L_a$ and $\iota(L_b) = L_b$. If w is a word over $\{R, F, L_a, L_b\}$, then $\widetilde{w} = \iota(w^R)$. A symmetric Grand Motzkin path is described by a word of the kind $w \, c \, \widetilde{w}$ where $c \in \{\varepsilon, L_a, L_b\}$.

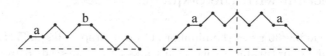

Fig. 1. Two Motzkin paths

Example 2. Two Motzkin paths colored with the letters a and b are represented in Figure 1. The left one is described by the word RL_aRFRL_bFFRF and the right one is symmetric and irreducible. It corresponds to the word $h(m(aaaababababb)) = RRL_aRFRFRFL_aFF$.

Lemma 3. *A word u starting with a is abelian unbordered if and only if $h(m(u))$ is a symmetric and irreducible Motzkin path.*

Proof. This is a reformulation of Lemma 2. □

Proposition 1. *The set of abelian unbordered words over $\{a, b\}$ starting with a and of length at least 2 is $a\pi_1(P \cap h^{-1}(\mathcal{M}_2))b$.*

Proof. Note that $h(P \cap h^{-1}(\mathcal{M}_2))$ is the set of all symmetric 2-colored Motzkin paths. Now observe that if u belongs to $a\pi_1(P \cap h^{-1}(\mathcal{M}_2))b$, then $h(m(u))$ starts with R and ends with F. So the corresponding 2-colored Motzkin path is irreducible. Conversely, if u is abelian unbordered and starts with a, then by Lemma 3, $h(m(u)) = RMF$, where M is a symmetric 2-colored Motzkin path. Thus, $u \in a\pi_1(P \cap h^{-1}(\mathcal{M}_2))b$. □

Remark 2. Any symmetric 2-colored Motzkin path can be built by reflecting a prefix of a 2-colored Motzkin path. Let $w \in \{R, F, L_a, L_b\}^*$ be a prefix of length $k - 1$ of a word in $\mathcal{M}_2$. By the previous proposition, we get that $a\pi_1[h^{-1}(w\,\widetilde{w})]b$, $a\pi_1[h^{-1}(w\,L_a\,\widetilde{w})]b$ and $a\pi_1[h^{-1}(w\,L_b\,\widetilde{w})]b$ are respectively an abelian unbordered word of length $2k$, of length $2k + 1$ having a as central letter, of length $2k + 1$ having b as central letter.

- The set of abelian unbordered words of length $2k$ starting with a is in one-to-one correspondence with the set of prefixes of length $k - 1$ of words in $\mathcal{M}_2$. Equivalently, the set of abelian unbordered words of length $2k$ starting with a is in one-to-one correspondence with the set of prefixes of length k of irreducible 2-colored Motzkin paths.
- The set of abelian unbordered words of length $2k + 1$ starting with a and having a central letter equal to a (resp. b) is in one-to-one correspondence with the set of prefixes of length $k - 1$ of words in $\mathcal{M}_2$. Equivalently, the set of abelian unbordered words of length $2k + 1$ starting with a is in one-to-one correspondence with the set of prefixes of length k of irreducible 2-colored Motzkin paths.

3   Connection with the Sequence A063886

The sequence A063886 gives the number $s(n)$ of n-step walks in $\mathbb{Z}$ starting from the origin but not returning to it. Such walks can be described by words over $\{r, \ell\}$ for right and left steps. The aim of this section is to show that the set of abelian unbordered words over a binary alphabet is in one-to-one correspondence with the n-step walks in $\mathbb{Z}$ starting from the origin but not returning to it. Let us first collect some well-known facts. The generating function for $s(n)$ is $\sqrt{\frac{1+2x}{1-2x}}$.

Consider a word $w = u_1 \cdots u_n v_n \cdots v_1 \in \{a, b\}^*$ of length $2n$. We consider the map c by

$$c : \begin{pmatrix} a \\ b \end{pmatrix} \mapsto rr, \quad \begin{pmatrix} b \\ a \end{pmatrix} \mapsto \ell\ell, \quad \begin{pmatrix} a \\ a \end{pmatrix} \mapsto \ell r, \quad \begin{pmatrix} b \\ b \end{pmatrix} \mapsto r\ell.$$

Applying c to the prefix of length n of $m(w)$ provides a unique path of length $2n$ in $\mathbb{Z}$. This path is denoted by $\mathfrak{p}(w)$. It is clear that $\mathfrak{p}$ is a one-to-one correspondence between the words of length $2n$ over $\{a, b\}$ and the paths of length $2n$ in $\mathbb{Z}$ starting from the origin. The following proposition follows immediately from Lemma 2.

Proposition 2. *A word w over $\{a, b\}$ of even length is abelian unbordered if and only if the path $\mathfrak{p}(w)$ does not return to the origin.*

We extend the definition of $\mathfrak{p}$ to words of odd length by

$$\mathfrak{p}(u_1 \cdots u_n \alpha v_n \cdots v_1) = \begin{cases} \mathfrak{p}(u_1 \cdots u_n v_n \cdots v_1)\, \ell \text{ , if } \alpha = a; \\ \mathfrak{p}(u_1 \cdots u_n v_n \cdots v_1)\, r \text{ , if } \alpha = b. \end{cases}$$

With this definition, $\mathfrak{p}$ is a one-to-one correspondence between the abelian unbordered words of length $2n + 1$ over $\{a, b\}$ and the paths of length $2n + 1$ in $\mathbb{Z}$ starting from the origin and not returning to it. It is therefore easy to get a result similar to the above proposition for words of odd length.

Proposition 3. *A word w over $\{a, b\}$ of odd length is abelian unbordered if and only if the path $\mathfrak{p}(w)$ does not return to the origin.*

The number of prefixes of Motzkin paths is well-known [10, Theorem 1]. Here, we have obtained the following.

Corollary 1. *The number of prefixes of length k of 2-colored Motzkin paths is equal to half the number of paths in $\mathbb{Z}$ of length $2k + 2$ starting from the origin but not returning to it. In particular, this number is equal to*

$$\frac{1}{2} \binom{2k + 2}{k + 1}.$$

4   Larger Alphabets

Let $k \geq 2$. Consider the alphabet $A = \{a_1, \ldots, a_k\}$, or simply $\{1, \ldots, k\}$, and $\mathbb{Z}^k$ equipped with the usual unit vectors $\mathbf{e}_1, \ldots, \mathbf{e}_k$, whose coordinates are all equal to zero except one which is equal to 1. To be able to define k-colored paths, we assume that at each point in $\mathbb{Z}^k$, there are exactly k loops colored with the k different letters.

We first consider a word $u_1 \cdots u_n v_n \cdots v_1$ of even length $2n$. Take the prefix of length n of $m(u_1 \cdots u_n v_n \cdots v_1)$ and apply to it the morphism $h_k : (A \times A)^* \to \{\mathbf{e}_i - \mathbf{e}_j \mid 1 \leq i, j \leq k\}^* \subset (\mathbb{Z}^k)^*$ defined by

$$h_k \begin{pmatrix} a_i \\ a_j \end{pmatrix} = \mathbf{e}_i - \mathbf{e}_j, \quad \forall i, j \in \{1, \ldots, k\}.$$

Therefore, to the word $w = u_1 \cdots u_n v_n \cdots v_1$ there corresponds the sequence of $n + 1$ points in $\mathbb{Z}^k$

$$p_0 = \mathbf{0}, \; p_1 = h_k \begin{pmatrix} u_1 \\ v_1 \end{pmatrix}, \; p_2 = h_k \begin{pmatrix} u_1 \\ v_1 \end{pmatrix} + h_k \begin{pmatrix} u_2 \\ v_2 \end{pmatrix}, \ldots, \; p_n = \sum_{j=1}^{n} h_k \begin{pmatrix} u_j \\ v_j \end{pmatrix},$$

where $\mathbf{0}$ denotes the origin $(0, 0, \ldots, 0)$. By the definition of h_k, note that all these points lie in the subspace $\mathbf{H}_k$ of $\mathbb{Z}^k$ satisfying the equation

$$x_1 + \cdots + x_k = 0.$$

Definition 3. *A path of length n in $\mathbf{H}_k$ is a sequence $p_0, \ldots, p_n$ of points in $\mathbf{H}_k$ such that, for all $j \geq 1$, $p_j - p_{j-1}$ belongs to $\{\mathbf{e}_i - \mathbf{e}_j \mid 1 \leq i, j \leq k, i \neq j\}$.*

A k-colored path of length n in $\mathbf{H}_k$ is a sequence $p_0, c_0, p_1, c_1, \ldots, p_{n-1}, c_{n-1}, p_n$ alternating points in $\mathbf{H}_k$ and elements belonging to $A \cup \{\varepsilon\}$ in such a way that, if $p_j \neq p_{j+1}$, then $p_{j+1} - p_j$ belongs to $\{\mathbf{e}_i - \mathbf{e}_j \mid 1 \leq i, j \leq k, i \neq j\}$ and $c_j = \varepsilon$, otherwise c_j belongs to A and can be interpreted as the color assigned to a loop on p_j. Note that paths are special cases of k-colored paths.

For the rest of this paper we will only consider paths that start at the origin.

Remark 3. For $k = 3$, $\mathbf{H}_3$ corresponds to the so-called *triangular lattice* (sometimes called hexagonal lattice) because a point $\mathbf{x}$ has exactly six neighbors. The set of neighbors of $\mathbf{x}$ is denoted by

$$N(\mathbf{x}) := \mathbf{x} + \{\mathbf{e}_1 - \mathbf{e}_2, \mathbf{e}_1 - \mathbf{e}_3, \mathbf{e}_2 - \mathbf{e}_1, \mathbf{e}_2 - \mathbf{e}_3, \mathbf{e}_3 - \mathbf{e}_1, \mathbf{e}_3 - \mathbf{e}_2\}.$$

Consider the word $w = 23321211$ over the alphabet $\{1, 2, 3\}$. The prefix of length 4 of $m(w)$ is

$$\begin{pmatrix} 2 \; 3 \; 3 \; 2 \\ 1 \; 1 \; 2 \; 1 \end{pmatrix}$$

and corresponds to the sequence of moves $p_1 - 0 = \mathbf{e}_2 - \mathbf{e}_1$, $p_2 - p_1 = \mathbf{e}_3 - \mathbf{e}_1$, $p_3 - p_2 = \mathbf{e}_3 - \mathbf{e}_2$ and $p_4 - p_3 = \mathbf{e}_2 - \mathbf{e}_1$ and the path represented in Fig. 2.

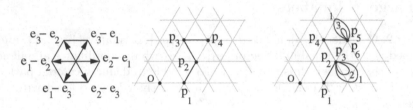

Fig. 2. In the triangular lattice, a path and a 3-colored path

The second path in Fig. 2 is colored and has four loops with labels $2, 1, 3$ and 1 respectively. It corresponds to $w' = 2321323113121211$. The prefix of length 8 of $m(w')$ is

$$\begin{pmatrix} 2\,3\,2\,1\,3\,2\,3\,1 \\ 1\,1\,2\,1\,2\,1\,3\,1 \end{pmatrix}.$$

Observe that in this prefix we have an occurrence of a repeated symbol in positions $3, 4$ and $7, 8$ corresponding to the four loops in the path.

The word $w = u_1 \cdots u_n v_n \cdots v_1$ is said to be *simple* if, for all $i \in \{1, \ldots, n\}$, $u_i \neq v_i$. In this case, in the sequence of points $p_0 = \mathbf{0}, p_1, \ldots, p_n$ corresponding to w, for all $j < n$, we have $p_j \neq p_{j+1}$. Therefore simple words w of length $2n$ correspond to paths of length n in $\mathbf{H}_k$. Such paths are denoted by $\mathfrak{p}(w)$. For a non-simple word w of length $2n$ there is a corresponding k-colored path $\mathfrak{p}(w)$ of length n in $\mathbf{H}_k$ (where at least one loop $p_j = p_{j+1}$ occurs along the path). Conversely, for each k-colored path of length n in $\mathbf{H}_k$, there is a corresponding word of length $2n$.

Proposition 4. *A word w over $\{a_1, \ldots, a_k\}$ of even length $2n$ is abelian unbordered if and only if the k-colored path $\mathfrak{p}(w)$ in $\mathbf{H}_k$ of length n does not return to the origin. Moreover, a simple word w over $\{a_1, \ldots, a_k\}$ of even length $2n$ is abelian unbordered if and only if $\mathfrak{p}(w)$ is a path in $\mathbf{H}_k$ of length n without loops that does not return to the origin.*

Proof. The proof is similar to the one of Proposition 2. □

If $w = u_1 \cdots u_n \alpha v_n \cdots v_1$ is a word of odd length, we can first consider the prefix of length n of $m(w)$ and then add an extra loop of color α to the end of the corresponding path $\mathfrak{p}(w)$. As for Proposition 3, we get the following.

Proposition 5. *A word w over $\{a_1, \ldots, a_k\}$ of odd length $2n + 1$ is abelian unbordered if and only if the k-colored path $\mathfrak{p}(w)$ of length $n + 1$ in $\mathbf{H}_k$ does not return to the origin. In particular, such a path ends with a loop whose color is the one corresponding to the central letter of w.*

Remark 4. The numbers of abelian unbordered words of length n over a 3-letter alphabet, for $1 \leq n \leq 10$, are: 3, 6, 18, 48, 144, 402, 1206, 3438, 10314, 29754 and for simple abelian unbordered words, we get 3, 6, 18, 30, 90, 168, 504, 954,

2862, 5508. As we can observe, over a 3-letter alphabet, the number of abelian unbordered words (resp. simple abelian unbordered words) of length $2n + 1$ is three times the number of abelian unbordered words (resp. simple abelian unbordered words) of length $2n$ because there are three available choices for the central letter. This observation extends trivially to an arbitrary alphabet.

From the discussion above and taking only entries of even index in the previous table, we also get the number of paths (resp. paths without loops) of length n in the triangular lattice $\mathbf{H}_3$ that do not return to the origin. We denote this quantity by $p_3(n)$ (resp. $s_3(n)$). The first few values of $p_3(n)$, $n \geq 1$, are 6, 48, 402, 3438, 29754, 259464, 2274462 and the first few values of $s_3(n)$ are 6, 30, 168, 954, 5508, 32016, 187200. The next statement means that one only needs to compute the sequence $(s_k(n))_{n \geq 1}$ to get $(p_k(n))_{n \geq 1}$ and thus the number of abelian unbordered words of length n.

Lemma 4. *We have*

$$p_k(n) = \sum_{i=1}^{n} s_k(i) \, k^{n-i} \binom{n-1}{n-i}.$$

Proof. By a (k-colored) path, we mean a path in $\mathbf{H}_k$ that does not return to the origin. Each such k-colored path of length n has a unique underlying path of length i, for some $i \in \{1, \ldots, n\}$. To get a k-colored path of length n, $n - i$ loops are added to this underlying path. Each loop can be placed independently at any point of the path, except the origin, and can be colored independently in one of k colors. So, the total number of ways to extend such a path of length i to a k-colored path of length n is $k^{n-i}\binom{n-1}{n-i}$.   □

4.1   Computation of $(s_3(n))_{n \geq 0}$ and Then $(s_k(n))_{n \geq 0}$

We show how to get a recurrence relation to compute the number $s_3(n)$, i.e., the number of paths in the triangular lattice $\mathbf{H}_3 = (V, E)$ that do not return to the origin; here V (resp. E) is the set of vertices (resp. edges) of $\mathbf{H}_3$. Consider the map

$$e : V \to \mathbb{N}, \quad \mathbf{x} \mapsto \begin{cases} 1 & \text{if } \mathbf{x} = \mathbf{0}, \\ 0 & \text{otherwise.} \end{cases}$$

If $f : V \to \mathbb{N}$ is a map, we denote by $\mathcal{S}f : V \to \mathbb{N}$ the map defined by

$$(\mathcal{S}f)(\mathbf{x}) = \sum_{\mathbf{y} \in N(\mathbf{x})} f(\mathbf{y})$$

where $N(\mathbf{x})$ is the set of neighbors of $\mathbf{x}$. In particular, if $f, g : V \to \mathbb{N}$ are maps, then $\mathcal{S}(f + g) = \mathcal{S}f + \mathcal{S}g$. A simple induction argument gives the following result.

Lemma 5. *With the above notation, $(\mathcal{S}^n e)(\mathbf{x})$ is equal to the number of paths of length n that end at $\mathbf{x}$.*

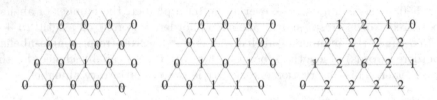

Fig. 3. values of e, $\mathcal{S}e$ and $\mathcal{S}^2e$ around $\mathbf{0}$

The values of the maps e, $\mathcal{S}e$ and $\mathcal{S}^2e$ around $\mathbf{0}$ are given in Figure 3. Let $r_{3,n} : V \to \mathbb{N}$ be defined as follows: $r_{3,n}(\mathbf{x})$ is the number of paths of length n that end at $\mathbf{x}$ and never return to the origin. Then

$$s_3(n) = \sum_{\mathbf{x} \in V} r_{3,n}(\mathbf{x})$$

where the sum is finite, since $r_{3,n}(\mathbf{x}) \neq 0$ implies that $\mathbf{x}$ is at distance at most n from the origin. If a map $f : V \to \mathbb{N}$ is constant on $N(\mathbf{0})$ (as is the case for $\mathcal{S}^n e$), then $\gamma(f)$ is a shorthand for $f(\mathbf{y})$ for any $\mathbf{y} \in N(\mathbf{0})$. By the symmetry of $\mathbf{H}_3$, we see that $r_{3,n}$ is constant on $N(\mathbf{0})$. Note that

$$s_3(n+1) = 6s_3(n) - \sum_{\mathbf{x} \in N(\mathbf{0})} r_{3,n}(\mathbf{x}) = 6s_3(n) - 6\gamma(r_{3,n})$$

because all paths except the ones that end in vertices adjacent to $\mathbf{0}$ have 6 prolongations, and the excluded ones have 5 possible prolongations. The same argument can be applied to maps: $r_{3,n+1} = \mathcal{S}r_{3,n} - 6\gamma(r_{3,n})\,e$ and, applied inductively, this leads to the following relation for $r_{3,n+1}$:

$$r_{3,n+1} = \mathcal{S}^{n+1}e - \sum_{i=0}^{n} 6\gamma(r_{3,i})\,\mathcal{S}^{n-i}e \ . \tag{1}$$

The sequence $((\mathcal{S}^n e)(\mathbf{0}))_{n \geq 0}$ counting the paths of length n starting and ending at $\mathbf{0}$ is well-known (A002898 gives the number of n-step closed paths on the hexagonal lattice). For instance, we have

$$(\mathcal{S}^n e)(\mathbf{0}) = \sum_{\ell=0}^{n} (-2)^{n-\ell} \binom{n}{\ell} \sum_{j=0}^{\ell} \binom{\ell}{j}^3 \tag{2}$$

and its first values are 1, 0, 6, 12, 90, 360, 2040, 10080, 54810, 290640,... Due to the 6-fold symmetry of the maps around the origin, note that

$$\gamma(\mathcal{S}^n e) = \frac{(\mathcal{S}^{n+1}e)(\mathbf{0})}{6}. \tag{3}$$

Taking into account both (1) and (3), for all $n \geq 0$, we have

$$\gamma(r_{3,n+1}) = \frac{(\mathcal{S}^{n+2}e)(\mathbf{0})}{6} - \sum_{i=0}^{n} \gamma(r_{3,i})(\mathcal{S}^{n-i+1}e)(\mathbf{0})$$

and $\gamma(r_{3,0}) = 0$. As a conclusion, using the sequence $((\mathcal{S}^n e)(\mathbf{0}))_{n\geq 0}$, we can compute inductively $(\gamma(r_{3,n}))_{n\geq 0}$ and therefore $(s_3(n))_{n\geq 0}$. Using the above formula, the first values of $(\gamma(r_{3,n}))_{n\geq 0}$ are 0, 1, 2, 9, 36, 172, 816, 4101, 20840, 108558, 572028,... Knowing that $s_3(0) = 1$ is enough to compute

$$s_3(1) = 6(s_3(0) - \gamma(r_{3,0})) = 6, \quad s_3(2) = 6(s_3(1) - \gamma(r_{3,1})) = 6(6-1) = 30,\ldots$$

Let $k \geq 3$. We now turn to the general case to compute $(s_k(n))_{n\geq 0}$. Consider the homomorphism of groups χ between $(\mathbf{H}_k, +)$ and $(\mathbb{Z}((z_1,\ldots,z_{k-1})), \cdot)$ defined by the images of a basis of $\mathbf{H}_k$

$$\chi : \mathbf{e}_1 - \mathbf{e}_k \mapsto z_1, \ \mathbf{e}_2 - \mathbf{e}_k \mapsto z_2, \ldots, \ \mathbf{e}_{k-1} - \mathbf{e}_k \mapsto z_{k-1}.$$

In particular, $\chi(-\mathbf{v}) = 1/\chi(\mathbf{v})$ and $\chi(\mathbf{v} + \mathbf{v}') = \chi(\mathbf{v}).\chi(\mathbf{v}')$. Any element of $\mathbf{H}_k$ can be decomposed as a linear combination with integer coefficients of $\mathbf{e}_1 - \mathbf{e}_k, \mathbf{e}_2 - \mathbf{e}_k, \ldots, \mathbf{e}_{k-1} - \mathbf{e}_k$ and corresponds by χ to a unique Laurent polynomial in $z_1, \ldots, z_{k-1}$. Each vertex $\mathbf{x}$ in $\mathbf{H}_k$ has exactly $k(k-1)$ neighbors:

$$\mathbf{x} + \{\mathbf{e}_i - \mathbf{e}_j \mid 1 \leq i, j \leq k, \ i \neq j\}$$

and these $k(k-1)$ translations of $\mathbf{x}$ are coded through χ by the terms

$$\left\{z_i + \frac{1}{z_i} \mid 1 \leq i \leq k-1\right\} \cup \left\{\frac{z_i}{z_j} \mid 1 \leq i, j \leq k-1, \ i \neq j\right\}.$$

Now consider the Laurent polynomial corresponding to these elementary translations:

$$T = \sum_{i=1}^{k-1}\left(z_i + \frac{1}{z_i}\right) + \sum_{i \neq j}\frac{z_i}{z_j} = \left(1 + \sum_{i=1}^{k-1} z_i\right)\left(1 + \sum_{i=1}^{k-1}\frac{1}{z_i}\right) - k.$$

Let $\mathbf{x} \in \mathbf{H}_k$ and $(j_1, \ldots, j_{k-1}) \in \mathbb{Z}^{k-1}$ be such that $\chi(\mathbf{x}) = z_1^{j_1} \cdots z_{k-1}^{j_{k-1}}$. The number of paths of length n from the origin to $\mathbf{x}$ in the lattice $\mathbf{H}_k$ is given by the coefficient of $z_1^{j_1} \cdots z_{k-1}^{j_{k-1}}$ in T^n. In particular, the constant term gives exactly the number of paths of length n returning to the origin. Furthermore, for $k = 3$ one can derive (2).

Example 3. For $k = 4$, the number of paths of length n in $\mathbf{H}_4$ starting and ending at the origin is Sloane's sequence A002899 and is given by

$$\sum_{\ell=0}^{n}(-4)^{n-\ell}\binom{n}{\ell}\sum_{j=0}^{\ell}\binom{\ell}{j}^2\binom{2\ell-2j}{\ell-j}\binom{2j}{j}.$$

The first values are $1, 0, 12, 48, 540, 4320, 42240, 403200, 4038300, \ldots$. For $k = 5$, we get $1, 0, 20, 120, 1860, 23280, 342200, 5115600, 79922500, \ldots$ and for $k = 6$: $1, 0, 30, 240, 4770, 82080, 1650900, 34524000, 758894850, \ldots$.

Being able to compute $(\mathcal{S}^n e)(\mathbf{0})$ for the lattice $\mathbf{H}_k$, we can proceed exactly as for the computation of $s_3(n)$ and get, for all $n \geq 0$,

$$\gamma(r_{k,n+1}) = \frac{(\mathcal{S}^{n+2}e)(\mathbf{0})}{k(k-1)} - \sum_{i=0}^{n} \gamma(r_{k,i})(\mathcal{S}^{n-i+1}e)(\mathbf{0})$$

with $\gamma(r_{k,0}) = 0$ and finally, $s_k(n+1) = k(k-1)(s_k(n) - \gamma(r_{k,n}))$.

5   About the Thue–Morse Word

Currie and Saari [4] proved that if $n \not\equiv 1 \pmod 6$, then the Thue–Morse word $\mathbf{t}$ has an unbordered factor of length n, but they left it open to decide for which lengths congruent to 1 modulo 6 does this property hold. Then Goč, Henshall and Shallit [6] showed that $\mathbf{t}$ has an unbordered factor of length n if and only if $(n)_2 \notin 1(01^*0)^*10^*1$, where $(n)_2$ denotes the base 2 expansion of n.

If we are interested in abelian unbordered factors of length n occurring in $\mathbf{t}$, we obtain a strict subset of the set described by the above theorem. For instance, for $n = 9$, $(n)_2 = 1001$ does not belong to $1(01^*0)^*10^*1$, so $\mathbf{t}$ contains an unbordered factor of length 9 but a direct inspection shows that all factors of length 9 occurring in $\mathbf{t}$ are abelian bordered. For instance, the factor 001100101 is unbordered but is abelian bordered. Obtained by a computer search, the first few values of $n \in \{0, \dots, 2000\}$ such that $\mathbf{t}$ has an abelian unbordered factor of length n are 0, 1, 2, 3, 5, 8, 10, 12, 14, 16, 22, 50, 54, 66, 70, 194, 198, 258, 262, 770, 774, 1026, 1030. We conjecture that the set of integers $n \geq 50$ such that $\mathbf{t}$ has an abelian unbordered factor of length n consists of those integers whose base 2 expansion belongs to $110(00)^*\{01, 11\}0 \cup 10(00)^+\{01, 11\}0$.

Generally, abelian properties of k-automatic sequences are not suited to be expressed in the extended Presburger arithmetic $\langle \mathbb{N}, +, V_k \rangle$. Nevertheless, we can take advantage of the fact that the Thue–Morse word is abelian periodic of period 2 and apply Büchi's theorem [2] with a technique similar to [3,6]. We take verbatim the statement of Büchi's theorem as formulated by Charlier, Rampersad and Shallit in [3], which states that the k-automatic sequences are exactly the sequences definable in the first order structure $\langle \mathbb{N}, +, V_k \rangle$.

Theorem 1. *[3] If we can express a property of a k-automatic sequence $\mathbf{x}$ using quantifiers, logical operations, integer variables, the operations of addition, subtraction, indexing into $\mathbf{x}$, and comparison of integers or elements of $\mathbf{x}$, then this property is decidable.*

The technique we are now describing can obviously be adapted to any k-automatic abelian periodic word. We will give in (4) below a first order formula $\varphi(n)$ in $\langle \mathbb{N}, +, V_2 \rangle$ that is satisfied if and only if an abelian unbordered factor of length n occurs in the Thue–Morse word $\mathbf{t}$. General procedures to obtain a finite automaton recognizing the base 2 expansions of the integers belonging to the set $\{n \in \mathbb{N} \mid \langle \mathbb{N}, +, V_2 \rangle \models \varphi(n)\}$ do exist (see for instance [2]). Hence a certified regular expression for the base 2 expansion of the elements in the above set will follow. Note that, since t is 2-automatic, we can define in $\langle \mathbb{N}, +, V_2 \rangle$ a unary function that maps i to $t(i)$. Such a formula is again described in [2]. Predicates $e(n)$ and $o(n)$ are simply shorthands to characterize even and odd integers, $e(n) \equiv (\exists x)(n = x + x)$, $o(n) = \neg e(n)$. We define a predicate $B(i, n, k)$ which is true if and only if the Thue–Morse word has an abelian bordered factor of length n occurring at i with a border of length k. Since the Thue–Morse word t is a concatenation of ab and ba, discussing only the parity of the position i, the length n of the factor and the length k of the border, the predicate $B(i, n, k)$ is defined by the disjunction of the following terms $(e(i) \wedge e(n) \wedge e(k))$, $(e(i) \wedge e(n) \wedge o(k) \wedge t(i+k-1) = t(i+n-k))$, $(e(i) \wedge e(n) \wedge e(k) \wedge t(i+n-k) \neq t(i+n-1))$, $(e(i) \wedge o(n) \wedge o(k) \wedge t(i+k-1) = t(i+n-1))$, $(o(i) \wedge e(n) \wedge o(k) \wedge t(i) = t(i+n-1))$, $(o(i) \wedge o(n) \wedge e(k) \wedge t(i) \neq t(i+k-1))$, $(o(i) \wedge o(n) \wedge o(k) \wedge t(i) = t(i+n-k))$ and $(o(i) \wedge e(n) \wedge e(k) \wedge [(t(i) = t(i+n-k) \wedge t(i+k-1) = t(i+n-1)) \vee (t(i) = t(i+n-1) \wedge t(i+k-1) = t(i+n-k))])$. As an example, if i is even, n and k are odd, we have the situation depicted in Figure 4. In that case, since all blocks ab and ba are abelian equivalent, one has just to check equality of two symbols in adequate positions corresponding to the parameters.

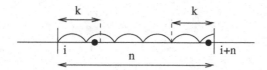

Fig. 4. A factor occurring in the Thue–Morse word

Now the Thue–Morse word has an abelian unbordered factor of length n if and only if the following formula holds true

$$\varphi(n) \equiv (\exists i)(\forall k)(k \geq 1 \wedge 2k \leq n) \to \neg B(i, n, k). \tag{4}$$

References

1. Barnabei, M., Bonetti, F., Silimbani, M.: Restricted involutions and Motzkin paths. Adv. in Appl. Math. 47, 102–115 (2011)
2. Bruyère, V., Hansel, G., Michaux, C., Villemaire, R.: Logic and p-recognizable sets of integers. Bull. Belg. Math. Soc. 1, 191–238 (1994)
3. Charlier, E., Rampersad, N., Shallit, J.: Enumeration and decidable properties of automatic sequences. Int. J. Found. Comput. Sci. 23, 1035–1066 (2012)
4. Currie, J.D., Saari, K.: Least periods of factors of infinite words. RAIRO Inform. Théor. App. 43, 165–178 (2009)
5. Ehrenfeucht, A., Silberger, D.M.: Periodicity and unbordered segments of words. Disc. Math. 26, 101–109 (1979)
6. Goč, D., Henshall, D., Shallit, J.: Automatic theorem-proving in combinatorics on words. In: Moreira, N., Reis, R. (eds.) CIAA 2012. LNCS, vol. 7381, pp. 180–191. Springer, Heidelberg (2012)
7. Graham, D., Knuth, D.E., Patashnik, O.: Concrete mathematics. A foundation for computer science, 2nd edn. Addison-Wesley Pub. Company (1994)

8. Harju, T., Nowotka, D.: Periodicity and Unbordered Words: A Proof of Duval's Conjecture. J. ACM 54 (2007)
9. Holub, S., Saari, K.: On highly palindromic words. Disc. Appl. Math. 157, 953–959 (2009)
10. Guibert, O., Pergola, E.: Enumeration of vexillary involutions which are equal to their mirror/complement. Disc. Math. 224, 281–287 (2000)
11. Sapounakis, A., Tsikouras, P.: On k-colored Motzkin words. J. Integer Seq. 7 (2004)
12. Sloane, N.J.A.: The On-Line Encyclopedia of Integer Sequences. The OEIS Foundation Inc., http://oeis.org/

Proof of a Phase Transition in Probabilistic Cellular Automata

Damien Regnault

IBISC, Université d'Évry Val-d'Essonne, 23 boulevard de France, 91037 Evry CEDEX
http://www.ibisc.fr/~dregnault/

Abstract. Cellular automata are a model of parallel computing. It is well known that simple deterministic cellular automata may exhibit complex behaviors such as Turing universality [3,13] but only few results are known about complex behaviors of probabilistic cellular automata.

Several studies have focused on a specific probabilistic dynamics: α-asynchronism where at each time step each cell has a probability α to be updated. Experimental studies [5] followed by mathematical analysis [2,4,7,8] have permitted to exhibit simple rules with interesting behaviors. Among these behaviors, most of these studies conjectured that some cellular automata exhibit a polynomial/exponential phase transition on their convergence time, *i.e.* the time to reach a stable configuration. The study of these phase transitions is crucial to understand the behaviors which appear at low synchronicity. A first analysis [14] proved the existence of the exponential phase in cellular automaton FLIP-IF-NOT-ALL-EQUAL but failed to prove the existence of the polynomial phase. In this paper, we prove the existence of a polynomial/exponential phase transition in a cellular automaton called FLIP-IF-NOT-ALL-0.

1 Introduction

Cellular automata are made of several cells which are characterized by a state. Time is discrete and at each iteration, the state of a cell evolves according to the states of its neighbors. On the one hand, cellular automata are used as a model of parallel computing. Programming synchronous cellular automata relies on signals and collisions. Such mechanism are hard to explain but rather well understood now (see the firing squad synchronization problem [11], the game of life [1] or universality of rule 110 [3]). On the other hand, they are also a common tool to model real life phenomena.

Theoretical studies of cellular automata as a computation model mainly focus on the synchronous dynamics, *i. e.* at each time step all cells are updated. Only few studies focus on other dynamics. For modeling real life phenomena, the assumption of synchronicity may be too restrictive. For example, boolean networks are commonly used to model the proteins-genes interaction and lots of works focus on different dynamics.

In this article, we will consider stochastic dynamics. At each time step, only the cells of a randomly chosen set are updated. The introduction of randomness has different motivations. It can be used:

M.-P. Béal and O. Carton (Eds.): DLT 2013, LNCS 7907, pp. 433–444, 2013.
© Springer-Verlag Berlin Heidelberg 2013

- to simulate faults in a system, *i.e.* as a difficulty to overcome;
- as an oracle, *i.e.* as a tool to develop more efficient algorithms;
- to accurately represent physical phenomena in models based on cellular automata.

Fault tolerant cellular automata are cellular automata where at each time step, each cell has a constant probability to be faultily updated and to switch to some other state. Toom [15] and Gács [9] have developed rules which are able to make reliable computation in the presence of random faults. Their results interlace two rules: a rule which is Turing universal for the parallel dynamics and a rule which detects and erases faults.

In algorithmic, it is well known that randomness can be useful in the development of efficient algorithms. Recently, Fatès [6] has shown that the density classification problem can be solved with arbitrary precision by two-state one-dimensional cellular automata under probabilistic dynamics. In this problem, the cells must all choose the overall majority state of the initial configuration. Land and Belew [10] have shown that this problem cannot be solved under deterministic dynamics.

Some theoretical studies [2,4,7,8] have focused on α-asynchronous cellular automata where at each time step, each cell has a probability α to be updated. The first theoretical analyses [2,7,8] mainly rely on simple stochastic processes such as random walks and coupon collectors. Later studies [4] have focused on the minority rule. The authors manage to analyze the very first and last steps of a classical evolution of Minority from a random configuration but they are not able to analyze the whole dynamics. They argue that even if Minority on a random configuration seems a "simple" process, some specific initial configurations lead to different dynamics. They conjectured that Minority can simulate some classical stochastic processes like percolation or TASEP on specific initial configurations. From these works, it seems that one simple cellular automaton may simulate several stochastic processes which were independently studied by different communities. If these conjectures are true, then these works may shed a new light on stochastic process simulation.

In this paper, we will prove the existence of a phase transition in the $1D$ cellular automaton FLIP-IF-NOT-ALL-0 depending on α. This is the first proof of a phase transition in a probabilistic cellular automaton even if the existence of a phase transition in α-asynchronous dynamics was conjectured and analyzed empirically [5]. Note that rules FLIP-IF-NOT-ALL-0 and FLIP-IF-NOT-ALL-EQUAL only differ on one neighborhood and rule $1D$ Minority is equivalent to rule FLIP-IF-NOT-ALL-EQUAL on even configuration by switching the state of one cell over two. A first study [14] proved the existence of one part of the phase transition on $1D$ Minority but failed to prove the other part. Only one last argument is necessary to prove the existence of a phase transition for Minority rule.

In Gács' $1D$ fault tolerant cellular automaton [9], the main difficulty was to develop a process which is able to detect faulty regions, *i.e.* to save one bit of information in presence of random faults. Since cells have only a local vision of the information, the positive rates conjecture states that it is not possible to keep one bit of information safe from the random faults and thus that reliable computation is impossible. Gács provided a counter-example but the community was unsatisfied with the size and the complexity of its rule and of the following proof. The quest for a simple fault-tolerant cellular automaton with a proof of reasonable size is still open. One phase of the transition phase of FLIP-IF-NOT-ALL-0 leads to the emergence of a white region and the other phase of the transition phase leads to the emergence of a checkerboard pattern. Thus this paper is another step to a better understanding of the emergence of homogeneous regions in $1D$ stochastic processes.

2   Asynchronous Cellular Automata

2.1   Definition

We give here a formal definition of FLIP-IF-NOT-ALL-0, also called ECA 50 (Wolfram encoding). The next part presents informally its behavior.

Definition 1 (Configuration). *Consider $n \in \mathbb{N}$, we denote by $\mathbb{Z}/n\mathbb{Z}$ the set of cells and $Q = \{0,1\}$ the set of states (white stands for 0 and black for 1 in the figures), n is the size of the configuration. The neighborhood of a cell i consists of the cells $i - 1$, i and $i + 1 \bmod n$. A configuration c is a function $c : \mathbb{Z}/n\mathbb{Z} \to Q$; c_i is the state of the cell i in configuration c.*

We consider configurations of size $n \in \mathbb{N}$ with periodic boundary conditions, thus all computation on the position of a cell are made modulo n.

Definition 2 (FLIP-IF-NOT-ALL-0). *The rule of a cellular automaton is a function which associates a new state to the states of a neighborhood. The rule δ of FLIP-IF-NOT-ALL-0 is defined as follows:*

$$\delta(c_{i-1}, c_i, c_{i+1}) = \begin{cases} c_i & \text{if } c_{i-1} = c_i = c_{i+1} = 0 \\ 1 - c_i & \text{otherwise} \end{cases}$$

Note that this rule is also known as ECA 50 (Wolfram coding). Time is discrete and in the classical deterministic synchronous dynamics all the cells of a configuration are updated at each time step according to the transition rule of the cellular automaton (see figure 1). Here we consider a stochastic asynchronous dynamics where only a random subset of cells is updated at each time step.

Definition 3 (Asynchronous dynamics). *Given $0 < \alpha \leqslant 1$, we call α-asynchronous dynamics the following process : time is discrete and c^t denotes the random variable for the configuration at time t. The configuration c^0 is the* initial configuration. *The configuration at time $t + 1$ is the random variable defined by the following process : each cell has independently a probability α to be updated according to the rule δ (we say that the cell* fires *at time t) and a probability $1 - \alpha$ to remain in its current state. A cell is said to be* active *if its state changes when fired.*

Figure 1 presents different space-time diagrams of ECA 50 for different values of α. By abuse of notation $\delta(c)$ is the probability distribution obtained after updating c one time with rule δ under asynchronous dynamics.

Definition 4 (Stable configuration). *A configuration c is stable if for all $i \in \mathbb{Z}/n\mathbb{Z}$, $\delta(c_{i-1}, c_i, c_{i+1}) = c_i$.*

FLIP-IF-NOT-ALL-0 admits only one stable configuration: $\bar{0} = 0^n$ where all cells are in state 0. Since any black cell is active, by firing all the black cells and no white cell, we have the following fact:

Fact 1 (One step convergence). *If $0 < \alpha < 1$, any configuration, evolving under FLIP-IF-NOT-ALL-0 and α-asynchronous dynamics, can reach the stable configuration $\bar{0}$ in one step.*

Definition 5 (Worst case convergence). *We say that a random sequence of configurations $(c^t)_{t \geqslant 0}$ evolving under FLIP-IF-NOT-ALL-0 and α-asynchronous dynamics converges from an initial configuration c^0 if the random variable $T = \min\{t : c^t$ is stable $\}$ is finite with probability 1. We say that the convergence occurs in expected polynomial (resp. exponential) time if and only if $\mathbb{E}[T] \leqslant p(n, 1/\alpha)$ (resp. $\mathbb{E}[T] \geqslant b^n$) for some polynomial p (resp. constant $b > 1$) and for any initial configuration (resp. for at least one initial configuration).*

From the definition of stable configuration, it follows that if there is t such that c^t is a stable configuration then for all $t' \geqslant t$ the configuration $c^{t'}$ is the same stable configuration. Since $(c^t)_{t \geqslant 0}$ is a finite Markow chain and since there is a path from any configuration to the stable configuration all white (see fact 1), any sequence of configuration converges with probability 1 when $0 < \alpha < 1$.

Theorem 1 (Main result). *Consider a sequence of configurations $(c^t)_{t \geqslant 0}$ evolving under rule FLIP-IF-NOT-ALL-0 and α-asynchronous dynamics then:*

- *if $0 < \alpha \leqslant 0.5$ then $\mathbb{E}[T] = O(n^2\alpha^{-1})$;*
- *if $1 - \epsilon \leqslant \alpha < 1$ (where $\epsilon = 0.187 \times 10^{-13} > 0$) then $\mathbb{E}[T] = \Omega(2^n)$.*

This is the first time that a phase transition is formally proved on a asynchronous cellular automata. This result shows that simple rules exhibit complex behavior and turn out to be hard to analyze. The following part exposes experimental results on the behavior of FLIP-IF-NOT-ALL-0. Section 3 is dedicated to the proof of the expected polynomial time convergence and section 4 is dedicated to the proof of the expected exponential time convergence.

2.2   Observations

In this section, we present empirical result on FLIP-IF-NOT-ALL-0 and the ideas behind Theorem 1 and Theorem 4. A detailed empirical study of this automata was published by Fatès [5]. We only give here an intuitive description of the behaviour of the automaton. Fatès gave experimental evidences of the

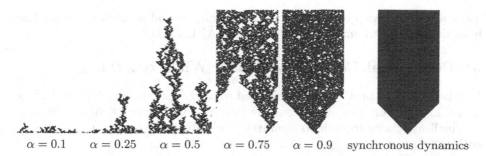

$\alpha = 0.1$ $\alpha = 0.25$ $\alpha = 0.5$ $\alpha = 0.75$ $\alpha = 0.9$ synchronous dynamics

Fig. 1. FLIP-IF-NOT-ALL-0 under different dynamics. The height of the diagrams is scaled according to α.

existence of a critical value $\alpha_c \approx 0.6282$. When $\alpha < \alpha_c$ then the stochastic process converges in polynomial time and when $\alpha > \alpha_c$ the stochastic process converges in exponential time.

Polynomial Time Convergence. If $\alpha < \alpha_c$, only black regions crumble and white regions expand on the whole configuration. The last isolated black cells manage to survive only for a little while and the dynamics quickly reaches the stable configuration $\bar{0}$. Lots of cells are inactive during these periods. We will prove in theorem 2 that the dynamics converges in expected polynomial time. This proof uses a potential function, a technique already used to analyze asynchronous cellular automata [4,7,8]. The proof is here more technical. Until now, this technique was ineffective to analyze FLIP-IF-NOT-ALL-EQUAL when $\alpha \leqslant \alpha'_c$. The analysis of FLIP-IF-NOT-ALL-0 is easier and we able are to conclude.

The idea of the result presented here is the following: consider a bi-infinite configuration with one semi-infinite white region on the left and one semi-infinite black region on the right. The border of the configuration corresponds to the position of the first black cell. Consider the limit case $\alpha = 0.5$: with probability $\frac{1}{2}$ the border move by one cell to the left, with probability $\frac{1}{4}$ the border move to the right by at least 1 cell, with probability $\frac{1}{8}$ the border move to the right by at least 2 cells, ..., with probability $\frac{1}{2^i}$ the border moves to the right by at least $i - 1$ cells. Thus, the expected movement of the border is 0 and it behaves as a non-biased random walk. For finite configurations, this means that the size of a white configuration behaves as a non-biased random walk in the worst case and thus reaches a size of n in quadratic time. The next section formalizes this idea and proves the convergence of the dynamics in expected polynomial time for $\alpha \leqslant 0.5$.

Exponential Time Convergence. Empirically, if $\alpha > \alpha_c$ then from any non-stable initial configuration, white and black regions crumble and a checkerboard pattern invades the space-time diagram. Most cells are active and switch their states at each time step. We will prove in theorem 4 that there exists an initial configuration such that the dynamics converges in expected exponential time by using a coupling with oriented percolation. This method is similar to what has been done in [14] to analyze FLIP-IF-NOT-ALL-EQUAL.

3   Polynomial Convergence Time When $\alpha \leqslant 0.5$

In this section, we prove that the expected convergence time is polynomial when $\alpha \leqslant 0.5$. Thus from now on, we assume that $\alpha \leqslant 0.5$. We will define a function F which associates to each configuration c a potential $F(c) \in \{0, \ldots, 2n+2\}$ such that:

- the expected variation of potential is negative at each time step and
- the stable configuration $\bar{0}$ is the only configuration of zero potential.

We will conclude by using the following lemma which is folklore (a proof can be found in [7]). Consider $\epsilon > 0$, an integer $m > 0$ and a sequence $(X^t)_{t \geqslant 0}$ of random variables with values in $\{0, \ldots, m\}$ and a suitable filtration $(\mathcal{F}^t)_{t \in \mathbb{N}}$:

Lemma 1. *Suppose that :*

- *if $0 < X^t < m$ then $\mathbb{E}[X^{t+1} - X^t | \mathcal{F}^t] \leqslant 0$ and $\Pr\{|X^{t+1} - X^t| \geqslant 1 | \mathcal{F}^t\} \geqslant \epsilon$;*
- *if $X^t = m$ then $\mathbb{E}[X^{t+1} - X^t | \mathcal{F}^t] \leqslant -\epsilon$.*

Let $T = \min\{t \in \mathbb{N} : X^t = 0\}$ and $x_0 = \mathbb{E}[X^0]$. Then:

$$\mathbb{E}[T] \leqslant \frac{x_0(2m+1-x_0)}{2\epsilon}$$

Now, we formalize the observations made in section 2.2. Consider a configuration c, a *white region* of c is a maximal set of consecutive cells which are all in the state 0. The value $W(c)$ is defined as the size of the largest white region of c: $W(c) = \max\{|W| : W \text{ is a white region of } c\}$. We can now introduce the following potential function:

Definition 6 (Potential function). *We consider the function $F : Q^{\mathbb{Z}/n\mathbb{Z}} \to \mathbb{N}$ defined as follows:*

$$F(c) = \begin{cases} 0 & \text{if } c = \bar{0} \\ 2n+2 - W(c) & \text{otherwise} \end{cases}$$

Note that for all configurations c, $F(c) \in \{0, \ldots, 2n+2\}$. Moreover, $\bar{0}$ is the only configuration of potential 0 and the configuration of potential $2n+2$ has all its cells in state 1. We denote by $\mathbb{E}[\Delta(F(c))] = \mathbb{E}[F(\delta(c)) - F(c)]$, *i.e.* the expected variation of potential of c after one update of FLIP-IF-NOT-ALL-0 under α-asynchronous dynamics. We introduce $p^{\bar{0}}(c) = \Pr\{\delta(c) = \bar{0}\}$ the probability that the dynamics converges in one step and for each cell i, we introduce $p_i^0(c) = \Pr\{\delta(c)_i = 0\}$ the probability that the cell is in state 0 at the next time step. The expected variation of potential can be expressed as follows:

$$\mathbb{E}[\Delta(F(c))] = -\mathbb{E}[\Delta(W(c))] - (2n+2 - W(c))p^{\bar{0}}(c)$$
$$\leqslant -\mathbb{E}[\Delta(W(c))] - (n+2)p^{\bar{0}}(c)$$

Lemma 2. *Consider a configuration c and a cell $i \in \mathbb{Z}/n\mathbb{Z}$, if $\alpha \leqslant 0.5$ then:*

$$p_i^0(c) = \begin{cases} 1 & \text{if cell } i \text{ is inactive in } c \\ \geqslant \alpha & \text{otherwise} \end{cases}$$

Proof. An inactive cell is in state 0 and will stay in state 0 with probability 1. An active cell in state 1 will switch to state 0 with probability α. An active cell in state 0 will stay in state 0 with probability $1-\alpha$. Since $0 < \alpha \leqslant 0.5$, $\alpha \leqslant 1-\alpha$. The probability that an active cell will be in state 0 at the next time step is at least α.

Fact 2. *Consider a configuration c, if $\alpha \geqslant 0.5$, then $p^0(c) \geqslant \alpha^{n-W(c)+2}$.*

Our aim is to apply lemma 1 on the potential function F. We start by analyzing some special configurations c where $W(c) \leqslant 1$.

Lemma 3. *Consider a configuration c such that $W(c) \leqslant 1$, then $\mathbb{E}[\Delta(F(c))] \leqslant -\alpha(1-\alpha)^2$ and $\Pr\{|\Delta(F(c))| \geqslant 1\} \geqslant \alpha(1-\alpha)^2$.*

Proof. If $W(c) = 0$ then all cells are in state 1 and $F(c)$ is maximum. Firing any cell leads to the creation of a white region of site at least 1 and thus $\mathbb{E}[\Delta(F(c))] \leqslant -\alpha$ and $\Pr\{|\Delta(F(c))| \geqslant 1\} \geqslant \alpha$.

If $W(c) = 1$ then consider a cell i of c such that $c_i = 0$. We have $c_{i-1} = c_{i+1} = 1$. If $W(\delta(c)) = 0$ then cells $i-1$ and $i+1$ are not fired while cell i fires, this event occurs with probability $\alpha^2(1-\alpha)$. If at least one of cells $i-1$ or $i+1$ fires while cell i is inactive then $W(\delta(c)) \geqslant 2$, this event occurs with probability $(1-(1-\alpha)^2)(1-\alpha)$. Thus,

$$\mathbb{E}[\Delta(F(c))] \leqslant \alpha^2(1-\alpha) - (1-(1-\alpha)^2)(1-\alpha)$$
$$\mathbb{E}[\Delta(F(c))] \leqslant -2\alpha(1-\alpha)^2.$$

Moreover,

$$\Pr\{|\Delta(F(c))| \geqslant 1\} \geqslant (1-(1-\alpha)^2)(1-\alpha)$$
$$\geqslant \alpha(1-\alpha).$$

Now, the difficulty lies in showing that $\mathbb{E}[\Delta(F(c))] \leqslant 0$ for all non stable configurations c of potential $F(c) \leqslant 2n$.

Lemma 4. *Consider a configuration c such that $2 \leqslant W(c) \leqslant n-1$ then $\mathbb{E}[\Delta(F(c))] \leqslant 0$ and $\Pr\{|\Delta(F(c))| \geqslant 1\} \geqslant \alpha(1-\alpha)^2$.*

Proof. Since a cellular automaton is shift invariant, we will now consider that $c_0 = 0$ and $\forall i \in \{n-W(c)+1,\ldots,n-1\}, c_i = 0$. Cells 0 and $n-W(c)+1$ are white active cells and cells of $\{n-W(c)+2,\ldots,n-1\}$ are white inactive cells. We consider the random variable M for the white region of $\delta(c)$ maximum for the inclusion and containing cell 0 if $\delta(c)_0 = 0$ and cell $n-1$ if $\delta(c)_{n-1} = 0$. If $\delta(c)_0 = 1$ and $\delta(c)_{n-1} = 1$ then $M = \emptyset$ (this case may only occurs when $W(c) = 2$). For $i \in \{0, \ldots, n-1\}$, cell i belongs to M if and only if:

- case A: $n - W(c) + 2 < i \leqslant n - 1$ or
- case B_1: $\forall j \in \{0, \ldots, i\}$, $\delta(c)_j = 0$ or
- case B_2: $\forall j \in \{i, \ldots, n - W(c) + 1\}$, $\delta(c)_j = 0$.

For $i \in \{1, 2\}$ we also call $B_i(j)$ the event: case B_i is true for cell j (and thus j belongs to M).

$$\mathbb{E}[|M|] = \mathbb{E}\left[\sum_{0 \leqslant i \leqslant n-1} \mathbf{1}_{i \in M}\right] = \sum_{0 \leqslant i \leqslant n-1} \mathbb{E}[\mathbf{1}_{i \in M}] = \sum_{0 \leqslant i \leqslant n-1} \Pr(i \in M).$$

$$= W(c) - 2 + \sum_{0 \leqslant i \leqslant n-W(c)+1} \Pr(i \in M)$$

$$= W(c) - 2 + \sum_{0 \leqslant i \leqslant n-W(c)+1} [\Pr(B_1(i)) + \Pr(B_2(i)) - \Pr(B_1(i) \cap B_2(i))].$$

Since $c_0 = 0$ and by lemma 2, we have $\Pr(B_1(i)) \geqslant (1 - \alpha)\alpha^i$. Then,

$$\sum_{0 \leqslant i \leqslant n-W(c)+1} \Pr(B_1(i)) \geqslant \sum_{0 \leqslant i \leqslant n-W(c)+1} (1 - \alpha)\alpha^i$$

$$\geqslant 1 - \alpha^{n-W(c)+2}$$

Similarly, we have $\Pr(B_2) \geqslant 1 - \alpha^{n-W(c)+2}$. Note that the event $B_1(i) \cap B_2(i)$ implies that $\delta(c) = \bar{0}$. Then, $\Pr(B_1(i) \cap B_2(i)) = p^{\bar{0}}(c)$. We have:

$$\mathbb{E}[|M|] \geqslant W(c) - 2\alpha^{n-W(c)+2} - (n - W(c) + 2)p^{\bar{0}}(c)$$

Then,

$$\mathbb{E}[\Delta(F(c))] \leqslant W(c) - \mathbb{E}[|M|] - (n+2)p^{\bar{0}}(c)$$

$$\leqslant 2\alpha^{n-W(c)+2} + (n - W(c) + 2)p^{\bar{0}}(c) - (n+2)p^{\bar{0}}(c)$$

$$\leqslant 2\alpha^{n-W(c)+2} - W(c)p^{\bar{0}}(c)$$

$$\leqslant 2\alpha^{n-W(c)+2} - 2p^{\bar{0}}(c)$$

$$\leqslant 0.$$

Note that if cells 0 and $n - W(c) + 1$ do not fire and cell 1 fires then $|M| \geqslant W(c) + 1$. This event occurs with probability $\alpha(1 - \alpha)^2$ and thus $\Pr\{|\Delta(F(c))| \geqslant 1\} \geqslant \alpha(1 - \alpha)^2$.

Theorem 2. *If $\alpha \leqslant 0.5$ then the expected convergence time of* FLIP-IF-NOT-ALL-0 *on any initial configuration of size n is $O(n^2\alpha^{-1})$.*

Proof. Using lemma 1, lemma 3 and 4, we obtain that if $\alpha \leqslant 0.5$, the stochastic process reaches a configuration of potential 0 after $O(n^2\alpha^{-1})$ iterations on expectation. Note that the factor $(1 - \alpha)^2$ is negligible since $\alpha \leqslant 0.5$. The only configuration of potential 0 is $\bar{0}$.

4   Exponential Convergence Time for α Large Enough

In this section, we will demonstrate that FLIP-IF-NOT-ALL-0 converges on expected exponential time when α is close enough to 1. To achieve this goal we will use a coupling with oriented percolation. This coupling will be done such that if the open cluster of the percolation is large enough then the dynamics of the automaton converges in expected exponential time. Since we will use percolation on a non-standard topology, we start by introducing this model. We will then construct the coupling.

4.1   Percolation

Consider a probability p, an integer n and the infinite randomly labeled oriented graph $\mathbb{L}(p, n) = (V, E)$ where $V = \{(i, j) \in \mathbb{N} \times \{0, \ldots, n\} : i + j \text{ is odd}\}$ is called the set of *sites* and E the set of *bonds*. For all sites $(i, j) \in V$, i is the *height* of the site. The height of a bond is the height of its origin. For all $(i, j) \in V$, there are oriented bonds between site (i, j) and sites $(i + 1, j - 1)$ (if $j \neq 0$) and $(i + 1, j + 1)$ (if $j \neq n$). Bonds have a probability p to be labeled *open* and a probability $1 - p$ to be labeled *closed*. These probabilities are independent and identically distributed. An open path of a randomly labeled graph is a path where all edges are open. We denote by C the open cluster of $\mathbb{L}(p, n)$: C contains all sites such that there exists an open path from a site of height 0 to this site.

Due to space constraints, we will admit the following theorem (a proof for a similar percolation process can be found in [12]).

Theorem 3. *If $p \geqslant \frac{16^2 - 1}{16^2}$ then there exists $\epsilon > 0$ such that $\forall n \in \mathbb{N}$, C contains a site of height 2^n with probability at least ϵ.*

4.2   Coupling

Consider a random sequence of configurations $(c^t)_{t \geqslant 0}$ evolving under FLIP-IF-NOT-ALL-0 and α-asynchronous dynamics. The size of the configuration is n. Consider a percolation grid $\mathbb{L}(p, n)$. Consider the following mapping:

Definition 7 (Mapping). *We define $g : V \to \{0, n\}^{\mathbb{N}}$ as the injection which associates the percolation site (i, j) to the cell j of configuration c^{2i}.*

Our aim is to design a coupling such that cells of $g(C)$ are active. The coupling will be defined recursively according to time and height. We denote by C^t the sites of height t which are in the open cluster C.

Definition 8 (Correspondence criterion). *We say that a space-time diagram $(c^t)_{t>0}$ and a labeled directed graph $\mathbb{L}(p, n)$ satisfy the* correspondence criterion *at step t if and only if the cells of $g(C^t)$ have at least one of their neighbors in a different state. We say that they satisfy the* correspondence criterion *if and only if they satisfy the correspondence criterion for all $t \geqslant 0$.*

Note that, satisfying the correspondence criterion implies that the cells of $g(C)$ are all active. The coupling will be defined such that if the correspondence criterion is true at time t, it remains true at time $t+1$. To achieve this goal efficiently, we will consider only local criteria.

Definition 9 (Candidate). *A site is a* candidate *of height $t + 1$ if and only if at least one of its predecessors is in C^t. We denote by $\hat{C}^{t+1}$ the set of candidates of height $t + 1$.*

Definition 10 (Constrained cells). *A cell c_i^t is* constrained *at time t if and only if $c_i^{t+2} \in g(V)$ and $g^{-1}(c_i^{t+2})$ is in $\hat{C}^{t+1}$.*

We have to find a way such that constrained cells possess a neighbor in a different state than themselves after two iterations of FLIP-IF-NOT-ALL-0. We will have to consider different patterns. For the rest of the paper, we will use the following kind of notation ▯◼ to represent the patterns. Here, ▯◼ designs a set of two consecutive cells i and $i+1$ such that $c_i = 0$, $c_{i+1} = 1$ and the arrowed cell $i + 1$ is constrained.

Definition 11 (block). *In a configuration c, a set of contiguous cells is a block if it belongs to the following set:*

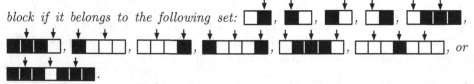

Lemma 5. *For any configuration, it is possible to compute a set of blocks such that any constrained cell of the configuration is in exactly one block.*

Proof. This proof is done by a long enumeration of cases. Due to space constraints, we will only explain how this enumeration is done. To prove that any constrained cell i can belong to a block, we enumerate all the possible neighborhoods of cells $i - 2$, $i - 1$, i, $i + 1$ and $i + 2$ and either i is in a block or i is not constrained. Now, consider a set B of blocks. If a cell belongs to two blocks of

B, the redundancy can be eliminated by removing a block, cutting a block into smaller ones or merging two blocks (and strictly decreasing the sum of the sizes of all blocks). For example, if the white constrained cell of a block ▢◼▮◼ is the same white constrained cell of a block ◼▮◼▢ , then these two blocks can be merged into block ◼▮◼▢◼▮◼▢ .

Lemma 6. *For any block, there is a probability at least α^{12} that all the constrained cells of the block possess a neighbor in a different state than themselves after two iterations of* FLIP-IF-NOT-ALL-0.

Proof. Figure 2 shows that for any block, all the constrained cells of the block can satisfy the correspondence criterion by firing at most 12 cells of the block during the next two steps.

Fig. 2. How to validate the correspondence criterion for any block. The cells drawn at time t and $t+1$ are either inactive or fired.

With the previous lemma, we can achieve our coupling.

Theorem 4. *If $\alpha \geqslant \sqrt[12]{1-(1-p)^6}$ then we can define a coupling between $(c^t)_{t \geqslant 0}$ and $\mathbb{L}(p,n)$ such that the correspondence criterion is true for all $t \in \mathbb{N}$.*

Proof. Consider $n \in \mathbb{N}$ and the initial configuration c^0 where $c_i^0 = 1$ if and only if $i = 0 \bmod 2$. The correspondence criterion is true at time 0. We suppose that the correspondence criterion is true at time t. Now, we should explain how to keep the criterion true at time $t+1$. Due to space constraints, we will only give the main idea of the proof (which is a straightforward adaptation of what has been done in [14]). If $\alpha \geqslant \sqrt[12]{1-(1-p)^6}$, a coupling can be defined such that for all sets of blocks B, if at least one bond ending at a constrained cell of B at time $t+1$ is open then all cells of B fire. Using lemma 6, if all cells of B fire than the criterion is still true at time $t+1$ for cells of B. This coupling can be done such that all cells fire independently from other cells and all bonds are open independently from other bonds.

Theorem 5. *If $\alpha \geqslant \sqrt[12]{1-(1-(\frac{16^2-1}{16^2})^2)^6}$, there exists an initial configuration c^0 such that the stochastic process $(c^t)_{t \geqslant 0}$ that evolves under rule* FLIP-IF-NOT-ALL-0 *and α-asynchronous dynamics converges in expected exponential time.*

Proof. Consider that $\alpha \geqslant \sqrt[12]{1 - (1 - \frac{16^2-1}{16^2})^6}$ then it is possible to define the coupling between FLIP-IF-NO-ALL-0 and $\mathbb{L}(p, n)$ with $p > \frac{16^2-1}{16^2}$. According to theorem 4, there exists an initial configuration such that the correspondence criterion is true for all $t \geqslant 0$. According to lemma 3, there exists a constant $\epsilon > 0$ such that there is a site of height 2^n in the open cluster with probability at least ϵ. According to the coupling definition, the probability that at least one cell is active in c^{2^n} is greater than ϵ.

Acknowledgements. Thanks to Nicolas Schabanel for the simulations of FLIP-IF-NOT-ALL-0 and his useful comments.

References

1. Adamatzky, A.: Collision-Based Computing. Springer (2002)
2. Chassaing, P., Gerin, L.: Asynchronous cellular automata and brownian motion. In: Proc. of AofA 2007. DMTCS Proceedings, vol. AH, pp. 385–402 (2007)
3. Cook, M.: Universality in elementary cellular automata. Complex System 15, 1–40 (2004)
4. Schabanel, N., Regnault, D., Thierry, É.: Progresses in the analysis of stochastic 2D cellular automata: A study of asynchronous 2D minority. Theoretical Computer Science 410, 4844–4855 (2009)
5. Fatès, N.: Asynchronism induces second order phase transitions in elementary cellular automata. Journal of Cellular Automata 4(1), 21–38 (2009)
6. Fatès, N.: Stochastic cellular automata solve the density classification problem with an arbitrary precision. In: Proc. of STACS 2011, pp. 284–295 (2011)
7. Fatès, N., Morvan, M., Schabanel, N., Thierry, É.: Fully asynchronous behavior of double-quiescent elementary cellular automata. Theoretical Computer Science 362, 1–16 (2006)
8. Fatès, N., Regnault, D., Schabanel, N., Thierry, É.: Asynchronous behavior of double-quiescent elementary cellular automata. In: Correa, J.R., Hevia, A., Kiwi, M. (eds.) LATIN 2006. LNCS, vol. 3887, pp. 455–466. Springer, Heidelberg (2006)
9. Gács, P.: Reliable cellular automata with self-organization. Journal of Statistical Physics 103(1/2), 45–267 (2001)
10. Land, M., Belew, R.K.: No perfect two-state cellular automata for density classification exists. Physical Review Letters 74, 5148–5150 (1995)
11. Mazoyer, J.: A six-state minimal time solution to the firing squad synchronization problem. Theoretical Computer Science 50(2), 183–238 (1987)
12. Mossel, E., Roch, S.: Slow emergence of cooperation for win-stay lose-shift on trees. Machine Learning 67(1-2), 7–22 (2006)
13. Ollinger, N., Richard, G.: 4 states are enough! Theoretical Computer Science 412, 22–32 (2011)
14. Regnault, D.: Directed percolation arising in stochastic cellular automata analysis. In: Ochmański, E., Tyszkiewicz, J. (eds.) MFCS 2008. LNCS, vol. 5162, pp. 563–574. Springer, Heidelberg (2008)
15. Toom, A.: Stable and attractive trajectories in multicomponent systems. Advances in Probability 6, 549–575 (1980)

Languages with a Finite Antidictionary: Growth-Preserving Transformations and Available Orders of Growth

Arseny M. Shur

Ural Federal University
arseny.shur@usu.ru

Abstract. We study FAD-languages, which are regular languages defined by finite sets of forbidden factors, together with their "canonical" recognizing automata. We are mainly interested in the possible asymptotic orders of growth for such languages. We analyze certain simplifications of sets of forbidden factors and show that they "almost" preserve the canonical automata. Using this result and structural properties of canonical automata, we describe an algorithm that effectively lists all canonical automata having a strong component isomorphic to a given digraph, or reports that no such automata exist. This algorithm can be used, in particular, to prove the existence of a FAD-language over a given alphabet with a given exponential growth rate. On the other hand, we provide an example showing that the algorithm cannot prove *non-existence* of a FAD-language having a given growth rate. Finally, we provide some examples of canonical automata with a nontrivial condensation graph and of FAD-languages with a "complex" order of growth.

Keywords: finite antidictionary, regular language, combinatorial complexity, growth rate.

1 Introduction

By growth properties of a language L over a finite alphabet Σ we mean the asymptotic behaviour of the function $C_L(n) = |L \cap \Sigma^n|$ called the *combinatorial complexity* of L. Growth properties of formal languages are intensively studied during the last decades. Such studies were initially motivated by the needs of symbolic dynamics (starting with [5]) and algebra (starting with [4]). In most cases, all considered languages were *factorial*, i.e., closed under taking factors of their words. In the context of dynamical systems, special attention was paid to *sofic subshifts*, which correspond to factorial regular languages, and to *subshifts of finite type*, which correspond to FAD-languages studied in this paper; see, e.g., [1]. In the algebraic context, FAD-languages correspond to finitely presented algebras of certain types; see, e.g., [3].

It is known [6] that the combinatorial complexity of a regular language is asymptotically equivalent to the function of the form $f(n) = p_{n \bmod r}(n)\alpha_{n \bmod r}^n$, where r is a positive integer, α_i are nonnegative algebraic numbers and $p_i(n)$

M.-P. Béal and O. Carton (Eds.): DLT 2013, LNCS 7907, pp. 445–457, 2013.
© Springer-Verlag Berlin Heidelberg 2013

are polynomials with algebraic coefficients. Recall that the *index* $\mathsf{Ind}(G)$ of a digraph G is the Frobenius root (or spectral radius) of the adjacency matrix of G. A finite automaton is considered as a labeled digraph and is *consistent* (or *trim*) if all its states are both accessible and coaccessible. If $\mathcal{A}$ is a consistent (partial) dfa recognizing L, then $\max_i \alpha_i = \mathsf{Ind}(\mathcal{A})$ (folklore). For factorial regular languages, there is a significant simplification: $\alpha_0 = \ldots = \alpha_{r-1}$ and the degrees of all polynomials p_i coincide. So, one has $C_L(n) = \Theta(n^{m-1}\alpha^n)$; here m is the maximum number of strong components of index $\alpha = \mathsf{Ind}(\mathcal{A})$ that are visited by a single directed walk in $\mathcal{A}$ [8]. We refer to the numbers α and $m-1$ as *growth rate* and *polynomial degree* of L, respectively, and use the notation $\mathsf{Gr}(L)$, $\mathsf{Pd}(L)$.

Any word from the complement of a factorial language L is a *forbidden word* for L. The set of all minimal (w.r.t. to the relation "to be a factor of") forbidden words for L is called the *antidictionary* of L. Antidictionaries are *antifactorial* languages in the sense that they are antichains w.r.t. to the relation "to be a factor of". Over a fixed alphabet, the function that maps any factorial language to its antidictionary obviously is a bijection between factorial and antifactorial languages. This bijection preserves regularity of a language. The languages with a finite antidictionary (*FAD-languages*) form a narrow and simple-looking subclass of regular languages. Besides symbolic dynamics and algebra, this class plays a special role in the study of combinatorial complexity for arbitrary factorial languages (see, e.g., [11, Sect. 3]).

In this paper we touch the following problem: *given a finite alphabet and a particular class $\Theta(n^m\alpha^n)$ of functions, find a FAD-language whose combinatorial complexity belongs to this class or prove that no such languages exist.*

For the related research in symbolic dynamics, see [1, Sect. 9] and the references therein. Note that it is essential to fix an alphabet. Indeed, if $\mathcal{A}$ is a consistent dfa, then it is enough to label all transitions in $\mathcal{A}$ with different letters to get a consistent dfa recognizing a FAD-language. Thus, without fixing an alphabet, the problem reduces to finding a strongly connected digraph with a given index. The latter problem is worthwhile, but does not involve languages at all.

We also mention that the above problem for the case $\alpha = 1$ was solved in [9]. So, from now on we assume that $\alpha > 1$. Approaching the above problem, we ask a natural question: *how to simplify an antidictionary preserving the Θ-class of combinatorial complexity of the corresponding factorial language?*

Studying FAD-languages, it is convenient to introduce "canonical" automata for them. We use a version [2] of Aho-Corasick's pattern matching automata. These automata are relatively small and very handy.

In Sect. 2 we recall necessary notation and definitions and prove basic properties of canonical automata. In Sect. 3, we introduce two reduction procedures for antidictionaries and prove that these procedures preserve the growth rate and the polynomial degree of the corresponding factorial languages as well as the essential elements of structure of their canonical automata. In Sect. 4 we provide Algorithm CC to reconstruct a canonical automaton from a given (unlabeled) strong component; this algorithm provides a way to prove the existence of a FAD-language with a given growth rate (Algorithm IT). In Sect. 5, the limitations of proposed approach are shown: in order to prove that an algebraic number of degree r is *not* the growth rate of a FAD-language, it is not enough to iterate Algorithm IT over the strong components of r vertices. The final Section 6 contains some results on the existence of FAD-languages with a given polynomial degree and of canonical automata with a "complex" condensation.

2   Canonical Automata and Their Properties

We consider deterministic finite automata (dfa's) with partial transition function and treat them as labeled digraphs. Such an automaton is *consistent* if any vertex can be reached by some accepting walk. A *trie* is a dfa which is a rooted tree with the root as the initial vertex and the set of leaves as the set of terminal vertices. Tries are used to represent prefix codes, e.g., antidictionaries.

Studying a class of languages, it is useful to have, for each language in the class, a unique "canonical" recognizing automaton. Since the languages with a finite antidictionary are usually described in terms of antidictionaries, it is natural to require that a canonical automaton is efficiently constructible from an antidictionary and keeps the information about it. As was shown in [2], a rather small (but not necessarily minimal) consistent dfa satisfying these two requirements can be built by a modification of Aho-Corasick's pattern matching algorithm. This modification is presented below. We assume that the input antidictionary is given by a trie. We write $\mathsf{Pre}(M)$ for the set of all prefixes of M and $\mathsf{L}(M)$ for the factorial language with the antidictionary M.

Algorithm CA (Canonical-automaton).
Input: a trie $\mathcal{T}$ recognizing an antidictionary M.
Output: a dfa $\mathcal{A}$ recognizing $\mathsf{L}(M)$.
Step 0. Associate each vertex in $\mathcal{T}$ with the word labeling the path from the root to this vertex. (Thus, the set of vertices of $\mathcal{T}$ is the set $\mathsf{Pre}(M)$.)
Step 1. For any nonterminal vertex u and any letter a, if no edge with the label a leaves u, add the edge $u \xrightarrow{a} v$ such that v is the longest suffix of the word ua among the vertices of $\mathcal{T}$. The resulting dfa is denoted by $\widehat{\mathcal{T}}$.
Step 2. Delete all terminal vertices from $\widehat{\mathcal{T}}$ and mark all remaining vertices as terminal to get $\mathcal{A}$.

The automaton $\mathcal{A}$ is deterministic and consistent, its set of vertices equals $\mathsf{Pre}(M) - M$, and accepting walks in $\mathcal{A}$ are exactly the walks from the initial vertex. The edges of $\mathcal{A}$ are naturally partitioned into two groups: *forward* edges belonging to $\mathcal{T}$ and *backward* edges added at step 1. Note that for an edge (u, v) one has $|v| = |u| + 1$ if the edge is forward, and $|v| \leq |u|$ if it is backward. We also make use of the *failure function* $f(u)$ which returns, for any given word, its longest *proper* suffix among the vertices of the trie. Throughout the paper, the notation $\mathcal{T}$, $\widehat{\mathcal{T}}$, and $\mathcal{A}$ is reserved for the automata described in Algorithm CA. Some properties of $\mathcal{A}$ are collected in the following lemma; the proof is omitted due to space constraints.

Lemma 1. *1) For any vertex u of $\mathcal{A}$ and any letter a, the word ua is forbidden if and only if $\mathcal{A}$ contains no outgoing edge from u labeled by a.*
2) If a walk with a label w ends in a vertex v, then the shorter word among v, w is the suffix of the longer one.
3) If an edge $u \xrightarrow{a} v$ was added to $\mathcal{T}$ at step 1, then the edge $f(u) \xrightarrow{a} v$ belongs to $\widehat{\mathcal{T}}$. In particular, if $\mathcal{A}$ has no outgoing edge from u with the label a, and $ua \notin M$, then $\mathcal{A}$ has no outgoing edge from $f(u)$ with the label a.

Remark 1. All assertions of Lemma 1 remain valid if we apply Algorithm CA to a broader class of tries. Namely, we can replace the condition "M is an antidictionary" by its weaker analog "the set $(\mathsf{Pre}(M) - M)$ has no factors in M".

We begin with several properties of canonical automata proved in [10]. Their proofs refer to Lemma 1 and to Algorithm CA, but not to the fact that $\mathcal{T}$ recognizes an antidictionary. Hence, all these properties can be extended to a wider class of automata according to Remark 1.

Proposition 1. *If a vertex v of a canonical automaton belongs to cycles of length p and length q, then $|v| < p + q - \gcd(p, q)$.*

Corollary 1. *Suppose that a canonical automaton contains a nontrivial strong component consisting of vertices that are words of length $\geq n$. Then this component contains a cycle of length $> n/2$. In particular, canonical automata over a fixed alphabet cannot have arbitrarily big number of isomorphic copies of any non-singleton strong component.*

Recall that a word is *primitive* if it is not an integral power of a shorter word.

Proposition 2. *The label of any cycle in a canonical automaton, read starting at any vertex, is a primitive word. If a cycle has the label u read starting at the vertex v, then the word v is a suffix of some power of u.*

Proposition 3. *A canonical automaton $\mathcal{A}$ has a loop at a vertex v if and only if there exist a letter c and an integer m such that $v = c^m$ and c^{m+1} is not a vertex of the trie $\mathcal{T}$.*

Corollary 2. *A canonical automaton $\mathcal{A}$ over the alphabet Σ has at most $|\Sigma|$ loops. There is no loop with a label c if and only if the antidictionary M contains a power of c.*

3   C-Graphs and Reduction Procedures for Antidictionaries

As was mentioned in the introduction, growth rate $\mathsf{Gr}(L)$ and polynomial degree $\mathsf{Pd}(L)$ describe the asymptotic behaviour of the combinatorial complexity of a FAD-language L up to a Θ-class. These two parameters are determined by the structure and mutual location of the strong components of the canonical automaton $\mathcal{A}(L)$. Since $\mathsf{Gr}(L) > 1$, it is sufficient to consider *nontrivial* strong components, i.e., the components that are neither singletons nor simple cycles. We define the *C-graph* of an automaton to be its subgraph generated by all vertices of nontrivial strong components. C-graph of a FAD-language L is the C-graph of $\mathcal{A}(L)$. The aim of this section is to learn how to transform finite antidictionaries in a way that (i) reduces the size of an antidictionary; (ii) either preserves the C-graph of the corresponding language L, or changes it slightly, preserving $\mathsf{Gr}(L)$ and $\mathsf{Pd}(L)$.

Recall that the vertices u and v of a finite automaton are *equivalent* if the set of words labeling all walks from u to terminal vertices coincides with such a set for v. Since all vertices in canonical automata are terminal, and the notion of equivalence does not use the initial vertex, we can extend the notion of equivalence to C-graphs. Vertices u and v of an arbitrary (not necessary labeled!) digraph G are called *duplicates* if their outgoing edges lead to the same vertices.

Let $\mathsf{Min}(K)$ be the set of minimal words of the language K w.r.t. the relation "to be a factor of". Then $\mathsf{Min}(K)$ is an antidictionary, and

$(\star)$ a word has no factors in K if and only if it has no factors in $\mathsf{Min}(K)$.

Now suppose that a trie $\mathcal{T}$ recognizes an antidictionary M, and the corresponding canonical automaton $\mathcal{A}$ recognizes the language $\mathsf{L}(M)$ of exponential growth. Arguing about subgraphs of $\mathcal{T}$, we distinguish between *subtries* (containing the root of $\mathcal{T}$) and *subtrees* (consisting of a vertex and all its descendants).

Consider the subtrees in $\mathcal{T}$ having no vertices from which the C-graph of $\mathcal{A}$ is reachable. Let $\mathcal{T}_1, \ldots, \mathcal{T}_s$ be all maximal subtrees with this property. Let M_1 [M_2] be the set of leaves [resp., roots] of the subtrees $\mathcal{T}_1, \ldots, \mathcal{T}_s$. We view the elements of M_1 and M_2 as words. The *reduction* of the antidictionary M is made in two steps. First, let $\mathsf{Trim}(M) = (M - M_1) \cup M_2$. Note that $\mathsf{Trim}(M)$ is a prefix code represented by the subtrie of $\mathcal{T}$ obtained by deleting all vertices of the subtrees $\mathcal{T}_1, \ldots, \mathcal{T}_s$ except for their roots. We denote this trie by $\mathsf{Trim}(\mathcal{T})$. Second, we apply the Min operation to obtain the *reduced* antidictionary $\mathsf{red}(M) = \mathsf{Min}(\mathsf{Trim}(M))$.

The following theorem shows that the reduction of an antidictionary preserves the parameters of growth of the language with this antidictionary and "almost" preserves the C-graph of this language.

Theorem 1. *Suppose that M is a finite antidictionary, $\mathsf{Gr}(\mathsf{L}(M)) > 1$, and C, C' are the C-graphs of $\mathsf{L}(M)$ and $\mathsf{L}(\mathsf{red}(M))$, respectively. Then*
1) $\mathsf{Gr}(\mathsf{L}(\mathsf{red}(M))) = \mathsf{Gr}(\mathsf{L}(M))$, $\mathsf{Pd}(\mathsf{L}(\mathsf{red}(M))) = \mathsf{Pd}(\mathsf{L}(M))$;
2) *The C-graph C' can be obtained from C by performing zero or more transformations of the following two types[1]:*
   (a) *merging two equivalent duplicates in the same strong component of C;*
   (b) *merging a vertex from C with its equivalent duplicate from outside C.*

---

[1] Note that (b) preserves edges inside the strong components of C and the reachability relation between these components.

Proof (sketched). Let $\mathcal{T}$ and $\mathcal{A}$ be the trie representing M and the corresponding canonical automaton. The subtrees $\mathcal{T}_1, \ldots, \mathcal{T}_s$ of $\mathcal{T}$ are defined above.

First we prove that the set $(\mathsf{Pre}(\mathsf{Trim}(M)) - \mathsf{Trim}(M))$ has no factors in $\mathsf{Trim}(M)$; in other words, if $w = xvy$ for some $v, w \in \mathsf{Trim}(M)$, $x, y \in \Sigma^*$, then $y = \lambda$. We show that in this case v is the root of some subtree $\mathcal{T}_i$ and then xv belongs to some $\mathcal{T}_j$. Since the only vertex of $\mathcal{T}_j$ in $\mathsf{Trim}(M)$ is its root, we have $xv = xvy = w$.

Let us take the trie $\mathsf{Trim}(\mathcal{T})$ and apply all steps of Algorithm CA to it, obtaining some automaton $\mathcal{A}^0$. We refer to Remark 1 and apply Lemma 1 (1) to the automaton $\mathcal{A}^0$, concluding that $\mathcal{A}^0$ accepts a word w if and only if w has no factors in $\mathsf{Trim}(M)$. Then $\mathcal{A}^0$ recognizes $\mathsf{L}(\mathsf{red}(M))$ by $(\star)$.

Next we check that $\mathcal{A}^0$ coincides with the automaton obtained from $\mathcal{A}$ by deleting all vertices from the sets $\mathcal{T}_1, \ldots, \mathcal{T}_s$. Since $\mathcal{A}^0$ can be obtained from $\mathcal{A}$ by deleting some vertices from which the C-graph cannot be reached, the C-graphs of $\mathcal{A}$ and $\mathcal{A}^0$ coincide. Statement 1 of the theorem follows from this.

Now let $\mathcal{A}'$ be the canonical automaton recognizing the language $\mathsf{L}(\mathsf{red}(M))$. The automata $\mathcal{A}^0$ and $\mathcal{A}'$ accept the same language, and $\mathcal{A}^0$ inherits the C-graph C from the automaton $\mathcal{A}$. Let us describe an iterative transformation of $\mathcal{A}^0$ into $\mathcal{A}'$, and watch the transformation of the C-graph in each iteration.

Let $K_0 = \mathsf{Trim}(M)$, $i = 0$, and proceed as follows until $K_i = \mathsf{red}(M)$:

- choose a pair $v, xv \in K_{i-1}$ $(x \neq \lambda)$;
- put $i = i + 1$ and $K_i = K_{i-1} - \{xv\}$;
- obtain $\mathcal{A}_i$ applying Algorithm CA to the trie $\mathcal{T}_i$ representing K_i
- compare the C-graph of $\mathcal{A}_i$ to the C-graph of $\mathcal{A}_{i-1}$.

Since each set $(\mathsf{Pre}(K_i) - K_i)$ has no factors in K_i, Remark 1 is applicable. In order to prove statement 2, it suffices to show that this statement holds for one iteration of the described procedure, i.e., for deleting one word from the set K_{i-1}. If the parent of the deleted word $w = xv$ in $\mathcal{T}_{i-1}$ has another child, then one can easily check that $\mathcal{A}_i = \mathcal{A}_{i-1}$. Otherwise, the vertex w and its ancestors up to the nearest branching point should be deleted from the trie $\mathcal{T}_{i-1}$. We perform this deletion in $\mathcal{A}_{i-1}$, one vertex at a time, starting from the longest vertex and redirecting the incoming backward edges of each deleted vertex z to $f(z)$. It is clear that deleting a vertex z with such a redirection is equivalent to merging z and $f(z)$.

Finally we show that, merging z and $f(z)$, we perform transformation (a) [resp., transformation (b); nothing] with the current C-graph if both [resp., only one of; neither of] z and $f(z)$ belong to this C-graph. $\square$

Corollary 3. *A C-graph having no duplicates is preserved by the reduction of the antidictionary. In particular, this is the case for C-graphs of minimal dfa's.*

We call an antidictionary M *reduced* if $\mathsf{red}(M) = M$. The reduction procedure can be "upgraded" to the *double reduction*, which results in the antidictionary $\mathsf{dred}(M) = \mathsf{red}(\overleftarrow{\mathsf{red}(\overleftarrow{M})})$, where $\overleftarrow{L}$ is the set of reversals of all words from L. For the double reduction, statement 1 of Theorem 1 holds true, because the languages

L and $\overleftarrow{L}$ have exactly the same combinatorial complexity, and $\overleftarrow{L(M)} = L(\overleftarrow{M})$. The statement 2 may fail: due to the asymmetric nature of automata, the C-graphs of $L(M)$ and $L(\overleftarrow{M})$ can be quite different.

Remark 2. The order of applying reductions in the double reduction procedure is irrelevant. Each word from a language L with a doubly reduced antidictionary is a prefix of exponentially many words from L and a suffix of exponentially many words from L; hence, no further reduction is possible.

Example 1. Let $\Sigma = \{a, b\}$, $M = \{a^3b, abba, babb, baabb\}$. The C-graph of the canonical automaton $\mathcal{A}$ is not reachable only from the vertex a^3 (see Fig. 1, left). So we have $\mathsf{Trim}(M) = \mathsf{Min}(\mathsf{Trim}(M)) = \{a^3, abba, babb, baabb\}$.

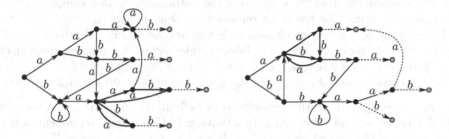

Fig. 1. Reducing canonical automata (for Example 1)

Now let us switch to the reversals: $\overleftarrow{\mathsf{red}(M)} = \{a^3, abba, bbab, bbaab\}$. In the obtained canonical automaton (see Fig. 1, right), the C-graph is not reachable from the vertices bb, bba, $bbaa$, and abb. Thus we obtain $\overleftarrow{\mathsf{red}(M)}_1 = \{abba, bbab, bbaab\}$, $\overleftarrow{\mathsf{red}(M)}_2 = \{abb, bb\}$, $\mathsf{Trim}(\overleftarrow{\mathsf{red}(M)}) = \{a^3, abb, b^2\}$, $\mathsf{dred}(M) = \{a^3, b^2\}$.   □

Remark 3. Reduction has an interesting interpretation in terms of symbolic dynamics. Recall that a (one-side) *subshift of finite type* coincides with the set of infinite words that label infinite walks in a dfa recognizing a FAD-language. A *periodicity marker* of such a subshift S is a finite word that appears only in ultimately periodic words from S (in Example 1, a^3 and b^2 are such markers in the left and the right automaton, respectively). Reduction of the antidictionary turns all periodicity markers into forbidden words and thus deletes the corresponding ultimately periodic words from S.

If we need to preserve (or almost preserve) C-graphs during the reduction, we cannot use the double reduction. But a single reduction can be enhanced by the *cleaning* procedure justified by the following proposition.

Proposition 4. *Suppose that a leaf w of a trie $\mathcal{T}$ is not reachable in the automaton $\widehat{\mathcal{T}}$ from the vertices of the C-graph C of the canonical automaton $\mathcal{A}$. Then the canonical automaton for the language $\mathsf{L}(M-w)$ has the C-graph C.*

Proof. Let us take the automaton $\mathcal{A}$ and restore the vertex w and its ingoing edges, deleted at step 2 of Algorithm CA. We add outgoing edges from w for all letters, following the rule of step 1. These edges cannot affect the C-graph C of $\mathcal{A}$, because their beginning w is not reachable from C. Thus, the obtained automaton $\mathcal{A}^0$ inherits the C-graph from $\mathcal{A}$. Now note that $\mathcal{A}^0$ recognizes $\mathsf{L}(M-w)$, because all missing edges initially lead to the vertices of the set $M - w$.

In order to obtain the canonical automaton $\mathcal{A}'$ for the language $\mathsf{L}(M-w)$, we cut the path representing w in $\mathcal{A}^0$ in the same fashion as in the proof of Theorem 1. All vertices that have to be deleted do not belong to C by the conditions of the proposition. As we have shown in the proof of Theorem 1, the deletion of such a vertex preserves C.   □

According to Proposition 4, we call an antidictionary M *clean* if all leaves of the trie $\mathcal{T}$ are reachable from the C-graph of the automaton $\mathcal{A}$. In Example 1, M is not clean: the word *abba* has to be removed (see Fig. 1, left).

Given a digraph G, its *condensation* is the acyclic digraph obtained by collapsing each strong component of G into a single vertex. A strong component of G is said to be a *sink component*, if it is a sink of the condensation of G. The following remark stems directly from the definition of the reduced antidictionary.

Remark 4. Suppose that the automaton $\mathcal{A}$ is built by Algorithm CA from a trie $\mathcal{T}$ that represents a reduced antidictionary. If H is a sink component of the C-graph of $\mathcal{A}$, then all edges missing in H lead in $\widehat{\mathcal{T}}$ to the leaves of $\mathcal{T}$.

4   Checking Candidate Strong Components

The following algorithm tries to build a canonical automaton containing a given digraph as a strong component.

Algorithm CC (check-component)

Input: a strongly connected digraph $G = (V, E)$, an alphabet Σ.
Output: an antidictionary $M \in \Sigma^*$ such that the canonical automaton for $\mathsf{L}(M)$ has a strong component isomorphic to G, or "NO".

Step 1. Return "NO" if G does not satisfy any of the following restrictions:
    1.1. $|V| < |E| < |V|\cdot|\Sigma|$;
    1.2. the outdegree of any vertex is at most $|\Sigma|$;
    1.3. the number of loops is at most $|\Sigma|$;
    1.4. multiple edges, if any, have the same endpoint and the multiplicity
        at most $|\Sigma|-1$; the endpoint has the outdegree $|\Sigma|$ and a loop.

Step 2. Calculate all labellings of the edges of G such that
    2.1. all edges with a common beginning have different labels;
    2.2. all edges with a common endpoint have the same label;
    2.3. all loops have different labels;
    2.4. the label of any cycle is a primitive word.

Step 3. For each labeling of edges built at step 2, calculate all labellings $\{v_1, \ldots, v_{|V|}\}$ of the vertices of G such that

3.1. if v_i belongs to cycles with labels w_1 and w_2 (read starting at v_i), then v_i is a common suffix of some words $w_1^{s_1}$ and $w_2^{s_2}$ ($s_1, s_2 \in \mathbb{N}$);

3.2. if $v_i \xrightarrow{c} v_j$ is a labeled edge, then v_j is a suffix of the word $v_i c$; moreover, $v_j = v_i c$ iff this edge is forward;

3.3. all forward edges have different endpoints;

3.4. a vertex has at most $|\Sigma|-1$ outgoing backward edges;

3.5. the vertex λ has a loop;

3.6. if $v_j = v_i c_1 \ldots c_{k+1}$, then the vertices $v_{i_1} = v_i c_1; \ldots; v_{i_k} = v_i c_1 \ldots c_k$ exist.

Step 4. For each labeling of vertices built at step 3,

4.1. calculate the set

$K = \{v_i c \mid$ no edge from v_i with the label $c\}$ of forbidden words;

4.2. calculate the antidictionary $M = \text{Min}(K)$ and the trie $\mathcal{T}$ for M;

4.3. build the canonical automaton $\mathcal{A}$ from $\mathcal{T}$;

4.4. if the vertices of $\mathcal{A}$ with the labels $\{v_1, \ldots, v_{|V|}\}$ form a strong component isomorphic to G, return M.

If no one of the obtained automata has the required component, return "NO".

Proposition 5. *Given an arbitrary strongly connected digraph G, Algorithm CC returns a clean reduced antidictionary M such that G is a sink component of the C-graph of the canonical automaton built from M, or "NO", if no reduced antidictionaries with this property exist.*

Proof (sketched). Restrictions 1.1–1.4 and 2.1–2.4 stem from the definitions, Corollary 2, Lemma 1 (3), and Proposition 2. Restriction 3.1 follows from Proposition 2 and implies $|v_i| < |w_1| + |w_2| - \gcd(|w_1|, |w_2|)$. Restriction 3.2 follows from the definition of Algorithm CA. Together, 3.1 and 3.2 guarantee finiteness of the set of possible vertex labelings and hence, the termination of the algorithm. Restrictions 3.3–3.6 stem from the definitions of Algorithm CA and strong connectedness.

By Proposition 4, the condition "M is clean" does not affect the class of possible C-graphs. Hence, if the graph G with the given labeling of edges and vertices form a strong component of a canonical automaton built from a reduced antidictionary, then by Remark 4 the set K built at step 4.1 consists of forbidden words, while all words from $\text{Pre}(K) - K$ are not forbidden. Thus, the only possible candidate antidictionary for this labeling is built at step 4.2. The last two steps show whether this candidate is suitable. If no labelings produce valid antidictionaries, then no such antidictionaries exist. $\square$

Proposition 6. *Given a strongly connected graph G without duplicates, Algorithm CC returns a clean reduced antidictionary M such that G is a sink component of the C-graph of the canonical automaton built from M, or "NO", if there exist no antidictionary for which G is any component of the C-graph of the canonical automaton.*

Proof (sketched). Let G be a strong component of a C-graph of a canonical automaton $\mathcal{A}$ built from an antidictionary M. By Corollary 3, M can be replaced by $\mathsf{red}(M) = \mathsf{Min}(\mathsf{Trim}(M))$, retaining the C-graph. By Proposition 4 we can consider $\mathsf{red}(M)$ as clean. Now we "cut" $\mathsf{red}(M)$ such that G becomes a sink component of the C-graph. In order to do this, it is enough to describe the transformation (of a clean reduced antidictionary) that deletes one sink component from the C-graph. Such a transformation is similar to the reduction; we omit the details here. Then we proceed with reduction–cleaning–deletion until G becomes a sink component, and apply Proposition 5.   □

Proposition 6 allows one to test candidate growth rates of FAD-languages, if these rates are given as indexes of strongly connected digraphs. Testing is performed by the following

Algorithm IT (index-testing)

Input: strongly connected digraph G without duplicates, an alphabet Σ.
Output: an antidictionary $M \in \Sigma^*$ such that its canonical automaton $\mathcal{A}$ contains G as a strong component and $\mathsf{Gr}(\mathsf{L}(M)) = \mathsf{Ind}(G)$, or "NO".

Step 1. Apply to G the modification of Algorithm CC, returning *all* suitable antidictionaries.
Step 2. For each antidictionary M, calculate the index of the canonical automaton; if this index coincides with $\mathsf{Ind}(G)$, return M.
Step 3. If no antidictionaries remain, return "NO".

Proving the correctness of Algorithm IT, it is sufficient to note that if G is a non-sink component of a canonical automaton $\mathcal{A}$ and $\mathsf{Ind}(\mathcal{A}) = \mathsf{Ind}(G)$, then there exists a canonical automaton with the same property, for which G is a sink component. Namely, this is the automaton obtained by the trimming procedure described in the proof of Proposition 6.

5   A Negative Example

It seems that it is impossible to prove, running Algorithm IT finitely many times, that the index of a given digraph is not the growth rate of any FAD-language over a given alphabet. This pessimistic forecast stems from the following statement. Note that this statement concerns about all consistent (not only canonical) dfa's.

Proposition 7. *There exists an algebraic number of degree 3 which is*
- a growth rate of a binary FAD-language but
- not an index of a 3-vertex strong component of a consistent dfa recognizing a binary FAD-language.

Proof (sketched). Consider the binary factorial language with the antidictionary $\{abbab, b^4, b^3a^2\}$. Its canonical automaton looks as follows.

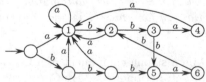

The index of this automaton is an algebraic number of degree 3, because the characteristic polynomial of the only nontrivial strong component (vertices 1–6) equals $x(x^2+1)(x^3-x^2-2x+1)$. For the binary alphabet, there is a unique strongly connected digraph G with three vertices and the same index:

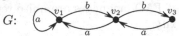

Next we show that if G is a strong component of a consistent dfa, then G is a component of the equivalent minimal dfa (we denote it by $\mathcal{B}$). Since FAD-languages are star-free languages, the label of *any* cycle in a minimal automaton is a primitive word by the Schützenberger theorem [7]. Then the labeling of G is unique. For this labeling, we find an infinite set of minimal forbidden words for the language $\mathsf{L}(\mathcal{B})$. These words have the form $a^j bb(ab)^i b$, and their existence proves that $\mathsf{L}(\mathcal{B})$ is not a FAD-language. □

6 Condensations of Canonical Automata

In this section we consider C-graphs consisting of several strong components. First, we are interested in FAD-languages having a nonzero polynomial degree. A priori, it is not obvious that such languages exist (cf. Corollary 1).

Proposition 8. *There exist k-ary FAD-languages having the combinatorial complexity $\Theta(n^{k-2}\alpha^n)$, where $\alpha > 1$.*

Proof. Let $\Sigma=\{1,\ldots,k\}$, $k\geq 3$. The canonical automaton for the antidictionary $\{ab \mid a,b \in \Sigma,\ a \neq 1, b \notin \{a-1,a,a+1\}\} \cup \{a(a-1)b \mid a,b \in \Sigma,\ a \neq 1, b \neq a\}$ is depicted below (for convenience, the labeling is given for the case $k = 9$). Each of the two-element components has the index ϕ (golden ratio), while the horizontal path meets all $k-1$ such components. □

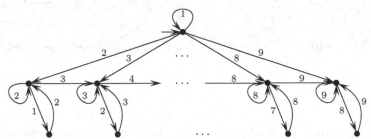

Proposition 9. *There exist binary FAD-languages having the combinatorial complexity $\Theta(n\alpha^n)$, where $\alpha > 1$.*

Proof. The canonical automaton for the antidictionary $\{a^2bab, a^2b^3, a^2b^2ab\}$ has two nontrivial components, each of index ϕ. □

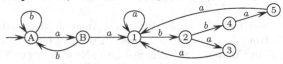

In the previous examples, C-graphs contain more than one strong component, but the condensation of the C-graph was the simplest possible: a chain. We now show that such a condensation can have a more complicated structure. Namely, it does not need to be weakly connected, and it does not need to be a forest. This shows once more that FAD-languages are quite representative in terms of the combinatorial complexity.

Proposition 10. *There exists a binary canonical automaton with a C-graph that is not weakly connected.*

Proof. Take the canonical automaton for $M = \{a^5, a^2b^2, a^2bab, a^2ba^2b\}$.

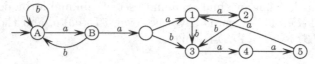

Proposition 11. *There exists a binary canonical automaton in which the condensation of the C-graph is not a forest.*

Proof. Take the canonical automaton for $M = \{a^4b^2a, a^4b^3a, a^4b^5, a^4b^4ab^4ab, abab, aba^2b, aba^3b, b^4a^2b, b^4a^3b, b^4a^5, b^4a^4ba^4ba, baba, bab^2a, bab^3a\}$. The condensation of the C-graph is depicted inside the dotted area.

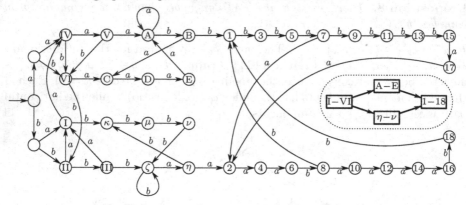

References

1. Béal, M.P., Perrin, D.: Symbolic dynamics and finite automata. In: Rozenberg, G., Salomaa, A. (eds.) Handbook of Formal Languages, vol. 2, pp. 463–505. Springer (1997)
2. Crochemore, M., Mignosi, F., Restivo, A.: Automata and forbidden words. Inform. Process. Lett. 67, 111–117 (1998)
3. Govorov, V.E.: Graded algebras. Math. Notes 12, 552–556 (1972)
4. Milnor, J.: Growth of finitely generated solvable groups. J. Diff. Geom. 2, 447–450 (1968)
5. Morse, M., Hedlund, G.A.: Symbolic dynamics. Amer. J. Math. 60, 815–866 (1938)
6. Salomaa, A., Soittola, M.: Automata-theoretic aspects of formal power series. Texts and Monographs in Computer Science. Springer, New York (1978)

7. Schützenberger, M.P.: On finite monoids having only trivial subgroups. Information and Computation 8, 190–194 (1965)
8. Shur, A.M.: Combinatorial complexity of regular languages. In: Hirsch, E.A., Razborov, A.A., Semenov, A., Slissenko, A. (eds.) CSR 2008. LNCS, vol. 5010, pp. 289–301. Springer, Heidelberg (2008)
9. Shur, A.M.: Polynomial languages with finite antidictionaries. RAIRO Inform. Théor. App. 43, 269–280 (2009)
10. Shur, A.M.: Languages with finite antidictionaries: growth index and properties of automata. Proc. Ural State Univ. 74, 220–245 (2010) (in Russian)
11. Shur, A.M.: Growth properties of power-free languages. Computer Science Review 6, 187–208 (2012)

Author Index